VOLUME FOUR HUNDRED AND SIXTY-EIGHT

METHODS IN ENZYMOLOGY

Biophysical, Chemical, and Functional Probes of RNA Structure, Interactions and Folding: Part A

METHODS IN ENZYMOLOGY

Editors-in-Chief

JOHN N. ABELSON AND MELVIN I. SIMON

Division of Biology
California Institute of Technology
Pasadena, California, USA

Founding Editors

SIDNEY P. COLOWICK AND NATHAN O. KAPLAN

VOLUME FOUR HUNDRED AND SIXTY-EIGHT

METHODS IN ENZYMOLOGY

Biophysical, Chemical, and Functional Probes of RNA Structure, Interactions and Folding: Part A

EDITED BY

DANIEL HERSCHLAG

Departments of Biochemistry and Chemistry
Stanford University
Stanford, California, USA

ELSEVIER

AMSTERDAM • BOSTON • HEIDELBERG • LONDON
NEW YORK • OXFORD • PARIS • SAN DIEGO
SAN FRANCISCO • SINGAPORE • SYDNEY • TOKYO
Academic Press is an imprint of Elsevier

Academic Press is an imprint of Elsevier
525 B Street, Suite 1900, San Diego, CA 92101-4495, USA
30 Corporate Drive, Suite 400, Burlington, MA 01803, USA
32 Jamestown Road, London NW1 7BY, UK

First edition 2009

For information on all Academic Press publications
visit our website at elsevierdirect.com

ISBN: 978-0-12-374399-2
ISSN: 0076-6879

Printed and bound in United States of America
09 10 11 12 10 9 8 7 6 5 4 3 2 1

Contents

Contributors

Tadepalli Adilakshmi
Weis Center for Research, Geisinger Medical Center, Danville, Pennsylvania, USA

Philip C. Bevilacqua
Department of Chemistry, The Pennsylvania State University, University Park, Pennsylvania, USA

Durga M. Chadalavada
Department of Chemistry, The Pennsylvania State University, University Park, Pennsylvania, USA

Eric L. Christian
Center for RNA Molecular Biology, Department of Biochemistry, Case Western Reserve University School of Medicine, Cleveland, Ohio, USA

Gloria M. Culver
Department of Biology, University of Rochester, Rochester, New York, USA

Victoria J. DeRose
Department of Chemistry, University of Oregon, Eugene, Oregon, USA

Caia D. S. Duncan
Department of Chemistry, University of North Carolina, Chapel Hill, North Carolina, USA

Martha J. Fedor
Departments of Chemical Physiology and Molecular Biology, The Skaggs Institute for Chemical Biology, The Scripps Research Institute, La Jolla, California, USA

Andrew L. Feig
Department of Chemistry, Wayne State University, Detroit, Michigan, USA

Marcello Forconi
Department of Biochemistry, Stanford University, Stanford, California, USA

John K. Frederiksen
The Pritzker School of Medicine, and Department of Biochemistry and Molecular Biology, University of Chicago, Chicago, Illinois, USA, and Department of Pathology and Laboratory Medicine, University of Rochester Medical Center, Rochester, New York, USA

Michael E. Harris
Center for RNA Molecular Biology, Department of Biochemistry, Case Western Reserve University School of Medicine, Cleveland, Ohio, USA

Daniel Herschlag
Departments of Biochemistry and Chemistry, Stanford University, Stanford, California, USA

James L. Hougland
Department of Chemistry, University of Michigan, Ann Arbor, Michigan, USA

Laura Hunsicker-Wang
Department of Chemistry, Trinity University, San Antonio, Texas, USA

Alain Laederach
Developmental Genetics and Bioinformatics, Wadsworth Center, Albany, New York, USA

Andreas Liebeg
Max F. Perutz Laboratories, Department of Biochemistry and Cell Biology, University of Vienna, Vienna, Austria

Joshua S. Martin
Developmental Genetics and Bioinformatics, Wadsworth Center, Albany, New York, USA

Jennifer L. McGinnis
Department of Chemistry, University of North Carolina, Chapel Hill, North Carolina, USA

David Mitchell III
Department of Chemistry and Biochemistry, Institute for Cellular and Molecular Biology, University of Texas at Austin, Austin, Texas, USA

Somdeb Mitra
Department of Biochemistry, Albert Einstein College of Medicine, Bronx, New York, USA

Tao Pan
Department of Biochemistry & Molecular Biology, University of Chicago, Chicago, Illinois, USA

Joseph A. Piccirilli
Departments of Chemistry and Biochemistry & Molecular Biology, University of Chicago, Chicago, Illinois, USA

Rick Russell
Department of Chemistry and Biochemistry, Institute for Cellular and Molecular Biology, University of Texas at Austin, Austin, Texas, USA

Susan J. Schroeder
Department of Chemistry and Biochemistry, University of Oklahoma, Norman, Oklahoma, USA

Inna Shcherbakova
Department of Biochemistry, Albert Einstein College of Medicine, Bronx, New York, USA, and Department of Biochemistry and Molecular Pharmacology, University of Massachusetts Medical School, Worcester, Massachusetts, USA

Katrina Simmons
Developmental Genetics and Bioinformatics, Wadsworth Center, Albany, New York, USA

Sarah F. C. Soper
Program in Cell, Molecular and Developmental Biology and Biophysics, Johns Hopkins University, Baltimore, Maryland, USA

Scott A. Strobel
Department of Molecular Biophysics and Biochemistry and Department of Chemistry, Yale University, New Haven, Connecticut, USA

Ian T. Suydam
Department of Molecular Biophysics and Biochemistry and Department of Chemistry, Yale University, New Haven, Connecticut, USA

Douglas H. Turner
Department of Chemistry, University of Rochester, Rochester, New York, USA

Matthew Vogt
Laboratory of Pathology, National Cancer Institute, Bethesda, Maryland, USA

Christina Waldsich
Max F. Perutz Laboratories, Department of Biochemistry and Cell Biology, University of Vienna, Vienna, Austria

Yaqi Wan
Department of Chemistry and Biochemistry, Institute for Cellular and Molecular Biology, University of Texas at Austin, Austin, Texas, USA

Peter Y. Watson
Departments of Chemical Physiology and Molecular Biology, The Skaggs Institute for Chemical Biology, The Scripps Research Institute, La Jolla, California, USA

Kevin M. Weeks
Department of Chemistry, University of North Carolina, Chapel Hill, North Carolina, USA

Terrence N. Wong
Department of Biochemistry & Molecular Biology, University of Chicago, Chicago, Illinois, USA

Sarah A. Woodson
T.C. Jenkins Department of Biophysics, Johns Hopkins University, Baltimore, Maryland, USA

Zhili Xu
Department of Biology, University of Rochester, Rochester, New York, USA

Preface

After the discovery of catalytic RNA nearly 30 years ago, and after the initial excitement wore off, RNA was viewed predominantly as an ancient biological macromolecule with vestigial, albeit critical, functions in modern-day biology. Thus, while important and informative, studies of RNA behavior and function took a back seat to the interrogation of transcription factors, protein kinases, and other molecules directly involved in the regulation of gene expression.

But more recent discoveries have clearly illuminated the central importance of RNA to modern-day biology—the discovery of fewer genes but vastly more alternative spliced gene products than anticipated in the human genome, the discovery of RNA regulatory elements in the form of riboswitches, the finding of many noncoded RNAs, the finding of functional groupings of RNAs by RNA binding proteins, and, of course, the discovery of RNA interference (RNAi) and its likely role in regulation of about half of all human genes.

Thus, it is now clear that we need to understand these molecules in order to understand how biology works, and applications to understanding and curing diseases, while still largely remote, are ultimately likely to become common. Fortunately, in the years since the discovery of catalytic RNA, many incisive and powerful chemical, biochemical, and biophysical tools have been developed to study the folding and conformational behavior of RNA. Here, we have assembled many of these together, in this two volume series. Descriptions of two of the most common and powerful techniques are not covered, NMR and X-ray crystallography (although there is a chapter on preparation of RNA for crystallography), as these approaches can warrant volumes on their own and have been treated separately (see, e.g., Volume 394 on Biological NMR).

I believe that this compilation is particularly important, not just because of the importance of the subject matter, but because the tools and contributors span previously distinct fields from molecular biology, biochemistry, chemistry, and physics. Now students and researchers in each of these areas can get a sense of, as well as detailed protocols for, the entire gambit of approaches. I hope that the efforts of the many contributors will inspire new students and investigators to join the search for understanding of these most fascinating macromolecules. And finally I want to thank a phenomenal group of contributors for agreeing to contribute and then coming through, even sometimes on time, with chapters of uniform high quality.

Dan Herschlag

METHODS IN ENZYMOLOGY

VOLUME 89. Carbohydrate Metabolism (Part D)
Edited by WILLIS A. WOOD

VOLUME 90. Carbohydrate Metabolism (Part E)
Edited by WILLIS A. WOOD

VOLUME 91. Enzyme Structure (Part I)
Edited by C. H. W. HIRS AND SERGE N. TIMASHEFF

VOLUME 92. Immunochemical Techniques (Part E: Monoclonal Antibodies and General Immunoassay Methods)
Edited by JOHN J. LANGONE AND HELEN VAN VUNAKIS

VOLUME 93. Immunochemical Techniques (Part F: Conventional Antibodies, Fc Receptors, and Cytotoxicity)
Edited by JOHN J. LANGONE AND HELEN VAN VUNAKIS

VOLUME 94. Polyamines
Edited by HERBERT TABOR AND CELIA WHITE TABOR

VOLUME 95. Cumulative Subject Index Volumes 61–74, 76–80
Edited by EDWARD A. DENNIS AND MARTHA G. DENNIS

VOLUME 96. Biomembranes [Part J: Membrane Biogenesis: Assembly and Targeting (General Methods; Eukaryotes)]
Edited by SIDNEY FLEISCHER AND BECCA FLEISCHER

VOLUME 97. Biomembranes [Part K: Membrane Biogenesis: Assembly and Targeting (Prokaryotes, Mitochondria, and Chloroplasts)]
Edited by SIDNEY FLEISCHER AND BECCA FLEISCHER

VOLUME 98. Biomembranes (Part L: Membrane Biogenesis: Processing and Recycling)
Edited by SIDNEY FLEISCHER AND BECCA FLEISCHER

VOLUME 99. Hormone Action (Part F: Protein Kinases)
Edited by JACKIE D. CORBIN AND JOEL G. HARDMAN

VOLUME 100. Recombinant DNA (Part B)
Edited by RAY WU, LAWRENCE GROSSMAN, AND KIVIE MOLDAVE

VOLUME 101. Recombinant DNA (Part C)
Edited by RAY WU, LAWRENCE GROSSMAN, AND KIVIE MOLDAVE

VOLUME 102. Hormone Action (Part G: Calmodulin and Calcium-Binding Proteins)
Edited by ANTHONY R. MEANS AND BERT W. O'MALLEY

VOLUME 103. Hormone Action (Part H: Neuroendocrine Peptides)
Edited by P. MICHAEL CONN

VOLUME 104. Enzyme Purification and Related Techniques (Part C)
Edited by WILLIAM B. JAKOBY

VOLUME 123. Vitamins and Coenzymes (Part H)
Edited by FRANK CHYTIL AND DONALD B. MCCORMICK

VOLUME 124. Hormone Action (Part J: Neuroendocrine Peptides)
Edited by P. MICHAEL CONN

VOLUME 125. Biomembranes (Part M: Transport in Bacteria, Mitochondria, and Chloroplasts: General Approaches and Transport Systems)
Edited by SIDNEY FLEISCHER AND BECCA FLEISCHER

VOLUME 126. Biomembranes (Part N: Transport in Bacteria, Mitochondria, and Chloroplasts: Protonmotive Force)
Edited by SIDNEY FLEISCHER AND BECCA FLEISCHER

VOLUME 127. Biomembranes (Part O: Protons and Water: Structure and Translocation)
Edited by LESTER PACKER

VOLUME 128. Plasma Lipoproteins (Part A: Preparation, Structure, and Molecular Biology)
Edited by JERE P. SEGREST AND JOHN J. ALBERS

VOLUME 129. Plasma Lipoproteins (Part B: Characterization, Cell Biology, and Metabolism)
Edited by JOHN J. ALBERS AND JERE P. SEGREST

VOLUME 130. Enzyme Structure (Part K)
Edited by C. H. W. HIRS AND SERGE N. TIMASHEFF

VOLUME 131. Enzyme Structure (Part L)
Edited by C. H. W. HIRS AND SERGE N. TIMASHEFF

VOLUME 132. Immunochemical Techniques (Part J: Phagocytosis and Cell-Mediated Cytotoxicity)
Edited by GIOVANNI DI SABATO AND JOHANNES EVERSE

VOLUME 133. Bioluminescence and Chemiluminescence (Part B)
Edited by MARLENE DELUCA AND WILLIAM D. MCELROY

VOLUME 134. Structural and Contractile Proteins (Part C: The Contractile Apparatus and the Cytoskeleton)
Edited by RICHARD B. VALLEE

VOLUME 135. Immobilized Enzymes and Cells (Part B)
Edited by KLAUS MOSBACH

VOLUME 136. Immobilized Enzymes and Cells (Part C)
Edited by KLAUS MOSBACH

VOLUME 137. Immobilized Enzymes and Cells (Part D)
Edited by KLAUS MOSBACH

VOLUME 138. Complex Carbohydrates (Part E)
Edited by VICTOR GINSBURG

Volume 157. Biomembranes (Part Q: ATP-Driven Pumps and Related Transport: Calcium, Proton, and Potassium Pumps)
Edited by Sidney Fleischer and Becca Fleischer

Volume 158. Metalloproteins (Part A)
Edited by James F. Riordan and Bert L. Vallee

Volume 159. Initiation and Termination of Cyclic Nucleotide Action
Edited by Jackie D. Corbin and Roger A. Johnson

Volume 160. Biomass (Part A: Cellulose and Hemicellulose)
Edited by Willis A. Wood and Scott T. Kellogg

Volume 161. Biomass (Part B: Lignin, Pectin, and Chitin)
Edited by Willis A. Wood and Scott T. Kellogg

Volume 162. Immunochemical Techniques (Part L: Chemotaxis and Inflammation)
Edited by Giovanni Di Sabato

Volume 163. Immunochemical Techniques (Part M: Chemotaxis and Inflammation)
Edited by Giovanni Di Sabato

Volume 164. Ribosomes
Edited by Harry F. Noller, Jr., and Kivie Moldave

Volume 165. Microbial Toxins: Tools for Enzymology
Edited by Sidney Harshman

Volume 166. Branched-Chain Amino Acids
Edited by Robert Harris and John R. Sokatch

Volume 167. Cyanobacteria
Edited by Lester Packer and Alexander N. Glazer

Volume 168. Hormone Action (Part K: Neuroendocrine Peptides)
Edited by P. Michael Conn

Volume 169. Platelets: Receptors, Adhesion, Secretion (Part A)
Edited by Jacek Hawiger

Volume 170. Nucleosomes
Edited by Paul M. Wassarman and Roger D. Kornberg

Volume 171. Biomembranes (Part R: Transport Theory: Cells and Model Membranes)
Edited by Sidney Fleischer and Becca Fleischer

Volume 172. Biomembranes (Part S: Transport: Membrane Isolation and Characterization)
Edited by Sidney Fleischer and Becca Fleischer

VOLUME 190. Retinoids (Part B: Cell Differentiation and Clinical Applications)
Edited by LESTER PACKER

VOLUME 191. Biomembranes (Part V: Cellular and Subcellular Transport: Epithelial Cells)
Edited by SIDNEY FLEISCHER AND BECCA FLEISCHER

VOLUME 192. Biomembranes (Part W: Cellular and Subcellular Transport: Epithelial Cells)
Edited by SIDNEY FLEISCHER AND BECCA FLEISCHER

VOLUME 193. Mass Spectrometry
Edited by JAMES A. MCCLOSKEY

VOLUME 194. Guide to Yeast Genetics and Molecular Biology
Edited by CHRISTINE GUTHRIE AND GERALD R. FINK

VOLUME 195. Adenylyl Cyclase, G Proteins, and Guanylyl Cyclase
Edited by ROGER A. JOHNSON AND JACKIE D. CORBIN

VOLUME 196. Molecular Motors and the Cytoskeleton
Edited by RICHARD B. VALLEE

VOLUME 197. Phospholipases
Edited by EDWARD A. DENNIS

VOLUME 198. Peptide Growth Factors (Part C)
Edited by DAVID BARNES, J. P. MATHER, AND GORDON H. SATO

VOLUME 199. Cumulative Subject Index Volumes 168–174, 176–194

VOLUME 200. Protein Phosphorylation (Part A: Protein Kinases: Assays, Purification, Antibodies, Functional Analysis, Cloning, and Expression)
Edited by TONY HUNTER AND BARTHOLOMEW M. SEFTON

VOLUME 201. Protein Phosphorylation (Part B: Analysis of Protein Phosphorylation, Protein Kinase Inhibitors, and Protein Phosphatases)
Edited by TONY HUNTER AND BARTHOLOMEW M. SEFTON

VOLUME 202. Molecular Design and Modeling: Concepts and Applications (Part A: Proteins, Peptides, and Enzymes)
Edited by JOHN J. LANGONE

VOLUME 203. Molecular Design and Modeling: Concepts and Applications (Part B: Antibodies and Antigens, Nucleic Acids, Polysaccharides, and Drugs)
Edited by JOHN J. LANGONE

VOLUME 204. Bacterial Genetic Systems
Edited by JEFFREY H. MILLER

VOLUME 205. Metallobiochemistry (Part B: Metallothionein and Related Molecules)
Edited by JAMES F. RIORDAN AND BERT L. VALLEE

Volume 256. Small GTPases and Their Regulators (Part B: Rho Family)
Edited by W. E. Balch, Channing J. Der, and Alan Hall

Volume 257. Small GTPases and Their Regulators (Part C: Proteins Involved in Transport)
Edited by W. E. Balch, Channing J. Der, and Alan Hall

Volume 258. Redox-Active Amino Acids in Biology
Edited by Judith P. Klinman

Volume 259. Energetics of Biological Macromolecules
Edited by Michael L. Johnson and Gary K. Ackers

Volume 260. Mitochondrial Biogenesis and Genetics (Part A)
Edited by Giuseppe M. Attardi and Anne Chomyn

Volume 261. Nuclear Magnetic Resonance and Nucleic Acids
Edited by Thomas L. James

Volume 262. DNA Replication
Edited by Judith L. Campbell

Volume 263. Plasma Lipoproteins (Part C: Quantitation)
Edited by William A. Bradley, Sandra H. Gianturco, and Jere P. Segrest

Volume 264. Mitochondrial Biogenesis and Genetics (Part B)
Edited by Giuseppe M. Attardi and Anne Chomyn

Volume 265. Cumulative Subject Index Volumes 228, 230–262

Volume 266. Computer Methods for Macromolecular Sequence Analysis
Edited by Russell F. Doolittle

Volume 267. Combinatorial Chemistry
Edited by John N. Abelson

Volume 268. Nitric Oxide (Part A: Sources and Detection of NO; NO Synthase)
Edited by Lester Packer

Volume 269. Nitric Oxide (Part B: Physiological and Pathological Processes)
Edited by Lester Packer

Volume 270. High Resolution Separation and Analysis of Biological Macromolecules (Part A: Fundamentals)
Edited by Barry L. Karger and William S. Hancock

Volume 271. High Resolution Separation and Analysis of Biological Macromolecules (Part B: Applications)
Edited by Barry L. Karger and William S. Hancock

Volume 272. Cytochrome P450 (Part B)
Edited by Eric F. Johnson and Michael R. Waterman

Volume 273. RNA Polymerase and Associated Factors (Part A)
Edited by Sankar Adhya

Volume 293. Ion Channels (Part B)
Edited by P. Michael Conn

Volume 294. Ion Channels (Part C)
Edited by P. Michael Conn

Volume 295. Energetics of Biological Macromolecules (Part B)
Edited by Gary K. Ackers and Michael L. Johnson

Volume 296. Neurotransmitter Transporters
Edited by Susan G. Amara

Volume 297. Photosynthesis: Molecular Biology of Energy Capture
Edited by Lee McIntosh

Volume 298. Molecular Motors and the Cytoskeleton (Part B)
Edited by Richard B. Vallee

Volume 299. Oxidants and Antioxidants (Part A)
Edited by Lester Packer

Volume 300. Oxidants and Antioxidants (Part B)
Edited by Lester Packer

Volume 301. Nitric Oxide: Biological and Antioxidant Activities (Part C)
Edited by Lester Packer

Volume 302. Green Fluorescent Protein
Edited by P. Michael Conn

Volume 303. cDNA Preparation and Display
Edited by Sherman M. Weissman

Volume 304. Chromatin
Edited by Paul M. Wassarman and Alan P. Wolffe

Volume 305. Bioluminescence and Chemiluminescence (Part C)
Edited by Thomas O. Baldwin and Miriam M. Ziegler

Volume 306. Expression of Recombinant Genes in Eukaryotic Systems
Edited by Joseph C. Glorioso and Martin C. Schmidt

Volume 307. Confocal Microscopy
Edited by P. Michael Conn

Volume 308. Enzyme Kinetics and Mechanism (Part E: Energetics of Enzyme Catalysis)
Edited by Daniel L. Purich and Vern L. Schramm

Volume 309. Amyloid, Prions, and Other Protein Aggregates
Edited by Ronald Wetzel

Volume 310. Biofilms
Edited by Ron J. Doyle

VOLUME 362. Recognition of Carbohydrates in Biological Systems (Part A)
Edited by YUAN C. LEE AND REIKO T. LEE

VOLUME 363. Recognition of Carbohydrates in Biological Systems (Part B)
Edited by YUAN C. LEE AND REIKO T. LEE

VOLUME 364. Nuclear Receptors
Edited by DAVID W. RUSSELL AND DAVID J. MANGELSDORF

VOLUME 365. Differentiation of Embryonic Stem Cells
Edited by PAUL M. WASSAUMAN AND GORDON M. KELLER

VOLUME 366. Protein Phosphatases
Edited by SUSANNE KLUMPP AND JOSEF KRIEGLSTEIN

VOLUME 367. Liposomes (Part A)
Edited by NEJAT DÜZGÜNEŞ

VOLUME 368. Macromolecular Crystallography (Part C)
Edited by CHARLES W. CARTER, JR., AND ROBERT M. SWEET

VOLUME 369. Combinational Chemistry (Part B)
Edited by GUILLERMO A. MORALES AND BARRY A. BUNIN

VOLUME 370. RNA Polymerases and Associated Factors (Part C)
Edited by SANKAR L. ADHYA AND SUSAN GARGES

VOLUME 371. RNA Polymerases and Associated Factors (Part D)
Edited by SANKAR L. ADHYA AND SUSAN GARGES

VOLUME 372. Liposomes (Part B)
Edited by NEJAT DÜZGÜNEŞ

VOLUME 373. Liposomes (Part C)
Edited by NEJAT DÜZGÜNEŞ

VOLUME 374. Macromolecular Crystallography (Part D)
Edited by CHARLES W. CARTER, JR., AND ROBERT W. SWEET

VOLUME 375. Chromatin and Chromatin Remodeling Enzymes (Part A)
Edited by C. DAVID ALLIS AND CARL WU

VOLUME 376. Chromatin and Chromatin Remodeling Enzymes (Part B)
Edited by C. DAVID ALLIS AND CARL WU

VOLUME 377. Chromatin and Chromatin Remodeling Enzymes (Part C)
Edited by C. DAVID ALLIS AND CARL WU

VOLUME 378. Quinones and Quinone Enzymes (Part A)
Edited by HELMUT SIES AND LESTER PACKER

VOLUME 379. Energetics of Biological Macromolecules (Part D)
Edited by JO M. HOLT, MICHAEL L. JOHNSON, AND GARY K. ACKERS

VOLUME 380. Energetics of Biological Macromolecules (Part E)
Edited by JO M. HOLT, MICHAEL L. JOHNSON, AND GARY K. ACKERS

VOLUME 418. Embryonic Stem Cells
Edited by IRINA KLIMANSKAYA AND ROBERT LANZA

VOLUME 419. Adult Stem Cells
Edited by IRINA KLIMANSKAYA AND ROBERT LANZA

VOLUME 420. Stem Cell Tools and Other Experimental Protocols
Edited by IRINA KLIMANSKAYA AND ROBERT LANZA

VOLUME 421. Advanced Bacterial Genetics: Use of Transposons and Phage for Genomic Engineering
Edited by KELLY T. HUGHES

VOLUME 422. Two-Component Signaling Systems, Part A
Edited by MELVIN I. SIMON, BRIAN R. CRANE, AND ALEXANDRINE CRANE

VOLUME 423. Two-Component Signaling Systems, Part B
Edited by MELVIN I. SIMON, BRIAN R. CRANE, AND ALEXANDRINE CRANE

VOLUME 424. RNA Editing
Edited by JONATHA M. GOTT

VOLUME 425. RNA Modification
Edited by JONATHA M. GOTT

VOLUME 426. Integrins
Edited by DAVID CHERESH

VOLUME 427. MicroRNA Methods
Edited by JOHN J. ROSSI

VOLUME 428. Osmosensing and Osmosignaling
Edited by HELMUT SIES AND DIETER HAUSSINGER

VOLUME 429. Translation Initiation: Extract Systems and Molecular Genetics
Edited by JON LORSCH

VOLUME 430. Translation Initiation: Reconstituted Systems and Biophysical Methods
Edited by JON LORSCH

VOLUME 431. Translation Initiation: Cell Biology, High-Throughput and Chemical-Based Approaches
Edited by JON LORSCH

VOLUME 432. Lipidomics and Bioactive Lipids: Mass-Spectrometry–Based Lipid Analysis
Edited by H. ALEX BROWN

VOLUME 433. Lipidomics and Bioactive Lipids: Specialized Analytical Methods and Lipids in Disease
Edited by H. ALEX BROWN

SECTION ONE

CHEMICAL AND FUNCTIONAL PROBING OF RNA STRUCTURE, INTERACTIONS, AND FOLDING

CHAPTER ONE

Nucleotide Analog Interference Mapping

Ian T. Suydam *and* Scott A. Strobel

Contents

Abstract

Nucleotide analog interference mapping (NAIM) is a powerful chemogenetic technique that rapidly identifies chemical groups essential for RNA function. Using a series of phosphorothioate-tagged nucleotide analogs, each carrying different modifications of nucleobase or backbone functionalities, it is possible to simultaneously, yet individually, assess the contribution of particular functional groups to an RNA's activity at every position within the molecule. In contrast to traditional mutagenesis, which modifies RNA on the nucleobase

Department of Molecular Biophysics and Biochemistry and Department of Chemistry, Yale University, New Haven, Connecticut, USA

Methods in Enzymology, Volume 468
ISSN 0076-6879, DOI: 10.1016/S0076-6879(09)68001-0

level, the smallest mutable unit in a NAIM analysis is a single atom, providing a detailed description of interactions at critical nucleotides. Because the method introduces modified nucleotides by *in vitro* transcription, NAIM offers a straightforward and efficient approach to study any RNA that has a selectable function, and it can be applied to RNAs of nearly any length.

1. Introduction

Structured RNAs rely on a large and diverse collection of functional group interactions to form folded states in active conformations (Noller, 2005). One approach to studying these interactions is the site-specific modification of functional groups, followed by a direct assay of activity. These experiments are most often performed using chimeric RNAs, typically generated by reassembling the full-length sequence from a substituted synthetic oligonucleotide and a truncated RNA transcript (Das and Piccirilli, 2005; Lafontaine *et al.*, 2002). Although powerful, these techniques are limited by the difficulty of assembling and measuring the effect of each singly substituted RNA. For this reason, site-specific substitution experiments are often restricted to a small number of modifications at a minimal number of sites. In contrast, the method described in this section provides a semiquantitative approach to rapidly assay the effect of a nucleotide modification at all sites simultaneously.

Nucleotide analog interference mapping (NAIM) utilizes a series of 5′-*O*-(1-thio)nucleoside triphosphate analogs, which are randomly incorporated at low frequency into an RNA with a selectable function. The method relies on two advantageous properties of nucleoside phosphorothioates. First is the observation that 5′-*O*-(1-thio)nucleoside triphosphates can be incorporated by T7 RNA polymerase (Eckstein, 1985). In these transcriptions, the polymerase accepts only the S_p diastereomer, a result explained by the Mg^{2+} coordination pattern observed in recent crystal structures (Fig. 1.1A) (Yin and Steitz, 2004). Because incorporation proceeds with inversion of configuration, the resulting RNA transcripts contain R_p phosphorothioate linkages (Griffiths *et al.*, 1987). When parental phosphorothioate-tagged nucleotides (NTPαS) are used, they are incorporated as readily as the corresponding nucleoside triphosphates (NTPs) (Christian and Yarus, 1992). For these phosphorothioate-tagged nucleotides, the level of incorporation is directly related to the molar ratios of NTPαS and NTP in the transcription reaction (Fig. 1.1B). The second property of phosphorothioates exploited by NAIM is the selective cleavage of phosphorothioate linkages by treatment with iodine (Fig. 1.1C). Although specific, the extent of cleavage at substituted sites is only 15%, which may indicate the formation of a phosphotriester intermediate, the

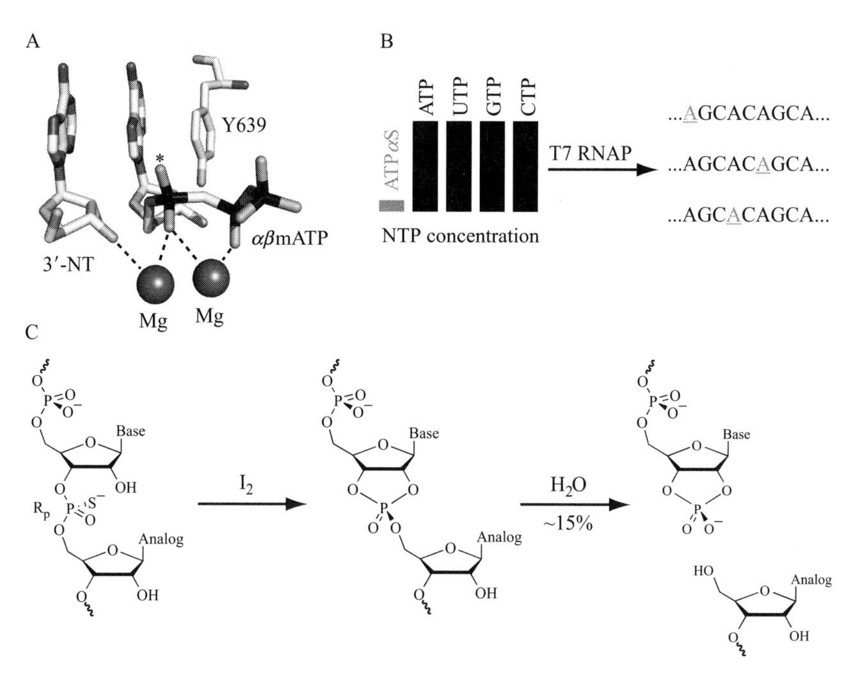

Figure 1.1 (A) Crystal structure of α-, β-methylene ATP bound to the active site of T7 RNA polymerase (PDB #1s76). The location of the sulfur substitution in S_p phosphorothioate-tagged triphosphates is indicated with an asterisk. (B) Schematic of ATPαS incorporation by T7 RNA polymerase leading to transcripts with a low level of substitution at random sites (underlined). (C) Cleavage of resulting R_p phosphorothioate linkage with iodine.

hydrolysis of which produces cleavage products or phosphodiester linkages (Gish and Eckstein, 1988). This selective cleavage has been used in RNA sequencing strategies (Gish and Eckstein, 1988), and provides a method of marking the site of functional group modification by the inclusion of a phosphorothioate tag.

The NAIM method expands on the technique of interference mapping by phosphorothioate substitution, and the initial work of Krupp and coworkers (Conrad *et al.*, 1995; Gaur and Krupp, 1993; Ruffner and Uhlenbeck, 1990). The method uses a series of 5′-*O*-(1-thio)nucleoside triphosphate analogs (δTPαS), each containing an incremental chemical modification in the nucleobase or ribose sugar, and consists of four basic steps. First, the parental and modified phosphorothioate-tagged triphosphates are randomly incorporated into the RNA by *in vitro* transcription (Fig. 1.2, step 1). In this step, the level of analog incorporation is adjusted such that any given transcript has approximately one site of substitution. Next, the pool of transcribed RNAs is subjected to a selection in which the active and inactive members are either physically separated or selectively

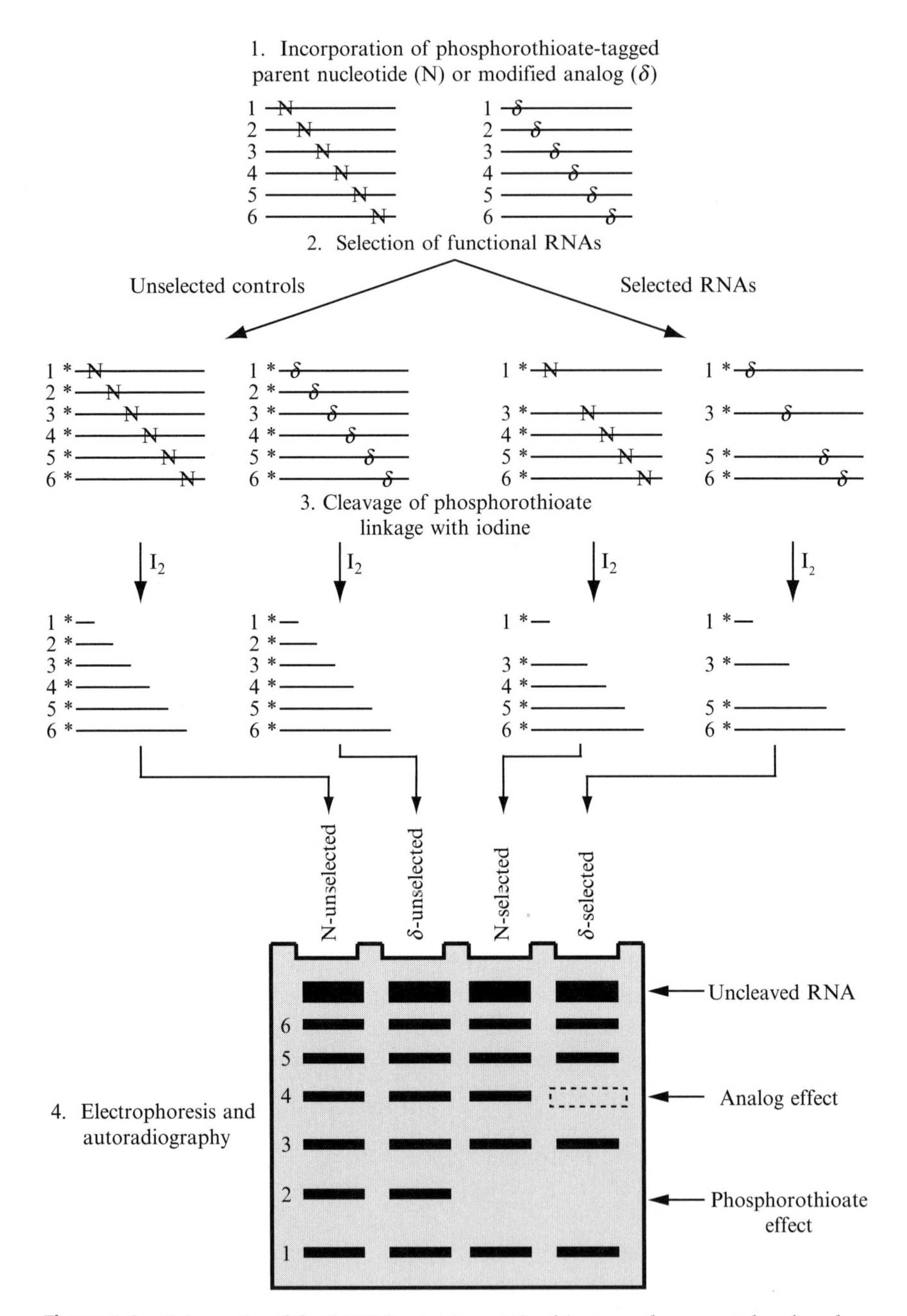

Figure 1.2 Schematic of the NAIM experiment. In this example, parental and analog-modified nucleotides are incorporated with equal efficiency at all sites. In the activity selected pool the R_p phosphorothioate interferes at position 2, while the modified analog interferes at position 4.

labeled (Fig. 1.2, step 2). Unselected RNAs from the original pool are also labeled to serve as controls. Each sample is then treated with iodine to selectively cleave the RNAs at the site of analog substitution (Fig. 1.2, step 3). Finally, the analog distribution at each site in the selected and unselected samples is revealed by resolving the cleavage products with polyacrylamide gel electrophoresis (PAGE) (Fig. 1.2, step 4). Sites of analog substitution that interfere with the selected function are identified as gaps in the sequencing ladder from the active pool of RNAs. Selections carried out with phosphorothioate-tagged parental nucleotides control for sites of interference that arise from the backbone sulfur substitution. Because every position in the sequence is resolved as a unique band on the sequencing gel, a single screen defines the effect of a particular modification at every site in the RNA.

NAIM can be expanded to use any analog that is incorporated by an RNA polymerase and can be applied to any RNA with a selectable function. Examples of RNA functions previously investigated with NAIM include RNA folding (Schwans *et al.*, 2003; Waldsich and Pyle, 2007), catalytic activity (Boudvillain and Pyle, 1998; Kazantsev and Pace, 1998; Oyelere *et al.*, 2002; Ryder and Strobel, 1999; Sood *et al.*, 2002; Strobel and Shetty, 1997), and the binding of proteins (Batey *et al.*, 2000), metal ions (Basu *et al.*, 1998), and small molecule metabolites (Jansen *et al.*, 2006; Kwon and Strobel, 2008). Many of these studies have identified specific interactions or RNA structural motifs that have been confirmed with subsequent atomic resolution structures (Adams *et al.*, 2004; Klein and Ferre-D'Amare, 2006; Ryder and Strobel, 2002; Toor *et al.*, 2008). More recently, the technique has been extended to study RNA interactions *in vivo* (Kolev and Steitz, 2006; Szewczak *et al.*, 2002). Several descriptions of the NAIM method are currently available (Ryder *et al.*, 2000; Strobel, 1999), including detailed protocols for specific applications (Cochrane and Strobel, 2004; Waldsich, 2008). Here, we focus on the methods required to synthesize and employ new analogs, including guidance in applying NAIM to new systems. Whenever multiple protocols have been used, we present those that have most efficiently produced high-quality results in a number of systems.

2. Materials and Reagents

2.1. Chemicals

- *5′-O-(1-thio)nucleoside triphosphates.* There are currently 20 commercially available phosphorothioate-tagged nucleoside triphosphates, including each parental NTPαS (Glen Research). A number of unprotected nucleosides are also available from a variety of sources that can be used to synthesize phosphorothioate-tagged nucleotide triphosphates (Sigma-Aldrich, RI Chemical, Toronto Research Chemical, TCI America, Berry & Associates).

- [γ-^{32}P]-adenosine 5′-triphosphate (6000 Ci/mmol, PerkinElmer #NEG-035C).
- All other chemicals and reagents were purchased from Sigma-Aldrich.

2.2. Equipment

- *Phosphor autoradiography.* All sequencing gels were visualized and quantitated with a Storm 820 Phosphorimager using ImageQuant 5.2 software (GE Healthcare).
- AKTA prime FPLC purification system (GE Healthcare).
- HiTrap Q HP anion-exchange columns (GE Healthcare #17-1153-01).
- QIAquick Nucleotide Removal Kit (Qiagen #28304).

2.3. Enzymes

- Appropriate Restriction Endonuclease (New England Biolabs).
- *T7 RNA polymerase.* Overexpressed from plasmid pT7-911Q (T.E. Shrader, personal communication) or purchased (New England Biolabs #M0251).
- *Y637F mutant form of T7 RNA polymerase.* Overexpressed from plasmid pDPT7-Y639F as described (Sousa and Padilla, 1995) or purchased (Epicentre Biotechnologies #D7P9201K).
- *Escherichia coli* Inorganic Pyrophosphatase (Sigma #I5907).
- Calf Intestinal Alkaline Phosphatase (New England Biolabs #M0290).
- T4 Polynucleotide Kinase (New England Biolabs #M0201).

2.4. Buffers

- *TEAB.* Triethylammonium bicarbonate, pH 8.0. Prepared by bubbling CO_2 through a suspension of triethylamine in water.
- *Transcription buffer.* 40 m*M* Tris–HCl (pH 8.0), 2.0 m*M* spermidine, 5.0 m*M* dithiothreitol (DTT), 0.005% Triton X-100, $MgCl_2$ (dependent on transcript).
- *Elution buffer.* 10 m*M* Tris–HCl (pH 7.5), 250 m*M* NaCl, 2% sodium dodecyl sulfate (SDS).
- *PCA.* Mixture of phenol/chloroform/isoamyl alcohol in ratio of 25:24:1, saturated with 10 m*M* Tris–HCl (pH 8.0), 1 m*M* EDTA.
- *TE.* 10 m*M* Tris–HCl (pH 7.5), 0.1 m*M* EDTA.
- *VS ribozyme reaction buffer.* 50 m*M* Tris–HCl (pH 7.5), 5 m*M* $MgCl_2$, 25 m*M* KCl.

- *2× loading buffer.* Prepared in deionized formamide, 20 m*M* EDTA, 0.05% (w/v) bromophenol blue, 0.05% (w/v) xylene cyanol FF.
- *Iodine/ethanol.* 100 m*M* iodine in 100% ethanol.

3. Methods

3.1. Synthesis of phosphorothioate-tagged nucleoside triphosphates

Several methods are available for the synthesis of 5′-*O*-(1-thio)nucleoside triphosphates, including enzymatic synthesis of pure S_p isomers (Eckstein, 1985). Chemical methods generally produce equimolar mixtures of R_p and S_p diastereomers, which can be separated by reverse-phase HPLC (Fischer *et al.*, 1999). However, these separations are not required for NAIM, since T7 polymerase is selective for the S_p isomer, and transcriptions for NAIM are often carried out using mixtures of both isomers. Although several high yielding chemical syntheses are now available starting from protected nucleosides, the simple one-pot method developed by Arabshahi and Frey (1994) is still preferred for most preparations. The modified procedure outlined below has been used to synthesize a wide range of purine and pyrimidine analogs in 5–40% yield starting from unprotected nucleosides. The synthesis is accomplished using limited equipment with commercially available reagents in less than 2 days. A typical reaction employs 100 mg of starting material, but we have routinely scaled the reaction down to as little as 25 mg of nucleoside. It is worth noting that even on this scale an overall yield of 5% produces sufficient analog triphosphate for 50–500 typical NAIM transcriptions (see below).

The conversion of unprotected nucleosides to 5′-*O*-(1-thio)nucleoside triphosphates is carried out in two steps using a single reaction flask (Fig. 1.3A). In the first step, the nucleoside is converted to the 5′-*O*-(1,1-dichloro-1-thio)phosphorylnucleoside by the addition of thiophosphoryl chloride ($PSCl_3$). This reaction has been shown to be regioselective for the 5′-hydroxyl when carried out in triethyl phosphate (TEP) (Arabshahi and Frey, 1994). In the second step, the initial product is converted directly to the triphosphate by the addition of tributylammonium pyrophosphate (TBAP). This reaction proceeds via a cyclic triphosphate intermediate which is subsequently hydrolyzed with water (Fig. 1.3A). Two factors greatly affect the overall yield of the reaction. The first is the concentration of reactants in the first step, so every effort should be made to dissolve the nucleoside in a minimal volume of TEP, aided by slight heating if necessary. The second and more important consideration is the strict exclusion of water in the initial steps to prevent the hydrolysis of $PSCl_3$ or the initially

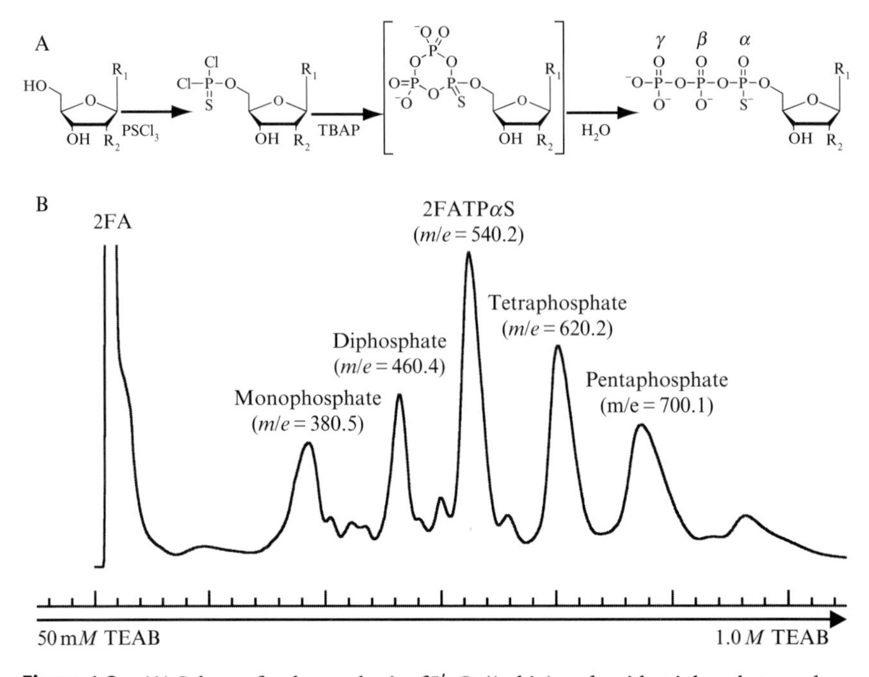

Figure 1.3 (A) Scheme for the synthesis of 5′-*O*-(1-thio)nucleoside triphosphates, where R_1 represents nucleobase analogs and R_2 represents 2′-ribose modifications. The unstable cyclic triphosphate intermediate is outlined in brackets. (B) Purification of the phosphorothioate-tagged 2-fluoroadenosine triphosphate by anion-exchange chromatography. The observed mass obtained with negative ion mode mass spectra of each elution peak indicates the formation of mono-, di-, tri-, tetra-, and pentaphosphate products.

formed 5′-*O*-(1,1-dichloro-1-thio)phosphorylnucleoside. For this reason, all glassware and syringes should be thoroughly dried, anhydrous solvents should be purchased in sealed containers, and reactions should be kept under an atmosphere of dry argon. Because TBAP is hygroscopic the introduction of water with this reagent is of particular concern. We have had good success purchasing a small quantity of the reagent and using it in its entirety after opening. The general procedure that follows has no strict timing requirements and is easily run in parallel, allowing the rapid synthesis of multiple analogs from a single stock of reagents.

The nucleoside (100 mg) is transferred to a 20-ml round-bottom flask and dried by coevaporating 3 × 10 ml of anhydrous pyridine and put under vacuum overnight. Once the nucleoside is dry the reaction flask is capped with a rubber septum and purged with dry argon. The nucleoside is then dissolved in a minimal volume (0.5–2.0 ml) of TEP. For less soluble nucleosides, the reaction flask can be heated gently with an air gun to obtain a clear solution. As was reported in the original description of the synthesis, we have observed reactivity differences between purine and pyrimidine nucleosides and have employed a slightly different procedure for the synthesis of their 5′-*O*-(1,1-dichloro-1-thio)phosphorylnucleosides.

For purine nucleoside analogs, 1.1 equivalents of trioctylamine ($M_W = 353.7$, $d = 0.809$ g/ml) is added to the stirred nucleoside solution, followed by 1.1 equivalents of $PSCl_3$ ($M_W = 169.4$, $d = 1.668$ g/ml). This solution is stirred at room temperature for 20–60 min until a significant amount of 5′-*O*-(1,1-dichloro-1-thio)phosphorylnucleoside has formed. The reaction progress can be monitored qualitatively by quenching a small aliquot (100–200 μl) in water and resolving the products with cellulose thin-layer chromatography using a solvent system of 0.5 *M* LiCl in water. The reaction products are visualized by shadowing with a hand held UV light, with the monophosphate migrating slower than the starting material. Alternatively, small aliquots (10–20 μl) of reaction can be quenched in 50 m*M* triethylammonium bicarbonate (TEAB) and loaded directly onto a prepacked anion-exchange column (see below). Elution with a steep gradient of TEAB easily separates the nucleoside from the monophosphate and provides a rapid quantitation of reaction progress.

For pyrimidine analogs the dry nucleoside is dissolved in a minimal volume (0.5–2.0 ml) of TEP as above, then cooled to 0 °C. To the cooled solution 1.2 equivalents of trioctylamine and 1.2 equivalents of collidine ($M_W = 121.2$, $d = 0.931$ g/ml) are added, followed by the dropwise addition of 1.3–1.5 equivalents of $PSCl_3$. After stirring at 0 °C for 30 min, the reaction is allowed to warm slowly to room temperature and stirred for an additional 45 min with the reaction progress monitored as above. For all subsequent chemical steps and purifications the procedure is the same for purine and pyrimidine analogs.

TBAP ($M_W = 548.7$) is dissolved in TEP to a final concentration of 0.1 g/ml by gently heating the solution with an air gun. Once a substantial quantity of 5′-*O*-(1,1-dichloro-1-thio)phosphorylnucleoside from the preceding step has formed the TBAP/TEP solution is added to the reaction until the final molar ratio of TBAP to $PSCl_3$ has reached 4:1. The addition of 4 equivalents of TBAP increases the overall yield of the triphosphate product but can also increase the quantity of branched higher order phosphate products (see below). The reaction is stirred at room temperature for 30 min and the reaction progress is monitored by rapid ion-exchange chromatography as above. Once a significant amount of triphosphate products are observed the reaction is exposed to air and transferred to a centrifuge tube. The nucleoside triphosphates are precipitated via amine exchange by the dropwise addition of 60 equivalents of triethylamine (TEA, $M_W = 101.2$, $d = 0.726$ g/ml) to the stirred reaction solution. The precipitate is collected by centrifugation, the supernatant is discarded and the white solid is dissolved in 5 ml of 50 m*M* TEAB. Incubation of this solution overnight at room temperature hydrolyzes the initially formed cyclic triphosphates to the linear form.

The 5′-*O*-(1-thio)nucleoside triphosphates are purified on prepacked anion-exchange columns using a linear gradient from 50 m*M* to 1.0 *M* TEAB while monitoring UV absorption at 260 nm. Poor binding to the matrix is often attributed to residual salt from the reaction and can be improved by further diluting the crude product in 50 m*M* TEAB before loading. The expected elution fraction can be identified by running a standard sample of the parental nucleoside triphosphate and is typically between 0.5 and 0.6 *M* TEAB. For some reactions several additional products are observed and these have been identified as the mono-, di-, tetra-, and pentaphosphates (Fig. 1.3B). The fractions containing the nucleoside triphosphate are pooled and lyophilized to a thin film, then redissolved in less than 1.0 ml TE. The pH of this solution should be checked and adjusted to near neutral if necessary to improve the stability of frozen stocks.

The 5′-*O*-(1-thio)nucleoside triphosphates are characterized by UV absorption, ^{31}P NMR, and mass spectroscopy. Because nucleobase modifications can produce significant changes in the UV spectrum it is important to measure extinction coefficients for the free nucleosides and to use these values when calculating nucleotide concentrations. NMR samples are prepared with 10% D_2O in buffered solutions (pH 8.0) and should contain approximately 1 m*M* nucleotide. The correct nucleotide has three resonances in the ^{31}P NMR spectrum at 42 to 43 (α), -5 to -10 (γ), and -20 to -25 (β) ppm relative to an external 85% phosphoric acid standard (Arabshahi and Frey, 1994). The distinctive downfield phosphorothioate resonance exists as a pair of doublets in a proton-decoupled spectrum, corresponding to the mixture of R_p and S_p diastereomers. Spectra taken without proton decoupling show additional splitting from the 5′-CH protons, which further confirms the 5′-*O*-(1-thio) linkage. An impurity that occasionally coelutes with the correct nucleotide and produces a ^{31}P NMR resonance between 33 and 38 ppm results from the reaction of unconsumed $PSCl_3$ with TBAP. Electrospray ionization mass spectra are obtained in negative ion mode from diluted nucleotide samples containing 95% methanol. A strong signal for the singly ionized parent ion is typically observed. Analogs can be stored as dry pellets or TE solutions at -20 °C for at least 2 years without substantial degradation.

3.2. Incorporation of phosphorothioate analogs by *in vitro* transcription

Nucleotide analogs are randomly incorporated into RNA by *in vitro* transcription using T7 RNA polymerase (Eckstein, 1985). For some analogs, particularly those that make modifications in the minor groove, the Y639F point mutant of T7 RNA polymerase greatly improves the level of incorporation (Sousa and Padilla, 1995) (Table 1.1). Transcription conditions are optimized to yield an analog incorporation level of approximately 5%. This level of incorporation provides iodine cleavage bands that are

Table 1.1 Conditions to incorporate nucleotide analogs for NAIM experiments

Analog δTPαS (S_p isomer only)	[δTPαS] (m*M*)	[Parent NTP] (m*M*)	Polymerase (WT/ Y639F)	References
AαS	0.05	1.0	WT	Christian and Yarus (1992)
DAPαS	0.025	1.0	WT	Strobel and Shetty (1997)
2AP	0.50	1.0	WT	Ortoleva-Donnelly *et al.* (1998b)
m^6AαS	0.10	1.0	WT	Ortoleva-Donnelly *et al.* (1998b)
PurαS	2.0	1.0	WT	Ortoleva-Donnelly *et al.* (1998b)
c^3AαS	1.0	0.5	Y639F	Ryder *et al.* (2001)
7dAαS	0.05	1.0	WT	Ortoleva-Donnelly *et al.* (1998b)
7F7dAαS	0.05	1.0	WT	Suydam and Strobel (2008)
2F7dAαS	0.2	1.0	WT	Suydam and Strobel (2008)
2FAαS	0.2	1.0	WT	Suydam and Strobel (2008)
FormAαS	0.05	1.0	WT	Ryder *et al.* (2001)
n^8AαS	0.25	1.0	WT	Ryder *et al.* (2001)
dAαS	0.75	1.0	Y639F	Ortoleva-Donnelly *et al.* (1998b)
^{OMe}AαS	2.0	0.2	Y639F	Ortoleva-Donnelly *et al.* (1998b)
^{F}AαS	0.25	1.0	Y639F	Ortoleva-Donnelly *et al.* (1998b)
^{SH}AαS	0.2	1.0	Y639F	Schwans *et al.* (2003)
GαS	0.05	1.0	WT	Christian and Yarus (1992)
IαS	0.1	1.0	WT	Strobel and Shetty (1997)
m^2GαS	0.75	0.5	Y639F	Ortoleva-Donnelly *et al.* (1998a)
S^6GαS (+Mn^{2+})	0.25	1.0	WT	Basu *et al.* (1998)
7dGαS	0.05	1.0	WT	Kazantsev and Pace (1998)
dGαS	0.25	1.0	Y639F	Szewczak *et al.* (1998)
^{OMe}GαS	2.0	0.1	Y639F	Conrad *et al.* (1995)
^{SH}GαS	0.25	1.0	Y639F	Schwans *et al.* (2003)
CαS	0.05	1.0	WT	Christian and Yarus (1992)
n^6CαS	0.5	0.5	WT	Oyelere and Strobel (2000)
f^5CαS	0.5	1.0	WT	Oyelere and Strobel (2000)
ΨiCαS	0.05	1.0	WT	Oyelere and Strobel (2000)
ZαS	0.50	0.5	Y639F	Szewczak *et al.* (1999)
dCαS	0.75	1.0	Y639F	Ryder and Strobel (1999)
^{OMe}CαS	2.0	0.05	Y639F	Conrad *et al.* (1995)
^{SH}CαS	0.2	1.0	Y639F	Schwans *et al.* (2003)
UαS	0.05	1.0	WT	Christian and Yarus (1992)

(*continued*)

Table 1.1 (*continued*)

Analog δTPαS (S_p isomer only)	[δTPαS] (m*M*)	[Parent NTP] (m*M*)	Polymerase (WT/ Y639F)	References
$m^5U\alpha S$	0.50	1.0	WT	Unpublished observation
$\Psi\alpha S$	0.05	1.0	WT	Unpublished observation
dUαS	0.25	1.0	Y639F	Szewczak *et al.* (1998)
$^{OMe}U\alpha S$	2.0	0.1	Y639F	Conrad *et al.* (1995)
$^{F}U\alpha S$	0.25	1.0	Y639F	Szewczak *et al.* (1998)
$^{SH}U\alpha S$	0.2	1.0	Y639F	Schwans *et al.* (2003)

intense enough to easily quantitate, but minimizes the possibility of cooperative interferences due to multiple substitutions within the same RNA. Incorporation above this level also increases the number of RNAs cleaved at more than one location by iodine, leading to an overrepresentation of short cleavage products in sequencing gels. The approximate conditions used to produce this level of incorporation for several analogs is provided in Table 1.1, but incorporation levels can be sequence dependent and may need to be reoptimized for a particular transcript. When incorporating new analogs the proper conditions are determined by carrying out multiple transcriptions with a range of analog concentrations, 5′-end labeling the transcripts, and comparing the intensity of the iodine cleavage pattern to the one obtained with the parent analog (see below).

Run-off transcriptions from DNA templates are carried out following standard procedures (Milligan and Uhlenbeck, 1989; Paschal *et al.*, 2008). If plasmid DNA is used the template is first linearized with the appropriate restriction enzyme. When designing these plasmids 3′-overhangs at the cut site should be avoided as these are known to lead to poor transcription (Milligan and Uhlenbeck, 1989). In many cases, the restriction enzyme can be heat-denatured and the cut plasmid used without further purification. Alternatively the reaction is PCA extracted, ethanol precipitated, and resuspended. It is essential that the DNA template is completely digested as even a small amount of undigested template drastically reduces the desired transcript. For short transcripts the DNA template can also be quickly assembled by PCR amplification of synthesized oligonucleotides. A typical transcription contains 1× transcription buffer, 10–20 m*M* $MgCl_2$ (optimized for transcript), 0.05 μg/μl linearized DNA template, 0.001 U/μl inorganic pyrophosphatase, 5 U/μl T7 RNA polymerase, 1.0 m*M* NTPs, and analog concentrations according to Table 1.1. The reactions are incubated for 2 h at 37 °C. If a self-cleaving ribozyme has been incorporated into the transcript for uniform end formation the transcription reaction is heated to 65 °C for 2 min and slow cooled to room temperature. The transcripts are purified by denaturing PAGE, visualized by UV shadowing, excised from the gel, and eluted by rocking in elution buffer for 1 h at room temperature. The filtered elution is PCA extracted, ethanol precipitated and redissolved in a minimal volume of TE. RNA concentrations are determined by UV absorption at 260 nm.

3.3. Incorporation efficiency controls

Before performing NAIM selections the level of analog incorporation at each site of the transcript must be determined. For new analogs these controls identify the concentration of analog required for an incorporation level of approximately 5% and ensures that the analog is incorporated specifically at coded positions. Incorporation controls also quantitate the level of incorporation at each site, which has been shown to be sequence specific, but reproducible for a given sequence. This sequence-dependent incorporation must be accounted for when analyzing sites of interference (see below). NAIM selections should be carried out using the same RNAs as the incorporation controls, so a sufficient quantity of RNA for the controls and activity selections should be transcribed once the incorporation efficiency is determined.

Incorporation levels are determined by 5′-end labeling analog-incorporated RNAs, cleaving with iodine and resolving the cleavage products by gel electrophoresis. The efficiency and distribution of analog incorporation is determined by comparing band intensities from the parental nucleotide and nucleotide analog cleavage patterns. For this reason incorporation controls for all analogs and the parent nucleotide should be included on a single gel. Transcripts retaining a 5′-terminal phosphate are first treated with calf intestinal alkaline phosphatase (CIP). Ten picomoles of each RNA are treated 50 units of CIP for 1 h at 37 °C. CIP is removed either by PCA extraction or more conveniently by the use of commercially available silica membrane spin columns (i.e., QIAquick® Nucleotide Removal Kit). The phosphatase step can be eliminated if a self-cleaving ribozyme is inserted in the leading sequence or if the transcription is initiated with guanosine. RNAs are 5′-end labeled with 20 units of T4 polynucleotide kinase (PNK) and 150 μCi of [γ-^{32}P]ATP by incubating at 37 °C for 1 h. Labeled RNAs are purified by denaturing PAGE, visualized by autoradiography, excised from the gel, and eluted by rocking in elution buffer for 1 h at room temperature. Short elution times in buffers containing SDS minimize the level of RNA degradation. RNAs are PCA extracted, ethanol precipitated and redissolved in a minimal volume of TE.

To normalize the loading of sequencing gels a small aliquot of each RNA solution is scintillation counted. Each labeled RNA is then combined with TE and 2× loading buffer to provide a 10 μl solution that contains 200,000 cpm, which is split into two 5 μl aliquots. One aliquot is treated with 0.5 μl of 100 mM I_2/ethanol, while the other is treated with 0.5 μl of ethanol alone to control for nonspecific degradation. Because the iodine cleavage reaction has an endpoint of approximately 15% at least 100,000 cpm should be loaded in each lane to accurately quantitate band intensities. To improve resolution in the sequencing gel samples are heated to 90 °C for 1 min and cooled on ice before loading. Gels are prerun for at least 2 h, loaded with iodine treated and untreated samples for the parent nucleotide and all analogs, and run for various lengths of time to maximally resolve different portions of the sequence. A phosphor storage screen is exposed to the dried gel overnight and imaged. Individual band intensities are quantitated by area integration of line profiles drawn through each lane (see below). The quality of the cleavage ladder is typically poorest for very short fragments and those with lengths closest to the uncleaved RNA. For difficult regions, the RNA can alternatively be labeled at its 3′-end with [^{32}P]pCp and T4 RNA ligase or additional noninteracting sequence can be added to the transcript at the 5′- or 3′-end (Kwon and Strobel, 2008).

3.4. Activity selections

NAIM studies require an assay to separate active and inactive RNAs from a pool of transcripts containing analog substitutions at random locations. A wide variety of selection strategies appropriate for NAIM studies have been developed, and detailed protocols for several of these methods have been published elsewhere (Cochrane and Strobel, 2004; Waldsich, 2008). In general, two broad categories of selections have been used. The first is to physically separate the active and inactive fractions, where the assay selects for a change in the RNAs' length, fold, or interaction with a binding partner. Separations are carried out with denaturing PAGE, native gel mobility shift assay, column chromatography, or filter binding. This approach has been used to investigate functional groups required for ribozyme activity, RNA folding, metabolite binding to riboswitches, and protein–RNA or RNA–RNA interactions (Cochrane *et al.*, 2003; Kwon and Strobel, 2008; Oyelere *et al.*, 2002; Waldsich and Pyle, 2007). In these studies, the analog-substituted RNAs are radiolabeled at their 3′- or 5′-end either before or after the selection, and band intensities can often be quantitated for both the active and inactive fractions. The second class of selections relies on the specific radiolabeling of the active fraction. This approach has been used to study ribozyme-mediated ligation reactions where a radiolabeled substrate is transferred to active members of the RNA pool (Jones and Strobel, 2003; Ryder and Strobel, 1999; Strobel *et al.*, 1998). A major advantage of these types of selections is that no separation step is required since only the active members become radiolabeled and contribute to the iodine cleavage pattern. Regardless of the type of selection employed the assay should be as stringent as possible. For equilibrium assays involving RNA binding or folding, the reaction conditions are typically set such that the active fraction is the minor component. For kinetic assays, stringent selections are produced by quenching the reactions as early in the time course as possible while still obtaining enough radiolabeled product for high-quality sequencing gels.

An example of a ribozyme-mediated ligation selection is provided below (Fig. 1.4). In this study, fluorine-substituted adenosine analogs were used to identify sites of functional ionization in a self-ligating form of the Varkud Satellite (VS) ribozyme (Suydam and Strobel, 2008). These analogs have N1 pK_a values spanning more than 4 pK_a units as measured for the free nucleosides; 7-deazaadenosine (7dA, pK_a 5.2), 7-fluoro-7-deazaadenosine

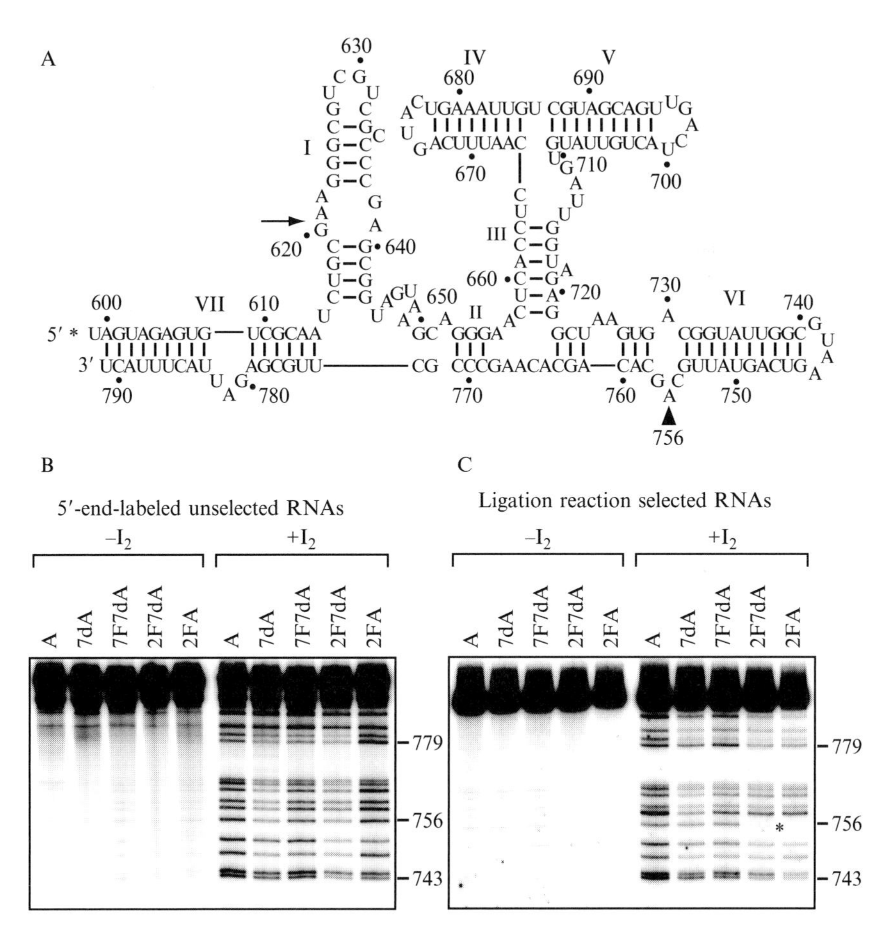

Figure 1.4 (A) Secondary structure of the Varkud Satellite ribozyme construct used for NAIM studies of ribozyme-catalyzed ligation. The site of ligation is marked with an arrow. (B) Sequencing gel of unselected RNAs labeled at their 5′-end and treated with or without iodine. (C) Sequencing gel of ligation selected RNAs treated with or without iodine. The site of pK_a-dependent interference is denoted with an asterisk. Additional gels resolved all sites of incorporation from A645 to A779.

(7F7dA, pK_a 4.5), adenosine (A, pK_a 3.5), 2-fluoro-7-deazaadenosine (2F7dA, pK_a < 1), and 2-fluoroadenosine (2FA, pK_a < 1). The activity selection in this system is the ligation of a 5′-end labeled substrate (NT 599–620) onto the ribozyme sequence (NT 621–791) (Jones and Strobel, 2003). An RNA sequence including a Hammerhead ribozyme at the 5′-end of the desired VS sequence is transcribed in the presence of analog nucleoside triphosphates according to Table 1.1. Hammerhead processing during transcription produces analog-substituted ribozyme RNA with a free 5′-hydroxyl at A621. A portion of these transcripts are labeled directly with PNK, purified by denaturing PAGE, iodine cleaved and loaded onto 8% denaturing sequencing gels for unselected incorporation efficiency controls (Fig. 1.4B).

Prior to the reaction, unlabeled ribozyme RNAs are incubated in reaction buffer at 50 °C for 15 min and cooled slowly to room temperature. Radiolabeled substrate RNA is produced with the appropriate 2′,3′-cyclic phosphate as described previously (Jones *et al.*, 2001). Ligation reactions containing 0.4 μM ribozyme and 0.05 μM substrate are initiated by the addition of 5′-end labeled substrate (>1,000,000 cpm). Reactions are allowed to proceed at room temperature until the fraction ligated reaches ~15%, then quenched with 1 volume of 2× loading buffer. The quenched reactions are iodine cleaved and loaded onto 8% denaturing sequencing gels as described for the unselected incorporation efficiency controls (Fig. 1.4C). Quantitating individual band intensities for these gels as described below identified a single adenosine, A756, whose interference pattern is consistent with function ionization (Suydam and Strobel, 2008).

3.5. Quantitating sites of interference

The analysis of NAIM data requires the inclusion of all relevant samples on a single gel. For every analog tested in a NAIM experiment, the relevant gel must also contain samples in which the parent nucleotide has been tested in a parallel experiment. Unselected controls can be run separately from the activity selected RNAs, but these gels must also contain lanes for all analogs and the parent nucleotide. Both types of gels also include lanes for parent and analog-incorporated RNAs that have not been iodine treated (Fig. 1.4B and C). These lanes serve as a control for nonspecific degradation of the RNA. Sites of nonspecific degradation often include unpaired sections of sequence as well as degradation products with lengths close to that of the uncleaved RNA. Positions of strong degradation are uninformative in NAIM experiments so every effort should be made to keep solutions RNAse free, to purify RNAs quickly, and to run sequencing gels soon after labeling reactions.

Sites of interference correspond to individual bands that are missing or underrepresented in the activity selected sequencing gels relative to the unselected controls. In some systems enhancements are also observed where bands are overrepresented in the activity selected lanes. The strongest interferences are easily identified as gaps in the activity selected sequencing ladder (Fig. 1.4C). To quantitate the level of interference at each site the peak intensities at every position of incorporation for both the parental nucleotide (NαS) and the nucleotide analog ($\delta\alpha$S) is calculated in both the activity selected and unselected sequencing gels. Line profiles are drawn through the appropriate lanes and peak areas are calculated by area integration. It is important to draw equivalent lines through each lane and to carefully define a consistent baseline for each before calculating the areas. Automated software has recently become available that may facilitate this process (Laederach *et al.*, 2008). The quality of the line profiles and subsequent integration depends on the overall intensity of the cleavage ladder, the even migration of the lanes and the resolution of the individual bands, so gels should be run to optimize these features. The extent of interference at each position is calculated by substituting the individual band intensities into Eq. (1.1):

$$\kappa = \mathrm{NF} \times \frac{(\mathrm{N}\alpha\mathrm{S\,selected}/\delta\alpha\mathrm{S\,selected})}{(\mathrm{N}\alpha\mathrm{S\,unselected}/\delta\alpha\mathrm{S\,unselected})}. \tag{1.1}$$

The normalization factor NF is introduced to account for uneven loading in the four lanes used to calculate the interference values. It is obtained by calculating the average of all interference values within two standard deviations of the mean prior to normalization, and is typically between 0.8 and 1.2. Interference values calculated with Eq. (1.1) normalize both for the extent of analog incorporation at a particular site and for the effect of the phosphorothioate substitution at that site. Sites of very weak incorporation or strong phosphorothioate interference cannot be quantitated in this way. Quantitation of interference values at over 100 sites in the group I intron provided an early experimental estimate for the noise in NAIM experiments (Ortoleva-Donnelly *et al.*, 1998b). A conservative interpretation of NAIM data defines sites of interference as those with κ greater than 2.0, sites of enhancements as those with κ less than 0.5, and sites with intervening values as showing no effect. Due to the difficulty in quantitating very weak bands in the background of minor degradation κ-values greater than 6.0 should be considered equivalent.

4. Properties of Analogs

Almost 40 nucleotide analogs have been utilized in NAIM to date (Figs. 1.5–1.8). While these analogs represent a great diversity of functional group modifications, there remains a wide variety of nucleoside analogs

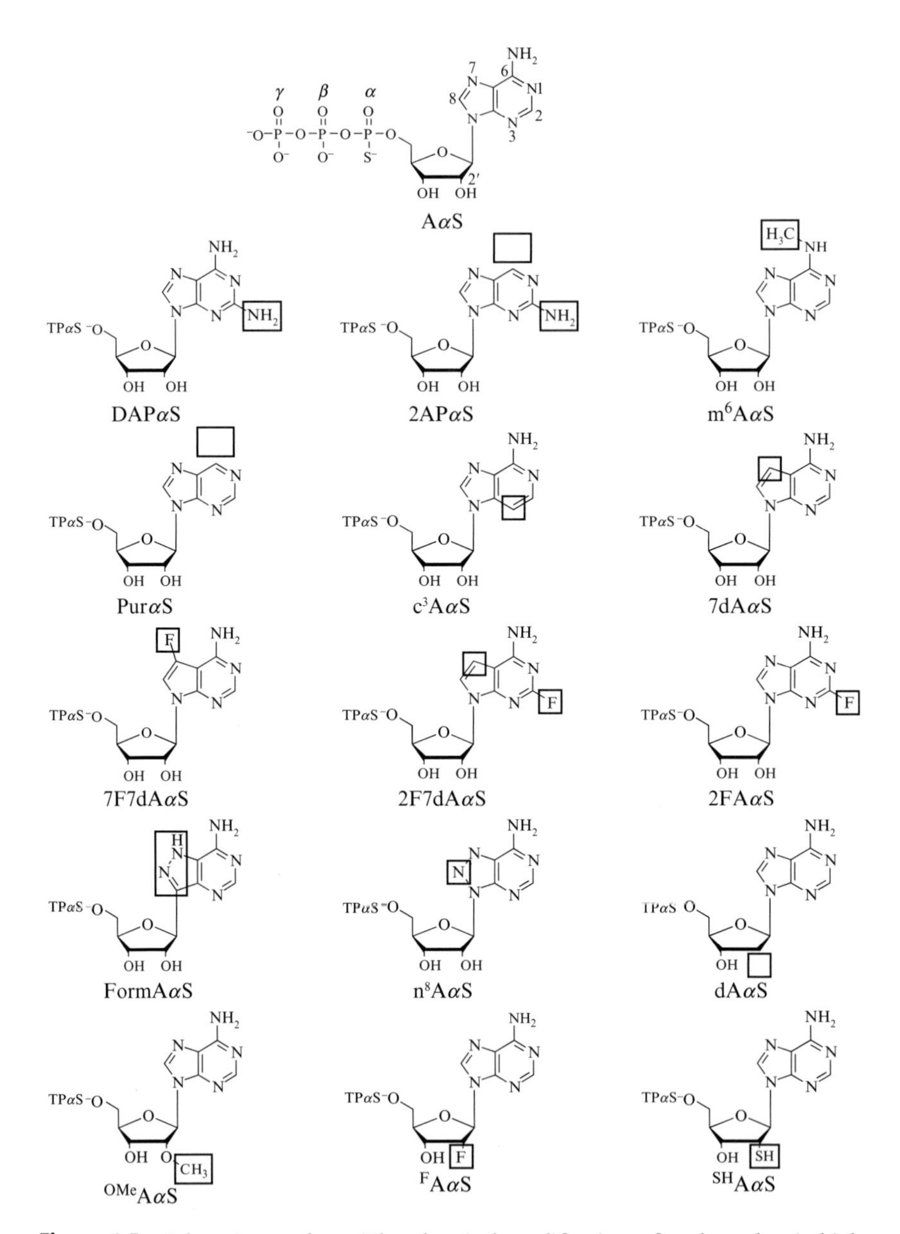

Figure 1.5 Adenosine analogs. The chemical modification of each analog is highlighted with a box.

currently untested in NAIM studies. Many of these nucleosides are commercially available or have been synthesized in sufficient quantities for the preparation of 5′-*O*-(1-thio)nucleoside triphosphates. The application of new analogs will continue to expand the types of RNA interactions

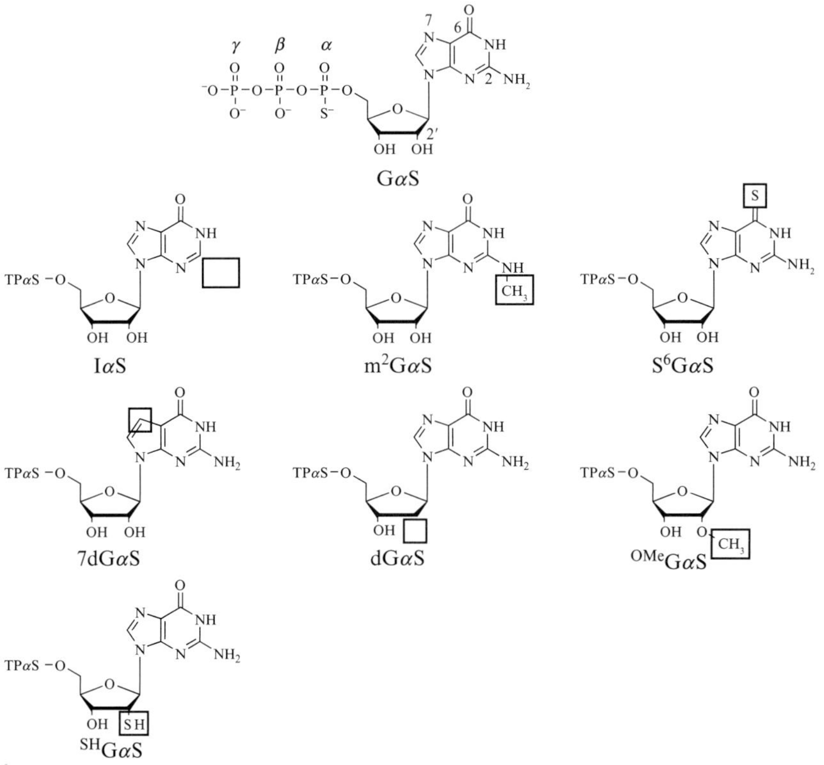

Figure 1.6 Guanosine analogs. The chemical modification of each analog is highlighted with a box.

revealed by the NAIM method, many of which are described below. Nucleotide analogs can be approximately classified into three groups: those deleting a functional group, those adding steric bulk to the functional group, and those altering the chemical nature of the functional group. Many analogs clearly modify more than one property of the nucleotide, so whenever possible a series of closely related analogs should be employed to dissect these effects. Nucleotide analog interference suppression (NAIS) has also been used to refine the role of functional groups at specific sites of interference (described below).

4.1. Ribose modifications

Four types of ribose modifications have been utilized in NAIM, each altering the 2′-position. These include 2′-deoxy modifications (dAαS, dGαS, dCαS, dUαS), 2′-*O*-methyl modifications (^{OMe}AαS, ^{OMe}GαS,

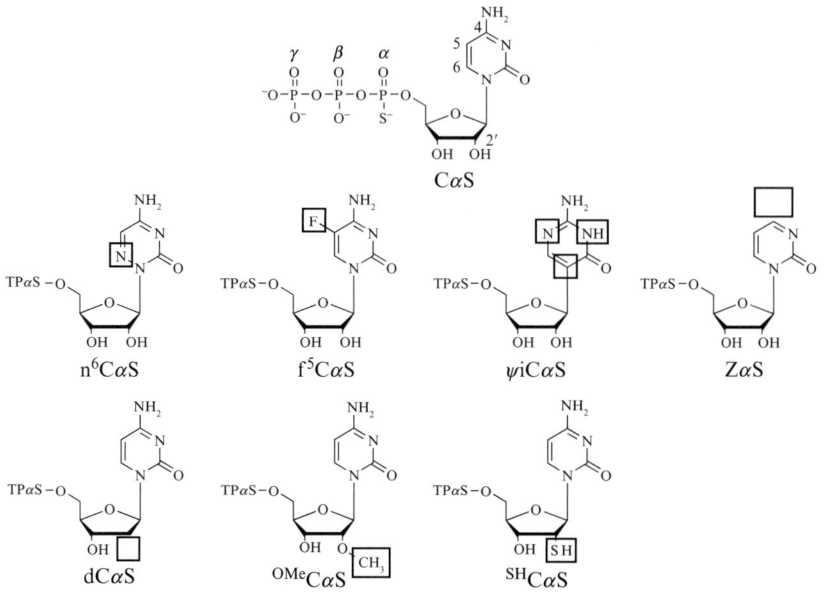

Figure 1.7 Cytosine analogs. The chemical modification of each analog is highlighted with a box.

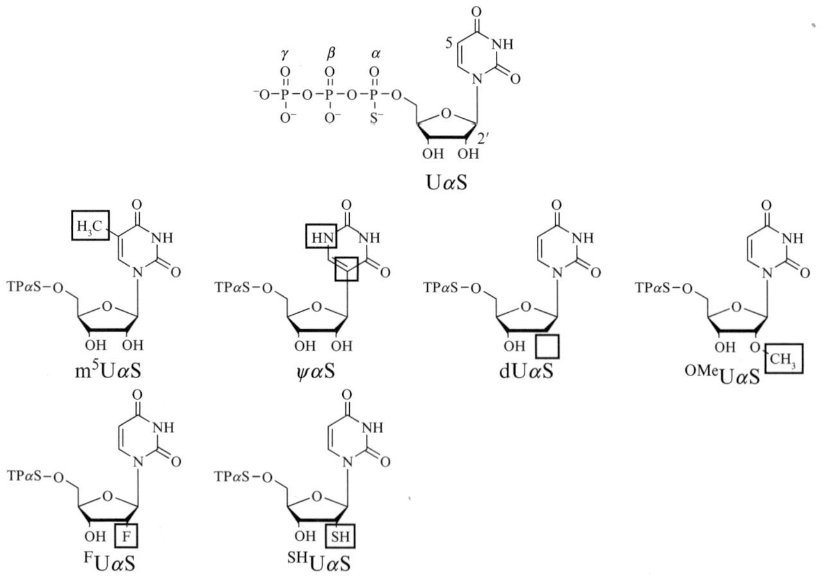

Figure 1.8 Uridine analogs. The chemical modification of each analog is highlighted with a box.

^{OMe}CαS, ^{OMe}UαS), 2′-deoxy-2′-fluoro modifications (^{F}AαS, ^{F}UαS), and 2′-deoxy-2′-thio modifications (^{SH}AαS, ^{SH}GαS, ^{SH}CαS, ^{SH}UαS). Each of these analogs is more readily incorporated by the Y639F point mutant of T7 RNA polymerase and may require increased analog concentration, decreased NTP concentration, or both for even incorporation (Table 1.1). These analogs probe the contribution of individual 2′-hydroxyls within an RNA (Conrad *et al.*, 1995; Gaur and Krupp, 1993; Ortoleva-Donnelly *et al.*, 1998b; Schwans *et al.*, 2003; Szewczak *et al.*, 1998). Interferences observed with dNαS substitutions identify those 2′-OH groups required for function, while the interference patterns observed with ^{OMe}NαS and ^{F}NαS analogs help define the functional role of these 2′-OH groups. Although 2′-*O*-methyl and 2′-deoxy-2′-fluoro analogs cannot act as hydrogen bond donors they have been shown to reproduce the hydrogen bond acceptor properties of 2′-hydroxyls (Ortoleva-Donnelly *et al.*, 1998b). For this reason a site that shows interference with dNαS, ^{OMe}NαS, and ^{F}NαS suggests a 2′-OH that acts as a hydrogen bond donor, while a site that shows interference with dNαS alone suggests the 2′-OH acts as a hydrogen bond acceptor.

Occasionally, ^{F}NαS interference is observed at sites that do not exhibit dNαS interference (Ortoleva-Donnelly *et al.*, 1998b). This interference pattern has been attributed to sites that require a 2′-C-endo ribose conformation, a sugar pucker that is destabilized by the electron withdrawing properties of the 2′-fluoro substituent in favor of the 3′-C-endo conformation (Uesugi *et al.*, 1979). Sites that exhibited this interference pattern in the group I intron map to ribose sugars refined as having 2′-C-endo conformations in subsequent crystal structures (Adams *et al.*, 2004; Cate *et al.*, 1996). Sites of ^{OMe}NαS interference that do not interfere with dNαS suggest tight packing in the minor groove that cannot accommodate the 2′-*O*-methyl substitution. These sites can be further investigated with ^{SH}NαS analogs which have been shown to cause interference at densely packed regions of RNA (Schwans *et al.*, 2003). In principle, ^{SH}NαS analogs should also produce interference at sites of Mg^{2+} coordination that could be rescued by the addition of thiophilic metals.

4.2. Base modifications

4.2.1. Adenosine analogs

The largest collection of nucleobase modifications developed for NAIM studies are those of adenosine (Fig. 1.5). The focus on adenosine analogs stems in part from the observation that adenosines are present at a disproportionately high frequency in several RNA structural motifs (Gutell *et al.*, 1985). Nucleobase-modified adenosine analogs include diaminopurine riboside (DAPαS), 2-aminopurine riboside (2APαS), *N*-methyladenosine (m^6AαS), purine riboside (PurαS), 3-deazaadenosine (c^3AαS),

7-deazaadenosine (7dAαS), 7-fluoro-7-deazaadenosine (7F7dAαS), 2-fluoro-7-deazaadenosine (2F7dAαS), 2-fluoroadenosine (2FAαS), formicin (FormαS), and 8-azaadenosine (n^8AαS). Even incorporation at low analog concentrations is observed for each analog except PurαS, 2APαS, and c^3AαS, which are incorporated with lower efficiency even at elevated concentrations (Table 1.1).

Each of the nucleobase analogs provides specific information on the role of adenosine functional groups in RNA activity. PurαS, 2APαS, and m^6AαS measure the effect of modifications at the *N*-6 exocyclic amine (Ortoleva-Donnelly *et al.*, 1998b). PurαS and 2APαS remove the amine, leading to interference at any site where this functional group is required for activity. m^6AαS replaces one proton of the amine with a methyl group. Interference with this analog indicates either that both N6 protons are required for function or that there is insufficient space in the local structure to accommodate the additional methyl group. Interference caused by the additional C2 amine of 2APαS can also arise from close packing in the minor groove. These sites can be identified by a similar interference with DAPαS or in very close contacts with 2FAαS (Ortoleva-Donnelly *et al.*, 1998b; Suydam and Strobel, 2008). 7dAαS replaces the N7 nitrogen with a C—H group. Interference with this analog indicates an important major groove contact with the N7 nitrogen (Kaye *et al.*, 2002; Ortoleva-Donnelly *et al.*, 1998b; Waldsich and Pyle, 2007). These sites will provide a similar result with 7F7dAαS and 2F7dAαS (Suydam and Strobel, 2008). The N7 nitrogen of adenosine also participates in metal binding in some systems and NAIM analysis has been used to identify these sites (Wrzesinski and Jozwiakowski, 2008). Interference with PurαS, m^6AαS, and 7dAαS is a strong indicator of Hoogsteen hydrogen bonding (Ortoleva-Donnelly *et al.*, 1998b; Waldsich and Pyle, 2007). c^3AαS replaces the N3 nitrogen with a C—H group and provides a similar test for interactions with this nitrogen. Interference with c^3AαS has identified a number of A-minor tertiary interactions involving the N3 nitrogen (Soukup *et al.*, 2002). In addition to structural interactions, NAIM has been used to identify sites of functional adenosine ionization. Analogs appropriate for these studies exhibit large N1 pK_a shifts with minimal structural perturbation and include FormAαS, 7dAαS, 7F7dAαS, n^8AαS, 2F7dAαS, and 2FAαS (Jones and Strobel, 2003; Ryder *et al.*, 2001; Suydam and Strobel, 2008).

4.2.2. Guanosine analogs

Four nucleobase-modified analogs of guanosine have been used in NAIM studies, including inosine (IαS), N^2-methylguanosine (m^2GαS), 6-thioguanosine (S^6GαS), and 7-deazaguanosine (7dGαS) (Fig. 1.6). IαS and 7dGαS are both incorporated efficiently and evenly by wild-type polymerase, whereas m^2GαS requires the Y639F point mutant of T7 RNA polymerase and decreased GTP concentrations (Table 1.1). S^6GαS is incorporated by the wild-type polymerase but requires the addition of 4 m*M* Mn^{2+} in the transcription buffer. RNAs containing $S^6G\alpha S$ are unusually unstable and should be used within 48 h of transcription (Basu *et al.*, 1998). Interferences arising from $I\alpha S$ and $m^2G\alpha S$ have been used extensively to distinguish between tertiary contacts in RNA (Heide *et al.*, 1999; Ortoleva-Donnelly *et al.*, 1998a; Strobel and Shetty, 1997; Szewczak *et al.*, 1998). $I\alpha S$ lacks the N2 exocyclic amine of guanosine and interference with this analog results from reduced duplex stability or loss of a tertiary contact to the N2 amine. In contrast, $m^2G\alpha S$ introduces a methyl group to the exocyclic amine and is isoenergetic with G in the context of G–C, G·U, and G·A base pairs (Rife *et al.*, 1998). Interference with $m^2G\alpha S$ has only been observed at sites of tertiary contact (Ortoleva-Donnelly *et al.*, 1998a). $7dG\alpha S$ like $7dA\alpha S$ replaces the N7 nitrogen with a C—H group and produces interference at sites with important major groove contacts to this nitrogen as well as at sites of metal binding (Heide *et al.*, 2001; Kazantsev and Pace, 1998; Kwon and Strobel, 2008; Wrzesinski and Jozwiakowski, 2008). $S^6G\alpha S$ replaces the O6 keto group of guanosine with sulfur and has been used to identify sites of monovalent metal binding (Basu *et al.*, 1998).

4.2.3. Pyridine analogs

The development of pyrimidine base-modified analogs has been much more limited than for the purines. Four cytidine analogs have been synthesized for use in NAIM studies, including 6-azacytidine ($n^6C\alpha S$), 5-fluorocytidine ($f^5C\alpha S$), pseudoisocytidine ($\Psi iC\alpha S$), and zebularine ($Z\alpha S$) (Fig. 1.7). Each of these analogs has altered N3 pK_a values and have been used to identify sites of functional cytidine ionization in catalytic RNAs (Oyelere and Strobel, 2000; Oyelere *et al.*, 2002). Two nucleobase-modified uridine analogs have also been used NAIM studies, including 5-methyluridine ($m^5U\alpha S$) and pseudouridine ($\Psi\alpha S$), both of which modify the 5-position in the major groove (Fig. 1.8). Nucleobase analogs of pyrimidines are generally incorporated evenly with wild-type polymerase (Table 1.1).

5. Nucleotide Analog Interference Suppression

Some of the interferences observed in traditional NAIM experiments arise from the deletion of a critical hydrogen bond and the subsequent destabilization of RNA folds, decrease in population of catalytically active conformations, or loss of affinity for small molecule ligands or proteins. Although NAIM identifies the RNA functional groups involved in these hydrogen bonds it does not identify the hydrogen bonding partners. NAIS is an extension of NAIM experiments that maps tertiary contacts by identifying pairs of interacting functional groups required for function (Boudvillain and Pyle, 1998; Fedorova and Pyle, 2005; Jansen *et al.*, 2006; Strobel and Ortoleva-Donnelly, 1999; Strobel *et al.*, 1998; Szewczak *et al.*, 1998, 1999). The approach combines a single mutation or functional group substitution at a specific site with random incorporation of phosphorothioate-tagged nucleotides throughout the remainder of the RNA. These chimeric RNA molecules are typically constructed by joining or annealing synthetic oligonucleotides carrying the site-specific deletion with the pool of transcribed RNAs carrying the randomly incorporated analogs. The resulting RNAs are then subjected to the same type of selection as the original NAIM study.

NAIS experiments rely on the expectation that the deletion of one functional group in a hydrogen bonding pair will eliminate the positive contribution made by the second, such that it can be deleted without an additional energetic penalty (Strobel and Ortoleva-Donnelly, 1999). In this way, the site-specific deletion of one functional group should lead to a suppression in interference caused by analog substitution at the interacting site, while leaving the remaining sites of interference unchanged. This pattern of interference suppression has been observed in several systems and the resulting predictions of tertiary contacts have agreed extremely well with subsequent crystal structures (Adams *et al.*, 2004; Toor *et al.*, 2008). The key to NAIS experiments is the ability to reconstruct the selection assay such that there is activity in the context of the site-specific substitution. This is often accomplished by modifying buffer conditions, lengthening reaction times, or increasing the concentration of substrates. Several additional controls are required to demonstrate that the suppression of interference is specific both to the nature of the site-specific substitution and to the site of suppression. First, site-specific substitutions with other functionalities or at other sites should not produce the same suppression pattern. Second, the observed suppression should be specific to a single residue among additional sites of interference. Finally, the observed suppression should be specific to a single or closely related set of analogs.

6. Conclusions

NAIM is a generalizable chemogenetic method that rapidly identifies the chemical groups important for RNA function. In principle, NAIM can be extended to any nucleotide analog that can be incorporated by *in vitro* transcription to study any RNA with a selectable function. The robustness of the method has been demonstrated by the excellent agreement between structural contacts predicted by NAIM and atomic resolution structures. Because a single NAIM experiment can assay several functional group modifications at every site in parallel the method provides an efficient route to define key features of RNA structure and function. As the number of functional RNAs identified continues to expand, NAIM should provide a straightforward method to determine which individual nucleotides or regions of sequence warrant additional study.

REFERENCES

Adams, P. L., Stahley, M. R., Kosek, A. B., Wang, J. M., and Strobel, S. A. (2004). Crystal structure of a self-splicing group I intron with both exons. *Nature* **430,** 45–50.

Arabshahi, A., and Frey, P. A. (1994). A simplified procedure for synthesizing nucleoside 1-thiotriphosphates—dATPS, dGTPS, UTPS, and dTTPS. *Biochem. Biophys. Res. Commun.* **204,** 150–155.

Basu, S., Rambo, R. P., Strauss-Soukup, J., Cate, J. H., Ferre-D'Amare, A. R., Strobel, S. A., and Doudna, J. A. (1998). A specific monovalent metal ion integral to the AA platform of the RNA tetraloop receptor. *Nat. Struct. Biol.* **5,** 986–992.

Batey, R. T., Rambo, R. P., Lucast, L., Rha, B., and Doudna, J. A. (2000). Crystal structure of the ribonucleoprotein core of the signal recognition particle. *Science* **287,** 1232–1239.

Boudvillain, M., and Pyle, A. M. (1998). Defining functional groups, core structural features and inter-domain tertiary contacts essential for group II intron self-splicing: A NAIM analysis. *EMBO J.* **17,** 7091–7104.

Cate, J. H., Gooding, A. R., Podell, E., Zhou, K. H., Golden, B. L., Kundrot, C. E., Cech, T. R., and Doudna, J. A. (1996). Crystal structure of a group I ribozyme domain: Principles of RNA packing. *Science* **273,** 1678–1685.

Christian, E. L., and Yarus, M. (1992). Analysis of the role of phosphate oxygens in the group-I intron from Tetrahymena. *J. Mol. Biol.* **228,** 743–758.

Cochrane, J. C., and Strobel, S. A. (2004). Current Protocols in Nucleic Acid Chemistry. Wiley, New York, NY.

Cochrane, J. C., Batey, R. T., and Strobel, S. A. (2003). Quantitation of free energy profiles in RNA–ligand interactions by nucleotide analog interference mapping. *RNA* **9,** 1282–1289.

Conrad, F., Hanne, A., Gaur, R. K., and Krupp, G. (1995). Enzymatic synthesis of 2′-modified nucleic acids—Identification of important phosphate and ribose moieties in RNAse-P substrates. *Nucleic Acids Res.* **23,** 1845–1853.

Das, S. R., and Piccirilli, J. A. (2005). General acid catalysis by the hepatitis delta virus ribozyme. *Nat. Chem. Biol.* **1,** 45–52.

Eckstein, F. (1985). Nucleoside phosphorothioates. *Annu. Rev. Biochem.* **54,** 367–402.

Fedorova, O., and Pyle, A. M. (2005). Linking the group II intron catalytic domains: Tertiary contacts and structural features of domain 3. *EMBO J.* **24,** 3906–3916.

Fischer, B., Chulkin, A., Boyer, J. L., Harden, K. T., Gendron, F. P., Beaudoin, A. R., Chapal, J., Hillaire-Buys, D., and Petit, P. (1999). 2-Thioether 5′-O-(1-thiotriphosphate)adenosine derivatives as new insulin secretagogues acting through P2Y-receptors. *J. Med. Chem.* **42,** 3636–3646.

Gaur, R. K., and Krupp, G. (1993). Modification interference approach to detect ribose moieties important for the optimal activity of a ribozyme. *Nucleic Acids Res.* **21,** 21–26.

Gish, G., and Eckstein, F. (1988). DNA and RNA sequence determination based on phosphorothioate chemistry. *Science* **240,** 1520–1522.

Griffiths, A. D., Potter, B. V. L., and Eperon, I. C. (1987). Stereospecificity of nucleases towards phosphorothioate-substituted RNA-stereochemistry of transcription by T7 RNA-polymerase. *Nucleic Acids Res.* **15,** 4145–4162.

Gutell, R. R., Weiser, B., Woese, C. R., and Noller, H. F. (1985). Comparative anatomy of 16-S-like ribosomal-RNA. *Prog. Nucleic Acid Res. Mol. Biol.* **32,** 155–216.

Heide, C., Pfeiffer, T., Nolan, J. M., and Hartmann, R. K. (1999). Guanosine 2-NH2 groups of *Escherichia coli* RNase P RNA involved in intramolecular tertiary contacts and direct interactions with tRNA. *RNA* **5,** 102–116.

Heide, C., Feltens, R., and Hartmann, R. K. (2001). Purine N7 groups that are crucial to the interaction of *Escherichia coli* RNase P RNA with tRNA. *RNA* **7,** 958–968.

Jansen, J. A., McCarthy, T. J., Soukup, G. A., and Soukup, J. K. (2006). Backbone and nucleobase contacts to glucosamine-6-phosphate in the glmS ribozyme. *Nat. Struct. Mol. Biol.* **13,** 517–523.

Jones, F. D., and Strobel, S. A. (2003). Ionization of a critical adenosine residue in the Neurospora Varkud satellite ribozyme active site. *Biochemistry* **42,** 4265–4276.

Jones, F. D., Ryder, S. P., and Strobel, S. A. (2001). An efficient ligation reaction promoted by a Varkud Satellite ribozyme with extended 5′- and 3′-termini. *Nucleic Acids Res.* **29,** 5115–5120.

Kaye, N. M., Christian, E. L., and Harris, M. E. (2002). NAIM and site-specific functional group modification analysis of RNase P RNA: Magnesium dependent structure within the conserved P1–P4 multihelix junction contributes to catalysis. *Biochemistry* **41,** 4533–4545.

Kazantsev, A. V., and Pace, N. R. (1998). Identification by modification-interference of purine N-7 and ribose 2′-OH groups critical for catalysis by bacterial ribonuclease P. *RNA* **4,** 937–947.

Klein, D. J., and Ferre-D'Amare, A. R. (2006). Structural basis of glmS ribozyme activation by glucosamine-6-phosphate. *Science* **313,** 1752–1756.

Kolev, N. G., and Steitz, J. A. (2006). *In vivo* assembly of functional U7 snRNP requires RNA backbone flexibility within the Sm-binding site. *Nat. Struct. Mol. Biol.* **13,** 347–353.

Kwon, M. Y., and Strobel, S. A. (2008). Chemical basis of glycine riboswitch cooperativity. *RNA* **14,** 25–34.

Laederach, A., Das, R., Vicens, Q., Pearlman, S. M., Brenowitz, M., Herschlag, D., and Altman, R. B. (2008). Semiautomated and rapid quantification of nucleic acid footprinting and structure mapping experiments. *Nat. Protoc.* **3,** 1395–1401.

Lafontaine, D. A., Wilson, T. J., Zhao, Z. Y., and Lilley, D. M. J. (2002). Functional group requirements in the probable active site of the VS ribozyme. *J. Mol. Biol.* **323,** 23–34.

Milligan, J. F., and Uhlenbeck, O. C. (1989). Synthesis of small RNAs using T7 RNA-polymerase. *Methods Enzymol.* **180,** 51–62.

Noller, H. F. (2005). RNA structure: Reading the ribosome. *Science* **309,** 1508–1514.

Ortoleva-Donnelly, L., Kronman, M., and Strobel, S. A. (1998a). Identifying RNA minor groove tertiary contacts by nucleotide analogue interference mapping with N-2-methylguanosine. *Biochemistry* **37,** 12933–12942.

Ortoleva-Donnelly, L., Szewczak, A. A., Gutell, R. R., and Strobel, S. A. (1998b). The chemical basis of adenosine conservation throughout the Tetrahymena ribozyme. *RNA* **4,** 498–519.

Oyelere, A. K., and Strobel, S. A. (2000). Biochemical detection of cytidine protonation within RNA. *J. Am. Chem. Soc.* **122,** 10259–10267.

Oyelere, A. K., Kardon, J. R., and Strobel, S. A. (2002). pKa perturbation in genomic hepatitis delta virus ribozyme catalysis evidenced by nucleotide analogue interference mapping. *Biochemistry* **41,** 3667–3675.

Paschal, B. M., McReynolds, L. A., Noren, C. J., and Nichols, N. M. (2008). Current Protocols in Molecular Biology. Wiley, New York, NY.

Rife, J. P., Cheng, C. S., Moore, P. B., and Strobel, S. A. (1998). N-2-methylguanosine is iso-energetic with guanosine in RNA duplexes and GNRA tetraloops. *Nucleic Acids Res.* **26,** 3640–3644.

Ruffner, D. E., and Uhlenbeck, O. C. (1990). Thiophosphate interference experiments locate phosphates important for the hammerhead RNA self-cleavage reaction. *Nucleic Acids Res.* **18,** 6025–6029.

Ryder, S. P., and Strobel, S. A. (1999). Nucleotide analog interference mapping of the hairpin ribozyme: Implications for secondary and tertiary structure formation. *J. Mol. Biol.* **291,** 295–311.

Ryder, S. P., and Strobel, S. A. (2002). Comparative analysis of hairpin ribozyme structures and interference data. *Nucleic Acids Res.* **30,** 1287–1291.

Ryder, S. P., Ortoleva-Donnelly, L., Kosek, A. B., and Strobel, S. A. (2000). Chemical probing of RNA by nucleotide analog interference mapping. *Methods Enzymol.* **317,** 92–109.

Ryder, S. P., Oyelere, A. K., Padilla, J. L., Klostermeier, D., Millar, D. P., and Strobel, S. A. (2001). Investigation of adenosine base ionization in the hairpin ribozyme by nucleotide analog interference mapping. *RNA* **7,** 1454–1463.

Schwans, J. P., Cortez, C. N., Olvera, J. M., and Piccirilli, J. A. (2003). 2′-mercaptonucleotide interference reveals regions of close packing within folded RNA molecules. *J. Am. Chem. Soc.* **125,** 10012–10018.

Sood, V. D., Yekta, S., and Collins, R. A. (2002). The contribution of 2′-hydroxyls to the cleavage activity of the Neurospora VS ribozyme. *Nucleic Acids Res.* **30,** 1132–1138.

Soukup, J. K., Minakawa, N., Matsuda, A., and Strobel, S. A. (2002). Identification of A-minor tertiary interactions within a bacterial group I intron active site by 3-deazaadenosine interference mapping. *Biochemistry* **41,** 10426–10438.

Sousa, R., and Padilla, R. (1995). Mutant T7 RNA-polymerase as a DNA-polymerase. *EMBO J.* **14,** 4609–4621.

Strobel, S. A. (1999). A chemogenetic approach to RNA function/structure analysis. *Curr. Opin. Struct. Biol.* **9,** 346–352.

Strobel, S. A., and Ortoleva-Donnelly, L. (1999). A hydrogen-bonding triad stabilizes the chemical transition state of a group I ribozyme. *Chem. Biol.* **6,** 153–165.

Strobel, S. A., and Shetty, K. (1997). Defining the chemical groups essential for Tetrahymena group I intron function by nucleotide analog interference mapping. *Proc. Natl. Acad. Sci. USA* **94,** 2903–2908.

Strobel, S. A., Ortoleva-Donnelly, L., Ryder, S. P., Cate, J. H., and Moncoeur, E. (1998). Complementary sets of noncanonical base pairs mediate RNA helix packing in the group intron active site. *Nat. Struct. Biol.* **5,** 60–66.

Suydam, I. T., and Strobel, S. A. (2008). Fluorine substituted adenosines as probes of nucleobase protonation in functional RNAs. *J. Am. Chem. Soc.* **130,** 13639–13648.

Szewczak, A. A., Ortoleva-Donnelly, L., Ryder, S. P., Moncoeur, E., and Strobel, S. A. (1998). A minor groove RNA triple helix within the catalytic core of a group I intron. *Nat. Struct. Biol.* **5,** 1037–1042.

Szewczak, A. A., Ortoleva-Donnelly, L., Zivarts, M. V., Oyelere, A. K., Kazantsev, A. V., and Strobel, S. A. (1999). An important base triple anchors the substrate helix recognition surface within the Tetrahymena ribozyme active site. *Proc. Natl. Acad. Sci. USA* **96,** 11183–11188.

Szewczak, L. B. W., DeGregorio, S. J., Strobel, S. A., and Steitz, J. A. (2002). Exclusive interaction of the 15.5 kD protein with the terminal box C/D motif of a methylation guide snoRNP. *Chem. Biol.* **9,** 1095–1107.

Toor, N., Keating, K. S., Taylor, S. D., and Pyle, A. M. (2008). Crystal structure of a self-spliced group II intron. *Science* **320,** 77–82.

Uesugi, S., Miki, H., Ikehara, M., Iwahashi, H., and Kyogoku, Y. (1979). Linear relationship between electronegativity of 2′-substituents and conformation of adenine nucleosides. *Tetrahedron Lett.* 4073–4076.

Waldsich, C. (2008). Dissecting RNA folding by nucleotide analog interference mapping (NAIM). *Nat. Protoc.* **3,** 811–823.
Waldsich, C., and Pyle, A. M. (2007). A folding control element for tertiary collapse of a group II intron ribozyme. *Nat. Struct. Mol. Biol.* **14,** 37–44.
Wrzesinski, J., and Jozwiakowski, S. K. (2008). Structural basis for recognition of Co^{2+} by RNA aptamers. *FEBS J.* **275,** 1651–1662.
Yin, Y. W., and Steitz, T. A. (2004). The structural mechanism of translocation and helicase activity in T7 RNA polymerase. *Cell* **116,** 393–404.

CHAPTER TWO

Hydroxyl-Radical Footprinting to Probe Equilibrium Changes in RNA Tertiary Structure

Inna Shcherbakova*,† *and* Somdeb Mitra*

Contents

Abstract

Hydroxyl-radical footprinting utilizes the ability of a highly reactive species to nonspecifically cleave the solvent accessible regions of a nucleic acid backbone. Thus, changes in a nucleic acids structure can be probed either as a function of time or of a reagent's concentration. When combined with techniques that allow single nucleotide resolution of the resulting fragments, footprinting experiments provide richly detailed information about local changes in tertiary structure of a nucleic acid accompanying its folding or ligand binding. In this chapter, we present two protocols of equilibrium hydroxyl-radical footprinting based on peroxidative and oxidative Fenton chemistry and discuss how to adjust the Fenton reagent concentrations for a specific experimental condition. We also discuss the choice of the techniques to separate the reaction products and specifics of the data analysis for equilibrium footprinting experiments. Protocols addressing the use of peroxidative Fenton chemistry for time-resolved

* Department of Biochemistry, Albert Einstein College of Medicine, Bronx, New York, USA
† Current address: Department of Biochemistry and Molecular Pharmacology, University of Massachusetts Medical School, Worcester, Massachusetts, USA

Methods in Enzymology, Volume 468
ISSN 0076-6879, DOI: 10.1016/S0076-6879(09)68002-2

studies have been published [Schlatterer and Brenowitz, 2009. *Methods*; Shcherbakova and Brenowitz, 2008. *Nat. Protoc.* 3(2), 288–302; Shcherbakova *et al.*, 2006. *Nucleic Acids Res.* 34(6), e48; Shcherbakova *et al.*, 2007. *Methods Cell Biol.* 84, 589–615].

1. Introduction

The term "footprinting" refers to an assay that probes changes in macromolecular structure by probing changes in the sensitivity of the individual residues to a modification or cleavage reagent. This technique was initially developed to probe specific DNA–protein interactions by DNase I (Galas and Schmitz, 1978), and the term arose from an association that a protein leaves "a footprint" on its cognate binding site on a DNA, visualized as a blank region lacking bands of cleaved DNA on the image of a sequencing gel. Quantitating changes in the band intensities within such "footprinted" regions as a function of the protein ligand concentration or as a function of time provided the binding affinities or interaction rates, respectively. Later, it was demonstrated that the same approach could be used with a variety of reagents to probe not only intermolecular (Tullius and Dombroski, 1986) but also intramolecular interactions (Latham and Cech, 1989) and was appropriate to probe changes in the macromolecular structure associated with macromolecular folding or assembly.

Among the wide variety of footprinting reagents used to probe RNA structure, small chemicals that do not confer significant steric hindrances proved to be most useful. One of the smallest one, the hydroxyl radical (•OH), has been used extensively to probe changes in the tertiary structure of RNA in kinetic (Adilakshmi *et al.*, 2005, 2008; Chauhan *et al.*, 2009; Deras *et al.*, 2000; Heilman-Miller and Woodson, 2003; Kwok *et al.*, 2006; Laederach *et al.*, 2007b; Lease *et al.*, 2007; Nguyenle *et al.*, 2006; Sclavi *et al.*, 1997, 1998; Shcherbakova and Brenowitz, 2005; Shcherbakova *et al.*, 2004; Silverman *et al.*, 2000; Uchida *et al.*, 2003) and equilibrium studies (Ralston *et al.*, 2000; Schlatterer *et al.*, 2008; Su *et al.*, 2003; Swisher *et al.*, 2002; Takamoto *et al.*, 2002, 2004; Uchida *et al.*, 2002, 2003). Falling only slightly short in its reactivity to elemental fluorine, this highly reactive species is capable of oxidizing virtually every biological molecule. In the case of nucleic acids, hydroxyl radicals cleave their backbones by abstracting hydrogens from the ribose or deoxyribose sugar rings (Hertzberg and Dervan, 1984). Because the sugars are ubiquitously located on the outer faces of base pairs in nucleic acid helices, the changes in •OH cleavage do not reflect changes in the RNA secondary structure but are associated with the formation of tertiary interactions. An additional advantage is that •OH radicals interact nonspecifically with nucleotides, thereby allowing us to monitor individual behavior of all residues that are involved in tertiary interactions.

When the products of the cleavage reaction are separated on a high-resolution gel or by capillary electrophoresis, this footprinting technique tracks changes in the reactivity of every nucleotide along the RNA chain. The small size of •OH and its properties comparable to that of a water molecule, allow us to link the changes in the •OH reactivity of a nucleotide to changes in its solvent accessibility, associated with conformational changes in an RNA molecule. This footprinting technique is appropriate for both *in vitro* and *in vivo* studies. The feasibility to footprint RNA structure *in vivo* was demonstrated in synchrotron footprinting experiments on 16S rRNA and RNase P (Adilakshmi *et al.*, 2006).

There are a wide variety of ways to produce •OH for footprinting. In the *in vivo* footprinting experiments mentioned above the •OH are produced by radiolysis of intracellular water by a high-energy X-ray beam generated by a synchrotron. However, for *in vitro* studies, the readily available chemical reagent hydrogen peroxide (H_2O_2) can generate •OH without the need for a synchrotron beam. A simple way to decompose H_2O_2 in a controlled manner is to use Fenton chemistry (Fenton, 1894) in its Udenfriend version (Udenfriend *et al.*, 1954) where EDTA-chelated iron(II) reduces peroxide to produce •OH and the resulting iron(III)–EDTA is recycled back to iron (II)–EDTA by ascorbate:

$$\mathrm{Fe(II)-EDTA} + \mathrm{H_2O_2} \rightarrow \mathrm{Fe(III)-EDTA} + \bullet\mathrm{OH} + \mathrm{OH^-}$$

$$\xleftarrow{\text{ascorbate}}$$

The concentrations of the reagents can be adjusted so that the cleavage occurs during several seconds. The reaction is quenched by addition of thiourea or ethanol. In this version this reaction can be used to probe equilibrium changes in the RNA structure, as well as to track slow kinetic processes taking minutes to hours. The protocols for millisecond time-resolved footprinting are published elsewhere (Schlatterer and Brenowitz, 2009; Shcherbakova and Brenowitz, 2008; Shcherbakova *et al.*, 2006, 2007).

It was recently noticed that minutes-long incubation of RNA with only Fe(II)–EDTA caused a small amount of RNA to be uniformly cleaved, presumably due to the production of the •OH by oxidative chemistry. This reaction, simpler in its execution (Takamoto *et al.*, 2004), requires longer reaction times. However, it can be a valuable tool in case of molecules or a macromolecular complex that are highly sensitive to hydrogen peroxide, as was reported for the TFIIIA–DNA complex (Tullius *et al.*, 1987).

In this chapter, we present two protocols of equilibrium •OH footprinting based on the peroxidative and the oxidative Fenton chemistry that we routinely use to probe changes in RNA tertiary structure during its cation-dependent folding. We discuss how the reaction conditions can be adjusted to accommodate a specific experimental requisite. We also note the available choices of techniques to separate and quantitate the reaction products and specifics of the data analysis of equilibrium footprinting experiments.

2. Sample Preparation

Our laboratory has over a decade's experience in studying the folding pathways of the group I intron ribozyme derived from *Tetrahymena thermophila*, and we shall use it as an example in the subsequent sections to highlight the applications of equilibrium •OH footprinting. The methods described here are very general and can be easily adapted for any structured RNA molecule. The RNA molecule under investigation can be directly end-labeled with a radioactive label, in which case the •OH cleavage products can be separated on a high-resolution denaturing polyacrylamide gel. The dried gel is exposed to a phosphor storage screen and visualized by scanning the screen in a phosphor-imaging system (STORM® phosphorimager from GE Healthcare in our case). The intensities of cleavage products, appearing as bands in the gel image, are quantitated using the software SAFA (Das *et al.*, 2005; Laederach *et al.*, 2007a; see also Chapter 3). Alternatively, to avoid radioactivity and increase the throughput of product separation and data analysis, we have recently adopted an indirect fluorescent labeling strategy, described in Section 5.

For the experiments with directly labeled RNA, we first separately label the 5′- and 3′-ends of RNA (Shcherbakova and Brenowitz, 2008) with ^{32}P. Several reaction aliquots of the RNA in a buffered solution are prepared so that each reaction tube contains enough radioactively labeled RNA for visualization and quantitation of the bands of the reaction products (usually 10–20 pmol RNA per tube). For investigating mono- or divalent cation mediated equilibrium folding of the RNA each reaction sample is heated to denature the RNA (90 °C for 3 min in our case) and gradually cooled to the desired folding temperature when a desired concentration of the cation is added to the RNA and the sample is allowed to equilibrate. Incubation at 50 °C is often a necessary intermediate step for large RNA molecules to avoid misfolding. Depending on the folding temperature and the type of the cations, equilibration can take from several minutes to hours. For example, in monovalent ions and/or at high temperatures, the *Tetrahymena* ribozyme folding has been shown to be completed within seconds. However, the same ribozyme can be trapped in a long lived misfolded state at lower temperatures or in high-divalent cation concentrations, so that reaching the catalytically active native state can take hours.

The RNA for the footprinting experiments should be prepared with a special care to minimize any prereaction cleavage of its backbone which generates high background noise that appears as high-intensity bands in sequencing gels or abnormally high peaks in automated sequencer readouts. The buffers should be chosen wisely to minimize quenching of •OH by buffer components and to prevent side reactions that might cause random

RNA cleavage. Inorganic buffers such as sodium cacodylate buffer are ideal due to their low propensity to quench •OH. (*Note.* Sodium phosphate buffer was reported to significantly increase nucleic acid cleavage; Tullius *et al.*, 1987.) Whereas monovalent cations do not interfere with the Fenton chemistry and hence does not affect the •OH cleavage reaction, high concentrations of divalent cations (higher than 50 m*M*) can reduce the extent of •OH cleavage by displacing iron (Fe^{2+}) from its complex with EDTA (Tullius *et al.*, 1987). Glycerol is a very potent •OH quencher (Tullius *et al.*, 1987) that should be omitted if possible. If glycerol must be present in a reaction buffer, its scavenging must be accommodated by a concomitant upscaling of •OH production by increasing the concentrations of the Fenton reagents. We discuss below how to troubleshoot such problems by scaling the •OH production.

3. Equilibrium •OH Footprinting Based on Peroxidative Fenton Chemistry

The standard set of Fenton reagents we use for footprinting experiments include ammonium iron(II) sulfate hexahydrate [$(NH_4)_2Fe(SO_4)_2 \cdot 6H_2O$] (Sigma Ultra, minimum 99% pure); 0.5 *M* EDTA solution, pH 8.0 (Ambion); (+)Sodium L-ascorbate (Sigma), 30% Hydrogen Peroxide (Sigma Aldrich), thiourea (Sigma), and RNase-free water (e.g., the DEPC-treated water from Ambion). A 100 m*M* stock solution of ferrous ammonium sulfate should be freshly prepared for every experiment or kept frozen in small aliquots. The color of this solution can change to slight yellowish due to the iron oxidation. Do not use the solution that turns brown or orange, with some precipitate. The Fe–EDTA solution is prepared by mixing the ferrous ammonium sulfate with EDTA to have 1.1 times excess of EDTA over iron. Solutions of sodium ascorbate and dilutions of hydrogen peroxide should be prepared fresh for each experiment.

Following equilibration of the RNA sample with the desired concentration of the cation at the folding temperature 2 μl drops of 5 m*M* Fe–EDTA, 50 m*M* sodium ascorbate, and 1.5% hydrogen peroxide solutions are separately placed on the inside wall of the microfuge tube containing the RNA. The Fenton reaction is initiated by vigorous vortexing of the tube, resulting in simultaneous addition of all the three components to the 100 μl of the RNA solution. After 15 s the reaction is quenched by addition of 20 μl of stop solution (prepared beforehand by mixing thiourea with EDTA to final concentrations of 100 and 200 m*M*, respectively). If a protein component was present in the reaction mix it can be removed by phenol/chloroform treatment of the samples with subsequent ethanol precipitation of nucleic acid from the aqueous phase.

Alternatively if only nucleic acids are present in solution, the reaction can be quenched by addition of three times the reaction volume of ethanol. Following addition of ethanol, the reaction products are kept for at least 30 min at −20 °C and then precipitated by spinning in a refrigerated centrifuge at 14,000 rpm. Formation of tertiary structure can require molar concentrations of monovalent cations that upon ethanol precipitation form large pellets. In the case of high concentrations of monovalent cations, the reaction solution should be diluted to adjust the monovalent ion concentration to 0.15–0.3 *M* prior to precipitation. We found that the salt pellets are significantly smaller if the samples are first diluted and then precipitated by an equal volume of isopropanol instead of ethanol. In the case of titration samples that contain less than 50 m*M* of monovalent ions additional monovalent salts should be added to 0.15–0.3 *M* prior to the precipitation step to facilitate precipitation of small RNA fragments generated during •OH cleavage. The precipitation of the small RNA fragments can be additionally facilitated by addition of glycogen solution (Ambion®) to 1 μg/μl.

Following precipitation, the pellets are dried in a SpeedVac and resuspended in a gel loading buffer containing formamide, EDTA, bromophenol blue, and xylene cyanol (we use Gel loading buffer II from Ambion®). The samples are then loaded into the wells of a denaturing sequencing gel and separated at constant power (70–80 W). In certain cases, when the reactions are performed in smaller volumes, the formamide gel loading buffer may be added directly to the reaction mix after quenching •OH production and a small aliquot (usually up to 10 μl) is loaded directly on the gel. However, in these situations the radioactive counts of the labeled RNA should be high enough to allow visualization of the cleavage products. We do not recommend direct sample loading for the cation titration experiments since the high amount of salt in some lanes will interfere with the electrophoretic separation of fragments, thereby significantly diminishing data quality. We have also noted improved signal to noise ratio as well as unbiased distribution of large and small cleavage fragments when the reaction products are completely precipitated, washed with 70% ethanol and resuspended in small volumes of gel loading buffer (usually 12 μl for loading 4 μl in each of the three sequencing gels) prior to separation by electrophoresis.

4. Equilibrium •OH Footprinting Based on Oxidative Fenton Chemistry

Radioactively labeled RNA samples are equilibrated in 100 μl buffered solution as described above. The following fresh solutions are prepared (or thawed from frozen stock aliquots): 250 m*M* $(NH_4)_2Fe(SO_4)_2$, 250 m*M*

EDTA, and 500 m*M* sodium ascorbate. Just before initiating the cleavage reaction, the Fenton reaction mix is prepared by mixing together 2 μl of the $(NH_4)_2Fe(SO_4)_2$ solution, 2.2 μl of the EDTA solution, 62.5 μl of the ascorbate solution, and 158 μl of RNase-free water. This Fenton reaction mix contains 0.1 m*M* Fe $(NH_4)_2(SO_4)_2$, 0.14 m*M* EDTA, and 6.94 m*M* ascorbate. •OH cleavage is initiated by adding 5 μl of the Fenton reaction mix to the RNA solution and allowing the cleavage reaction to proceed for 30 min at a desired temperature (we have not noticed significant temperature dependence of •OH production). The reaction is quenched by addition of three times the volume of ethanol or equal volume of isopropanol and precipitated as described for the peroxidative Fenton reaction.

The concentrations of the reagents listed in both protocols are used to probe structural changes during equilibrium RNA folding under ideal non-radical scavenging condition, namely, in the presence of sodium or potassium cacodylate buffer and various concentrations of mono- or divalent cations. Usually a small concentration of EDTA (0.1 m*M*) is present in the solution to bind contaminant transition metals, thereby preventing them from binding to the RNA and cleaving its backbone. We recommend minimizing the amount of •OH scavengers (glycerol, DTT, β-mercaptoethanol and organic buffers) in the buffer. If their presence is unavoidable and/or a relatively high concentration of biological components is present, the concentrations of the Fenton reagents should be scaled appropriately.

The optimal •OH cleavage under nonoptimal scavenging condition can be found by a "dose response" experiment. For this experiment, the RNA sample is prepared as described above and aliquoted into several tubes containing the reaction mix in question. Then Fenton reaction is performed such that the concentration of any one of the Fenton reagents increases from one test tube to another. The products of the reaction are separated by denaturing gel electrophoresis, while taking care that the smallest cleavage fragments do not run out of the bottom of the gel, so that they can be visualized with the uncut RNA band (at the top) on the same gel. The fraction of cleaved RNA is calculated from the ratio of the intensity for the contour enclosing all the cleaved fragments, to the total intensity of cleaved and uncut samples (Shcherbakova and Brenowitz, 2008). Single-hit kinetics is typically achieved when no less than 70% of the RNA remains intact (Tullius *et al.*, 1987). The product profile should be relatively uniform. When the intensity is visibly increased from longer to shorter fragments it means that the RNA is cleaved multiple times and the concentration of •OH should be decreased to achieve the single-hit kinetics condition. The reagent concentrations determined by the dose response experiments are then used for equilibrium folding studies of the RNA under the desired experimental condition.

5. Cleavage Product Separation

The cleavage products of the peroxidative and oxidative Fenton reaction are processed identically. Following ethanol/isopropanol precipitation as described above, the pellets are washed with 70% ethanol, dried in a SpeedVac, dissolved in a formamide gel loading buffer, heated for 30 s at 95 °C and loaded onto high-resolution 8% acrylamide/bisacrylamide sequencing gel containing 7 *M* urea (Tijerina *et al.*, 2007). More than a single gel may be required to clearly separate the regions of interest. For every gel it is important to run a ladder for assigning nucleotide position numbers to the bands on the gel. We typically use a G ladder generated by T1 ribonuclease which, in the presence of 7.5 *M* urea at 55 °C, cleaves after every guanosine residue along the RNA backbone.

We found that one-meter-long electrophoresis gel plate can save some time especially in the case of long RNA molecules. In this case, the special long-range gel solutions (e.g., LongRanger Gel solution from Cambrex®) are recommended. It is critical to siliconize one of the long gel plates. The long gels have to be cut in half before transferring them to a gel paper. Gels are then dried, exposed to a phosphor storage screen and visualized by scanning the screen by a phosphor imager.

A typical equilibrium footprinting gel obtained for a Mg^{2+} titration of the *Tetrahymena* ribozyme bearing the UUCG mutation of the L5b tetraloop is shown in Fig. 2.1. The "no •OH" lane demonstrates a single RNA band of a low electrophoretic mobility that corresponds to the full-length RNA without any detectable degradation. The same band of a diminished intensity is present in all other lanes on this gel along with additional bands corresponding to the shorter cleavage products separated at single nucleotide resolution. For the lanes corresponding to the RNA incubated in the absence of Mg^{2+}, the backbone cleavage is nearly uniform. The decrease in the intensity of cleaved bands, corresponding to certain nucleotides, correlate with the decreased solvent accessibility of those nucleotides due to tertiary interactions formed upon addition of higher concentrations of mono- or divalent cations. Some bands display a reverse trend, that is, increase in intensity with increasing ionic strength. Increased reactivity to •OH can be generally rationalized as an increase in the solvent exposure of some regions in the RNA molecule upon its folding (e.g., the increase in the cleavage of the nucleotide 122 reports on the formation of the characteristic structural feature of the hinge region in the P4–P6 domain).

We have recently begun to routinely implement separation of the products of the cleavage reaction by capillary electrophoresis (Mitra *et al.*, 2008). This method involves footprinting unlabeled RNA molecules followed by reverse transcription (RT) of the products into fluorophore-labeled cDNA

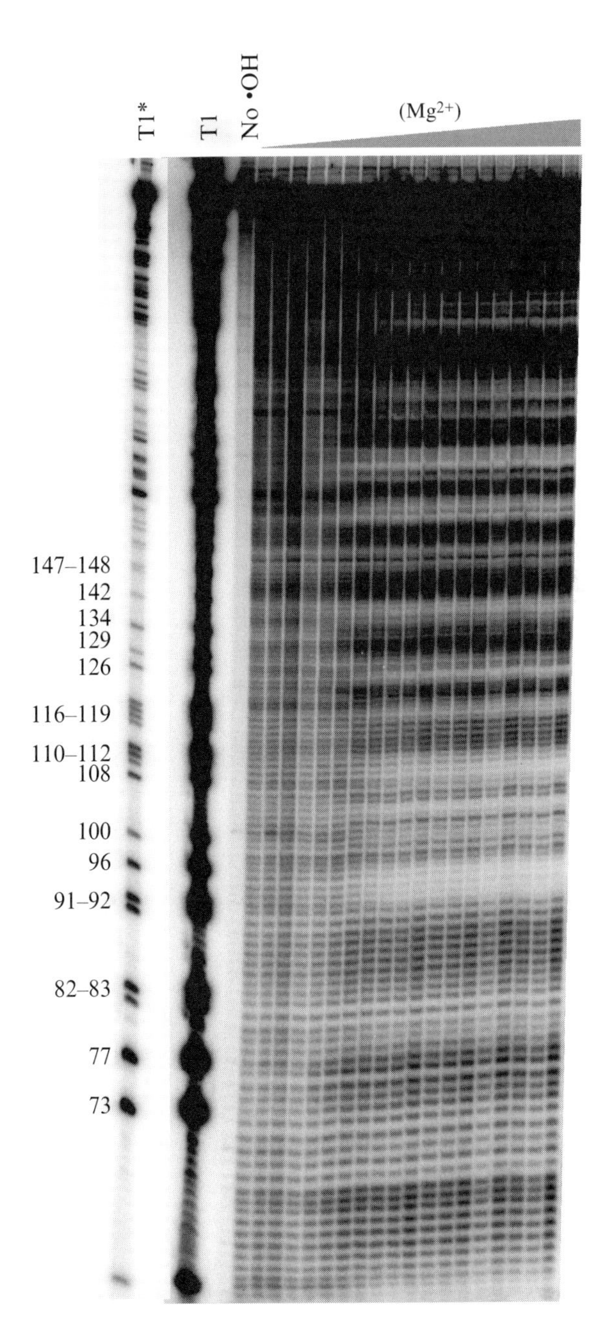

Figure 2.1 Hydroxyl-radical footprinting of the mutant of the *Tetrahymena* ribozyme bearing UUCG mutation of the L5b tetraloop. The equilibrium folding experiment was performed as a function of the increasing concentration of Mg^{2+} in the background of 100 m*M* KCl at 25 °C buffered by 10 m*M* of potassium cacodylate, pH 7.0.

fragments using DNA primers labeled by a fluorescent dye (Cy5 in our case) that anneal to the 3′-end of the RNA molecules. Briefly, the unlabeled RNA molecule is subjected to •OH cleavage subsequent to its cation mediated folding reaction. The cleaved products are precipitated, dried, and resuspended in 9 μl of a buffer containing 50 m*M* Tris–HCl pH 8.3, 60 m*M* NaCl, and 10 m*M* DTT. 0.5 μl of a 10 μ*M* Cy5-labeled primer stock solution is added to each sample, heated at 85 °C for 1 min and slowly cooled to 25 °C for primer annealing.

We then add 9 μl of reverse transcription mix (4 μl of 5× RT buffer supplied with Superscript III (Invitrogen®), 1 μl of 0.1 *M* DTT, 2 μl of RNase Inhibitor, 2 μl of 10 m*M* dNTP mix) to each tube. The solutions are incubated at 55 °C for 5 min before addition of 1 μl (200 units) of Superscript III to initiate RT reaction that occurs at 55 °C for 15–30 min. Upon completion of RT extension, we degrade the RNA by adding 2 μl of 2 N NaOH and incubating at 95 °C for 3 min and neutralize the solution by 2 μl of 2 N HCl. Finally, 3 μl of 3 *M* sodium acetate and 80 μl of ethanol is added and the cDNA is precipitated by centrifugation at 16,100×*g* for 30 min. The pellets are washed with 70% ethanol, dried, and resuspended in 40 μl of the Sample Loading Solution (Beckman®). The dideoxy sequencing reactions (for ladder lanes) are performed in the same way except that we add 0.25 m*M* of one of the ddNTPs.

The resuspended cDNA fragments are then separated in an automated DNA sequencing apparatus (Beckman® CEQ 8000 in our case). The optimized parameters that, in our hands, produce high-resolution peak traces at single nucleotide resolution for about 300 nucleotides are electrokinetic injection voltage, 2 kV; electrokinetic injection time, 7 s; denaturation, 90 °C for 150 s; separation voltage, 3 kV; and capillary temperature, 60 °C. The fragment of a sequencer profile that demonstrate visible differences for unfolded and folded RNA are shown in Fig. 2.2.

6. Quantitation of the Changes in the Reactivity and Data Analysis

The changes in the extent of cleavage for individual or grouped nucleotides can be teased out by quantitating the changes in the intensities of the corresponding bands. When the RNA fragments are separated by

---

The oxidative Fenton chemistry was used to produce hydroxyl radicals. The radioactively labeled RNA fragments were separated on 8% gel electrophoresis with 7 *M* urea. Each band represents a single nucleotide. Location of the nucleotide in the RNA sequence is assigned from the G ladder generated by RNase T1 digestion of the same RNA, run on the same gel. Changes in solvent accessibility of individual nucleotides appear as decrease or increase in intensity of separate bands or their groups.

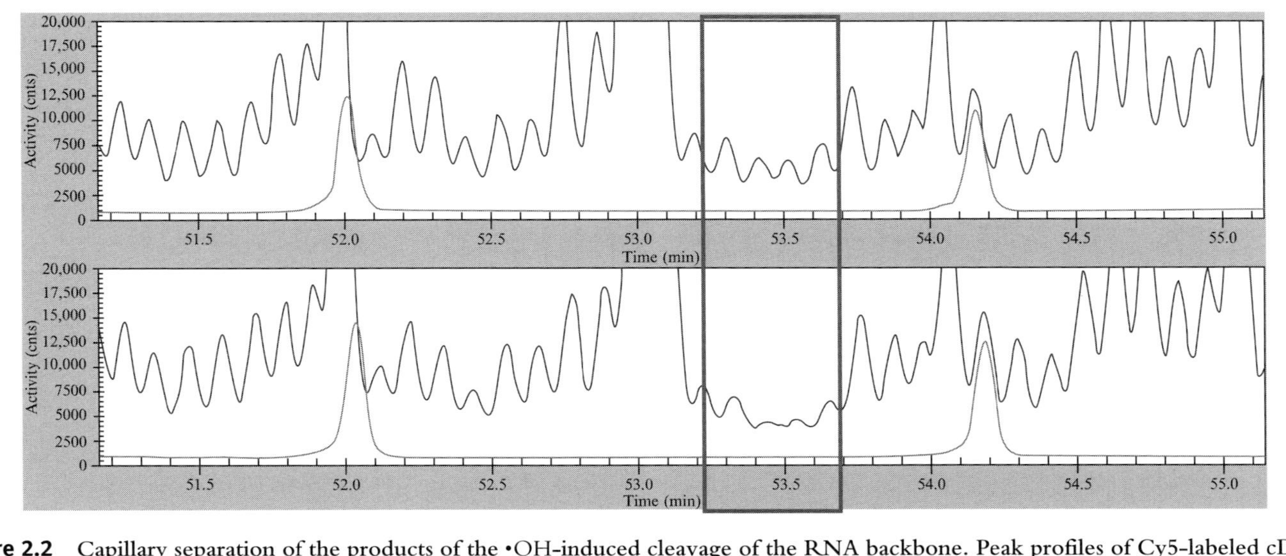

Figure 2.2 Capillary separation of the products of the •OH-induced cleavage of the RNA backbone. Peak profiles of Cy5-labeled cDNAs corresponding to cleaved unfolded (upper panel) and folded (lower panel) *T. thermophila* ribozyme are shown by blue lines. Red peaks corresponding to the Beckman® size standard ladder are used to correlate the peaks with the RNA nucleotides. The peaks that demonstrate significant decrease in reactivity upon folding are enclosed within gray contour. (See Color Insert.)

denaturing gel electrophoresis, we quantitate these changes either semiautomatically using the SAFA software (Chapter 3, Laederach *et al.*, 2008) or manually using a standard gel analysis software like ImageQuant® in which the bands demonstrating a visible change in intensity are enclosed within a contour whose area is reported (Shcherbakova and Brenowitz, 2008). When the cleaved fragments are separated by capillary electrophoresis, the peak traces of the sequencer readout (Fig. 2.2), stored as numerical values in arbitrary units, are exported in the form of text files (.txt), are then semiautomatically processed and analyzed using the CAFA software (Mitra *et al.*, 2008).

The data normalization procedure is described in detail in Chapter 3. Briefly, the data for each sample is normalized to account for the variations in sample loading and scaled to reflect the observed changes between 0 (initial state) and 1 (final state). This normalization can be executed either manually or automatically by gel or capillary electrophoresis data analysis software.

In the case of the equilibrium studies the normalized data can be fit to the Hill equation:

$$\bar{Y}_i = \frac{K_{\mathrm{di}}^{n_{\mathrm{H}}}[\mathrm{M}^+]^{n_{\mathrm{H}}}}{1 + K_{\mathrm{di}}^{n_{\mathrm{H}}}[\mathrm{M}^+]^{n_{\mathrm{H}}}},$$

where K_{di} is the equilibrium dissociation constant, $[\mathrm{M}^+]$ refers to the concentration of the reagent that caused the analyzed structural changes (e.g., Mg^{2+} in our studies of Mg^{2+}-induced folding), and n_{H} is the Hill coefficient. An example of the equilibrium curves generated for several local reporters of folding from the gel electrophoresis shown in Fig. 2.1 is presented in Fig. 2.3. We report lower and upper limits calculated at the 65% confidence interval that corresponds to one standard deviation of the data.

Depending on the length of the RNA and the extent of its tertiary structure, multiple titration curves can be generated for a single RNA molecule (e.g., for the 385 nucleotide *Tetrahymena* ribozyme we usually generate about 30 titration curves). A simple visual inspection of the curves can answer some important questions about equilibrium folding of the RNA, for example: How concerted is the equilibrium folding? Can several distinct groups of titration curves be distinguished? Are there any obvious equilibrium intermediates present? Which structural elements are involved in different phases of the process?

In some cases, it is beneficial to complement this richly detailed local information by additional information about global changes in the nucleic acid structure (Kwok *et al.*, 2006; Laederach *et al.*, 2007b; Schlatterer *et al.*, 2008; Takamoto *et al.*, 2002, 2004). Analytical ultracentrifugation (see chapter 10 in volume 469 of these series) or small angle X-ray scattering can help set constrains on the conformational space for generating coarse-grained

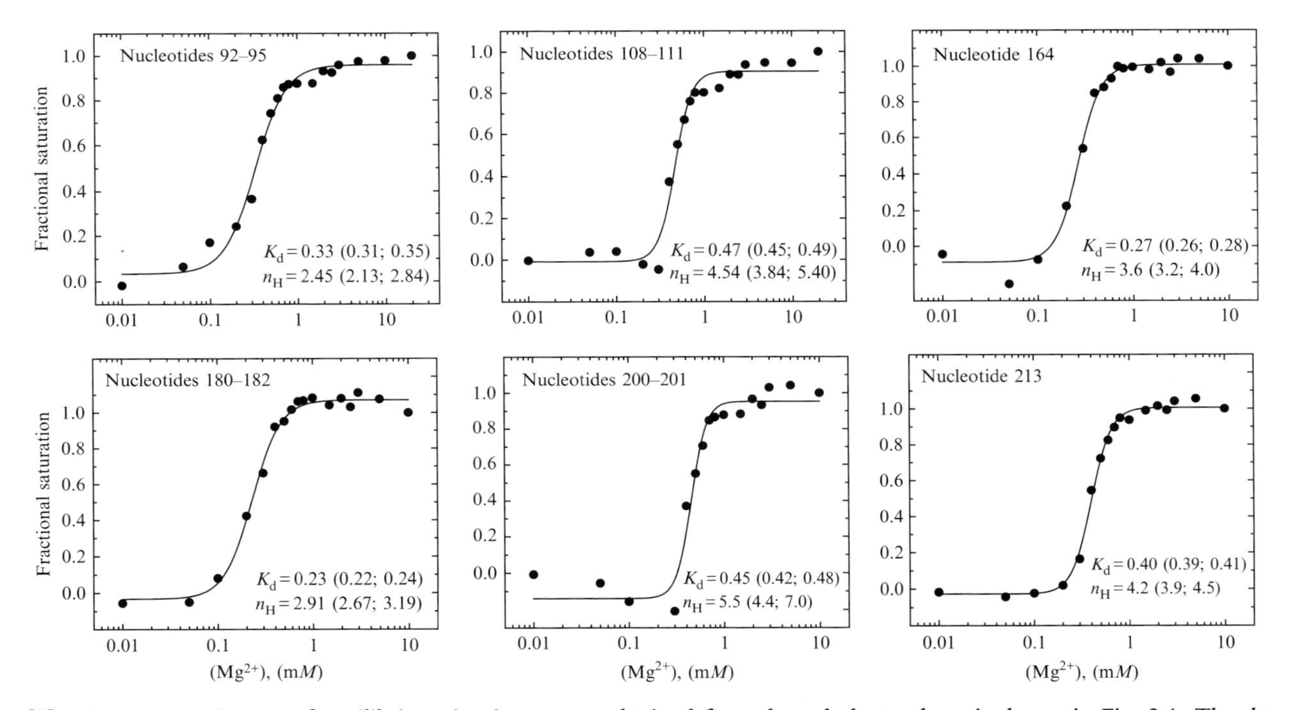

Figure 2.3 A representative set of equilibrium titration curves obtained from the gel electrophoresis shown in Fig. 2.1. The data were normalized and scaled as described in Chapter 3. The Hill coefficient n_H and equilibrium dissociation constant K_d (in m*M*) for each of the sites that demonstrate change in the reactivity associated with increase in the concentration of Mg^{2+} were determined when the data were fit to the Hill equation.

structural models of equilibrium folding intermediates (Baird *et al.*, 2005; Laederach *et al.*, 2007b; Lipfert *et al.*, 2007).

7. Conclusions

Hydroxyl-radical footprinting provides vast local information about formation of tertiary structure during equilibrium RNA folding. The same approach can be used to probe specific nucleic acid–protein interactions. This simple experiment combined with a separation technique capable of single nucleotide resolution can provide information about changes in solvent accessibility of every nucleotide along the nucleic acid backbone as a function of a reagent concentration. This information can be used by itself to characterize the sequence of events and intermediate structures during equilibrium folding (or molecular association) or can be melded with data obtained by other techniques, under identical, or closely related experimental conditions. The recent advances in the data analysis (see Chapter 3) and modeling of the RNA structure to be compatible with the experimental data (Jonikas *et al.*, 2009) provide a valuable set of tools to interpret this information in terms of coarse-grained models of the molecular structure.

Acknowledgments

The authors are members of the laboratory of Michael Brenowitz within which the studies described in this chapter were conducted. The writing of this chapter was supported by grant RO1-GM085130 from the National Institute of General Medical Sciences of the National Institutes of Health. We are thankful to Michael Brenowitz for critical reading of the manuscript and his valuable suggestions.

References

Adilakshmi, T., Ramaswamy, P., *et al.* (2005). Protein-independent folding pathway of the 16S rRNA 5′ domain. *J. Mol. Biol.* **351**(3), 508–519.

Adilakshmi, T., Lease, R. A., *et al.* (2006). Hydroxyl radical footprinting *in vivo*: Mapping macromolecular structures with synchrotron radiation. *Nucleic Acids Res.* **34**(8), e64.

Adilakshmi, T., Bellur, D. L., *et al.* (2008). Concurrent nucleation of 16S folding and induced fit in 30S ribosome assembly. *Nature* **455,** 1268–1272.

Baird, N. J., Westhof, E., *et al.* (2005). Structure of a folding intermediate reveals the interplay between core and peripheral elements in RNA folding. *J. Mol. Biol.* **352**(3), 712–722.

Chauhan, S., Behrouzi, R., *et al.* (2009). Structural rearrangements linked to global folding pathways of the Azoarcus group I ribozyme. *J. Mol. Biol.* **386**(4), 1167–1178.

Das, R., Laederach, A., *et al.* (2005). SAFA: Semi-automated footprinting analysis software for high-throughput quantification of nucleic acid footprinting experiments. *RNA* **11**(3), 344–354.

Deras, M. L., Brenowitz, M., *et al.* (2000). Folding mechanism of the Tetrahymena ribozyme P4–P6 domain. *Biochemistry* **39**(36), 10975–10985.

Fenton, H. J. H. (1894). The oxidation of tartaric acid in presence of iron. *J. Chem. Soc.* **65,** 899–910.

Galas, D. J., and Schmitz, A. (1978). DNAse footprinting: A simple method for the detection of protein–DNA binding specificity. *Nucleic Acids Res.* **5**(9), 3157–3170.

Heilman-Miller, S. L., and Woodson, S. A. (2003). Perturbed folding kinetics of circularly permuted RNAs with altered topology. *J. Mol. Biol.* **328**(2), 385–394.

Hertzberg, R. P., and Dervan, P. B. (1984). Cleavage of DNA with methidiumpropyl–EDTA–iron(II): Reaction conditions and product analyses. *Biochemistry* **23**(17), 3934–3945.

Jonikas, M. A., Radmer, R. J., *et al.* (2009). Coarse-grained modeling of large RNA molecules with knowledge-based potentials and structural filters. *RNA* **15**(2), 189–199.

Kwok, L. W., Shcherbakova, I., *et al.* (2006). Concordant exploration of the kinetics of RNA folding from global and local perspectives. *J. Mol. Biol.* **355**(2), 282–293.

Laederach, A., Das, R., *et al.* (2008). Rapid, quantitative semi-automated analysis of "footprinting" gels. *Nat. Protoc.* **3**(9), 1395–1401.

Laederach, A., Shcherbakova, I., *et al.* (2007b). Distinct contribution of electrostatics, initial conformational ensemble, and macromolecular stability in RNA folding. *Proc. Natl. Acad. Sci. USA* **104**(17), 7045–7050.

Latham, J. A., and Cech, T. R. (1989). Defining the inside and outside of a catalytic RNA molecule. *Science* **245**(4915), 276–282.

Lease, R. A., Adilakshmi, T., *et al.* (2007). Communication between RNA folding domains revealed by folding of circularly permuted ribozymes. *J. Mol. Biol.* **373**(1), 197–210.

Lipfert, J., Das, R., *et al.* (2007). Structural transitions and thermodynamics of a glycine-dependent riboswitch from *Vibrio cholerae*. *J. Mol. Biol.* **365**(5), 1393–1406.

Mitra, S., Shcherbakova, I. V., *et al.* (2008). High-throughput single-nucleotide structural mapping by capillary automated footprinting analysis. *Nucleic Acids Res.* **36**(11), e63.

Nguyenle, T., Laurberg, M., *et al.* (2006). Following the dynamics of changes in solvent accessibility of 16 S and 23 S rRNA during ribosomal subunit association using synchrotron-generated hydroxyl radicals. *J. Mol. Biol.* **359**(5), 1235–1248.

Ralston, C. Y., He, Q., *et al.* (2000). Stability and cooperativity of individual tertiary contacts in RNA revealed through chemical denaturation. *Nat. Struct. Biol.* **7**(5), 371–374.

Schlatterer, J. C., and Brenowitz, M. (2009). Complementing global measures of RNA folding with local reports of backbone solvent accessibility by time resolved hydroxyl radical footprinting. *Methods* **49**(2), 142–147.

Schlatterer, J. C., Kwok, L. W., *et al.* (2008). Hinge stiffness is a barrier to RNA folding. *J. Mol. Biol.* **379**(4), 859–870.

Sclavi, B., Woodson, S., *et al.* (1997). Time-resolved synchrotron X-ray "footprinting", a new approach to the study of nucleic acid structure and function: Application to protein-DNA interactions and RNA folding. *J. Mol. Biol.* **266**(1), 144–159.

Sclavi, B., Woodson, S., *et al.* (1998). Following the folding of RNA with time-resolved synchrotron X-ray footprinting. *Methods Enzymol.* **295,** 379–402.

Shcherbakova, I., and Brenowitz, M. (2005). Perturbation of the hierarchical folding of a large RNA by the destabilization of its Scaffold's tertiary structure. *J. Mol. Biol.* **354**(2), 483–496.

Shcherbakova, I., and Brenowitz, M. (2008). Monitoring structural changes in nucleic acids with single residue spatial and millisecond time resolution by quantitative hydroxyl radical footprinting. *Nat. Protoc.* **3**(2), 288–302.

Shcherbakova, I., Gupta, S., *et al.* (2004). Monovalent ion-mediated folding of the *Tetrahymena thermophila* ribozyme. *J. Mol. Biol.* **342**(5), 1431–1442.
Shcherbakova, I., Mitra, S., *et al.* (2006). Fast Fenton footprinting: A laboratory-based method for the time-resolved analysis of DNA, RNA and proteins. *Nucleic Acids Res.* **34**(6), e48.
Shcherbakova, I., Mitra, S., *et al.* (2007). Following molecular transitions with single residue spatial and millisecond time resolution. *Methods Cell Biol.* **84,** 589–615.
Silverman, S. K., Deras, M. L., *et al.* (2000). Multiple folding pathways for the P4–P6 RNA domain. *Biochemistry* **39**(40), 12465–12475.
Su, L. J., Brenowitz, M., *et al.* (2003). An alternative route for the folding of large RNAs: Apparent two-state folding by a group II intron ribozyme. *J. Mol. Biol.* **334**(4), 639–652.
Swisher, J. F., Su, L. J., *et al.* (2002). Productive folding to the native state by a group II intron ribozyme. *J. Mol. Biol.* **315**(3), 297–310.
Takamoto, K., He, Q., *et al.* (2002). Monovalent cations mediate formation of native tertiary structure of the *Tetrahymena thermophila* ribozyme. *Nat. Struct. Biol.* **9**(12), 928–933.
Takamoto, K., Das, R., *et al.* (2004). Principles of RNA compaction: Insights from the equilibrium folding pathway of the P4–P6 RNA domain in monovalent cations. *J. Mol. Biol.* **343**(5), 1195–1206.
Tijerina, P., Mohr, S., *et al.* (2007). DMS footprinting of structured RNAs and RNA–protein complexes. *Nat. Protoc.* **2**(10), 2608–2623.
Tullius, T. D., and Dombroski, B. A. (1986). Hydroxyl radical "footprinting": High-resolution information about DNA–protein contacts and application to lambda repressor and Cro protein. *Proc. Natl. Acad. Sci. USA* **83**(15), 5469–5473.
Tullius, T. D., Dombroski, B. A., *et al.* (1987). Hydroxyl radical footprinting: A high-resolution method for mapping protein–DNA contacts. *Methods Enzymol.* **155,** 537–558.
Uchida, T., He, Q., *et al.* (2002). Linkage of monovalent and divalent ion binding in the folding of the P4–P6 domain of the Tetrahymena ribozyme. *Biochemistry* **41**(18), 5799–5806.
Uchida, T., Takamoto, K., *et al.* (2003). Multiple monovalent ion-dependent pathways for the folding of the L-21 *Tetrahymena thermophila* ribozyme. *J. Mol. Biol.* **328**(2), 463–478.
Udenfriend, S., Clark, C. T., *et al.* (1954). Ascorbic acid in aromatic hydroxylation I. A model system for aromatic hydroxylation. *J. Biol. Chem.* **208**(2), 731–739.

CHAPTER THREE

Rapid Quantification and Analysis of Kinetic ·OH Radical Footprinting Data Using SAFA

Katrina Simmons,* Joshua S. Martin,* Inna Shcherbakova,† *and* Alain Laederach*

Contents

Abstract

The use of highly reactive chemical species to probe the structure and dynamics of nucleic acids is greatly simplified by software that enables rapid quantification of the gel images that result from these experiments. Semiautomated footprinting analysis (SAFA) allows a user to quickly and reproducibly quantify a chemical footprinting gel image through a series of steps that rectify, assign, and integrate the relative band intensities. The output of this procedure is raw band intensities that report on the relative reactivity of each nucleotide with the

* Developmental Genetics and Bioinformatics, Wadsworth Center, Albany, New York, USA
† Department of Biochemistry and Molecular Pharmacology, University of Massachusetts Medical School, Worcester, Massachusetts, USA

Methods in Enzymology, Volume 468
ISSN 0076-6879, DOI: 10.1016/S0076-6879(09)68003-4

chemical probe. We describe here how to obtain these raw band intensities using SAFA and the subsequent normalization and analysis procedures required to process these data. In particular, we focus on analyzing time-resolved hydroxyl radical (•OH) data, which we use to monitor the kinetics of folding of a large RNA (the L-21 *T. thermophila* group I intron). Exposing the RNA to bursts of •OH radicals at specific time points during the folding process monitors the time progress of the reaction. Specifically, we identify protected (nucleotides that become inaccessible to the •OH radical probe when folded) and invariant (nucleotides with constant accessibility to the •OH probe) residues that we use for monitoring and normalization of the data. With this analysis, we obtain time–progress curves from which we determine kinetic rates of folding. We also report on a data visualization tool implemented in SAFA that allows users to map data onto a secondary structure diagram.

1. Introduction

Chemical probes such as the •OH radical (Latham and Cech, 1989), dimethyl sulfate (DMS) (Tijerina *et al.*, 2007), *N*-methylisatoic anhydride (NMIA) (Wilkinson *et al.*, 2006), and various RNA modifying and cleaving enzymes (Donis-Keller *et al.*, 1977) are commonly used to determine aspects of RNA structure and dynamics. The general principle behind all of these probes is their selective reactivity with nucleotides in different conformations or structural states (Felden *et al.*, 1996). The information provided by these experiments indicates how the local structure of the nucleic acid affects the reactivity of the chemical probe. For example, •OH radicals will react more readily with nucleotides that are solvent exposed, reporting on the solvent accessibility of particular residues (Latham and Cech, 1989). DMS reacts selectively with adenines and cytosines that are not involved in base-pairing reactions (Lempereur *et al.*, 1985), whereas NMIA creates 2′-*O*-adducts with highly flexible nucleotides in the RNA (Badorrek and Weeks, 2005).

Chemical mapping experiments are relatively simple to carry out experimentally, and readout is often accomplished using gel electrophoresis (Laederach *et al.*, 2008). Nucleic acids are readily separated on electrophoretic gels with single nucleotide resolution. Large electrophoretograms such as that illustrated in Fig. 3.1A will often contain several thousand individual bands, each representative of a nucleotide's relative chemical reactivity. Each lane in the gel can represent the RNA's reactivity under different conditions, or as is the case in Fig. 3.1A, at different times during a folding reaction. The ability to probe RNA structure under different conditions and at different time points is a major motivation for obtaining quantitative band intensities from the gel image shown in Fig. 3.1A.

The gel image shown in Fig. 3.1A reports on the Mg^{2+}-induced folding reaction of a mutant (L5b) of the L-21 *Tetrahymena thermophila* group I

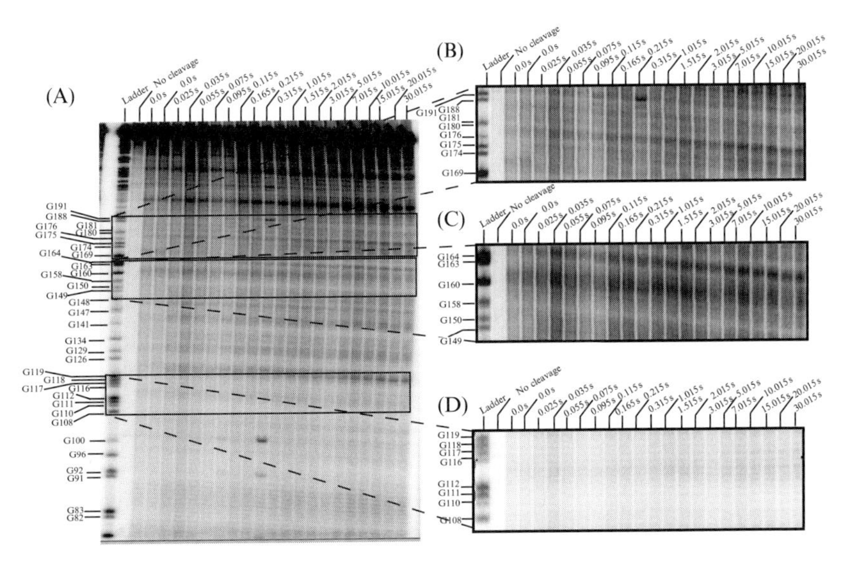

Figure 3.1 (A) Gel image of a typical chemical mapping experiment on the L5b mutant of the *T. thermophila* group I intron analyzed by •OH radical cleavage. Each band in the gel corresponds to a specific nucleotide and each lane, in this case, a different time point along the folding reaction. Nucleotides where a visual change in band intensity is seen as a function of time are magnified in (B) and (C). (D) A zoomed in version of the gel image on the invariant residues (113–115) used for normalization. This chapter focuses on the SAFA software, which is designed to rapidly quantify the relative band intensities from such a gel image (Das *et al.*, 2005).

intron (Laederach *et al.*, 2007). We will use these data (along with a similar gel reporting on the folding reaction of the wild-type RNA) to illustrate the process of obtaining accurate and reproducible time–progress curves for analysis. We analyze the Mg^{2+}-induced folding reaction as measured by •OH radical footprinting for both a wild-type and mutant RNA, which have different folding rates (Laederach *et al.*, 2007; Shcherbakova and Brenowitz, 2005). •OH radicals measure the solvent accessibility of nucleotides by selectively reacting with highly accessible residues, indicating the sites of protection, and accessibility in the molecule (Latham and Cech, 1989). The radicals are generated either using synchrotron radiation or the Fenton chemical reaction (Fenton, 1894; Shcherbakova and Brenowitz, 2008; Shcherbakova *et al.*, 2006). The basic premise of the data we are analyzing is a relative change in the accessibility of nucleotides as the RNA folds upon the addition of Mg^{2+}. As shown in Fig. 3.1B, certain bands become lighter (or darker) as a function of time, indicating folding of the RNA. By sequentially cleaving the RNA with •OH radicals at different time points during the reaction, we measure the fractional saturation of the reaction. We will use these data to generate time–progress curves, from which we can then obtain local folding rates.

SAFA is software designed to facilitate the analysis of RNA footprinting gels. We describe here the major steps in analyzing these data using the software, which can be downloaded along with the example files at http://simtk.org/home/safa. SAFA runs on Mac OS X, Windows XP, and is written using the Matlab (The Mathworks) programming language. Heavy users of the system should consider acquiring the Matlab software (which is licensed campus-wide at most academic institutions) as performance of SAFA is significantly improved when used in conjunction with Matlab. When users choose to use SAFA in conjunction with Matlab, they must download the "source" version of the software independently of whether they are running Matlab on a Windows or Macintosh computer. Furthermore, if users have the Windows Vista or Home edition, using Matlab will solve many compatibility issues.

2. Using SAFA

Using SAFA is relatively straightforward. Users will recognize familiar menus (such as the File Menu, indicated by red arrow in Fig. 3.2A). The basic workflow of SAFA is controlled through a series of buttons (marked by green arrows) on the right side of the program in Fig. 3.2A. Clicking on each button successively leads the user through the different steps of the procedure. Initially, the software requires as input both a gel image (either with a .gel or .tiff extension, 16-bit images are generally easier to analyze with the software, although 8-bit images are supported) and a sequence file (in FASTA, .fas format). The gel image will appear in the central SAFA window (Fig. 3.2A) while reading in a sequence file will open the sequence selection dialog box (Fig. 3.2B). In general, keyboard commands or keyboard shortcuts are displayed above the main gel window in the software. Several example data sets are available for download from http://simtk.org/home/safa including all the data presented in this chapter.

Our goal is to obtain quantitative intensities for all bands shown in Fig. 3.1. As can be seen in Fig. 3.1B and C, certain bands (in this case corresponding to nucleotides C177 and G160, respectively) become significantly lighter as the folding reaction progresses. This indicates that as the molecule is folding these regions are becoming progressively more protected. By plotting the relative change in band intensity as a function of time, we can generate time–progress curves that report on the local rate of conformational change in the RNA molecule. Several steps are required to obtain these data from the gel image. The first step we describe here

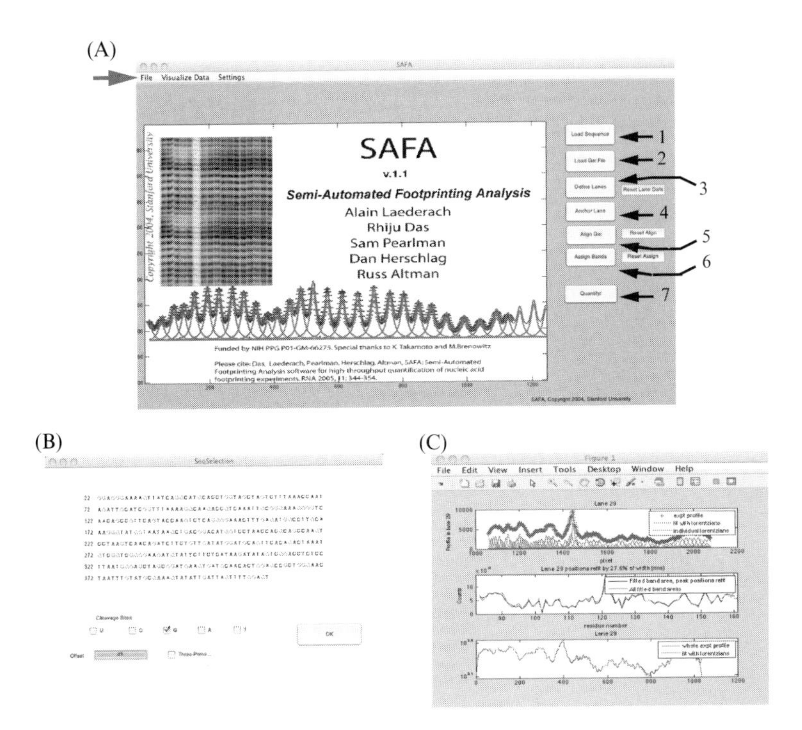

Figure 3.2 Screenshots of the SAFA software. (A) The main SAFA window used to control all aspects of the software. The analysis procedure is carried out by successively clicking on each of the buttons to the right side the main window in which the gel image is displayed. (B) The sequence selection tool allows the user to define how the semi-automated procedure of band assignment will be carried out. In this case, the sequence selection tool is setup to use a T1 ladder, which indicates the positions of guanosines on the gel. (C) The quantification progress window appears when the software is automatically fitting lanes and is used to monitor the progress and accuracy of the single-peak fitting procedure (Takamoto *et al.*, 2004).

involves integrating the band intensity using a procedure called "single-band peak fitting" (Takamoto *et al.*, 2004).

Single-band peak fitting fits a model to the raw image data, but requires accurate initial guesses of the relative positions of the bands on the gel. The inevitable imperfections in the gel matrix yield gel images with nonparallel lanes and "smiles" (or "frowns") that make the problem of estimating band positions quite difficult. SAFA includes functionality to "rectify" the gel image, effectively correcting the imperfections so as to have parallel lanes, and no smiles (or frowns) in the gel image. Details of the procedure for gel rectification can be found in previous publications (Laederach *et al.*, 2008) and a video demonstration of the process can be downloaded from http://simtk.org/home/safa. We recommend viewing the video tutorials, as the gel rectification process is fundamentally a visual process, requiring accurate user input. For the purposes of this method, we will briefly outline the procedure for the example data set we use here.

2.1. Gel cropping

The gel image must be initially cropped. In Fig. 3.3A, we show a screenshot of the main SAFA window in which we have loaded the gel image into SAFA. Using the cursor, we have drawn a cropping box around the right half of the gel image, which reports on the folding progress of the L5b mutant of the *T. thermophila* group I intron. We are careful to include in our cropping the T1 digest lane, which will help us assign the bands to the sequence in later steps. In general, a large enough portion of the gel image should be kept so that all lanes are encompassed.

2.2. Lane definition and gel rectification

The next step in the SAFA procedure involves defining lanes. This procedure is used to vertically align the gel by identifying lane boundaries. Such lane boundaries are generally trivial to identify by eye, but present a significant computational challenge when attempted in a fully automatic way. SAFA therefore uses a semiautomated approach to resolve this problem. The user initially defines at least one lane by drawing the boundaries of that lane using a click and draw procedure, analogous to a line tool in a computer graphics program. The user can then invoke an automated procedure for "guessing" the next lane boundaries, which uses the previous lane definitions as a starting point. This automated procedure is invoked by typing "G."

The procedure of guessing lane boundaries is continued until all lanes are correctly selected, at which point the user invokes a vertical rectification procedure by typing "Z." The vertical rectification procedure uses a linear interpolation routine that preserves the total pixel count in the gel image. The result of this procedure is illustrated in Fig. 3.3B, where each lane is properly defined by red vertical lines. SAFA then identifies the center of the lanes as illustrated by dotted lines in Fig. 3.3C. Once the lanes have been defined, the gel now requires horizontal alignment.

The horizontal alignment procedure in SAFA is analogous to the vertical one. In this case, horizontal lines are drawn through corresponding bands in the gel image as shown in Fig. 3.3D. These lines should trace through the bands in the gel and will act as anchor lines for the horizontal rectification procedure. This procedure is manual, and the user should choose several well-resolved bands throughout the gel image as shown in Fig. 3.3D. The user can zoom in and out by right clicking to more accurately trace through

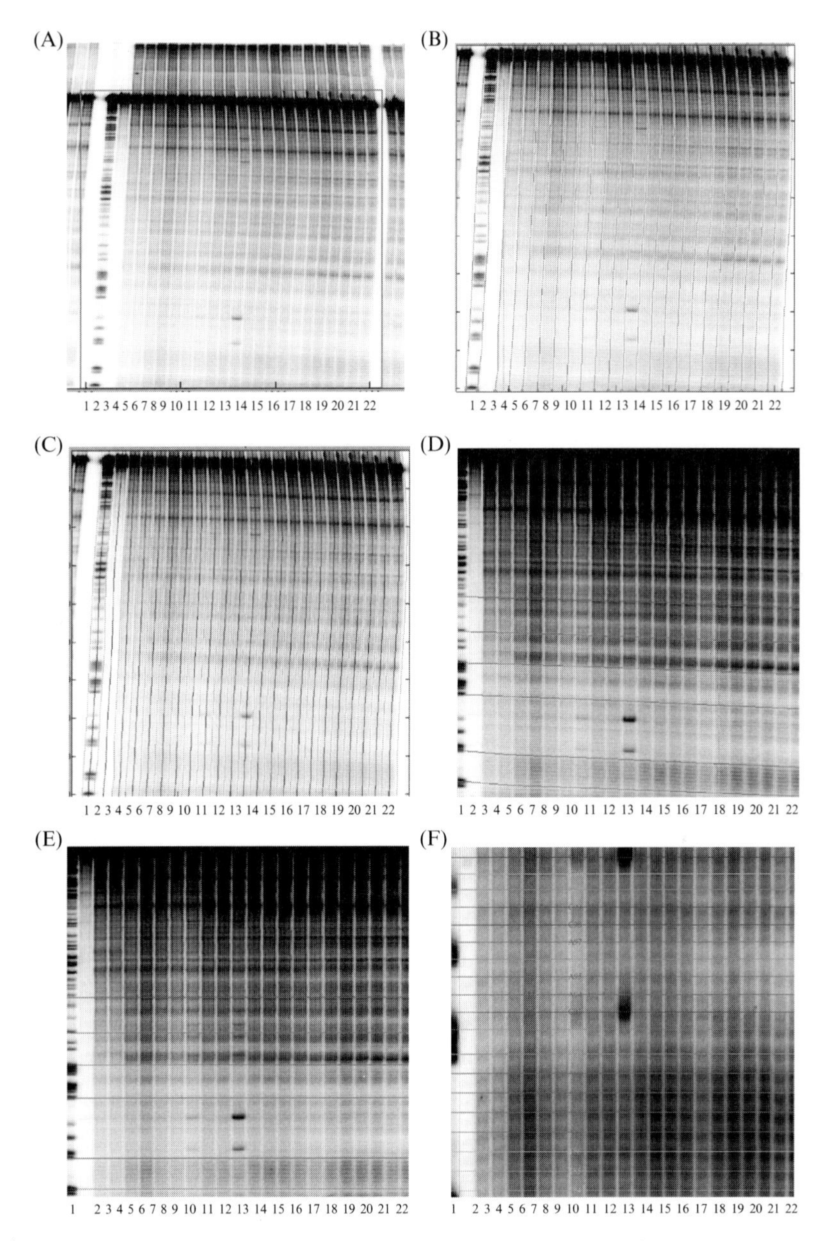

Figure 3.3 Illustration of the gel rectification procedure implemented in SAFA. (A) Screenshot of the main SAFA window with a gel image loaded; red box delineates the user-defined cropping boundary for subsequent analysis. (B) Screenshot of the lane definition procedure. Red line is manually drawn that marks the boundaries of each lane. Once a few lanes have been manually defined, SAFA is able to "guess" the outline of the following lanes, which is done by typing the letter "G." (C) Screenshot of the completed and recorded defined lanes that occur after the user types "Z" or "Q."

the bands in the gel. The number of such anchor lines required and their position is dependent on the extent of horizontal deformation within the gel image. The ultimate result (Fig. 3.3E) is a gel that is both vertically and horizontally aligned, making the process of band assignment straightforward.

2.3. Band assignment and use of a sequence file

As noted above, SAFA requires two input files, a gel image and a FASTA sequence file. The FASTA file for the L-21 *T. thermophila* group I intron is shown below:

>Group I Intron, T. thermophila
GGAGGGAAAAGUUAUCAGGCAUGCACCUGGUAGCUAGU
CUUUAAACCAAUAGAUUGCAUCGGUUUAAAAGGCAAGACC
GUCAAAUUGCGGGAAAGGGGUCAACAGCCGUUCAGUACCA
AGUCUCAGGGGAAACUUUGAGAUGGCCUUGCAAAGGGUAU
GGUAAUAAGCUGACGGACAUGGUCCUAACCACGCAGCCAA
GUCCUAAGUCAACAGAUCUUCUGUUGAUAUGGAUGCAGU
UCACAGACUAAAUGUCGGUCGGGGAAGAUGUAUUCUUCU
CAUAAGAUAUAGUCGGACCUCUCCUUAAUGGGAGCUAGCG
GAUGAAGUGAUGCAACACUGGAGCCGCUGGGAACUAAUUU
GUAUGCGAAAGUAUAUUGAUUAGUUUUGGAGU

In general, the first line of the FASTA file contains some descriptive text of the sequence preceded by a "greater than" character (>). It is also important to input the RNA sequence (replacing T with U) if the molecule being studied is in fact RNA. SAFA can also be used to study DNA, and therefore distinguishes between T and U in the sequence. Most importantly, the file should be saved as a "Western ASCII" text file, especially if it is being edited in a word processing application. SAFA assumes that the first nucleotide in the file corresponds to the first nucleotide in the sequence, unless an offset is set in the sequence selection tool (Fig. 3.2B).

The sequence selection tool (Fig. 3.2B) defines how the user will assign bands on the gel. It is invoked by clicking on "load sequence" (Fig. 3.2A, green arrow marked 1) in the main SAFA window. The general concept

The color of the lines changes from red to green and a dotted line marks the middle of the lane when the procedure in completed. (D) Screenshot of the horizontal alignment procedure. The gel is aligned horizontally to ensure the bands are parallel. Similar to the lane definition process, the user identifies bands across the gel image with a horizontal line. Only one line should be drawn per band. The left button of the mouse begins the line and marks sequential points throughout the row and the right button ends the line. The number of rows to mark is at the users' discretion; however, the more rows used will result in a more accurate an adjustment of the image. (E) Screenshot of the finished horizontally aligned gel. (F) Screenshot of the band assignment procedure. (See Color Insert.)

behind the sequence selection tool is that one or several ladder lanes will be run in a separate lane on the gel being analyzed. In the case of RNA, T1 digests are often used to identify guanosine nucleotides; this is the case in Fig. 3.1. Alternatively, sequencing ladders can also be run to identify any (or all) of the other nucleotides (Tijerina *et al.*, 2007). Ultimately, with the gel rectified, the user will click on the bands they have chosen with the sequence selection tool. As is illustrated in Fig. 3.2B, the user has selected "G" and set an offset of 21, since the first nucleotide in the sequence file corresponds to the 21st residue of the *T. thermophila* group I intron. The user will only need to identify bands corresponding to G residues, and SAFA will guess the positions of all the residues in-between.

This semi-automated procedure allows the user to precisely define bands in the gel across all lanes rapidly. Furthermore, by adjusting the parameters in the sequence selection tool, the user has the option of selecting one or multiple nucleotides for manual definition during the assignment process. In general, selecting one nucleotide is sufficient, as SAFA is able to extrapolate the positions of the other nucleotides accurately during the assignment process. The sequence selection tool also allows the user to specify that the RNA is 3′-labeled, indicating that the 5′-band will be at the bottom of the gel. This option can also be used if the gel to be analyzed is the result of reverse transcription and the user wishes to use the RNA sequence. SAFA will use a single-peak fitting algorithm to fit the data, which can handle overlapping bands. As a result, the user can choose to continue to assign bands in the overlapping region of the gel. How "high" to go in the gel will depend greatly on the quality of the gel image, desired quality of the data, and complexity of the experiment. Users will need to evaluate the reproducibility of the procedure on their own data to evaluate the degree to which they can analyze highly overlapping regions in the gel. In the example we provide along with this chapter, we analyze the gel conservatively, obtaining data for the first 100 bands.

2.4. Single-peak fitting with SAFA

Peak fitting is a fully automated process in SAFA and is activated by clicking on the "quantify" button in the main SAFA window (Fig. 3.2A, green arrow marked 7). SAFA will only perform the peak-fitting analysis if all the previous steps have been completed. Progress in the peak-fitting procedure is monitored through a series of three graphs that appear when the peak-fitting procedure is run (Fig. 3.2C). SAFA proceeds through each lane of the gel successively, allowing the user to visualize the quality of the fit. Visualizing the resulting fit is key when troubleshooting the quantification procedure, as potential errors in the fitting procedure can easily be seen when looking at these results. SAFA will plot all of the fits again once the fitting procedure is complete, to encourage users to look carefully at the traces.

We illustrate in Fig. 3.4A a successful single-peak fitting procedure on Lane 4 of the gel shown in Fig. 3.1A. In Fig. 3.4A, SAFA plots the individual peaks in red, and the overall lane profile in blue. In this case, the individual Lorentzian peaks are evenly spaced, and have uniform widths. In general, SAFA optimizes the fit so as to have regularly spaced peaks with linearly varying widths. However, if the initial positions of the peaks as defined by the band assignment procedure (Fig. 3.3F) are inaccurate, SAFA will find a solution with large differences in peak widths, as is illustrated in Fig. 3.4B. In this case, the user has incorrectly assigned bands and the solution found by SAFA has irregularly spaced peaks and highly variable

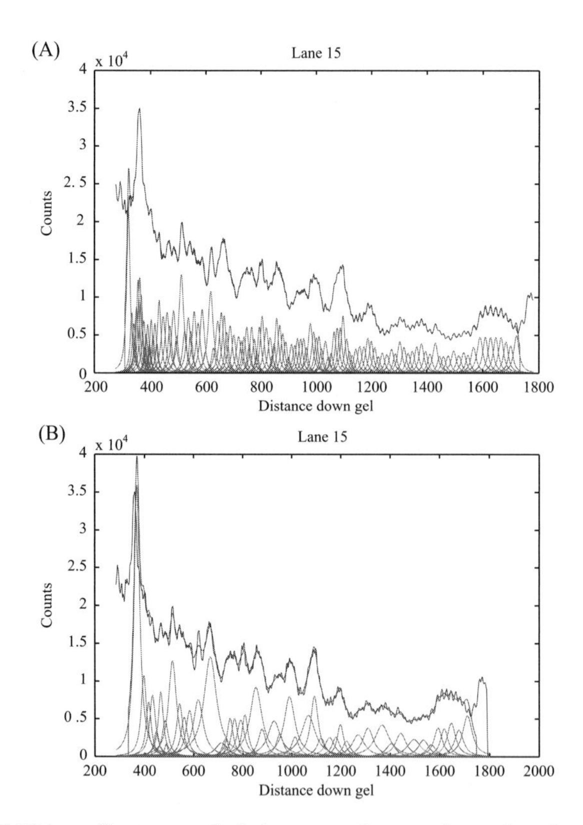

Figure 3.4 SAFA performs a peak-fitting procedure to determine the relative band intensity in the gel image. When the fitting procedure has converged, SAFA will plot the results as shown here. Red lines represent individual Lorentzians that when summed yield the lane profile shown in blue. (A) A successful and accurate single-peak fitted graph for Lane 15 of the gel shown in Fig. 3.1. (B) Visual inspection of the fitted profile can be used to determine if the user was not accurate in assigning bands (Fig. 3.3F). In this example, there are large differences in the peak widths and irregularly spaced peaks. These are signs of inaccurate results. (See Color Insert.)

peak widths. SAFA will replot the fitted data as illustrated in Fig. 3.4 for each lane after the fitting procedure, and it is recommended to visually inspect these plots to identify any fits that look like Fig. 3.4B, as these will indicate potential inaccuracy in the results. Although SAFA will check for convergence of the fitting procedure and provide an error message if the convergence criteria are not met, it is often still possible to obtain convergence on an incorrect solution as is illustrated in Fig. 3.4B. Visual inspection is thus an important final step of the fitting procedure.

2.5. Data output

At this stage, SAFA will output a text file with the raw band intensities determined through the single-peak fitting procedure. The text file is tab delimited, with the first column containing the residue numbers, and the next columns the actual raw band areas, as determined by peak fitting. It should be noted that the data are output in scientific notation, and when the file is opened in a spreadsheet program like Excel, the nucleotide numbers are displayed in scientific notation as well. Interpretation of the raw data, however, requires further analysis, normalization, and visualization. Although some users at this stage may prefer to perform these analyses in other software, SAFA offers several other tools for data visualization and normalization. The automated normalization functionality of SAFA is described in detail in a previous publication (Laederach *et al.*, 2008); we choose here instead to analyze the time-resolved data manually to better illustrate the role of normalization in obtaining time–progress curves from these data (Shcherbakova *et al.*, 2006). We also turn our focus to a previously unpublished tool built into SAFA for the visualization/projection of chemical mapping data on RNA secondary structure diagrams. These two approaches are critical for the correct interpretation of chemical mapping data.

3. Data Normalization

3.1. Data normalization using invariant residues

The first step of data normalization requires correcting for nonuniform loading of the RNA into each lane. This procedure involves identifying invariant residues, that is, nucleotides where the relative density changes reflect the variations in the amount of RNA loaded per lane. The choice of the invariant residues is based on visual inspection of the gel image to identify bands that demonstrate no visible systematic change in intensity. Furthermore, any available structural information can be used to ensure that the selected residues belong to a region where structural changes are not expected. To verify whether the choice of invariant bands was correct, we plot the invariant band intensities as a function of reaction time to rule out a systematic increase or decrease in the band intensities. Usually, a single set of invariant residues can be used for all the protections and sites of the enhanced reactivity on the same gel. In the case of our example, we chose nucleotides 113–115 (Fig. 3.1D), which are in the P4 helix of L-21 *T. thermophila*, a highly stable region of the molecule that does not change conformation during folding. In general, helical segments work best as invariant residues.

To normalize the time-resolved data, we divide the intensities of the all the bands by the average intensity of the invariant residues. The result of this procedure is illustrated in Fig. 3.5 in which the raw data (Fig. 3.5A) and invariant normalized data (Fig. 3.5B) for the wild-type *Tetrahymena* ribozyme (black) and L5b mutant is plotted (red symbols). The effect of this simple procedure is dramatic, resulting in uniform time–progress curves which can be subjected to further analysis. We provide the Microsoft® Excel spreadsheet used to normalize these data at http://simtk.org/home/safa. It should be noted that small variations in sample loading are easily corrected in this way. However, if large variations in sample loading are observed, this will lead to much greater noise in the data regardless of which invariant residues are chosen.

The second step of data normalization aims to determine the extent of protection as a fractional value between 0 (initial state) and 1 (final state). This is accomplished by dividing the band intensity by the absolute in band intensity between the final and initial states. Following this transformation, the intensities are individually scaled to fractional saturation and this is illustrated in Fig. 3.5C. This transformation creates uniform kinetic data, which allows us to compare changes in intensity between different molecules. In this case, it is clear that the wild-type and mutant RNA (black and red in Fig. 3.5, respectively) have different rates of formation. For the three protections, we illustrate in Fig. 3.5, the L5b mutant appears to fold slower.

3.2. Generating time–progress curves and obtaining rate constants

The normalized data are best visualized in Fig. 3.5C by plotting the fractional saturation of the individual sites that demonstrate changes in the reactivity as a function of time. The reaction time is calculated as a sum of the quench-flow delay time and half of the cleavage (or modification) reaction time. To visualize the details of the fastest events the *X*-axis is often set to a logarithmic scale.

The collection of the individual time points can be fit to an exponential function to discern rate constants for the processes that are reflected in the time-resolved changes in the local reactivity to the footprinting reagent

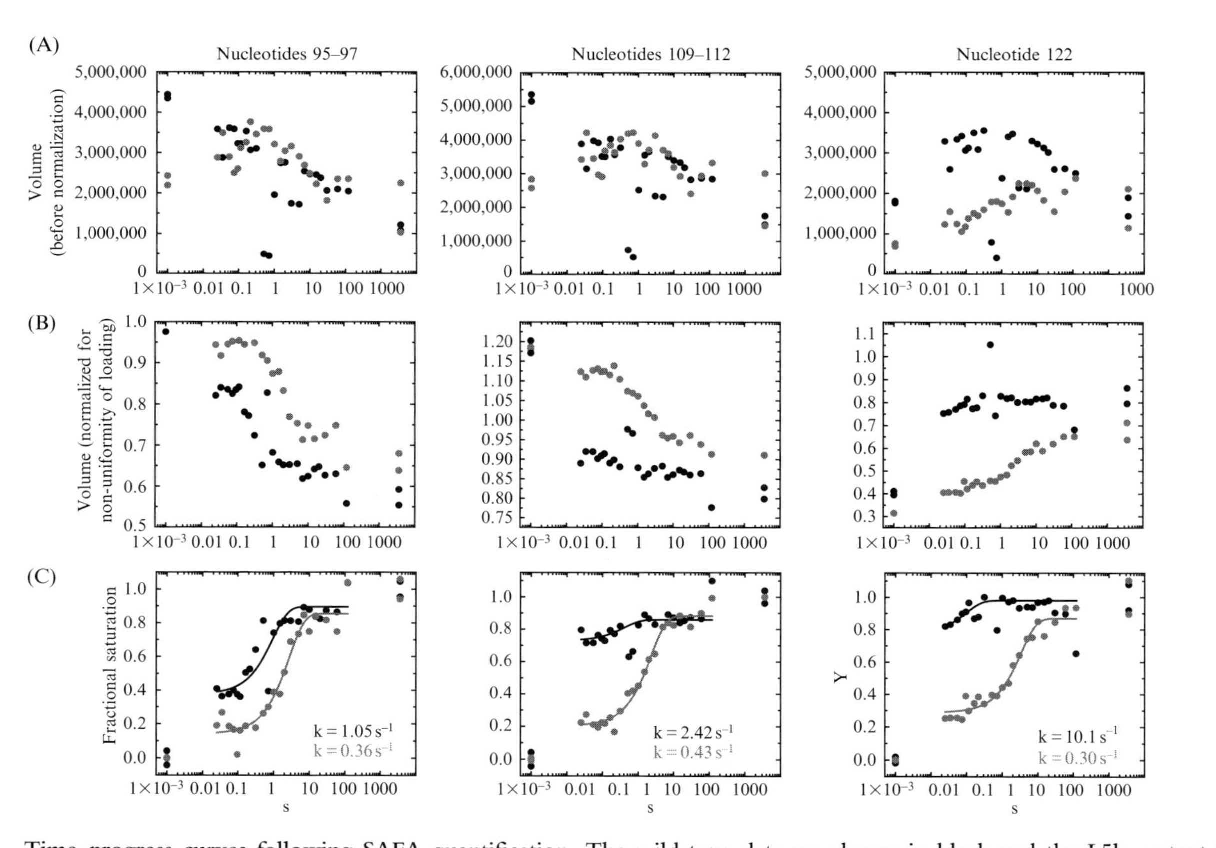

Figure 3.5 Time–progress curves following SAFA quantification. The wild-type data are shown in black and the L5b mutant in red. (A) When the data that are not normalized are plotted as a function of time, it is difficult to identify any trends in the changes in reactivity.

(Fig. 3.5C, lines). These changes are best described in terms of an exponential function:

$$\bar{Y} = 1 - \sum_{i=1} \alpha_i \exp(-k_i t),$$

where α_i and k_i are the amplitude and rate constant, respectively, of the *i*th kinetic phase. When fitting the data we start with a single exponential parameter ($i = 1$). If a systematic error arises as a result of the fitting two rate constants are fit. We usually report lower and upper limits calculated at the 65% confidence interval that corresponds to one standard deviation of the data.

3.3. Data interpretation by inspection of the time–progress curves and comparison of rate constants

In many cases, a working hypothesis can be discerned from a simple inspection of the time–progress curves and the comparison of the rate constants. For example, for the limited number of time–progress curves illustrated in Fig. 3.5, we conclude that not all the time–progress curves demonstrate the same kinetic behavior. The significant offset in the reactivity (or burst) between initial state and the first data point for the wild-type data (Fig. 3.5, black curves) indicates that the onset of the folding process is faster than the dead time of the experiment. In this case better time resolution would be beneficial to characterize the fastest events of the folding under this condition. Improved time resolution is achieved by Fast Fenton footprinting (Shcherbakova *et al.*, 2006) and will reveal the multiple kinetic phases of the local changes in the reactivity at shorter time points. The data we present here, however, were collected using synchrotron radiation to generate •OH radicals, which did not have the time resolution to resolve the fastest events in the folding of the wild-type molecule (Ralston *et al.*, 2000). We can nonetheless conclude from this analysis that the mutation of the L5b tetraloop causes significant changes in the folding mechanism, which yield slower rates (Fig. 3.5C). The presence of a distinct hierarchical folding mechanism for the wild-type ribozyme is gleaned from the comparison of the rate constants and the amplitudes of the fastest folding events: formation of the hinge region of the P4–P6 domain, reported by increase in the reactivity for nucleotide 122, precedes to the formation of junction region

(B) When invariant residue normalization is applied (in this case based on invariant residues 113–115), much clearer changes in reactivity are revealed. (C) When further normalization procedures are applied based on setting the initial and final fractional saturation to zero for $t = 0$ and one for the final state, we are able to compare the data between experiments and determine rate constants for the transitions. (See Color Insert.)

between P4–P6 domain and the catalytic core (nucleotides 109–112), whereas the catalytic core (nucleotides 95–97) is the slowest to fold. A complete analysis of these data is described in Laederach *et al.* (2007) where we model the folding pathways of the RNA based on these experiments. An automated approach to modeling this data kinetically is the Kinfold software described in detail by Martin *et al.* (2009).

4. Data Visualization

SAFA outputs a large table of numbers, which alone can be difficult to interpret. If the data measure time progress of the folding reaction, time–progress curves can be generated and visualized as described above (Fig. 3.5). Often, however, the ability to project the data onto the two-dimensional structure of the RNA is valuable in identifying regions of the molecule that are highly reactive to a particular probe. We have therefore developed a way to quickly and clearly map the quantified data generated through SAFA analysis onto the molecule's secondary structure. Our idea is based on the fact that each lab has their favorite secondary structure diagram, and that a tool able to map data directly onto this diagram will be most valuable in interpreting the data within the structural context of the experiment. Other tools such as VARNA (Ferre *et al.*, 2007) and S2S (Jossinet and Westhof, 2005) generate automated secondary structure representations on which users may map data. These representations, however, do not always correspond to the context in which the data are being visualized, and as such we decided a tool that can facilitate plotting of data onto a user-defined diagram would have value. This tool can be accessed from the Secondary Structure Plot menu item under Visualize Data in SAFA.

The visualization tool requires two inputs, SAFA data and a high-resolution (>150 dpi) image of the secondary structure. SAFA reads either JPEG, GIF, or TIFF formatted pictures with the preferred format TIFF (without any form of compression). The image should have some space reserved on one side for a color legend. This procedure allows the user to manually define the relative positions on the image once and plot any data onto the image thereafter. It is therefore worthwhile obtaining a high-quality image before carrying out the manual parts of the procedure.

There are four steps required to generate a figure:

1. The image (in JPEG, GIF, or TIFF format) of the molecule's secondary structure is loaded using the *SecStruct Image* button (Fig. 3.6A, red arrow) When the image loads correctly, "loaded" should appear to the right of the *SecStruct Image* button, and the image will appear in the main window as shown in Fig. 3.6B.

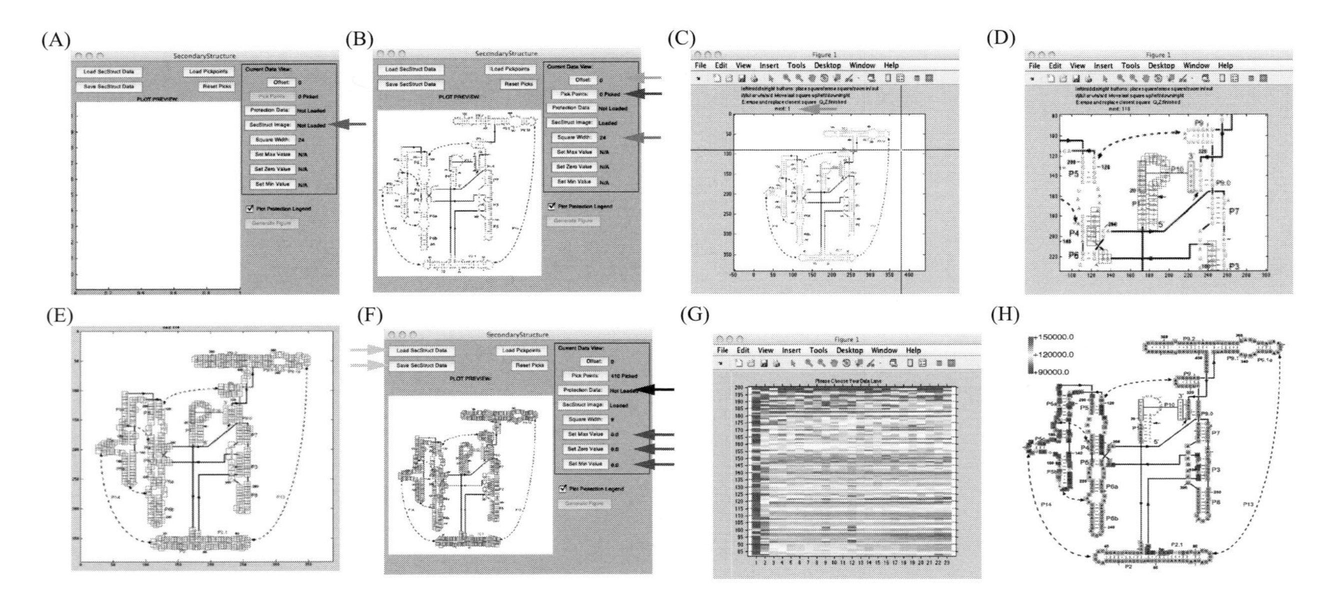

Figure 3.6 The data visualization component allows users to map their data onto secondary structure diagrams. (A) Screenshot of the starting window that opens when *Visualize Data* → *Secondary Structure Plot* is selected. Selecting the *SecStruct Image* button (highlighted by red arrow) allows the user to load a JPEG, GIF, or TIFF image of their secondary structure. (B) Once the image is loaded, it will appear in the window and the next step is to pick points on the image. The *Pick Points* button indicated by blue arrow begins the procedure where the user defines the location of each individual nucleotide on the diagram. Magenta arrow indicates the *Square Width* button, which determines the size of the box that is placed over the selected residues, and green arrow indicates the *Offset* button that allows the user to account for any offset in the start nucleotide of the nucleic acid. (C) A new window appears when the *Pick Points* button is pushed. Indications for keyboard shortcuts are printed above the upper *X*-axis along with the next box number indicated by orange arrow. (D) An example of pick points procedure on the *T. thermophila* group I intron secondary structure diagram. (E) An example of a completed procedure on the same figure. Each nucleotide in the diagram is covered by a square. (F) Once the procedure of picking points is completed, the image and corresponding boxes are saved

2. The *Pick Points* (Fig. 3.6B, blue arrow) button will become active once an image is loaded. Clicking the *Pick Points* button will begin the procedure for indicating the location of nucleotides on the secondary structure diagram. The user can place squares over each residue in the corresponding region of the molecule. SAFA assumes that the first square to be placed corresponds to nucleotide number one, unless the user specifies an offset (Fig. 3.6B, green arrow). Each box added by a left click corresponds to a particular numbered nucleotide starting from the offset and increasing by increments of one. To assist in determining which box corresponds to what nucleotide, the next box number is indicated by orange arrow in Fig. 3.6C. To place successive squares, the left button is clicked. The right button on the mouse zooms in and out and inaccurately placed boxes are removed by either clicking the middle button on the mouse or moving the crosshairs over the box and typing "E" on the keyboard. The size of the squares can be set with the *Square Width* button (magenta arrow in Fig. 3.6B). The value must be a whole number. An example of the pick points procedure on the 5′-region of the *T. thermophila* group I intron is shown in Fig. 3.6D. Once all the boxes have been matched with their corresponding nucleotides (as illustrated in Fig. 3.6E), clicking either Z or Q will end the pick points procedure. The figure will return to its original window. Here, red highlighted boxes will become blue, and this signifies the picking points procedure is done. Figure 3.6F illustrates the result of the completed procedure.
3. Once all the points have been picked, clicking the *Protection Data* button (indicated by black arrow in Fig. 3.6F) will invoke a dialog to input the text file. A new window will appear with a color plot of the data (Fig. 3.6G). The user can then choose the lane corresponding to the data that he/she wishes to visualize on the secondary structure diagram. The colors used to represent the data are defined in the following manner: White nucleotides correspond to "zero" intensity. The value of the zero level is initially set to be the average of the minimum and maximum intensities in the data. Each residue that has higher reactivity will be colored red with intensity on a linear scale from this zero level to a maximum value. The maximum value is also a default value and is plotted in solid red. Similarly, residues with

and can also be reloaded. The boxes change color from red to blue signifying the points have been saved. The light blue arrows mark the *Load SecStruct Data* and *Save SecStruct Data* buttons that allow the user to save their progress and load their data. Black arrow highlights the *Protection Data* button that initiates the next step in the process where the user chooses the data to be displayed on the figure. The purple arrows indicate the *Set Max/Zero/Min Value* buttons that allow the user to adjust these values. (G) Representation of the SAFA quantified data for the gel image shown in Fig. 3.1 in a color plot allowing the user to choose which lane to plot on the secondary structure. (H) Mapping of Lane 15 onto the secondary structure of the *T. thermophila* group I intron indicating nucleotides that are protected in blue and cleaved by •OH radicals in red. (See Color Insert.)

reactivity values lower than the zero value will be colored blue with intensity on a linear scale from the zero level to a minimum value. The default minimum value is plotted as solid blue. The maximum, minimum, and zero values are adjusted by clicking the appropriate *Set Max/Zero/Min Value* buttons (Fig. 3.6F, purple arrows).

4. When the structure image, residue locations, and accessibility data have all been loaded or assigned, the *Generate Figure* button will become active. This generates a full-sized figure showing the reactivity levels plotted on the secondary structure of the molecule (Fig. 3.6H). The plot protection legend checkbox controls whether the color legend is displayed in the preview plot as well as in the full-sized figure. If a legend is being displayed, its location is indicated before saving the figure or exporting to the file format of choice. Any nucleotides where no data are available will be covered in a gray box (as illustrated in Fig. 3.6H).

The secondary structure image and the points picked for the residue locations on that image can be loaded independently from previously saved data files using the *SecStruct Image* and *Load Pickpoints* buttons, and all fields can be saved and later loaded together from those data files using the *Load/Save SecStruct Data* buttons (Fig. 3.6F, light blue arrows).

5. Conclusion

SAFA is a software package specifically designed for obtaining quantitative and reproducible data from a gel image of a chemical mapping reaction. It was originally intended for use mostly with •OH radical footprinting gels (Brenowitz *et al.*, 2002; Gross *et al.*, 1998; Shcherbakova *et al.*, 2006; Woodson, 2008), but is now being used for many other applications in nucleic acid chemical mapping (Vicens *et al.*, 2007; Wilkinson *et al.*, 2005, 2006). Since most chemical mapping experiments use similar principles to •OH radical footprinting for the analysis of the reactions, it is easy to adapt SAFA to a wide range of applications. The sequence browser (Fig. 3.2B) is instrumental in defining how the band assignment procedure will behave, and by using the correct sequence and selection of nucleotides; the user can define any arbitrary combination of nucleotides. This flexibility, along with the gel rectification procedure, provide a series of tools for the rapid and accurate quantification of chemical mapping gels (Das *et al.*, 2005).

Acknowledgment

This work is supported the US National Institutes of Health (NIGMS) through grants R00-79953 and R21-87336 to A. L.

REFERENCES

Badorrek, C. S., and Weeks, K. M. (2005). RNA flexibility in the dimerization domain of a gamma retrovirus. *Nat. Chem. Biol.* **1,** 104–111.

Brenowitz, M., Chance, M. R., Dhavan, G., and Takamoto, K. (2002). Probing the structural dynamics of nucleic acids by quantitative time-resolved and equilibrium hydroxyl radical "footprinting". *Curr. Opin. Struct. Biol.* **12,** 648–653.

Das, R., Laederach, A., Pearlman, S. M., Herschlag, D., and Altman, R. B. (2005). SAFA: Semi-automated footprinting analysis software for high-throughput quantification of nucleic acid footprinting experiments. *RNA* **11,** 344–354.

Donis-Keller, H., Maxam, A. M., and Gilbert, W. (1977). Mapping adenines, guanines, and pyrimidines in RNA. *Nucleic Acids Res.* **4,** 2527–2538.

Felden, B., Florentz, C., Westhof, E., and Giege, R. (1996). Usefulness of functional and structural solution data for the modeling of tRNA-like structures. *Pharm. Acta Helv.* **71,** 3–9.

Fenton, H. J. H. (1894). What species is responsible for strands scission in the reaction of Fe(II) EDTA2 with H_2O_2 with DNA? *J. Am. Chem. Soc.* **6,** 899.

Ferre, F., Ponty, Y., Lorenz, W. A., and Clote, P. (2007). DIAL: A web server for the pairwise alignment of two RNA three-dimensional structures using nucleotide, dihedral angle and base-pairing similarities. *Nucleic Acids Res.* **35,** W659–W668.

Gross, P., Arrowsmith, C. H., and Macgregor, R. B. Jr. (1998). Hydroxyl radical footprinting of DNA complexes of the Ets domain of PU.1 and its comparison to the crystal structure. *Biochemistry* **37,** 5129–5135.

Jossinet, F., and Westhof, E. (2005). Sequence to structure (S2S): Display, manipulate and interconnect RNA data from sequence to structure. *Bioinformatics* **21,** 3320–3321.

Laederach, A., Shcherbakova, I., Jonikas, M. A., Altman, R. B., and Brenowitz, M. (2007). Distinct contribution of electrostatics, initial conformational ensemble, and macromolecular stability in RNA folding. *Proc. Natl. Acad. Sci. USA* **104,** 7045–7050.

Laederach, A., Das, R., Vicens, Q., Pearlman, S. M., Brenowitz, M., Herschlag, D., and Altman, R. B. (2008). Semiautomated and rapid quantification of nucleic acid footprinting and structure mapping experiments. *Nat. Protoc.* **3,** 1395–1401.

Latham, J. A., and Cech, T. R. (1989). Defining the inside and outside of a catalytic RNA molecule. *Science* **245,** 276–282.

Lempereur, L., Nicoloso, M., Riehl, N., Ehresmann, C., Ehresmann, B., and Bachellerie, J. P. (1985). Conformation of yeast 18S rRNA. Direct chemical probing of the 5′ domain in ribosomal subunits and in deproteinized RNA by reverse transcriptase mapping of dimethyl sulfate-accessible. *Nucleic Acids Res.* **13,** 8339–8357.

Martin, J. S., Simmons, K., and Laederach, A. (2009). Exhaustive enumeration of kinetic model topologies for the analysis of time-resoved RNA folding. *Algorithms* **2,** 200–214.

Ralston, C. Y., Sclavi, B., Sullivan, M., Deras, M. L., Woodson, S. A., Chance, M. R., and Brenowitz, M. (2000). Time-resolved synchrotron X-ray footprinting and its application to RNA folding. *Methods Enzymol.* **317,** 353–368.

Shcherbakova, I., and Brenowitz, M. (2005). Perturbation of the hierarchical folding of a large RNA by the destabilization of its Scaffold's tertiary structure. *J. Mol. Biol.* **354,** 483–496.

Shcherbakova, I., and Brenowitz, M. (2008). Monitoring structural changes in nucleic acids with single residue spatial and millisecond time resolution by quantitative hydroxyl radical footprinting. *Nat. Protoc.* **3,** 288–302.

Shcherbakova, I., Mitra, S., Beer, R. H., and Brenowitz, M. (2006). Fast Fenton footprinting: A laboratory-based method for the time-resolved analysis of DNA, RNA and proteins. *Nucleic Acids Res.* **34,** e48.

Takamoto, K., Chance, M. R., and Brenowitz, M. (2004). Semi-automated, single-band peak-fitting analysis of hydroxyl radical nucleic acid footprint autoradiograms for the quantitative analysis of transitions. *Nucleic Acids Res.* **32,** E119.

Tijerina, P., Mohr, S., and Russell, R. (2007). DMS footprinting of structured RNAs and RNA–protein complexes. *Nat. Protoc.* **2,** 2608–2623.

Vicens, Q., Gooding, A. R., Laederach, A., and Cech, T. R. (2007). Local RNA structural changes induced by crystallization are revealed by SHAPE. *RNA* **13,** 536–548.

Wilkinson, K. A., Merino, E. J., and Weeks, K. M. (2005). RNA SHAPE chemistry reveals nonhierarchical interactions dominate equilibrium structural transitions in tRNA(Asp) transcripts. *J. Am. Chem. Soc.* **127,** 4659–4667.

Wilkinson, K. A., Merino, E. J., and Weeks, K. M. (2006). Selective 2′-hydroxyl acylation analyzed by primer extension (SHAPE): Quantitative RNA structure analysis at single nucleotide resolution. *Nat. Protoc.* **1,** 1610–1616.

Woodson, S. A. (2008). RNA folding and ribosome assembly. *Curr. Opin. Chem. Biol.* **12,** 667–673.

CHAPTER FOUR

High-Throughput SHAPE and Hydroxyl Radical Analysis of RNA Structure and Ribonucleoprotein Assembly

Jennifer L. McGinnis, Caia D. S. Duncan, *and* Kevin M. Weeks

Contents

Abstract

RNA folds to form complex structures vital to many cellular functions. Proteins facilitate RNA folding at both the secondary and tertiary structure levels. An absolute prerequisite for understanding RNA folding and ribonucleoprotein (RNP) assembly reactions is a complete understanding of the RNA structure at

Department of Chemistry, University of North Carolina, Chapel Hill, North Carolina, USA

Methods in Enzymology, Volume 468
ISSN 0076-6879, DOI: 10.1016/S0076-6879(09)68004-6

each stage of the folding or assembly process. Here we provide a guide for comprehensive and high-throughput analysis of RNA secondary and tertiary structure using SHAPE and hydroxyl radical footprinting. As an example of the strong and sometimes surprising conclusions that can emerge from high-throughput analysis of RNA folding and RNP assembly, we summarize the structure of the bI3 group I intron RNA in four distinct states. Dramatic structural rearrangements occur in both secondary and tertiary structure as the RNA folds from the free state to the active, six-component, RNP complex. As high-throughput and high-resolution approaches are applied broadly to large protein–RNA complexes, other proteins previously viewed as making simple contributions to RNA folding are also likely to be found to exert multifaceted, long-range, cooperative, and nonadditive effects on RNA folding. These protein-induced contributions add another level of control, and potential regulatory function, in RNP complexes.

1. Introduction

RNA is actively involved in diverse cellular processes, including protein synthesis, gene regulation, and maintenance of genome stability (Gesteland *et al.*, 2006). In many critical examples, RNA function is directly related to its ability to fold back on itself to form complex higher order structures, or in some cases, the ability to achieve multiple structures. For example, during mRNA processing, catalytic RNAs such as group I and II introns, RNase P, small ribozymes, and some riboswitches fold into highly ordered three-dimensional structures to cleave RNA (Cochrane and Strobel, 2008; Scott, 2007; Torres-Larios *et al.*, 2006). Other RNAs, especially riboswitches, sample multiple stable conformations that then allow these RNAs to function as metabolite sensors that regulate gene expression (Edwards *et al.*, 2007; Winkler and Breaker, 2005).

Many RNAs, especially larger RNAs like catalytic introns and the ribosome, also recruit proteins to facilitate fast and stable folding in the cell. The proteins that facilitate RNA folding are generally classified into two categories, chaperones and cofactors (Herschlag, 1995; Rajkowitsch *et al.*, 2007; Weeks, 1997). Chaperone proteins interact transiently with RNA and function to allow an RNA to refold and achieve a thermodynamically favored structure. Cofactor proteins are defined by their ability to bind stably to an RNA to create a structurally well-defined ribonucleoprotein (RNP) complex. Cofactors often stabilize a specific, active, tertiary structure. The mechanisms by which proteins facilitate RNA folding can be complex and emerging work has begun to emphasize that some proteins do not fit clearly into the traditional definitions of either chaperones or cofactors.

An absolute prerequisite for understanding RNA folding and RNP assembly reactions is a complete understanding of the RNA structure at each stage of the predominant folding or assembly pathway. Information regarding the final, stable, structure of an RNA or RNA–protein complex can be obtained by high-resolution NMR or by crystallography or, in a few cases, by molecular modeling. However, for most initial and intermediate states, RNA secondary and tertiary structure must be inferred from either comparative sequence analysis or chemical probing approaches.

The most successful method for determining an RNA secondary structure has been by phylogenetic comparative sequence analysis (Gutell *et al.*, 2002; Michel and Westhof, 1990). However, covariation analysis only demonstrates that a pairing is likely to exist and not that a physical pairing occurs (Woese *et al.*, 1980). Covariation analysis also does not reveal whether a specific physical pairing actually exists in any given RNA state or folding intermediate.

Alternatively, RNA structure information can be inferred experimentally by treating an RNA with small molecule or enzyme reagents that are sensitive to local RNA structure. Two broad classes of experiments have proven especially useful. First, many chemical and enzymatic reagents react with partial selectivity toward single-stranded nucleotides. Reactivity patterns obtained using these reagents provide information useful for inferring base-pairing interactions (Brunel and Romby, 2000; Ehresmann *et al.*, 1987). Second, Fenton chemistry can be used to generate short-lived hydroxyl radicals in solution, which then react with the RNA backbone in a way that is sensitive to solvent accessibility (Tullius and Greenbaum, 2005; Tullius *et al.*, 1987). Identification of RNA backbone positions that are protected from the hydroxyl radical reagent can provide strong evidence for higher order tertiary interactions (Latham and Cech, 1989).

Conventional approaches for analyzing RNA base pairing or tertiary interactions often lead to a view of an RNA structure that is incomplete in key features. Conventional secondary structure-selective chemical and enzymatic RNA mapping reagents tend to have a narrow dynamic range and typically yield information for one-half, or less, of the nucleotides in an RNA. In addition, both secondary structure-sensitive and hydroxyl radical-mediated solvent accessibility experiments have been analyzed predominantly by sequencing gel electrophoresis. The effort required to analyze long, intact, RNAs by this approach is very large. A useful approach can be to focus on short RNAs, on simplified models of larger RNAs, or on a small region within a large RNA. The challenge in focusing on a segment of a large RNA is that long-range and unanticipated interactions, critical for understanding how an RNA functions or how proteins affect an RNA folding process, will be missed. These challenges can be addressed by new chemistries and high-throughput analysis approaches.

Two methods that, in principle, yield comprehensive single nucleotide resolution information about RNA secondary and tertiary structure are SHAPE (selective 2′-hydroxyl acylation analyzed by primer extension) (Merino *et al.*, 2005; Wilkinson *et al.*, 2005, 2006) and the aforementioned hydroxyl radical footprinting (Brenowitz *et al.*, 2002; Latham and Cech, 1989; Shcherbakova and Brenowitz, 2008; Shcherbakova *et al.*, 2008; Tullius and Greenbaum, 2005; Tullius *et al.*, 1987). SHAPE chemistry interrogates local nucleotide flexibility, while hydroxyl radical footprinting assesses solvent accessibility and, thus, the global RNA fold. Both techniques are adaptable to a variety of reaction conditions, including the presence of proteins. Additionally, both techniques are insensitive to nucleotide identity and can potentially provide structural information at nearly every nucleotide position in an RNA.

When used in tandem, SHAPE and hydroxyl radical footprinting comprehensively probe multiple levels of RNA structure and provide an impressively detailed view of an RNA or RNP folding state or assembly intermediate. Combining these sensitive structural probes with an experimental readout using capillary electrophoresis makes it possible to characterize 350–600 nucleotides (nt) in a single high-throughput experiment. Thus, large RNAs and complex RNP structures can be readily studied in detail. These developments in high-throughput in-solution RNA structure analysis are revealing unanticipated, and complex, mechanisms for how RNAs fold and interact with proteins to achieve a final active structure. In this review, we will focus on the bI3 group I intron RNA and its RNP complexes as an example of how these technological advances are leading to new insights into RNA folding and RNP assembly reactions.

2. Theory

SHAPE chemistry is a robust and comprehensive method for analyzing local nucleotide dynamics in RNA (Merino *et al.*, 2005; Wilkinson *et al.*, 2005, 2006). In a SHAPE reaction, the RNA is treated with an electrophilic reagent like NMIA or 1M7 (Merino *et al.*, 2005; Mortimer and Weeks, 2007) that selectively reacts with the ribose 2′-hydroxyl group at conformationally flexible nucleotides (Fig. 4.1A). SHAPE reactivity reports local nucleotide flexibility because unconstrained nucleotides (like those in single-stranded structures) are more likely to sample relatively rare conformations that make the 2′-hydroxyl more nucleophilic and promote the reaction that forms a 2′-*O*-adduct. All four RNA nucleotides show nearly identical intrinsic reactivities when not constrained by base pairing or tertiary interactions (Wilkinson *et al.*, 2009). In contrast, nucleotides constrained by base pairing or other interactions react poorly with SHAPE reagents. SHAPE chemistry is not strongly sensitive to solvent accessibility and, for example, conformationally dynamic but solvent inaccessible nucleotides are generally still reactive (Gherghe *et al.*, 2008; Merino *et al.*, 2005). SHAPE reagents both modify flexible sites in RNA and also undergo self-inactivating hydrolysis.

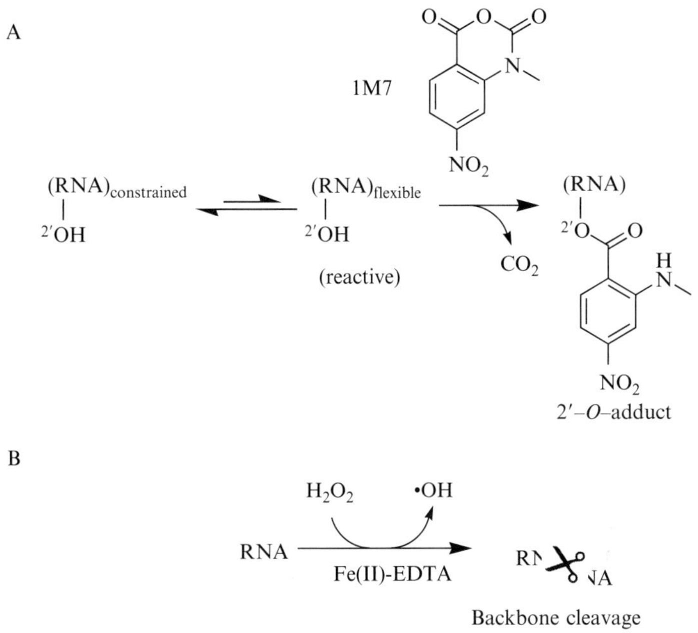

Figure 4.1 Schemes for (A) RNA SHAPE chemistry and (B) RNA hydroxyl radical footprinting chemistries.

As a practical matter, this autoinactivation means that reagent reactivity does not need to be quenched, as long as the 2′-*O*-adduct forming reaction is allowed to proceed to completion (Merino *et al.*, 2005). These bulky RNA adducts can be detected as stops to primer extension.

The current gold standard for mapping solvent accessibility at the RNA backbone is hydroxyl radical footprinting. Short-lived hydroxyl radicals (•OH) are generated *in situ* in a reaction between Fe^{2+} and H_2O_2 (Brenowitz *et al.*, 2002; Latham and Cech, 1989; Shcherbakova and Brenowitz, 2008; Shcherbakova *et al.*, 2008; Tullius and Greenbaum, 2005; Tullius *et al.*, 1987). The iron ion is chelated by EDTA, which prevents the ion from binding directly to RNA. Hydroxyl radicals are then generated at the periphery of the RNA, leading to RNA cleavage (Fig. 4.1B). Cleavage occurs preferentially at RNA positions accessible to solvent, although other factors also contribute to reactivity (Balasubramanian *et al.*, 1998; Lu *et al.*, 1990). Backbone cleavage sites can also be detected by primer extension.

Both SHAPE and the hydroxyl radical experiment can be converted to high-throughput formats by detecting sites of 2′-*O*-adduct formation or of backbone cleavage, respectively, by reverse transcriptase-mediated primer extension, resolved by capillary electrophoresis (Fig. 4.2A–D). In addition to the RNA modification reaction, a control no-reagent experiment is used to assess background. These reactions are compared to dideoxy sequencing

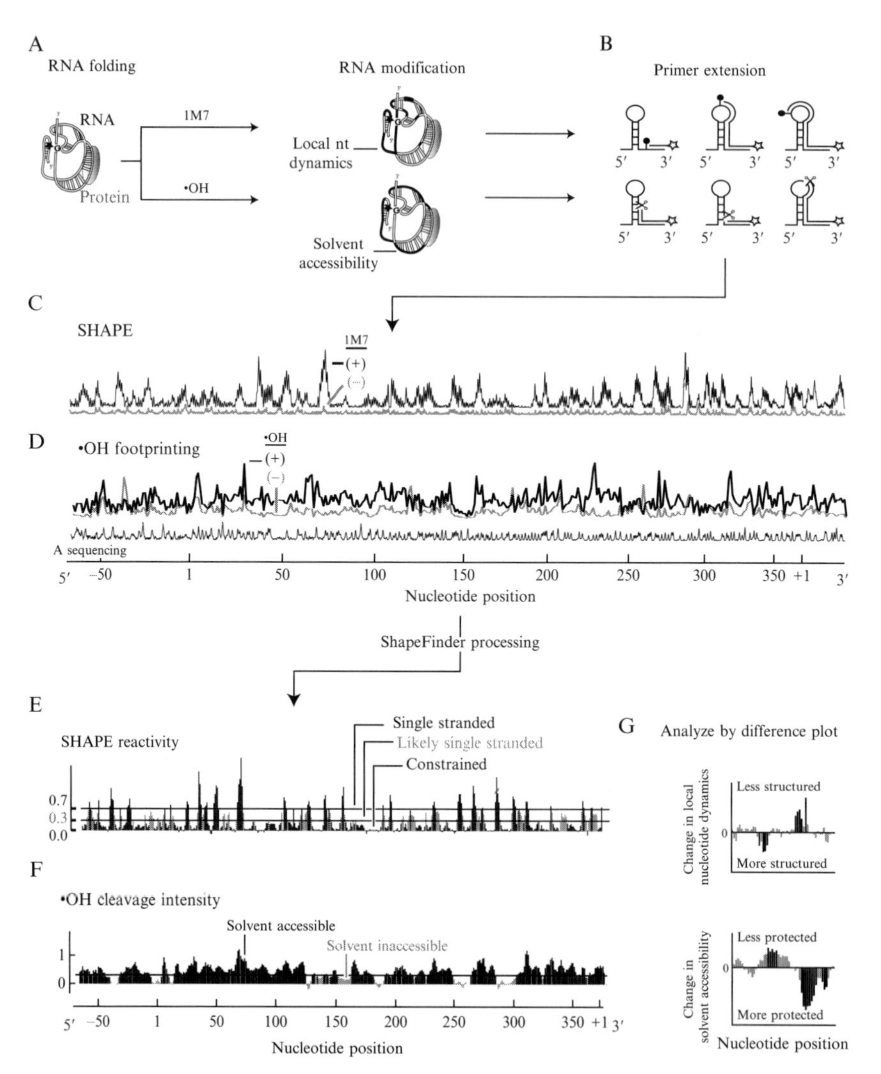

Figure 4.2 High-throughput RNA structure mapping. (A) The RNA is folded and modified with a SHAPE reagent (1M7 or NMIA) or cleaved by hydroxyl radical chemistry. (B) RNA modifications or cleavages yield stops to primer extension, which are detected using color-coded, fluorescently labeled primers. (C, D) Resulting cDNAs are resolved by automated capillary electrophoresis. (E, F) After quantifying the net reactivity at each nucleotide position, the data are normalized on a scale spanning 0–∼2, where 1.0 is defined as the average intensity at highly reactive positions. (G) Changes in local nucleotide dynamics and solvent accessibility that distinguish any two RNA folding or RNP assembly states are readily detected at single nucleotide resolution by difference plot analysis.

markers to assign nucleotide positions. After data processing, each nucleotide is characterized by a quantitative SHAPE reactivity or hydroxyl radical cleavage intensity that directly reflects the underlying RNA structure at single nucleotide resolution (Fig. 4.2E and F). In the SHAPE experiment, single-stranded nucleotides are reactive while base paired nucleotides are generally unreactive (Fig. 4.2E). Among other applications, SHAPE data can be converted to pseudo-free energy change terms to constrain the output of a thermodynamic structure prediction program. This blended approach generally produces highly accurate RNA secondary structure models (Deigan *et al.*, 2009; Wilkinson *et al.*, 2008). For the hydroxyl radical experiment, inspection of the final reactivity versus position histogram makes it possible to identify compact tertiary interactions as solvent inaccessible elements in the context of a large RNA (Fig. 4.2F).

3. Practice

Both SHAPE and hydroxyl radical footprinting advantageously yield a comprehensive and single nucleotide resolution view of an RNA state in a concise set of experiments. In outline, both SHAPE and hydroxyl radical footprinting involve folding the RNA, adding protein (if appropriate), incubating the RNA with either the SHAPE electrophile or the hydroxyl radical reagents, recovering the RNA, and mapping the sites of modified or cleaved nucleotides by primer extension (Fig. 4.2). The following experimental overview is based on the analysis of the structure of the free bI3 group I intron RNA and its complexes with protein facilitators.

3.1. RNA folding

SHAPE and hydroxyl radical footprinting experiments are most commonly performed with RNAs that have been generated by *in vitro* transcription (Duncan and Weeks, 2008; Latham and Cech, 1989; Merino *et al.*, 2005; Tullius and Greenbaum, 2005; Wilkinson *et al.*, 2006). These RNAs require purification by denaturing gel electrophoresis and then must be renatured to achieve a biologically relevant conformation. RNA folding reactions involving the renatured RNA can be initiated in several ways, including a change in ionic strength or addition of a protein or small molecule ligand. This protocol describes a general method in which an RNA is first heated and snap cooled in a low ionic strength buffer to eliminate multimeric forms. RNA folding is initiated by adding mono- and divalent ions to this solution. The RNA is folded in a single step and then separated into (+) and (−) reagent reactions for either SHAPE or hydroxyl radical footprinting.

1. Add 4 pmol RNA in 10 μl sterile water to a 0.65 ml (Eppendorf) reaction tube.

Note: This amount of RNA is based on using 2 pmol per (+) and (−) reagent reaction. The amount of RNA needed for complete analysis may vary with the RNA. The useful range is 0.5–8 pmol per reaction.

2. Heat the RNA to 95 °C for 1 min. Place the RNA on ice for 1 min.
3. Add 24 μl 5× folding buffer [5× = 200 m*M* MOPS (pH 8.0), 400 m*M* potassium acetate (KOAc) (pH 8.0), 100 m*M* $MgCl_2$]. Adjust final volume to 108 μl with water.
4. Incubate at 37 °C for 10 min. This is the "free RNA reaction mix."

Note: For protein-binding experiments, add proteins and incubate at appropriate concentrations, temperature, and time for the system. Adjustment of the volume of added water will be necessary to ensure a final 30 μl reaction volume in the chemical probing step.

3.2. Chemical probing of RNA secondary and tertiary structure

Modification and cleavage reactions should take place under conditions that yield single hits over the length of RNA to be analyzed by primer extension (~1 modification per 300 nt). For SHAPE, too high or low a concentration of reagent results in a steep signal decay in the fluorescence trace or in a low signal-to-noise ratio in the (+) reagent lane, respectively. For hydroxyl radical cleavage, excessive cleavage yields a complex cleavage pattern that is biased toward short RNA fragments. For both reactions, fresh reagents are imperative.

3.2.1. SHAPE

1. Aliquot 3 μl 50 m*M* 1M7 (or NMIA) in DMSO and 3 μl neat DMSO into two 0.65 ml reaction tubes. These will be the (+) and (−) reagent reactions. We strongly recommend 1M7 for routine SHAPE analysis of RNA. 1M7 can be synthesized as described (Mortimer and Weeks, 2007); alternatively, the NMIA reagent can be used.

Note: The optimal concentration of 1M7 can vary with RNA length. The useful range is 0.5–8 m*M*; 5 m*M* final is a good starting concentration. For longer RNAs, or very AU rich RNAs, use the lower end of these 1M7 concentrations.

2. Add 27 μl of the free RNA reaction mix to 3 μl 1M7 in DMSO and an additional 27 μl to the tube containing 3 μl neat DMSO (final volume = 30 μl).

3. Allow reaction mixtures to react for five hydrolysis half lives. This time is 70 s for 1M7 and 35 min for NMIA (Merino *et al.*, 2005; Mortimer and Weeks, 2007). No additional quench step is necessary.

Note: For protein-binding experiments, remove proteins by proteolysis and phenol–chloroform extraction (Duncan and Weeks, 2008) before proceeding to step 4.

4. Adjust volume to 96 μl with water and recover RNA by ethanol precipitation: add 1 μl 3 *M* NaCl, 1 μl 20 mg/ml glycogen, 2 μl 0.5 *M* EDTA (pH 8.0), and 400 μl ethanol to both (+) and (−) reactions, mix, and incubate at −80 °C for 30 min. Sediment the RNA by spinning at maximum speed in a microfuge at 4 °C for 30 min.
5. Redissolve RNA in 10 μl sterile water.

3.2.2. Hydroxyl radical footprinting

Note: Fresh solutions are imperative for efficient cleavage reactions. Weigh out ammonium iron(II) sulfate hexahydrate [$(NH_4)_2Fe(SO_4)_2 \cdot 6H_2O$] and ascorbic acid (sodium salt) into dry reaction tubes; wait to make the solutions until just before addition to the RNA.

1. Prepare a solution of 7.5 m*M* iron(II)/11.25 m*M* EDTA (pH 8.0).
2. Prepare a 0.3% H_2O_2 solution in water.
3. Prepare a 150 m*M* sodium ascorbate solution in water.
4. Transfer two 27 μl aliquots of the free RNA reaction mix to two sterile 0.65 ml reaction tubes. One tube will be used for the (+) •OH reaction and the other for the (−) •OH reaction.
5. For the (+) •OH reaction, sequentially add 1 μl each of the iron(II)–EDTA, hydrogen peroxide, and ascorbate solutions to the reaction solution.
6. Repeat step 5 for the (−) •OH reaction, substituting 1 μl water for the iron(II)–EDTA solution.
7. Quench the reactions after 5 min at 37 °C with 10.5 μl 75% glycerol (v/v).

Note: For protein-binding experiments, remove proteins by proteolysis and phenol–chloroform extraction (Duncan and Weeks, 2008) before proceeding to step 8.

8. Recover RNA as in steps 4–5 for the SHAPE reaction procedure.

3.3. Primer extension to map modification and cleavage sites

Long RNA regions can be analyzed in a single experiment by performing each primer extension using a primer labeled with a color-coded fluorophore. The resulting cDNA products (from the (+) and (−) reagent reactions plus one or two dideoxy sequencing reactions) are combined and resolved in one multifluor run by automated capillary electrophoresis.

In a single read, quantitative RNA structural information at single nucleotide resolution can be routinely obtained for 350–600 nt. Analysis is simplified by choosing fluorescent dyes with similar electrophoretic mobilities. The dyes 6-FAM, TET, HEX, and NED require very little correction for fluorophore-induced mobility shifts. This correction is readily performed using the ShapeFinder software (Vasa *et al.*, 2008). Primer extension protocols are identical for RNA analyzed by SHAPE or by hydroxyl radical footprinting.

1. Add 3 μl of 0.3 μM fluorescently labeled primer to the RNA solutions from (+) and (−) reagent reactions.
2. For sequencing reactions, add 3 μl 0.3 μM fluorescently labeled primer to 1–2 pmol of RNA in 8 μl sterile water.
3. For all reactions, anneal the primer to the RNA by heating at 65 °C for 5 min, reducing the temperature to 37 °C for 1 min, and placing on ice.
4. Add 6 μl of Superscript III enzyme mix (the solution is a 4:1:1 mixture of 5× first strand buffer (from Invitrogen), 100 m*M* DTT, and a solution that is 10 m*M* in each dNTP) to the (+) and (−) reagent reactions and to the sequencing reactions. Also add 1 μl of 5 m*M* of a selected ddNTP to each sequencing reaction.

Note: For AU rich RNAs, better extension can be achieved by decreasing the deoxynucleotide concentration used for sequencing. For example, if using ddATP to sequence at T, reduce the dATP concentration in the reverse transcriptase enzyme mix to 5 m*M*.

5. Add 1 μl SuperScript III Reverse Transcriptase (Invitrogen) to each tube. Mix well and incubate at 37 °C for 5 min, 52 °C for 20 min, 60 °C for 5 min, and then place on ice.
6. Quench the primer extension reactions by adding 4 μl of a 1:1 mixture of 4 *M* NaOAc and 100 m*M* EDTA (pH 8.0) and place on ice.
7. Combine 22 μl from each (+) and (−) reaction and one or two sequencing reactions into a 1.5 ml reaction tube and recover cDNA products by ethanol precipitation (240 μl 100% ethanol). Incubate at −80 °C for 15 min. Sediment the cDNA by spinning at maximum speed in a microfuge at 4 °C for 15 min.
8. Wash with 800 μl 70% ethanol to remove excess salt (which leads to poor resolution during capillary electrophoresis).
9. Repeat step 8 and dry pellet by vacuum for 10 min.
10. Resuspend in 10 μl deionized formamide.

3.4. cDNA analysis by capillary electrophoresis

1. Load each 10 μl sample into separate input wells on a capillary electrophoresis DNA sequencing instrument and run.
2. Export raw traces into ShapeFinder (Vasa *et al.*, 2008).

3.5. Data analysis

Data analysis using ShapeFinder has been outlined in detail elsewhere (Vasa *et al.*, 2008; Wilkinson *et al.*, 2008). Most tools are straightforward to use. The Align and Integrate tool in ShapeFinder performs a whole trace Gaussian integration to quantify the intensity of every peak in the (+) and (−) reagent lanes. This tool works best for data in which traces have been adjusted to remove fluorescence background and in which peaks corresponding to the same positions have been aligned to overlap consistently.

1. Analyze SHAPE and hydroxyl radical data using ShapeFinder, as described (Duncan and Weeks, 2008; Wilkinson *et al.*, 2008).
2. ShapeFinder is available at: http://bioinfo.unc.edu/Downloads/index.html.

In brief, adjust the fluorescent baseline (window of 40 pixels), perform a mobility shift to account for the effect of different dyes on cDNA mobility, correct for signal decay, and scale the (+) and (−) reagent traces to make them equal to each other in regions in which reagent-induced reactivities are low. Integrate all peaks in the (+) and (−) reagent traces.

3. Subtract the (−) from the (+) reagent reactivities to create a table of net reactivity as a function of nucleotide position.

3.6. Normalization

A consistent approach to normalization is important for comparing data sets and for making accurate secondary structure predictions. We normalize SHAPE and hydroxyl radical data to a scale starting at zero (no reactivity) and in which 1.0 is defined as the average intensity of highly reactive positions.

1. Identify the (usually small number of) highly reactive outliers.
2. For large data sets (>300 measurements), use a model-free box plot analysis (Deigan *et al.*, 2009). Reactivities greater than 1.5 times the interquartile range are outliers (Chernick and Friis, 2003). If the box plot approach is used for small RNA data sets (<~100 nt), the maximum number of outliers should be capped at 5%. After eliminating outliers, compute the mean of the next (highest) 10% of intensities and divide by this value.
3. For smaller data sets or for RNAs with a relatively small number of reactive peaks, we use a "2–8%" rule (Duncan and Weeks, 2008; Wilkinson *et al.*, 2008). The most reactive 2% of peaks are taken to be outliers. Then, calculate the mean reactivity for the next 8% most reactive nucleotides and divide all intensities by this value.
4. Using either a box plot or the 2–8% rule to identify outliers, the resulting reactivities typically span a scale from 0 to ~2.

5. For hydroxyl radical cleavage data, it is usually helpful to algebraically smooth the data over a three nucleotide window to account for the structurally heterogeneous reactivity of this reagent.

3.7. SHAPE-directed RNA structure prediction

SHAPE reactivities can be converted to pseudo-free energy change constraints to enable highly accurate prediction of RNA secondary structure (Deigan *et al.*, 2009). SHAPE energies combined with conventional thermodynamic parameters typically yield secondary structure models in which 95% of all accepted base pairs are predicted correctly, for RNAs that do not contain pseudoknots (Deigan *et al.*, 2009).

1. Use RNA structure, available at: http://rna.urmc.rochester.edu/rnastructure.html.
2. Set the maximum base-pairing distance to 600 nt.
3. Input SHAPE data as pseudo-free energy change terms using 2.6 and −0.8 for the slope and intercept parameters, respectively.
4. For a well-supported structure with complete SHAPE data, a small number of similar structures will be obtained. We focus on the first (lowest energy) structure.

4. Examples and Interpretation

4.1. High-throughput structure analysis of protein-assisted RNA folding of the bI3 RNA

High-throughput structure probing yields a wealth of structural information about an RNA in an efficiently performed set of experiments. The methods outlined here typically yield in-solution structural information for ≥95% of the nucleotides in intact, full length RNAs (Deigan *et al.*, 2009; Duncan and Weeks, 2008; Wilkinson *et al.*, 2008). Because all regions in an RNA can be probed comprehensively, many fewer structural assumptions have to be made about unprobed regions of an RNA. In addition, we generally find that resolving chemical modification or backbone cleavage information by capillary electrophoresis and quantification using the tools in ShapeFinder yield reactivity information that is of much higher quality than we are able to obtain using sequencing gel approaches. Because the high-throughput data are so quantitative, the RNA structural changes that differentiate two states can be identified rapidly by simply subtracting one reactivity profile from another (Fig. 4.2G). As an example of the impact comprehensive structural information has for developing models for RNA structure and RNP assembly, we probed the bI3 group I intron complex at four distinct folding stages using both SHAPE and hydroxyl radical footprinting.

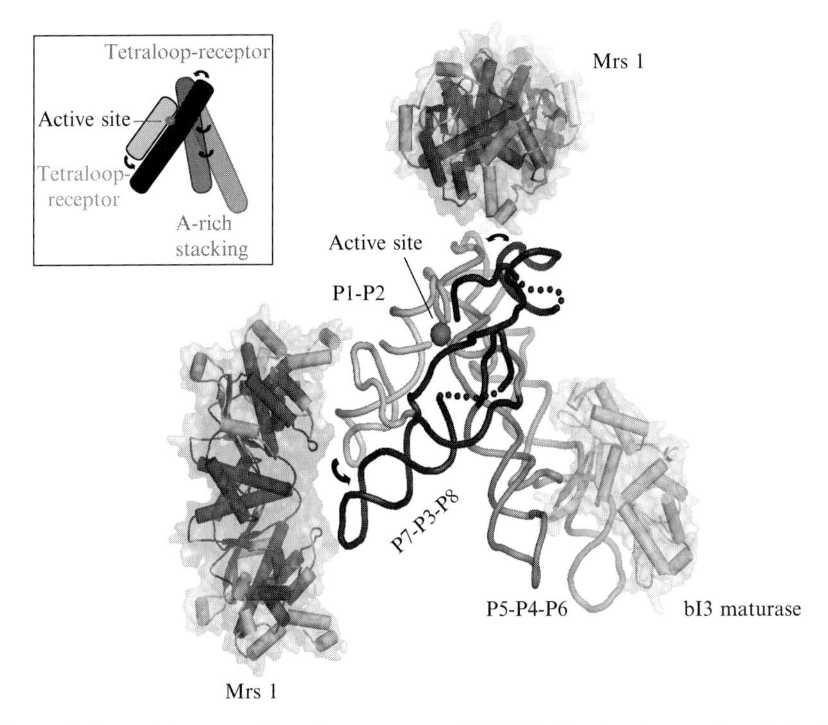

Figure 4.3 Architecture of the six-component bI3 RNP. Tertiary interactions, including tetraloop–receptor motifs (arrows) and A-minor interactions, mediate long-range interactions and stabilize a compact and tight RNA structure. The splice site is highlighted by gray sphere; the three group I intron domains are labeled. For clarity, the P7.1 and P9.1 helices are replaced by a dashed backbone. This model summarizes results from high-throughput SHAPE and hydroxyl radical footprinting, interpreted in the context of a molecular model for the bI3 group I intron.

The *Saccharomyces cerevisiae* bI3 group I intron is a good model for complex protein-mediated RNA folding processes. It is a large (~370 nt) catalytic RNA composed of two coaxially stacked domains that interact through extensive tertiary interactions to form a highly structured catalytic core that then docks with a third helical domain to form a sophisticated active site (Cech, 1990; Michel and Westhof, 1990; Vicens and Cech, 2006; Woodson, 2005) (inset, Fig. 4.3). The intron RNA requires binding by two proteins, the bI3 maturase and Mrs1, to splice *in vivo* (Kreike *et al.*, 1987; Lazowska *et al.*, 1989) or *in vitro* (Bassi and Weeks, 2003; Bassi *et al.*, 2002; Longo *et al.*, 2005). The active complex consists of six subunits: the intron RNA, a single monomer of the bI3 maturase protein, and two dimers of the Mrs1 protein (Fig. 4.3). The complete bI3 RNP is ~420 kDa and splices with a k_{cat} of 0.3 min (Bassi *et al.*, 2002). Each protein binds the free RNA independently and each shows modest cooperativity relative to binding of the second protein (Bassi and Weeks, 2003; Bassi *et al.*, 2002).

The maturase protein binds in a peripheral helix (termed the P5c helix) in the P5–P4–P6 domain. The maturase binds across consecutive RNA minor grooves with a K_d of 1.0 nM (Bassi and Weeks, 2003; Longo *et al.*, 2005). Maturase binding stabilizes tertiary structure folding in the entire P5–P4–P6 domain which then stabilizes critical interactions in the catalytic active site, which lie over 50 Å away (Fig. 4.3). Mrs1 is a dimer and two dimers bind cooperatively to the bI3 RNA with a $K_{1/2}$ of 11 nM and a Hill coefficient of ~2 (Bassi and Weeks, 2003). Until the development of high-throughput RNA footprinting, the RNA binding sites for Mrs1 remained unknown.

The use of high-throughput chemical probing methods has, in conjunction with three-dimensional structure modeling, enabled the development of a model for the free bI3 RNA, identified the RNA binding site for Mrs1, made it possible to visualize the global three-dimensional architecture of this RNP complex, and revealed new "nonhierarchical" contributions of protein-facilitated RNA folding (Duncan and Weeks, 2008, 2009 and see Fig. 4.3).

4.2. Identification of a misfolded free RNA state using SHAPE

When studying RNA–protein interactions, often the initial secondary structure of an RNA is taken to be similar to that seen in the final complex. For example, it would be easy to assume that, prior to protein binding, the free bI3 RNA folds into the same secondary structure as established by comparative sequence analysis. Both the bI3 maturase and Mrs1 proteins have characteristics of protein cofactors and, therefore, might have only a limited effect on the bI3 RNA secondary structure.

The structure of the entire free bI3 RNA was analyzed in a single experiment using high-throughput SHAPE, as outlined above. Inspection of the SHAPE reactivity data indicated that about one-half of the RNA folds into a structure consistent with the evolutionarily conserved secondary structure common to all group I introns (Duncan and Weeks, 2008; Michel and Westhof, 1990). For example, in the P5–P4–P6 domain, all base-paired regions are unreactive toward SHAPE whereas the single-stranded nucleotides that link these regions are reactive (Fig. 4.4A). In contrast, SHAPE data also indicate that about one-half of the free bI3 RNA does not fold in a manner consistent with this consensus group I intron secondary structure (Duncan and Weeks, 2008). For example, nucleotides in the P1 helix are predicted to be base paired in the active RNA state and should therefore be unreactive toward SHAPE; in contrast, these nucleotides are reactive (Fig. 4.4A). The conserved P3 and P7 helices also show significant reactivity. In a single experiment, high-throughput SHAPE readily identified a misfolded free RNA state.

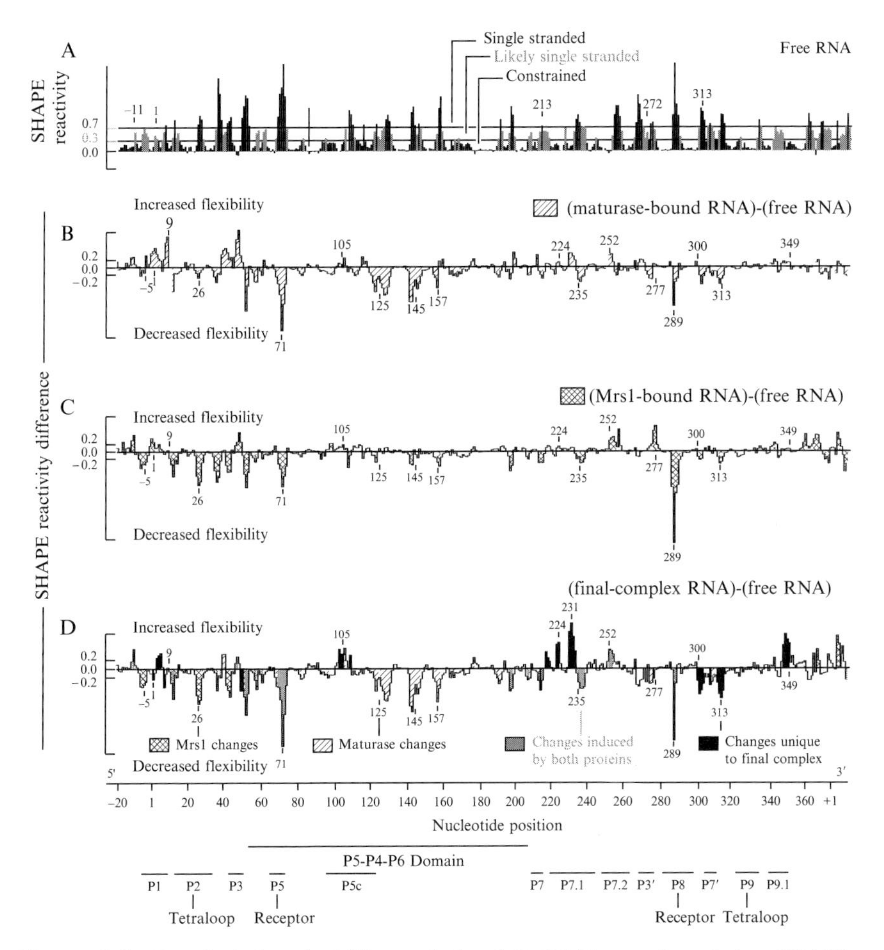

Figure 4.4 Protein-induced folding of the bI3 RNA monitored by SHAPE. (A) Reactivity as a function of nucleotide position for the free RNA. Horizontal lines indicate reactivity cut-offs characteristic of base paired and single-stranded nucleotides. (B–D) Difference histograms for the maturase-bound, Mrs1-bound, and final complex RNAs. Nucleotides with significant changes in reactivity are shaded to indicate the protein that induces a given effect. Structural landmarks are highlighted below the axis.

A new secondary structure model for the RNA was developed using SHAPE-directed secondary structure prediction (Deigan *et al.*, 2009; Duncan and Weeks, 2008). This model supported the existence of canonical pairings in regions where SHAPE reactivities were consistent with the phylogenetic model, including in the P5–P4–P6 domain and in the P2, P8, P9, and P7.2 helices (Fig. 4.6A). However, in this new model, roughly half of the RNA is folded in a conformation different from that of the phylogenetic structure, including in the P1 helix, in the P7–P3 pseudoknot, and in the peripheral P7.1 and P9.1 helices (termed the P7.1/P9.1 alternative helix). Additionally, an unanticipated helix between 5′ and 3′ exons was predicted (Fig. 4.6A). SHAPE data are almost exactly consistent with this alternate secondary structure model.

The working model for the free bI3 RNA could be tested and was ultimately confirmed through SHAPE analysis of point mutations. SHAPE analysis of point mutations is proving to be a powerful approach for evaluating unconventional secondary structure models. The basic idea is to make a precise sequence change in one region of a large RNA and then evaluate the structural consequences in the RNA element that is postulated to interact with the mutated region. The second site can be located hundreds of nucleotides away in the primary sequence. For example, mutations were designed in the bI3 exon helix, in the P1 splice site helix, and in the large P7.1/P9.1 alternative helix. Single nucleotide mutations that either disrupted or stabilized the proposed alternative secondary structure or that induced small helical defects were easily detected by the single nucleotide resolution SHAPE experiment and provided strong confirmation of widespread misfolding in the free bI3 RNA state (Duncan and Weeks, 2008).

Comparing the structure of the free RNA to the protein-bound, catalytically active RNA showed extensive differences in SHAPE reactivity, confirming that dramatic structural rearrangements occurred as the RNA formed the intact RNP complex. Interestingly, SHAPE analysis of the bI3 RNA, after proteins were removed, showed that the RNA relaxed back to a structural state similar to that of the free RNA (Duncan and Weeks, 2008). These experiments indicate that the catalytically active secondary structure is not the thermodynamically most stable structure for the bI3 RNA. High-throughput SHAPE thus made it possible to establish whether an individual large RNA folds to the structure characteristic of its phylogenetic family and to evaluate the extent of protein-induced secondary structure rearrangement.

4.3. Identification of a protein cofactor that binds the ubiquitous RNA tetraloop–receptor motif

Identification of the RNA interaction site is a crucial step in understanding the role a protein plays in RNP function. High-throughput hydroxyl radical footprinting identified two tetraloop–receptor interactions as the binding sites for the Mrs1 protein on the bI3 RNA (Duncan and Weeks, 2009). The tetraloop–receptor interaction is a common long-range tertiary interaction in RNA. The motif consists of a GNRA tetraloop in which the RA (purine–adenosine) nucleotides form A-minor type hydrogen bonds in the minor groove of a receptor helix (Cate *et al.*, 1996; Jaeger *et al.*, 1994; Nissen *et al.*, 2001).

In the free bI3 RNA, only 20% of nucleotides are protected from hydroxyl radical cleavage prior to binding by the Mrs1 protein

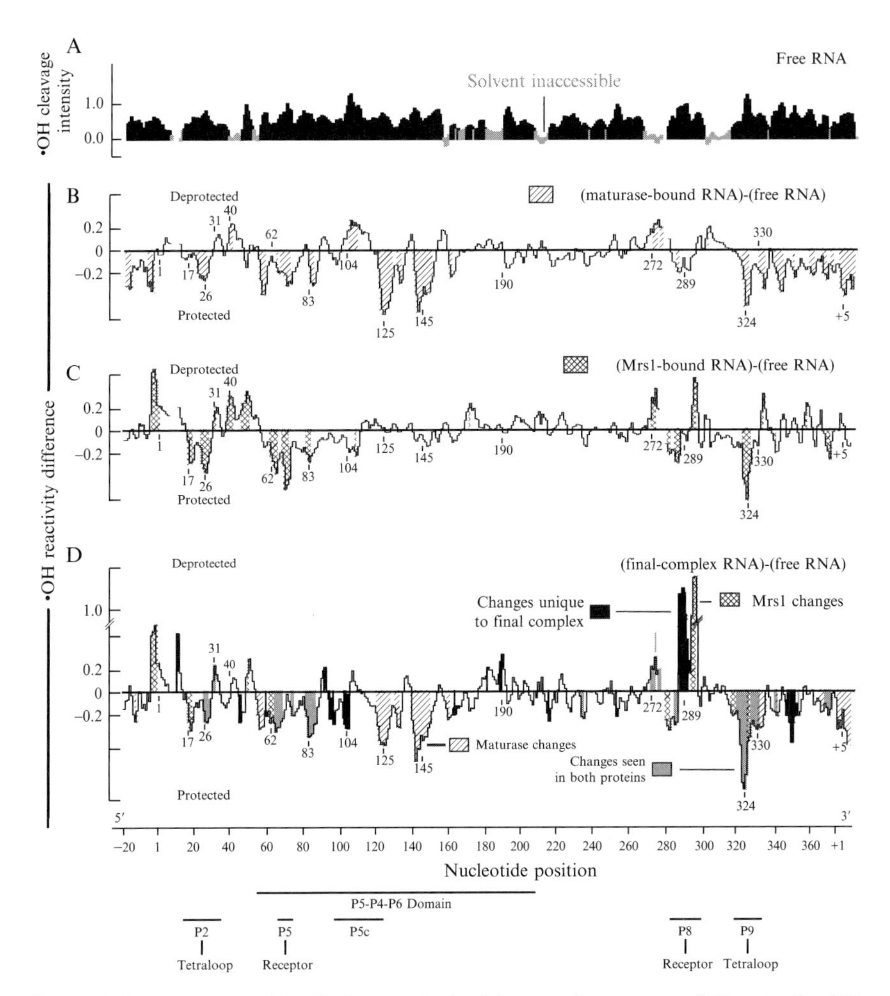

Figure 4.5 Protein-mediated changes in backbone solvent accessibility in the bI3 RNA. (A) Hydroxyl radical (•OH) cleavage intensity versus nucleotide position for the free RNA. Protected nucleotides (reactivities less than one-half the mean) are emphasized in gray. (B–D) Difference histograms for the maturase-bound, Mrs1-bound, and final complex RNAs. Nucleotide regions with significant protein-induced changes in reactivity are shaded as in Fig. 4.4. Structural landmarks are highlighted below the axis.

(Fig. 4.5A). Most RNA elements expected to form tertiary contacts are solvent accessible, including the entire P5–P4–P6 domain, the GNRA tetraloops at the ends of the P2 and P9 helices, and their respective receptors in the P8 and P5 helices.

Upon addition of Mrs1, extensive regions in the RNA become protected from cleavage. The effects are readily quantified by creating a difference plot in which the hydroxyl radical cleavage intensities for the free RNA are subtracted from those for the Mrs1-bound RNA (Fig. 4.5C). Most significant differences in cleavage intensity are localized at or near the two distinct GNRA tetraloop–receptor interactions in the bI3 RNA. These tetraloop–receptor interactions are the L2 loop with the P8 helix and the L9 loop with the P5 helix (see Fig. 4.5C; indicated with arrows on the secondary structure in Fig. 4.6B). These results provide strong evidence that Mrs1 binds to and stabilizes the two independent GNRA tetraloop–receptor interactions in the otherwise misfolded bI3 RNA.

4.4. Mechanism of cooperative folding of the bI3 RNP

Folding for some RNAs is strongly hierarchical, such that the secondary structure forms first, followed by tertiary structure (Brion and Westhof, 1997; Tinoco and Bustamante, 1999). However, there are substantial and accumulating examples in which RNA tertiary structure either folds in tandem with the secondary structure or induces changes to the secondary structure during folding (Buchmueller *et al.*, 2000; Chauhan and Woodson, 2008; Chauhan *et al.*, 2005; Gluick and Draper, 1994; LeCuyer and Crothers, 1993; Wang *et al.*, 2008; Wilkinson *et al.*, 2005; Wu and Tinoco, 1998; Zheng *et al.*, 2001). A protein cofactor that stabilizes an RNA tertiary structure element might therefore also have a dramatic impact on secondary structure. The use of high-throughput SHAPE and hydroxyl radical footprinting, as applied to the bI3 RNP, has identified a new and complex nonhierarchical mechanism for protein-facilitated RNA folding. The net contributions of the bI3 maturase and Mrs1 on protein-facilitated folding of the bI3 group I intron are distinct from the conventional definitions of RNA cofactor and chaperone functions.

The secondary and tertiary structures of the bI3 RNA were probed in four conformational states: the free RNA, the maturase- and Mrs1-bound complexes, and the final six-component complex (Figs. 4.4 and 4.5). High-throughput SHAPE and hydroxyl radical experiments indicate that binding by the individual maturase and Mrs1 proteins primarily stabilize long-range RNA tertiary structures but, nonetheless, do not repair the misfolded secondary structure (Figs. 4.4B, C and 4.5B, C). However, upon binding all protein components to form the final six-component complex, the RNA undergoes large-scale secondary structure rearrangements to form the active group I intron structure (Figs. 4.4D and 4.6). Strikingly, dramatic structural rearrangements occur in RNA regions located long distances from the protein-binding sites. These data imply secondary structure rearrangements in the bI3 RNA involve highly cooperative interactions and that, apparently, binding by both the maturase and Mrs1 proteins "pulls" the RNA secondary structure to the native conformation.

The cooperative rearrangement of secondary structure in the bI3 RNA supports a new, and ambiguous, classification for the bI3 maturase and Mrs1

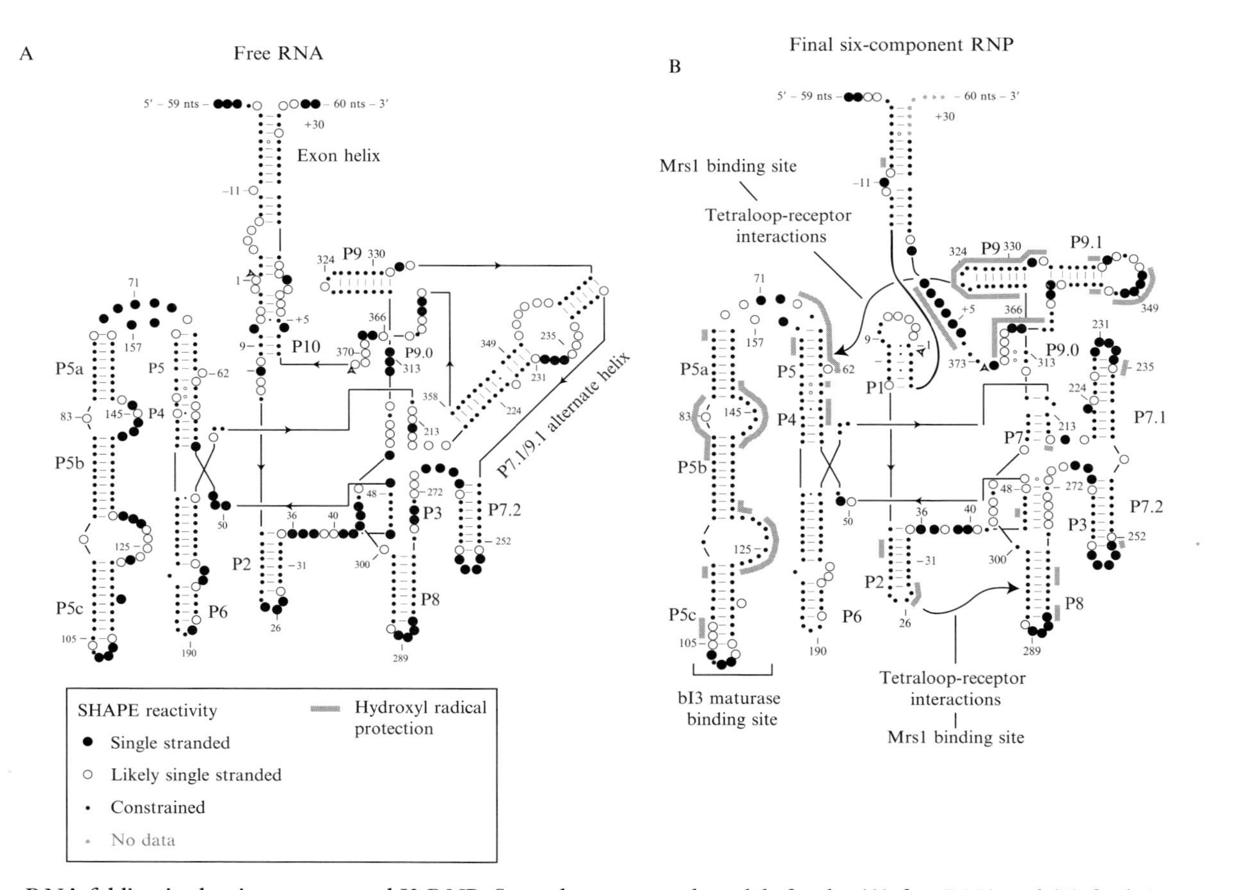

Figure 4.6 RNA folding in the six-component bI3 RNP. Secondary structural models for the (A) free RNA and (B) final six-component RNP. SHAPE reactivities are indicated by spheres. Protection from hydroxyl radical cleavage is emphasized with gray lines.

proteins. Individually, each seems to function as a cofactor: each binds tightly to the bI3 RNA to stabilize specific long-range and higher order tertiary structures (Figs. 4.4B, C and 4.5B, C). In concert, these proteins also induce additional large-scale structural rearrangements in the RNA secondary structure (Figs. 4.4D and 4.5D), a role usually attributed to proteins with chaperone-like activities. However, because the active secondary structure relaxes back to the inactive structure when proteins are removed by proteolysis, these proteins must remain bound to the RNA to exert their structure rearranging functions, which is not consistent with the traditional definition of a chaperone. The detailed information provided by high-throughput RNA structure analysis has revealed unexpectedly rich and complex consequences of protein binding on RNA folding.

5. Perspectives and Conclusion

Intricate and elegant relationships are emerging between the functional roles of RNA in the cell and its ability to form multiple, complex structures, often in concert with proteins. Only by obtaining detailed structural information for almost every nucleotide in each folded state of an RNA can the relationships that govern folding, assembly, conformational changes, and function be fully understood.

Together, SHAPE and hydroxyl radical footprinting comprehensively probe many levels of RNA structure and RNA–protein interactions at single nucleotide resolution. These techniques, in combination with resolution by high-throughput capillary electrophoresis, yield quantitative, single nucleotide resolution structural information for large and complex RNAs and RNPs in a concise and efficiently performed experiments. In our experience, this level of detail consistently yields new insights and structural surprises, even in intensively studied RNAs and RNPs.

As the high-throughput and high-resolution approaches described here are applied more broadly to large RNA–protein complexes, many proteins previously thought to make simple contributions to RNA folding will also likely to be found to exert highly complex, cooperative, and nonadditive effects on RNA folding and RNP function. These sophisticated protein-induced affects add another level of control, and potential regulatory function, in RNP complexes.

ACKNOWLEDGMENTS

This research program on high-throughput RNA structure analysis is supported by the US National Institutes of Health (AI068462 to K. M. W.). Experimental work on the bI3 intron RNP was supported by GM056222. We dedicate this paper to the memory of our friend and colleague Gurminder S. Bassi.

REFERENCES

Balasubramanian, B., Pogozelski, W. K., and Tullius, T. D. (1998). DNA strand breaking by the hydroxyl radical is governed by the accessible surface areas of the hydrogen atoms of the DNA backbone. *Proc. Natl. Acad. Sci. USA* **95,** 9738–9743.

Bassi, G. S., and Weeks, K. M. (2003). Kinetic and thermodynamic framework for assembly of the six-component bI3 group I intron ribonucleoprotein catalyst. *Biochemistry* **42,** 9980–9988.

Bassi, G. S., de Oliveira, D. M., White, M. F., and Weeks, K. M. (2002). Recruitment of intron-encoded and co-opted proteins in splicing of the bI3 group I intron RNA. *Proc. Natl. Acad. Sci. USA* **99,** 128–133.

Brenowitz, M., Chance, M. R., Dhavan, G., and Takamoto, K. (2002). Probing the structural dynamics of nucleic acids by quantitative time-resolved and equilibrium hydroxyl radical "footprinting" *Curr. Opin. Struct. Biol.* **12,** 648–653.

Brion, P., and Westhof, E. (1997). Hierarchy and dynamics of RNA folding. *Annu. Rev. Biophys. Biomol. Struct.* **26,** 113–137.

Brunel, C., and Romby, P. (2000). Probing RNA structure and RNA-ligand complexes with chemical probes. *Methods Enzymol.* **318,** 3–21.

Buchmueller, K. L., Webb, A. E., Richardson, D. A., and Weeks, K. M. (2000). A collapsed non-native RNA folding state. *Nat. Struct. Biol.* **7,** 362–366.

Cate, J. H., Gooding, A. R., Podell, E., Zhou, K., Golden, B. L., Kundrot, C. E., Cech, T. R., and Doudna, J. A. (1996). Crystal structure of a group I ribozyme domain: Principles of RNA packing. *Science* **273,** 1678–1685.

Cech, T. R. (1990). Self-splicing of group I introns. *Annu. Rev. Biochem.* **59,** 543–568.

Chauhan, S., and Woodson, S. A. (2008). Tertiary interactions determine the accuracy of RNA folding. *J. Am. Chem. Soc.* **130,** 1296–1303.

Chauhan, S., Caliskan, G., Briber, R. M., Perez-Salas, U., Rangan, P., Thirumalai, D., and Woodson, S. A. (2005). RNA tertiary interactions mediate native collapse of a bacterial group I ribozyme. *J. Mol. Biol.* **353,** 1199–1209.

Chernick, M. R., and Friis, R. H. (2003). Introductory Biostatistics for the Health Sciences: Modern Applications Including Bootstrap. Wiley-Interscience, Hoboken, NJ.

Cochrane, J. C., and Strobel, S. A. (2008). Catalytic strategies of self-cleaving ribozymes. *Acc. Chem. Res.* **41,** 1027–1035.

Deigan, K. E., Li, T. W., Mathews, D. H., and Weeks, K. M. (2009). Accurate SHAPE-directed RNA structure determination. *Proc. Natl. Acad. Sci. USA* **106,** 97–102.

Duncan, C. D. S., and Weeks, K. M. (2008). SHAPE analysis of long-range interactions reveals extensive and thermodynamically preferred misfolding in a fragile group I intron RNA. *Biochemistry* **47,** 8504–8513.

Duncan, C. D. S., and Weeks, K. M. (2009). The Mrs1 splicing factor binds the RNA tetraloop-receptor motif, in preparation.

Edwards, T. E., Klein, D. J., and Ferré-D'Amaré, A. R. (2007). Riboswitches: Small-molecule recognition by gene regulatory RNAs. *Curr. Opin. Struct. Biol.* **17,** 273–279.

Ehresmann, C., Baudin, F., Mougel, M., Romby, P., Ebel, J. P., and Ehresmann, B. (1987). Probing the structure of RNAs in solution. *Nucleic Acids Res.* **15,** 9109–9128.

Gesteland, R. F., Cech, T., and Atkins, J. F. (2006). The RNA World: The Nature of Modern RNA Suggests a Prebiotic RNA World. Cold Spring Harbor Laboratory Press, Cold Spring Harbor, NY.

Gherghe, C. M., Shajani, Z., Wilkinson, K. A., Varani, G., and Weeks, K. M. (2008). Strong correlation between SHAPE chemistry and the generalized NMR order parameter (S^2) in RNA. *J. Am. Chem. Soc.* **130,** 12244–12245.

Gluick, T. C., and Draper, D. E. (1994). Thermodynamics of folding a pseudoknotted mRNA fragment. *J. Mol. Biol.* **241,** 246–262.

Gutell, R. R., Lee, J. C., and Cannone, J. J. (2002). The accuracy of ribosomal RNA comparative structure models. *Curr. Opin. Struct. Biol.* **12,** 301–310.

Herschlag, D. (1995). RNA chaperones and the RNA folding problem. *J. Biol. Chem.* **270,** 20871–20874.

Jaeger, L., Michel, F., and Westhof, E. (1994). Involvement of a GNRA tetraloop in long-range RNA tertiary interactions. *J. Mol. Biol.* **236,** 1271–1276.

Kreike, J., Schulze, M., Ahne, F., and Lang, B. F. (1987). A yeast nuclear gene, MRS1, involved in mitochondrial RNA splicing: Nucleotide sequence and mutational analysis of two overlapping open reading frames on opposite strands. *EMBO J.* **6,** 2123–2129.

Latham, J. A., and Cech, T. R. (1989). Defining the inside and outside of a catalytic RNA molecule. *Science* **245,** 276–282.

Lazowska, J., Claisse, M., Gargouri, A., Kotylak, Z., Spyridakis, A., and Slonimski, P. P. (1989). Protein encoded by the third intron of cytochrome b gene in *Saccharomyces cerevisiae* is an mRNA maturase. Analysis of mitochondrial mutants, RNA transcripts proteins and evolutionary relationships. *J. Mol. Biol.* **205,** 275–289.

LeCuyer, K. A., and Crothers, D. M. (1993). The Leptomonas collosoma spliced leader RNA can switch between two alternate structural forms. *Biochemistry* **32,** 5301–5311.

Longo, A., Leonard, C. W., Bassi, G. S., Berndt, D., Krahn, J. M., Hall, T. M., and Weeks, K. M. (2005). Evolution from DNA to RNA recognition by the bI3 LAGLI-DADG maturase. *Nat. Struct. Mol. Biol.* **12,** 779–787.

Lu, M., Guo, Q., Wink, D. J., and Kallenbach, N. R. (1990). Charge dependence of Fe(II)-catalyzed DNA cleavage. *Nucleic Acids Res.* **18,** 3333–3337.

Merino, E. J., Wilkinson, K. A., Coughlan, J. L., and Weeks, K. M. (2005). RNA structure analysis at single nucleotide resolution by selective 2′-hydroxyl acylation and primer extension (SHAPE). *J. Am. Chem. Soc.* **127,** 4223–4231.

Michel, F., and Westhof, E. (1990). Modelling of the three-dimensional architecture of group I catalytic introns based on comparative sequence analysis. *J. Mol. Biol.* **216,** 585–610.

Mortimer, S. A., and Weeks, K. M. (2007). A fast-acting reagent for accurate analysis of RNA secondary and tertiary structure by SHAPE chemistry. *J. Am. Chem. Soc.* **129,** 4144–4145.

Nissen, P., Ippolito, J. A., Ban, N., Moore, P. B., and Steitz, T. A. (2001). RNA tertiary interactions in the large ribosomal subunit: The A-minor motif. *Proc. Natl. Acad. Sci. USA* **98,** 4899–4903.

Rajkowitsch, L., Chen, D., Stampfl, S., Semrad, K., Waldsich, C., Mayer, O., Jantsch, M. F., Konrat, R., Blasi, U., and Schroeder, R. (2007). RNA chaperones, RNA annealers and RNA helicases. *RNA Biol.* **4,** 118–130.

Scott, W. G. (2007). Ribozymes. *Curr. Opin. Struct. Biol.* **17,** 280–286.

Shcherbakova, I., and Brenowitz, M. (2008). Monitoring structural changes in nucleic acids with single residue spatial and millisecond time resolution by quantitative hydroxyl radical footprinting. *Nat. Protoc.* **3,** 288–302.

Shcherbakova, I., Mitra, S., Beer, R. H., and Brenowitz, M. (2008). Following molecular transitions with single residue spatial and millisecond time resolution. *Methods Cell Biol.* **84,** 589–615.

Tinoco, I. Jr., and Bustamante, C. (1999). How RNA folds. *J. Mol. Biol.* **293,** 271–281.

Torres-Larios, A., Swinger, K. K., Pan, T., and Mondragón, A. (2006). Structure of ribonuclease P—A universal ribozyme. *Curr. Opin. Struct. Biol.* **16,** 327–335.

Tullius, T. D., and Greenbaum, J. A. (2005). Mapping nucleic acid structure by hydroxyl radical cleavage. *Curr. Opin. Chem. Biol.* **9,** 127–134.

Tullius, T. D., Dombroski, B. A., Churchill, M. E., and Kam, L. (1987). Hydroxyl radical footprinting: A high-resolution method for mapping protein-DNA contacts. *Methods Enzymol.* **155,** 537–558.

Vasa, S. M., Guex, N., Wilkinson, K. A., Weeks, K. M., and Giddings, M. C. (2008). ShapeFinder: A software system for high-throughput quantitative analysis of nucleic acid reactivity information resolved by capillary electrophoresis. *RNA* **14,** 1979–1990.

Vicens, Q., and Cech, T. R. (2006). Atomic level architecture of group I introns revealed. *Trends Biochem. Sci.* **31,** 41–51.

Wang, B., Wilkinson, K. A., and Weeks, K. M. (2008). Complex ligand-induced conformational changes in tRNAAsp revealed by single nucleotide resolution SHAPE chemistry. *Biochemistry* **47,** 3454–3461.

Weeks, K. M. (1997). Protein-facilitated RNA folding. *Curr. Opin. Struct. Biol.* **7,** 336–342.

Wilkinson, K. A., Merino, E. J., and Weeks, K. M. (2005). RNA SHAPE chemistry reveals nonhierarchical interactions dominate equilibrium structural transitions in $tRNA^{Asp}$ transcripts. *J. Am. Chem. Soc.* **127,** 4659–4667.

Wilkinson, K. A., Merino, E. J., and Weeks, K. M. (2006). Selective 2′-hydroxyl acylation analyzed by primer extension (SHAPE): Quantitative RNA structure analysis at single nucleotide resolution. *Nat. Protoc.* **1,** 1610–1616.

Wilkinson, K. A., Gorelick, R. J., Vasa, S. M., Guex, N., Rein, A., Mathews, D. H., Giddings, M. C., and Weeks, K. M. (2008). High-throughput SHAPE analysis reveals structures in HIV-1 genomic RNA strongly conserved across distinct biological states. *PLoS Biol.* **6,** e96.

Wilkinson, K. A., Vasa, S. M., Deigan, K. E., Mortimer, S. A., Giddings, M. C., and Weeks, K. M. (2009). Influence of nucleotide identity on ribose 2′-hydroxyl reactivity in RNA. *RNA* **15,** 1314–1321.

Winkler, W. C., and Breaker, R. R. (2005). Regulation of bacterial gene expression by riboswitches. *Annu. Rev. Microbiol.* **59,** 487–517.

Woese, C. R., Magrum, L. J., Gupta, R., Siegel, R. B., Stahl, D. A., Kop, J., Crawford, N., Brosius, J., Gutell, R., Hogan, J. J., and Noller, H. F. (1980). Secondary structure model for bacterial 16S ribosomal RNA: Phylogenetic, enzymatic and chemical evidence. *Nucleic Acids Res.* **8,** 2275–2293.

Woodson, S. A. (2005). Structure and assembly of group I introns. *Curr. Opin. Struct. Biol.* **15,** 324–330.

Wu, M., and Tinoco, I. Jr. (1998). RNA folding causes secondary structure rearrangement. *Proc. Natl. Acad. Sci. USA* **95,** 11555–11560.

Zheng, M., Wu, M., and Tinoco, I. Jr. (2001). Formation of a GNRA tetraloop in P5abc can disrupt an interdomain interaction in the Tetrahymena group I ribozyme. *Proc. Natl. Acad. Sci. USA* **98,** 3695–3700.

CHAPTER FIVE

Metal Ion-Based RNA Cleavage as a Structural Probe

Marcello Forconi* *and* Daniel Herschlag†

Contents

Abstract

It is well established that many metal ions accelerate the spontaneous degradation of RNA. This property has been exploited in several ways to garner information about RNA structure, especially in regards to the location of site-specifically bound metal ions, the presence of defined structural motifs, and the occurrence of conformational changes in structured RNAs. In this chapter, we review this information, briefly giving strengths and limitations for each of these approaches. Finally, we provide a general protocol to perform metal ion-mediated cleavage of RNA.

* Department of Biochemistry, Stanford University, Stanford, California, USA
† Departments of Biochemistry and Chemistry, Stanford University, Stanford, California, USA

Methods in Enzymology, Volume 468
ISSN 0076-6879, DOI: 10.1016/S0076-6879(09)68005-8

1. Introduction

RNA molecules fold in intricate three-dimensional structures that are crucial for their biological functions. Metal ions play key roles in this process. Diffusely bound metal ions (often referred to as "the ion atmosphere") neutralize the negative charge present on the phosphodiester backbone, allowing nucleic acids to adopt globular structure. More specifically bound metal ions help to bring distant residues together, and shape certain motifs allowing the nucleic acid to adopt a well-defined three-dimensional structure. In addition, site-specifically bound metal ions may be involved in catalysis by nucleic acid enzymes; in particular, several RNA enzymes (ribozymes) use strategically positioned Mg^{2+} ions to help catalysis (Fedor, 2002; Frederiksen *et al.*, 2009). A key part in understanding RNA structure and function is to identify the interactions made by the specifically bound metal ions with the RNA, and how these interactions change in response to different conditions and reaction steps.

Mg^{2+} and other metal ions can promote cleavage of nucleic acids when added in micro- to millimolar quantities to solution (Breslow and Huang, 1991; Dimroth *et al.*, 1950; Farkas, 1968; Huff *et al.*, 1964). In principle, this ability of metal ions to promote RNA cleavage has the potential to reveal regions of the RNA molecule in close proximity to metal ions. For example, this ability may reveal sites of the molecule exposed to ion-atmosphere metal ions, and thus lying on the outer part of the folded RNA molecule, or sites in proximity of tightly bound metal ions.

However, as described in this chapter, the cleavage pattern of RNA molecules in the presence of metal ions is affected by many factors, and it is often impossible to interpret this pattern in terms of a single structural parameter, such as proximity between metal ions and the RNA backbone. Nevertheless, comparison of cleavage patterns in different conditions, for example, upon binding of a substrate to ribozyme, can provide information about local changes, complementing and expanding structural information that can be obtained using other techniques.

2. Mechanisms of Metal Ion-Based Cleavage of Nucleic Acids

Metal ions are known to promote cleavage of nucleic acid in aqueous solution. Typically, simple phosphate esters or dinucleotides are used in model studies to determine the rate acceleration provided by metal ions (Mikkola *et al.*, 2001; Oivanen *et al.*, 1998). Lanthanide ions, such as Eu^{3+}, Tb^{3+}, and Yb^{3+}, are particularly efficient in catalyzing cleavage of simple dinucleotides, with rate acceleration of 3–4 orders of magnitude over the uncatalyzed reaction (Breslow and Huang, 1991). Pb^{2+}, Zn^{2+}, and to some extent Mg^{2+} also display rate acceleration over the uncatalyzed reaction (Breslow and Huang, 1991).

In general, cations with a low pK_a of their hydrates (Table 5.1) cleave RNA better than cations with a high pK_a. This is consistent with the proposal (Brown *et al.*, 1985) that hydrated metal ions cleave RNA acting as Brönsted bases, abstracting a proton from the 2′-OH group of the ribose (Fig. 5.1). This base generates a 2′-O$^-$ group that attacks the phosphorous atom, with departure of the 5′-hydroxyl group. This reaction is greatly facilitated by a particular geometry, referred to as "in-line" geometry,

Table 5.1 Properties of some multivalent cations

Ion	Ionic radius (Å)	Coordination number	First pK_a of $[M(H_2O)_x]$
Mg^{2+}	0.57–0.89	6	11.4
Ca^{2+}	0.99	8	12.6
Mn^{2+}	0.66–0.96	4,6,8	10.6
Zn^{2+}	0.60–0.90	4,6,8	8.2–9.8
Ni^{2+}	0.55–0.69	4,6	6.5–10.2
Fe^{2+}	0.63	4	6.0–6.7
Co^{2+}	0.72	6	7.6–9.9
Pb^{2+}	0.98–1.5	4,6,8,10,12	6.5–8.4
UO_2^{2+}	0.8 (U^{6+})	6,8	5.7
Eu^{3+}	0.95	9	4.8–8.5
Tb^{3+}	0.92	8–9	8.2
Yb^{3+}	0.86	6–9	

Values are from Dallas *et al.* (2004), Frederiksen *et al.* (2009), and references therein.

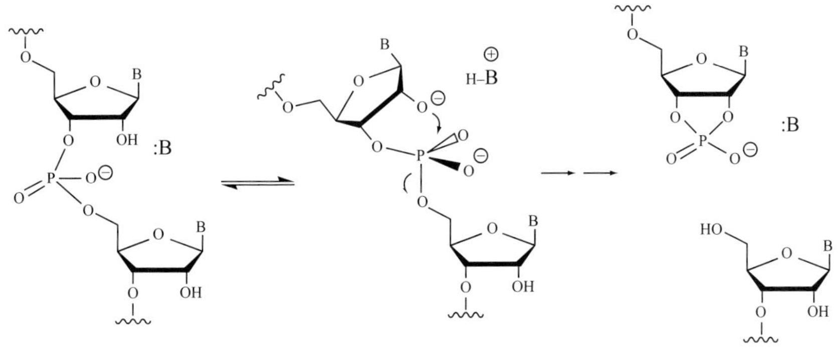

Figure 5.1 Possible mechanism of base-induced cleavage of RNA. Figure adapted from Kirsebom and Ciesiolka (2005).

whereby the attacking nucleophile and the leaving group are positioned about 180° from one another (Westheimer, 1968). Although the mechanism in which a metal ion facilitates abstraction of a proton from the 2′-OH group of the ribose is consistent with the experimental observations, it is not the only possibility (Arnone *et al.*, 1971; Butzow and Eichhorn, 1971). Indeed, alternative mechanisms of metal ion-mediated cleavage of RNA occur in natural RNA enzymes (ribozymes), as shown in Fig. 5.2 (Frederiksen *et al.*, 2009). In particular, metal ions can contribute to stabilization of the developing negative charge on the leaving group (Fig. 5.2A), on the nucleophile (Fig. 5.2B), and on the nonbridging phosphoryl oxygen atom (Fig. 5.2C). Further, they may coordinate both the nucleophile and the nonbridging phosphoryl oxygen (Fig. 5.2D), stabilizing the in-line geometry required for phosphoryl transfer reactions.

In addition, metal ions may impact catalysis through indirect effect. For example, they may constrain the RNA backbone so that the in-line attack of a distal 2′-hydroxyl is facilitated. Electrostatic stabilization of the transition state of the cleavage reaction through outer-sphere or long-range interactions is also possible, although not established.

Finally, it is important to point out that single-stranded regions in RNA are spontaneously cleaved by the mechanism in Fig. 5.1 ~100-fold more effectively than double-stranded regions, because single-stranded regions can sample the in-line conformation required for cleavage more often than double-stranded regions (Soukup and Breaker, 1999). According to this observation, metal ions are expected to generally be more effective in cleaving single-stranded regions of RNA (Hall *et al.*, 1996; Husken *et al.*, 1996; Kolasa *et al.*, 1993; Zagorowska *et al.*, 1998).

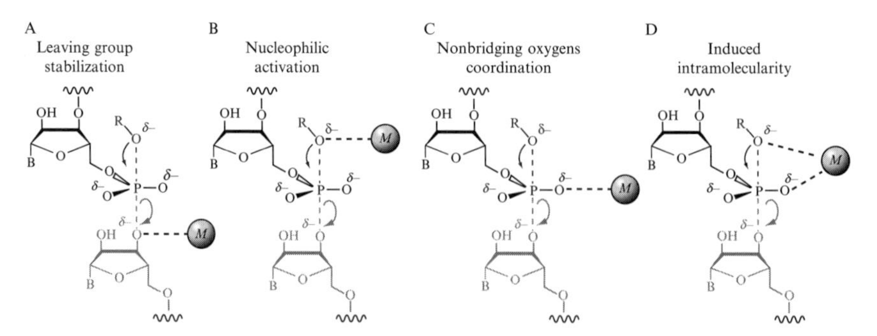

Figure 5.2 Additional strategies used by divalent metal ions to accelerate nucleophilic phosphotransesterification. (A) Lewis acid stabilization of the leaving group 3′-oxygen. (B) Activation of the incoming nucleophile. (C) Coordination of nonbridging oxygen atoms. (D) Induced intramolecularity, in which the external nucleophile and a nonbridging oxygen are coordinated by a metal ion. Figure adapted from Frederiksen *et al.* (2009).

3. Metal Ion-Based Cleavage of RNA as a Structural Probe

In this section, we briefly outline how metal ion-based cleavage of RNA has been used to garner structural information. We have used the excellent review by Kirsebom and Ciesiolka (2005) as a starting point for our considerations, and the reader is encouraged to read this original work. As outlined in that review, metal ion-mediated cleavage of RNA has been used in structural analysis of RNA and RNA complexes in two main ways: (i) to probe metal ion binding sites and (ii) to probe RNA structure and RNA–ligand interactions.

3.1. Probing metal ion binding sites

Metal ion-mediated cleavage of RNA to probe metal ion binding sites has been used in various RNAs, including tRNA (Ciesiolka *et al.*, 1989; Krzyzosiak *et al.*, 1988; Sampson *et al.*, 1987), group I (Rangan and Woodson, 2003; Streicher *et al.*, 1993, 1996) and group II (Sigel *et al.*, 2000) ribozymes, RNase P RNA (Brannvall *et al.*, 2001; Kaye *et al.*, 2002; Kazakov and Altman, 1991), the HDV ribozyme (Lafontaine *et al.*, 1999; Matysiak *et al.*, 1999; Rogers *et al.*, 1996), and the 16S and 23S subunits in the 70S ribosome (Dorner and Barta, 1999; Winter *et al.*, 1997). Because of their high affinity for RNA, lanthanide ions (especially Tb^{3+}) are commonly used in these experiments, although Pb^{2+}, Mn^{2+}, Fe^{2+}, and other metal ions (including Mg^{2+}) have also been used. The idea that underlies these experiments is that a strong metal ion-induced cleavage suggests the presence of a specific metal-ion binding site and that the residues involved in the coordination of the metal ion are close to the site of cleavage. This idea relies on several assumptions.

First, it implies that metal ions other than Mg^{2+} can displace Mg^{2+} from its high-affinity binding sites. This assumption is supported by the observation of relaxed-specificity of metal ion binding sites in two model RNAs, the P4–P6 domain of the *Tetrahymena* group I ribozyme (Travers *et al.*, 2007) and an engineered allosteric ribozyme (Zivarts *et al.*, 2005). However, it is not clear whether trivalent metal ions, such as Tb^{3+}, can effectively replace Mg^{2+} in its binding sites and in some cases the sites of bound metal ions are known to vary (e.g., Basu and Strobel, 1999; Jack *et al.*, 1977). Disappearance of the metal ion-induced cleavage when increasing quantities of Mg^{2+} are added is often taken as further support for this assumption. However, this interpretation is not unequivocal, as the disappearance of a band can also be due to indirect effects, such as changes in structure or simply replacement of the ion-atmosphere metal ions.

The second assumption is that the cleavage event can be ascribed to the abstraction of a proton from a 2′-OH group positioned near to the metal ion. Although this is reasonable, it is important to consider also the alternative mechanisms presented above, including the ones without direct involvement of the metal ion in the cleavage event. Further, it is possible that regions of the RNA only transiently sample the space near a metal ion, but nevertheless get cleaved more than other regions because of favorable geometry.

The third assumption is that the sites of cleavage are responsible for coordination of metal ions. However, there is no guarantee that just because a 2′-OH is nearby to a metal ion, it will act as a ligand for that metal ion or allow proton abstraction via a metal ion-coordinated hydroxide.

Finally, some metal ions may not be able to coordinate a water molecule, or coordinate it in the right place for the cleavage event, and rigid binding sites may be limited in cleavage efficiency. Secondary structure formation may complicate this analysis. As mentioned above, single-stranded regions are cleaved by metal ions more effectively than double-stranded regions (Hall *et al.*, 1996; Husken *et al.*, 1996; Kolasa *et al.*, 1993; Zagorowska *et al.*, 1998), and therefore metal ions bound to or near double-stranded regions may not cleave the RNA backbone efficiently. In addition, rigidity of metal ion binding sites may affect cleavage of metal ions located within the active sites of ribozymes. Because catalysis requires precise positioning of the catalytic residues, including metal ions, rigidity of active site metal ion binding sites may be used by ribozymes to contribute to precise positioning of the metal ion. However, metal ion-mediated cleavage of RNA may be more effective in regions more flexible, which presumably reside outside the active site. In this respect, it is interesting to note that the strongest sites of metal ion-induced cleavage for two ribozymes, the group I intron (Rangan and Woodson, 2003; Streicher *et al.*, 1993, 1996) and the RNase P RNA (Brannvall *et al.*, 2001; Kaye *et al.*, 2002; Kazakov and Altman, 1991), correspond to regions near metal ions located outside the active site, as shown by subsequent structural studies (Kazantsev *et al.*, 2009; Stahley and Strobel, 2007).

In summary, in the absence of other structural data, metal ion-induced cleavage of RNA may provide hints about the location of some metal ions, but it cannot establish the atoms involved in coordination of these metal ions or the importance of these metal ion in folding and catalysis. This technique can be used more powerfully to provide other structural information, as explained in Section 3.2.

3.2. Probing RNA structure and RNA–ligand interactions

Pb^{2+} and Tb^{3+} have been extensively used to probe RNA structure, usually employing higher quantities (millimolar) compared to experiments aimed at probing metal ion binding sites.

Pb^{2+}-mediated cleavage has been historically used with the assumption that more "flexible" regions (such as single-stranded regions) are cleaved better than more "rigid" regions (such as paired regions), as described above. Determining the cleavage pattern of an RNA in the presence of Pb^{2+} may reveal such regions. This approach is complementary to well-established approaches that use nucleases to detect such regions, has similar drawbacks, and presents additional complications due to the possibility that additional cleavage events may arise at high Pb^{2+} concentration from altered RNA conformations or from nonspecific Pb^{2+}-induced cleavage. Pb^{2+}-mediated cleavage has also been proposed as a technique to identify structural motifs in RNA molecules of unknown structure, based on the observation that the cleavage pattern of several structural motifs is identical independent of the rest of the RNA molecule (Ciesiolka *et al.*, 1998). These motifs are shown in Fig. 5.3.

Further, Pb^{2+}-mediated cleavage has been used in RNA–ligands complexes to identify regions of contacts between RNA and ligands (see Kirsebom and Ciesiolka, 2005, and references therein). This technique can match the information obtained with more established footprinting

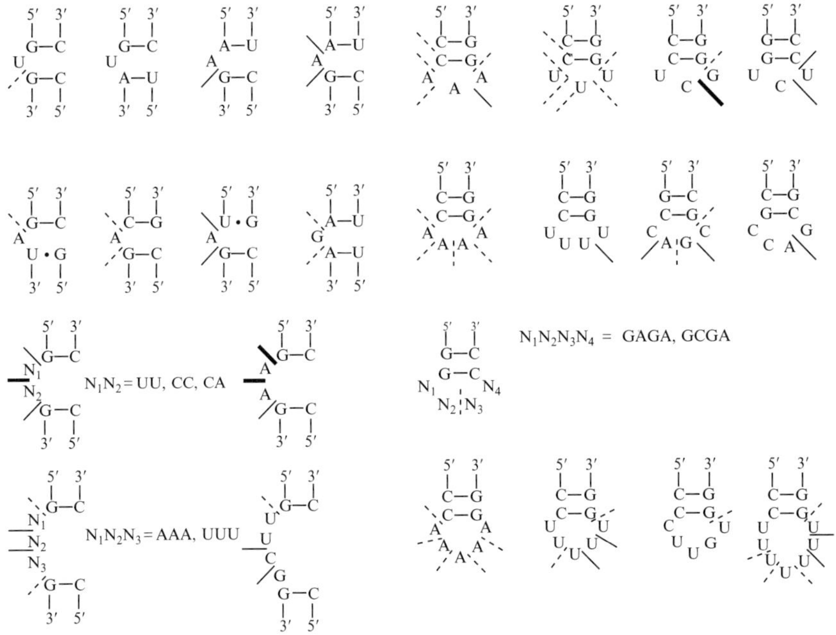

N_1N_2 = UU, CC, CA

$N_1N_2N_3N_4$ = GAGA, GCGA

$N_1N_2N_3$ = AAA, UUU

Figure 5.3 Cleavages induced by Pb^{2+} in RNA bulge regions (left panel) and terminal hairpin loops (right panel). Cleavage sites and their relative intensities are marked by lines—dotted lines: weak cleavages; normal lines: strong cleavages; thick lines: very strong cleavages. Figure adapted from Ciesiolka *et al.* (1998).

techniques, such as hydroxyl-radical footprinting. Because of the ambiguity of the metal ion-induced cleavage in comparison to hydroxyl-radical-mediated cleavage, the use of Pb^{2+} (or other metal ions) to probe RNA–ligands interactions seems to be at most complementary to other techniques.

Tb^{3+}-mediated cleavage of RNA has been introduced more recently as a technique to probe dynamics in RNA. In particular, structural changes in tRNA upon binding of HIV neurocapsid protein (Hargittai *et al.*, 2001) and in the *trans*-acting version of the HDV ribozyme upon binding of an oligonucleotide product or substrate (Jeong *et al.*, 2003) have been detected using this technique. In case of the HDV ribozyme, the differences in cleavage pattern with and without bound oligonucleotides were consistent with a structural change proposed on the basis of other techniques, such as fluorescence resonance energy transfer (Pereira *et al.*, 2002), fluorescence quenching (Harris *et al.*, 2002), and NMR spectroscopy (Luptak *et al.*, 2001; Tanaka *et al.*, 2002), but provided a local view of the regions that undergo structural changes. In conjunction with other metal ion-mediated cleavages of the RNA backbone (e.g., to control for effects given by the different charge on Tb^{3+} compared to Mg^{2+}), in-line probing (Regulski and Breaker, 2008; Soukup and Breaker, 1999), SHAPE (Chapter 4 of this volume), hydroxyl-radical footprinting (Chapters 2 and 3 of this volume), and structural information derived from X-ray crystallography, this approach has the potential to significantly contribute to the understanding of the structural changes associated to binding events or changes in conditions (such as pH, concentration of monovalent cations, etc.).

Finally, UO^{2+} has been proposed as a probe to monitor solvent exposure of the phosphodiester backbone of nucleic acids (Mollegard and Nielsen, 2000). According to this proposal, cleavage by UO^{2+} could be used in conjunction with hydroxyl-radical footprinting, which reports on the solvent exposure of the carbon atoms of the sugar ring, to determine residues with the phosphoryl group and the sugar ring in a different environment. However, little information is present about the fidelity of this cationic probe of RNA structure.

4. Protocols

4.1. General overview

Metal ion-mediated cleavage of RNA is fairly simple in principle. The RNA is radiolabeled, gel purified, subjected to cleavage in the appropriate condition, and the cleavage pattern is detected by running the cleavage reactions on a denaturing gel and detecting the radioactivity by phosphorimaging. However, these simple experiments can be complicated by several factors.

First, it is essential to work in RNase-free conditions; therefore, it is critical to always run a blank experiment, with no metal ions and Mg^{2+} only added to the RNA, to ensure that cleavage is due to the added metal ions and not to some RNase present in solution. However, remember that, as noted above, also Mg^{2+} can induce some cleavage in the RNA backbone.

Second, appropriate controls to ensure the homogeneity of the radiolabeled RNA should also be performed: at minimum, a T1 digestion and an alkaline hydrolysis ladder should always be run in gel lanes next to those for the reactions.

Third, different phosphodiester bonds are cleaved at different rates, because of both the multitude of mechanisms of nucleic acid cleavage and the idiosyncratic nature of local environments within a complex macromolecule such as RNA. This means that in these experiments there is sometimes no such thing as "single-hit kinetics." In other words, by the time one cleavage event happens, another cleavage event may have happened 10 times, generating fragments that, at least in principle, can be further cleaved. These fragments from secondary cleavage events may be erroneously treated as primary cleavages based on their mobility on the gels. Primary and secondary cleavage events can often be distinguished at least in principle, based on the kinetics of product formation (Fersht, 1999). In practice, many experiments are not performed under "single-hit kinetics," but this is not a problem if the structure of the RNA does not change significantly after the first cleavage event.

Finally, parameters such as temperature, pH, buffer concentration, and reaction time can significantly affect the cleavage pattern. Therefore, it is helpful to perform an initial set of experiments to optimize the conditions, avoiding, for example, extensive degradation of the RNA due to high pH or long reaction times.

4.2. RNA radiolabeling

Radiolabeling is usually performed using ^{32}P. RNA can be radiolabeled at the 5′- or 3′-end. In general, it is advisable to perform both radiolabeling and compare the results, because the cleavage pattern should not be affected by the particular label chosen. Here, we describe simple protocols that are generally applicable to many RNAs, but other techniques (Gimple and Schon, 2005) can also be used.

4.2.1. 5′-Radiolabeling

5′-Radiolabeling is achieved using the bacteriophage T4 polynucleotide kinase (PNK), which transfers the gamma phosphate from γ-^{32}P-ATP to the 5′-hydroxyl terminus of the RNA. RNA obtained from *in vitro* transcription has a 5′-triphosphate, which must be removed prior to 5′-radiolabeling with γ-^{32}P-ATP. Typical yields from a 20 μl reaction are 150 μl of radiolabeled RNA, 300,000 cpm/μl.

4.2.1.1. *Removal of the 5′-triphosphate*

1. Set up the following reaction:

RNA (~12 μ*M*)	8.0 μl
10× Antarctic phosphatase buffer (500 m*M* bis-tris-propane–HCl, pH 6.0; 100 m*M* $MgCl_2$; 10 m*M* $ZnCl_2$)	1.0 μl
Antarctic phosphatase (5000 units/ml)	1.0 μl
Total volume	10.0 μl

2. Incubate at 37 °C for 30 min.
3. Terminate the reaction by adding 0.8 μl of 250 m*M* EDTA and incubate at 65 °C for 5 min to denature the phosphatase.
4. Spin down the tubes and proceed to the 5′-radiolabeling reactions.

The quantity of RNA to be radiolabeled may be scaled up or down as needed.

4.2.1.2. *Radiolabeling the RNA 5′-end*

1. Set up the following reaction:

RNA (from triphosphate removal reaction)	10.8 μl
10× PNK buffer (700 m*M* Tris–HCl, pH 7.6; 100 m*M* $MgCl_2$; 50 m*M* dithiothreitol)	2.0 μl
100 m*M* $MgCl_2$	1.0 μl
[γ-^{32}P] ATP (160 μCi/μl)	1.0 μl
T4 PNK (10,000 units/ml)	2.0 μl
Water	3.2 μl
Total volume	20.0 μl

2. Incubate at 37 °C for 30 min.
3. (Optional step) Remove unincorporated short fragments by passing the sample through a size-exclusion column such as Bio-Rad P30 microspin column. This step is optional, but will significantly reduce the amount of "junk" radioactivity in the sample. Make sure to use a column with a cutoff compatible to the radiolabeled RNA.
4. Add 0.5–1 volumes of loading buffer (8 *M* urea; 50 m*M* EDTA, pH 8.0; 0.01% bromophenol blue; 0.01% xylene cyanol) to the resulting solution and proceed to gel purification.

4.2.2. 3′-Radiolabeling

3′-Radiolabeling can be performed using a number of different techniques (Gimple and Schon, 2005). We describe the protocol that uses the Klenow Fragment (3′→5′-exo$^-$) of DNA polymerase I (KF), to incorporate a radiolabeled deoxynucleotide at the 3′-terminus of RNA. The RNA to be labeled is annealed to a short DNA template, which is complementary to the 3′-end of the RNA and a two-nucleotide overhang at its 5′-end. The Klenow Fragment is then used to extend the 3′-end of RNA by one nucleotide that is complementary to the annealed DNA template. The DNA template should have a dideoxy terminator at its 3′-end to prevent labeling of the primer, leading to decrease in the labeling yield of the RNA. Typical yields from an 80 μl reaction are 100 μl of radiolabeled RNA, 200,000 cpm/μl.

1. Set up the annealing reaction as follows:

10× annealing buffer (140 m*M* Tris–HCl, pH 7.5; 400 m*M* NaCl; 2 m*M* EDTA)	4.0 μl
RNA to be labeled, 10 μ*M*	10.0 μl
DNA template, 30 μ*M*	4.2 μl
Water	25.8 μl
Total volume	40.0 μl

2. Incubate the annealing reaction at 60 °C for 1 min.
3. Cool the reaction for 10 min at room temperature. Spin down the tube.
4. During the 10 min cooling step, prepare the labeling reaction, adding the enzyme last.

20× Klenow Fragment buffer (140 m*M* $MgCl_2$, 20 m*M* DTT)	4.0 μl
Annealing reaction	40.0 μl
[α-^{32}P]-dATP (20 μCi/μl)	8.0 μl
Water	20.0 μl
Klenow Fragment (3′→5′-exo^{-}), 5000 units/ml	8.0 μl
Total volume	80.0 μl

5. Incubate the reaction for 50–60 min at 37 °C. Avoid longer times to reduce the possible incorporation of multiple nucleotides.
6. Concentrate the sample and remove unincorporated short fragments by passing the sample through a size-exclusion column such as Bio-Rad P30 microspin column.
7. Add 0.5–1 volumes of loading buffer (8 *M* urea; 50 m*M* EDTA, pH 8.0; 0.01% bromophenol blue; 0.01% xylene cyanol) to the resulting solution and proceed to gel purification.

4.3. RNA purification

1. Purify the radiolabeled RNA on a denaturing polyacrylamide gel containing 7 *M* urea in TBE buffer (90 m*M* Tris–borate, pH 8.5, 2.5 m*M* EDTA). Run a 0.5 mm thick gel at 50 W using a metal plate for an appropriate amount of time (use bromophenol blue and xylene cyanol as markers).
2. Remove the top plate. Cover with saran wrap.

3. Detect the position of the radioactive bands by exposing a film to the gel in a dark room. Punch holes in the gel using a needle to mark gel and film. Make sure not to poke holes in the radioactive RNA bands.
4. Place the film on a light box and use the punched holes to align the gel.
5. Cut out the gel area corresponding to the desired bands on film. Use sterile scalpels or ethanol-flamed razor blades. Place each band in a different tube.
6. Place the tubes on dry ice and freeze them. Freeze and thaw the gel slices three times.
7. When slices are thawed for the third time, add 150 μl of water or TEN buffer (50 m*M* Tris–HCl, pH 7.5; 10 m*M* EDTA; 0.3 *M* NaCl). Rotate the tubes overnight at 4 °C.
8. Remove traces of acrylamide, concentrate the RNA, and/or exchange the RNA in the appropriate buffer (or water) by using a size-exclusion column such as a Bio-Rad P30 column.
9. (Optional) Perform a phenol/chloroform extraction to remove possible protein contaminants and precipitate the RNA.
10. Dilute the RNA to less than 300,000 cpm/μl to minimize autoradiolysis and store at −20 °C.

4.4. Cleavage reactions

1. Prepare a stock solution of radiolabeled RNA and denature it by heating at 90 °C for 2 min. The amount of radiolabeled RNA should be sufficient for aliquots at each desired metal ion concentration; each reaction should have at least 30,000 cpm/μl.
2. Prefold the RNA by incubating at temperature, salt concentration, and pH optimized for the RNA under investigation.
3. Prepare reaction tubes with aliquots of Mg^{2+} and other components (*except for the cleavage-inducing metal ion*) to obtain the desired final concentrations. Include also a sample with buffer only and one with 10 m*M* EDTA (final concentration).
4. Prepare serial sets of cleaving-inducing metal ion dilutions in water, ranging from micro- to millimolar concentrations. Chloride salts are usually used, except for Pb^{2+}, whose acetate salt is used for cleavage experiments. ($PbCl_2$ is insoluble in water.)
5. Aliquot the prefolded RNA into the tubes previously prepared. Allow 5–10 min to equilibrate. If a potentially reacting substrate is present, ensure that the reaction does not occur in the experiments' time scale.
6. Initiate the metal ion-induced cleavage by adding 2 μl of the proper metal ion solution to the aliquoted RNA in (5).
7. Quench the reaction by adding the appropriate volume of stop solution (90% formamide, 50 m*M* EDTA, pH 8.0, 0.01% bromophenol blue, 0.01% xylene cyanol), so that the concentration of EDTA is at least

twofold the sum of the concentrations of the multivalent metal ions in the reaction aliquot.

8. Separate the cleavage products on a denaturing gel, with the percentage of acrylamide depending on the RNA in study. Use markers, such as RNase T1-digested RNA and alkaline hydrolysis ladder of the RNA in question to identify cleavage sites. In case of long RNAs, use two gels running for different times to resolve different regions of the molecule.
9. Dry the gel under vacuum, 80 °C, for 30–60 min and expose it to a phosphor screen overnight.
10. Align the gel using dedicated software such as SAFA (Das *et al.*, 2005). Quantify the intensity of the cleavage bands using the same program.

4.5. Troubleshooting

- No cleavage detected.
  - The metal ion solution could be too old: prepare a new solution of the metal ion salts.
  - The cleavage conditions are not optimized: vary incubation time, concentration of the metal ions, pH. It may be useful to run a control with single-stranded RNA as standard, to ensure that the conditions used can result in RNA cleavage.
  - The RNA contains a contaminant that interferes with cleavage: try to exchange the RNA in the cleavage buffer or prepare new RNA.
- Complete degradation detected.
  - The concentration of cleaving metal ion is too high: reduce the concentration.
  - The time of incubation is too long: decrease the incubation time.
  - The pH is too high: lower the pH.
  - One or more solutions are contaminated with nucleases. Change all solutions or perform a scan to determine which one(s) is responsible.

ACKNOWLEDGMENT

This research was supported by a grant from the NIH (GM 49243) to D. H.

REFERENCES

Arnone, A., Bier, C. J., Cotton, F. A., Day, V. W., Hazen, E. E. Jr., Richardson, D. C., Yonath, A., and Richardson, J. S. (1971). A high resolution structure of an inhibitor complex of the extracellular nuclease of *Staphylococcus aureus*. I. Experimental procedures and chain tracing. *J. Biol. Chem.* **246,** 2302–2316.

Basu, S., and Strobel, S. A. (1999). Thiophilic metal ion rescue of phosphorothioate interference within the *Tetrahymena* ribozyme P4–P6 domain. *RNA* **5,** 1399–1407.

Brannvall, M., Mikkelsen, N. E., and Kirsebom, L. A. (2001). Monitoring the structure of *Escherichia coli* RNase P RNA in the presence of various divalent metal ions. *Nucleic Acids Res.* **29,** 1426–1432.

Breslow, R., and Huang, D.-L. (1991). Effects of metal ions, including Mg^{2+} and lanthanides, on the cleavage of ribonucleotides and RNA model compounds. *Proc. Natl. Acad. Sci. USA* **88,** 4080–4083.

Brown, R. S., Dewan, J. C., and Klug, A. (1985). Crystallographic and biochemical investigation of the lead(II)-catalyzed hydrolysis of yeast phenylalanine tRNA. *Biochemistry* **24,** 4785–4801.

Butzow, J. J., and Eichhorn, G. L. (1971). Interaction of metal ions with nucleic acids and related compounds. XVII. On the mechanism of degradation of polyribonucleotides and oligoribonucleotides by zinc(II) ions. *Biochemistry* **10,** 2019–2027.

Ciesiolka, J., Wrzesinski, J., Gornicki, P., Podkowinski, J., and Krzyzosiak, W. J. (1989). Analysis of magnesium europium and lead binding sites in methionine initiator and elongator tRNAs by specific metal ion-induced cleavages. *Eur. J. Biochem.* **186,** 71–77.

Ciesiolka, J., Michalowski, D., Wrzesinski, J., Krajewski, J., and Krzyzosiak, W. J. (1998). Pattern of cleavages induced by lead ions in defined RNA secondary structure motifs. *J. Mol. Biol.* **275,** 211–229.

Dallas, A., Vlassov, A. V., and Kazakov, S. A. (2004). Principles of nucleic acid cleavage by metal ions. *In* "Nucleic Acids and Molecular Biology," (M.A Zenkova, ed.), Vol. 13, pp. 61–88. Springer-Verlag, Berlin Heidelberg.

Das, R., Laederach, A., Pearlman, S. M., Herschlag, D., and Altman, R. B. (2005). SAFA: Semi-automated footprinting analysis software for high-throughput identification of nucleic acid footprinting experiments. *RNA* **11,** 344–354.

Dimroth, K., Jaenicke, L., and Heinzel, D. (1950). Die Spaltung Der Pentose-Nucleinsaure Der Hefe Mit Bleihydroxyd. 1. Uber Nucleinsauren. *Annalen Der Chemie-Justus Liebig* **566,** 206–210.

Dorner, S., and Barta, A. (1999). Probing ribosome structure by europium-induced RNA cleavage. *Biol. Chem.* **380,** 243–251.

Farkas, W. R. (1968). Depolymerization of ribonucleic acid by plumbous ion. *Biochim. Biophys. Acta* **155,** 401–409.

Fedor, M. J. (2002). The role of metal ions in RNA catalysis. *Curr. Opin. Struct. Biol.* **12,** 289–295.

Fersht, A. (1999). Structure and Mechanism in Protein Science. W.H. Freeman and Company, New York.

Frederiksen, J. K., Fong, R., and Piccirili, J. A. (2009). Metal ions in RNA catalysis. *In* "Nucleic Acid-Metal Ion Interactions," (N. V. Hud, ed.), Royal Society of Chemistry, Cambridge, UK.

Gimple, O., and Schon, A. (2005). Direct determination of RNA sequence and modification by radiolabeling methods. *In* "Handbook of RNA Biochemistry," (R. K. Hartmann, ed.), Vol. 1, pp. 132–150. Wiley-VCH Verlag GmbH & Co., Weinheim.

Hall, J., Husken, D., and Haner, R. (1996). Towards artificial ribonucleases: The sequence-specific cleavage of RNA in a duplex. *Nucleic Acids Res.* **24,** 3522–3526.

Hargittai, M. R. S., Mangla, A. T., Gorelick, R. J., and Musier-Forsyth, K. (2001). HIV-1 neurocapsid protein zinc finger structures induce tRNALys, 3 structural changes but are not critical for primer/template annealing. *J. Mol. Biol.* **312,** 985–997.

Harris, D. A., Rueda, D., and Walter, N. G. (2002). Local conformational changes in the catalytic core of the trans-acting hepatitis delta virus ribozyme accompany catalysis. *Biochemistry* **41,** 12051–12061.

Huff, J. W., Sastry, K. S., Gordon, M. P., and Wacker, W. E. (1964). The action of metal ions on tobacco mosaic virus ribonucleic acid. *Biochemistry* **3,** 501–506.

Husken, D., Goodall, G., Blommers, M. J., Jahnke, W., Hall, J., Haner, R., and Moser, H. E. (1996). Creating RNA bulges: Cleavage of RNA in RNA/DNA duplexes by metal ion catalysis. *Biochemistry* **35,** 16591–16600.

Jack, A., Ladner, J. E., Rhodes, D., Brown, R. S., and Klug, A. (1977). A crystallographic study of metal-binding to yeast phenylalanine transfer RNA. *J. Mol. Biol.* **111,** 315–328.

Jeong, S., Sefcikova, J., Tinsley, R. A., Rueda, D., and Walter, N. G. (2003). Trans-acting Hepatitis Delta Virus ribozyme, catalytic core and global structure are dependent on the 5′ substrate sequence. *Biochemistry* **42,** 7727–7740.

Kaye, N. M., Zahler, N. H., Christian, E. L., and Harris, M. E. (2002). Conservation of helical structure contributes to functional metal ion interactions in the catalytic domain of ribonuclease P RNA. *J. Mol. Biol.* **324,** 429–442.

Kazakov, S., and Altman, S. (1991). Site-specific cleavage by metal ion cofactors and inhibitors of M1 RNA, the catalytic subunit of RNase P from *Escherichia coli*. *Proc. Natl. Acad. Sci. USA* **88,** 9193–9197.

Kazantsev, A. V., Krivenko, A. A., and Pace, N. R. (2009). Mapping metal-binding sites in the catalytic domain of bacterial RNase P RNA. *RNA* **15,** 266–276.

Kirsebom, L. A., and Ciesiolka, J. (2005). Pb2+-induced cleavage of RNA. *Handb. RNA Biochem.* **1,** 214–228.

Kolasa, K. A., Morrow, J. R., and Sharma, A. P. (1993). Trivalent lanthanide ions do not cleave RNA in DNA–RNA hybrids. *Inorg. Chem.* **32,** 3983–3984.

Krzyzosiak, W. J., Marciniec, T., Wiewiorowski, M., Romby, P., Ebel, J. P., and Giege, R. (1988). Characterization of the lead (II)-induced cleavages in tRNA in solution and effect of the Y-base removal in yeast $tRNA^{Phe}$. *Biochemistry* **27,** 5771–5777.

Lafontaine, D. A., Ananvoranich, S., and Perreault, J.-P. (1999). Presence of a coordinated metal ion in a trans-acting antigenomic *delta* ribozyme. *Nucleic Acids Res.* **27,** 3236–3243.

Luptak, A., Ferre-D'Amare, A. R., Zhou, K. H., Zilm, K. W., and Doudna, J. A. (2001). Direct pK(a) measurement of the active-site cytosine in a genomic hepatitis delta virus ribozyme. *J. Am. Chem. Soc.* **123,** 8447–8452.

Matysiak, M., Wrzesinski, J., and Ciesiolka, J. (1999). Sequential folding of the genomic ribozyme of the Hepatitis Delta Virus: Structural analysis of RNA transcription intermediates. *J. Mol. Biol.* **291,** 283–294.

Mikkola, S., Kaukinen, U., and Lonnberg, H. (2001). The effect of secondary structure on cleavage of the phosphodiester bonds of RNA. *Cell Biochem. Biophys.* **34,** 95–119.

Mollegard, N. E., and Nielsen, P. E. (2000). Application of uranyl cleavage mapping of RNA structure. *Methods Enzymol.* **318,** 43–47.

Oivanen, M., Kuusela, S., and Lonnberg, H. (1998). Kinetics and mechanisms for the cleavage and isomerization of the phosphodiester bonds of RNA by Bronsted acids and bases. *Chem. Rev.* **98,** 961–990.

Pereira, M. J. B., Harris, D. A., Rueda, D., and Walter, N. G. (2002). Reaction pathway of the trans-acting hepatitis delta virus ribozyme: A conformational change accompanies catalysis. *Biochemistry* **41,** 730–740.

Rangan, P., and Woodson, S. A. (2003). Structural requirement for Mg^{2+} binding in the group I intron core. *J. Mol. Biol.* **329,** 229–238.

Regulski, E. E., and Breaker, R. R. (2008). In-line probing analysis of riboswitches. *Methods Mol. Biol.* **419,** 53–67.

Rogers, J., Chang, A. H., Von Ahsen, U., Schroeder, R., and Davies, J. (1996). Inhibition of the self-cleaving reaction of the human Hepatitis Delta Virus ribozyme by antibiotics. *J. Mol. Biol.* **259,** 916–925.

Sampson, J. R., Sullivan, F. X., Behlen, A. B., DiRenzo, O. C., and Uhlenbeck, O. C. (1987). Characterization of two RNA-catalyzed RNA cleavage reactions. *Cold Spring Harbor Symp. Quant. Biol.* **52,** 267–275.

Sigel, R. K. O., Vaidya, A., and Pyle, A. M. (2000). Metal ion binding sites in a group II intron core. *Nat. Struct. Biol.* **7,** 111–116.

Soukup, G. A., and Breaker, R. R. (1999). Relationship between internucleotide linkage geometry and the stability of RNA. *RNA* **5,** 1308–1325.

Stahley, M. R., and Strobel, S. A. (2007). Structural metals in the group I intron: A ribozyme with a multiple metal ion core. *J. Mol. Biol.* **372,** 89–102.

Streicher, B., von Ahsen, U., and Schroeder, R. (1993). Lead cleavage sites in the core structure of group I intron-RNA. *Nucleic Acids Res.* **21,** 311–317.

Streicher, B., Westhof, E., and Schroeder, R. (1996). The environment of two metal ions surrounding the splice site of a group I intron. *EMBO J.* **15,** 2256–2264.

Tanaka, Y., Tagaya, M., Hori, T., Sakamoto, T., Kurihara, Y., Katahira, M., and Uesugi, S. (2002). Cleavage reaction of HDV ribozymes in the presence of Mg^{2+} is accompanied by a conformational change. *Genes Cells* **7,** 567–579.

Travers, K. J., Boyd, N., and Herschlag, D. (2007). Low specificity of metal ion binding in the metal ion core of a folded RNA. *RNA* **13,** 1205–1213.

Westheimer, F. H. (1968). Pseudo-rotation in the hydrolysis of phosphate esters. *Acc. Chem. Res.* **1,** 70–78.

Winter, D., Polaceck, N., Halama, I., Streicher, B., and Barta, A. (1997). Lead-catalysed specific cleavage of ribosomal RNAs. *Nucleic Acids Res.* **25,** 1817–1824.

Zagorowska, I., Kuusela, S., and Lonnberg, H. (1998). Metal ion-dependent hydrolysis of RNA phosphodiester bonds within hairpin loops. A comparative kinetic study on chimeric ribo/2′-*O*-methylribo oligonucleotides. *Nucleic Acids Res.* **26,** 3392–3396.

Zivarts, M., Liu, Y., and Breaker, R. R. (2005). Engineered allosteric ribozymes that respond to specific divalent metal ions. *Nucleic Acids Res.* **33,** 622–631.

CHAPTER SIX

2′-Amino-Modified Ribonucleotides as Probes for Local Interactions Within RNA

James L. Hougland* *and* Joseph A. Piccirilli†

Contents

Abstract

The 2′-hydroxyl group plays an integral role in RNA structure and catalysis. This ubiquitous component of the RNA backbone can participate in multiple interactions essential for RNA function, such as hydrogen bonding and metal ion coordination, but the multifunctional nature of the 2′-hydroxyl renders identification of these interactions a significant challenge. By virtue of their versatile physicochemical properties, such as distinct metal coordination preferences, hydrogen bonding properties, and ability to be protonated,

* Department of Chemistry, University of Michigan, Ann Arbor, Michigan, USA

† Departments of Chemistry and Biochemistry & Molecular Biology, University of Chicago, Chicago, Illinois, USA

Methods in Enzymology, Volume 468

ISSN 0076-6879, DOI: 10.1016/S0076-6879(09)68006-X

2′-amino-2′-deoxyribonucleotides can serve as tools for probing local interactions involving 2′-hydroxyl groups within RNA. The 2′-amino group can also serve as a chemoselective site for covalent modification, permitting the introduction of probes for investigation of RNA structure and dynamics. In this chapter, we describe the use of 2′-aminonucleotides for investigation of local interactions within RNA, focusing on interactions involving 2′-hydroxyl groups required for RNA structure, function, and catalysis.

1. Introduction

The 2′-hydroxyl group plays an integral role in RNA structure and catalysis. This multifunctional component of the RNA backbone influences ribose conformation and helix geometry (Saenger, 1984), coordinates metal ions (Gordon *et al.*, 2000; Shan and Herschlag, 1999; Smith and Pace, 1993), provides a scaffold for protein or solvent interactions (Dertinger *et al.*, 2001; Egli *et al.*, 1996; Hou *et al.*, 2001; Juneau *et al.*, 2001), mediates tertiary interactions and catalysis via hydrogen bonding (Hermann and Patel, 1999; Herschlag *et al.*, 1993b; Klostermeier and Millar, 2002; Knitt *et al.*, 1994; Silverman and Cech, 1999a; Yoshida *et al.*, 2000), and serves as a nucleophile in cleavage reactions catalyzed by both RNA- and protein-based enzymes (Cochrane and Strobel, 2008; Gordon *et al.*, 2007; Raines, 1998; Steyaert, 1997). While this multitude of potential functions demonstrates the versatility of the 2′-hydroxyl group, it also means that deconvoluting the roles played by a given 2′-hydroxyl group within a structured RNA is inherently difficult. The requirement for a 2′-hydroxyl at a given site within RNA can be readily determined through 2′-deoxyribonucleotide substitution (Abramovitz *et al.*, 1996; Herschlag *et al.*, 1993b; Klostermeier and Millar, 2002; Schwans *et al.*, 2003; Silverman and Cech, 1999a), but any resulting loss of function provides little information about the specific functional contribution(s) of the 2′-hydroxyl group. Thus, determining the interactions of a 2′-hydroxyl that contribute to RNA structure and function requires a detailed dissection of the potential roles played by this functional group.

2′-Amino-2′-deoxyribonucleotides have received increased attention in recent years as probes for investigating interactions within RNA involving 2′-hydroxyl groups (Earnshaw and Gait, 1998; Gordon *et al.*, 2000, 2004; Hougland *et al.*, 2004; Persson *et al.*, 2003; Shan and Herschlag, 1999; Verma and Eckstein, 1998). The substitution of a 2′-amino group for a 2′-hydroxyl group introduces minimal changes in steric volume and polarity at the 2′-position, thereby minimizing undesired changes in the local environment of the 2′-substituent (Gordon *et al.*, 2004). The introduction of 2′-aminonucleotides into RNA or DNA can destabilize duplex structure (Aurup *et al.*, 1994), but this destabilization can be minimized or compensated for by appropriate choice of experimental conditions or modification chemistry (Chamberlin and Weeks, 2000; Hougland *et al.*, 2004; Narlikar *et al.*, 1997; Pham *et al.*, 2004; Yoshida *et al.*, 2000). The distinct physicochemical properties of the 2′-amino group provide for simultaneous investigation of multiple potential roles played by 2′-hydroxyl groups. For example, metal coordination to the 2′-hydroxyl can be probed using "metal ion rescue" experiments that exploit the different metal coordination preferences exhibited by 2′-hydroxyl and 2′-amino groups (Gordon *et al.*, 2000; Shan and Herschlag, 1999; Sjogren *et al.*, 1997). The 2′-amino group also displays a different array of hydrogen bonding sites compared to the 2′-hydroxyl group, which can be harnessed to identify 2′-hydroxyl groups that donate functionally significant hydrogen bonds (Gordon *et al.*, 2004; Hougland *et al.*, 2004, 2008; Yoshida *et al.*, 2000). Finally, the 2′-amino group exhibits a pK_a of ~6 and can be protonated under biologically relevant pH conditions to yield a 2′-ammonium group (Aurup *et al.*, 1994; Dai *et al.*, 2007), which exhibits distinct electrostatic, hydrogen bonding, and metal coordination properties compared to both 2′-hydroxyl and 2′-amino groups (Shan *et al.*, 1999a; Yoshida *et al.*, 2000). With judicious choice of experimental conditions and careful comparison of the activity or binding of RNA bearing 2′-aminonucleotide modifications to that of an unmodified RNA, interactions involving the native 2′-hydroxyl group can be functionally identified and energetically dissected.

Use of 2′-aminonucleotide substitution for exploration of direct contacts, such as hydrogen bonding and metal ion coordination, is consistent with the strictest definition of "local" wherein only direct interactions with the 2′-hydroxyl are investigated (interaction distances <5 Å). However, chemoselective covalent modification of 2′-amino groups can be used to probe the larger "local" environment within RNA, providing information on RNA structure and dynamics within close proximity to the introduced 2′-aminonucleotide. The 2′-amino group exhibits much higher nucleophilicity than the nucleobase amino groups and sugar hydroxyl groups in unmodified nucleic acids, allowing for site-specific attachment of a wide variety of tags (e.g., fluorescent groups, spin labels, and proximity cleavage reagents), as described below. The 2′-position can also be isotopically labeled for NMR studies by synthetic incorporation of ^{15}N (Dai *et al.*, 2007), providing a spectroscopic probe of the local environment of a specific 2′-hydroxyl. These approaches allow for investigation of an expanded "local" environment within RNA, providing information on RNA structure and dynamics in regions within close proximity to an introduced 2′-aminonucleotide.

In this chapter, we discuss the use of 2′-amino-2′-deoxyribonucleotides to probe local interactions in RNA. We first describe the current methods for incorporating 2′-amino modifications into RNA, followed by a survey of 2′-amino group covalent modifications that introduce site-specific probes of the local environment within RNA. We then discuss the general strategies used to test for direct interactions with the 2′-hydroxyl group, focusing on the conditions required for interpretation of changes in reactivity or binding resulting from 2′-amino group incorporation. Finally, we describe experimental setups and analysis wherein 2′-amino modification can be used to probe metal ion coordination, hydrogen bonding, and the local electrostatic environment.

2. 2′-Amino-2′-Deoxynucleotide Synthesis and Incorporation

2′-Amino-2′-deoxynucleotides can be incorporated into RNA by one of two approaches, either site-specific incorporation through use of solid-phase oligoribonucleotide synthesis or random incorporation via enzymatic transcription (reviewed in Das *et al.*, 2005). The syntheses of all four 2′-amino-2′-deoxynucleosides and their corresponding phosphoramidites required for solid-phase synthesis have been reported (Dai *et al.*, 2006; Karpeisky *et al.*, 2002); 2′-amino-2′-deoxynucleoside phosphoramidites are generally compatible with standard solid-phase RNA synthesis and subsequent deprotection protocols (Dai *et al.*, 2006; Tuschl *et al.*, 1993). The 2′-amino-2′-deoxypyrimidines are also currently available for incorporation into oligoribonucleotides from several commercial vendors. In the event that a site-specific 2′-amino modification is desired within a RNA that is too large for chemical synthesis, 2′-amino modifications can be incorporated by enzymatic ligation (Moore and Query, 2000; Moore and Sharp, 1992). In this approach, a synthetic oligonucleotide bearing the 2′-amino-2′-deoxynucleotide can be attached to larger transcribed RNA(s) by splint-mediated ligation using T4 DNA ligase to yield a semisynthetic construct bearing a single site-specific 2′-amino modification. Although this approach can suffer from low yields and significant dependence of ligation efficiency on the location of the ligation site, development of improved ligation catalysts and protocols may help overcome this limitation (Baum and Silverman, 2008).

In contrast to the size limitations encountered with chemical synthesis of 2′-amino containing oligoribonucleotides, enzymatic synthesis of RNA in the presence of 2′-amino-2′-deoxynucleoside triphosphates (2′-amino-NTPs) allows for incorporation of 2′-amino groups throughout RNA without appreciable restrictions on RNA transcript size or sequences. Syntheses of both 2′-amino-NTPs and α-thio-2′-amino-NTPs have been reported (Aurup *et al.*, 1992; Das *et al.*, 2008), and 2′-amino-NTPs are commercially available (Trilink Biotechnologies). Transcription does not permit site-specific incorporation of modified nucleotides, but enzymatic incorporation of 2′-modified nucleotides allows for population-wide scanning for functionally important 2′-hydroxyl groups within RNA, using approaches such as nucleotide analogue interference mapping (NAIM) (Ryder *et al.*, 2000). To optimize the yield of transcribed RNAs containing 2′-amino substitutions, a mutant form of T7 RNA polymerase should be used as wild-type T7 RNA polymerase does not incorporate 2′-amino-NTPS efficiently (Guillerez *et al.*, 2005; Padilla and Sousa, 2002).

3. 2′-Amino-2′-Deoxynucleotides as Sites for Covalent Modification

Covalent modification of a 2′-amino-2′-deoxynucleotide allows site-specific introduction of probes for studying RNA structure and dynamics, and can therefore serve as a valuable tool for investigating the local environment and surrounding architecture of 2′-hydroxyl groups within RNA. For example, Silverman and coworkers have utilized a pyrene fluorophore attached at the 2′-position to probe folding of the P4–P6 domain of the *Tetrahymena* group I ribozyme (Silverman and Cech, 1999b; Silverman *et al.*, 2000; Young and Silverman, 2002). In these studies, RNA folding leads to increased pyrene fluorescence, and this approach appears compatible with a variety of fluorophore tether lengths and attachment positions within a RNA of interest (Smalley and Silverman, 2006). 2′-Amino-modified ribonucleotides have also been used to incorporate nitroxide spin labels that allow investigation of RNA structure and dynamics using electron paramagnetic resonance (EPR) spectroscopy (Edwards and Sigurdsson, 2007; Edwards *et al.*, 2002; Schiemann *et al.*, 2007). In addition to their use as probes for studying RNA folding, structure, and dynamics, 2′-aminonucleotides can also be utilized to probe RNA flexibility and solvent accessibility in the vicinity of the 2′-hydroxyl of interest. An example of this approach comes in the work of Weeks and coworkers, wherein the acylation efficiency of a 2′-aminonucleotide (i.e., the covalent modification reaction itself) was shown to be modulated by the flexibility of the residue bearing the 2′-amino group (Chamberlin and Weeks, 2000; John and Weeks, 2000, 2002; John *et al.*, 2004). This approach has also been used to monitor changes in RNA flexibility upon formation of structural elements in the bI5 group I intron (Chamberlin and Weeks, 2003).

In a second example, Herschlag and coworkers demonstrated the use of covalent modification of transcriptionally incorporated 2′-aminonucleotides to probe RNA structure and generate information on the proximity of different regions of a folded RNA. This approach used multiplexed hydroxyl-radical cleavage analysis (MOHCA) in the context of the P4–P6 domain on the *Tetrahymena* group I intron (Das *et al.*, 2008). Transcription with α-thio-2′-amino-NTPs results in incorporation of nucleotides bearing both a 2′-amino group and a phosphorothioate linkage, which allow attachment of a hydroxyl-radical cleavage reagent and specific cleavage at the site of the modified nucleotide, respectively. In combination with two-dimensional gel electrophoresis, the orthogonal cleavage chemistries permitted by these modifications allow high-throughput contact mapping of a structured RNA, yielding distance constraints ($\leq$25 Å) between the sites of the modified nucleotides and the regions of the RNA cleaved by the reagent attached to the 2′-amino group (Das *et al.*, 2008).

4. General Strategy for Investigating 2′-Hydroxyl Interactions Using 2′-Deoxy and 2′-Aminonucleotides

2′-Aminonucleotides can be used in combination with 2′-deoxynucleotide incorporation to provide a qualitative probe of the potential functional roles played by a specific 2′-hydroxyl, as shown in Table 6.1. Initially, inhibition of function upon 2′-deoxynucleotide incorporation indicates that a given 2′-hydroxyl is required for function, whether that function be stable folding of a RNA or catalysis by a ribozyme. If replacement of the 2′-hydroxyl group with a 2′-amino group also inhibits function in the presence of Mg^{2+}, this suggests the possibility of a metal ion coordination at this position; restoration of function upon addition of a softer metal ion (e.g., Mn^{2+}, Zn^{2+}, or Cd^{2+}) provides further evidence for metal ion coordination. When incorporation of a 2′-aminonucleotide does not significantly impact function, this indicates that the 2′-hydroxyl under investigation may donate or accept a functionally important hydrogen bond. Protonation of the 2′-amino group can provide further information regarding the role of a 2′-hydroxyl, as the 2′-ammonium group cannot coordinate a metal ion or accept a hydrogen bond but can donate a hydrogen bond. Inhibition upon protonation of the 2′-amino group can also be due to electrostatic interference with the binding of a metal ion required for function, such as that observed in the group I ribozyme (Shan *et al.*, 1999a). Thus, a reactivity profile developed from comparison of four 2′-substituents (2′-OH, 2′-H, 2′-NH_2, and 2′-NH_3^+) provides a functional signature of the potential roles played by a given 2′-hydroxyl. However, as described in detail below, this signature can be complicated if a 2′-hydroxyl is involved in multiple interactions, such as simultaneous metal ion coordination and hydrogen bonding. Furthermore, as incorporation of a 2′-aminonucleotide can destabilize RNA and DNA duplexes (Aurup *et al.*, 1994; Pham *et al.*, 2004), care must be taken to ensure that such destabilization is not interpreted as evidence for the inability of the 2′-amino group to functionally substitute for the 2′-hydroxyl group (Hougland *et al.*, 2004, 2008; Narlikar *et al.*, 1997).

Table 6.1 Qualitative analysis of the role of a 2′-hydroxyl group using 2′-deoxynucleotide and 2′-aminonucleotide modifications

	Predicted activity with various 2′-substituents[a] (background metal ion[b])					
Potential role of 2′-hydroxyl	2′-OH (Mg^{2+})	2′-H (Mg^{2+})	2′-NH_2 (Mg^{2+})	2′-NH_2 (Mn^{2+})	2′-NH_3^+ (Mg^{2+})	2′-NH_3^+ (Mn^{2+})
Donates a hydrogen bond	+	−	+	+	+	+
Accepts a hydrogen bond	+	−	+	+	−	−
Coordinates metal ion	+	−	−	+	−	−
Serves as nucleophile	+	−	−	−	−	−

[a] Activity denoted as + (active) or − (inhibited/inactive).
[b] Mn^{2+} can represent any of several divalent metal ions (e.g., Mn^{2+}, Zn^{2+}, Cd^{2+}) with higher affinity than Mg^{2+} for the 2′-NH_2 group.

5. Studies of RNA Catalysis Using 2′-Amino-2′-Deoxynucleotides

2′-Aminonucleotides have been used to probe the role of 2′-hydroxyl groups during catalysis in several RNA-based systems, including peptidyl transfer reaction catalyzed by the ribosome as well as all three large naturally occurring ribozymes (RNase P, the group I ribozyme, and the group II ribozyme). In the ribosome, removal of the 2′-hydroxyl at the A76 site of the P-site tRNA results in a 10^6-fold reduction in the rate of peptide bond formation (Weinger *et al.*, 2004). Strobel and coworkers introduced a 2′-amino modification at the A76 position of a peptidyl transfer transition state analogue to serve as an electrostatic probe, testing whether the ribosome active site stabilizes an ionized form of the A76 2′-hydroxyl group (Huang *et al.*, 2008). Such stabilization could aid in a proton-shuttle mechanism involving this 2′-hydroxyl by lowering the energetic cost of localizing charge on the A76 2′-hydroxyl (Beringer and Rodnina, 2007). By measuring the binding affinity of the analogue as the 2′-NH_2 group transitioned to a 2′-NH_3^+ group as a function of pH, they determined that the ribosome does not preferentially bind an inhibitor with a positive charge at the 2′-position of A76. This result suggests that the 2′-hydroxyl remains neutral in the transition state of the peptidyl transfer reaction, potentially supporting a model wherein this 2′-hydroxyl simultaneously accepts a proton from the incoming amine nucleophile while donating a proton to the departing 3′-oxygen leaving group during the peptidyl transfer reaction (Beringer and Rodnina, 2007).

In RNase P, substitution of the 2′-hydroxyl at the cleavage site of the pre-tRNA substrate impacts ribozyme activity, with a 2′-aminonucleotide substitution inhibiting cleavage to approximately the same degree as 2′-deoxy substitution (Brannvall *et al.*, 2004; Persson *et al.*, 2003). Addition of soft metal ions such as Mn^{2+} activates cleavage of the pre-tRNA with a 2′-aminonucleotide at the cleavage site but also enhances cleavage of the substrate containing a 2′-deoxynucleotide, suggesting that this rescue does not reflect direct metal ion coordination to the cleavage site 2′-hydroxyl. Protonation of the 2′-aminonucleotide impacts cleavage site selection in the pre-tRNA substrate, suggesting the potential for electrostatic interference with a metal ion within the ribozyme active site, but the specific effects of 2′-aminonucleotide incorporation at the cleavage site of RNase P are difficult to interpret. Taken in total, studies employing 2′-aminonucleotide substitution at the cleavage site of RNase P indicate that the 2′-hydroxyl at this position plays an important role in RNase P-catalyzed cleavage, but the specific interactions involving this 2′-hydroxyl are not yet well defined or energetically characterized.

In the group I ribozyme, 2′-aminonucleotides have been extensively utilized to investigate aspects of substrate binding and the ribozyme catalytic mechanism. These studies have focused on three regions of the ribozyme–substrate complex: (1) interactions between the P1 helix formed from the oligonucleotide substrate and the internal guide sequence (IGS) of the ribozyme that stabilize "docking" of this helix into the catalytic core of the ribozyme (Herschlag *et al.*, 1993a; Narlikar *et al.*, 1997); (2) the role of the 2′-hydroxyl group at the cleavage site (U_{-1}) of the oligonucleotide substrate (Herschlag *et al.*, 1993b; Hougland *et al.*, 2004; Strobel and Ortoleva-Donnelly, 1999; Yoshida *et al.*, 2000); and (3) the interactions with the 2′-hydroxyl of the guanosine nucleophile (Hougland *et al.*, 2008; Kuo *et al.*, 2004; Shan and Herschlag, 1999; Sjogren *et al.*, 1997). 2′-Aminonucleotide substitutions within the P1 helix revealed that a 2′-NH_3^+ group can substitute for a 2′-hydroxyl group at the -3 position of the substrate, indicating that this position donates a hydrogen bond important for stabilizing the docking of the P1 helix (Narlikar *et al.*, 1997). This study is particularly notable in that the docking stabilization provided by the 2′-NH_3^+ group was only detected once the destabilizing effect of 2′-aminonucleotide protonation on P1 helix stability was compensated for, illustrating the need to characterize the effect of 2′-aminonucleotide incorporation and protonation on each step of the reaction under investigation. At the cleavage site, 2′-amino substitution was instrumental in identifying that this 2′-hydroxyl donates a hydrogen bond to the adjacent 3′-oxygen leaving group but does not coordinate a metal ion in the course of the ribozyme-catalyzed reaction (Yoshida *et al.*, 2000). Further studies of this 2′-hydroxyl using an array of 2′-modifications in the framework of an atomic mutation cycle (AMC, described in detail below), including 2′-amino substitution, provided both an estimate for the energetic cost of removing this hydrogen bond and the energetic benefit of coupling the cleavage site 2′-hydroxyl into a larger hydrogen bonding network within the ribozyme active site (Hougland *et al.*, 2004; Strobel and Ortoleva-Donnelly, 1999). Finally, studies involving 2′-amino-2′-deoxyguanosine revealed that the 2′-hydroxyl of the guanosine nucleophile coordinates a catalytic metal ion in the group I ribozyme active site (Kuo *et al.*, 2004; Shan and Herschlag, 1999; Shan *et al.*, 1999a; Sjogren *et al.*, 1997), an interaction subsequently observed in group I intron crystal structures (Stahley and Strobel, 2006). Protonation of the 2′-amino group of 2′-amino-2′-deoxyguanosine interferes with the binding of this catalytic metal ion, illustrating the utility of the 2′-NH_3^+ group as an electrostatic probe within a large RNA (Shan *et al.*, 1999a). Investigation of the guanosine nucleophile 2′-hydroxyl using AMC analysis revealed that it also donates a functionally significant hydrogen bond in both the ground state and the transition state of the group I ribozyme reaction, indicating that this 2′-hydroxyl forms multiple interactions within an active-site network important for catalysis by this ribozyme (Hougland *et al.*, 2008).

In the group II intron, 2′-aminonucleotides have been employed to investigate interactions in both the forward- and reverse-splicing reactions catalyzed by this ribozyme (Gordon *et al.*, 2000, 2004). In an assay that mimics the second step of forward splicing wherein the exons that flank the group II intron are ligated, analysis of 2′-amino-substituted substrates provided evidence that the 2′-hydroxyl at the cleavage site coordinates a catalytic metal ion and may also donate a hydrogen bond (Gordon *et al.*, 2000). In contrast, in the spliced exons reopening (SER) reaction that mimics the first step of reverse splicing, analysis of the effects of 2′-amino and other substitutions of the cleavage site 2′-hydroxyl indicates that this 2′-hydroxyl group does not coordinate a catalytic metal ion, nor does the removal of hydrogen bond donation ability at the cleavage site 2′-substituent block catalysis (Gordon *et al.*, 2004). Protonation of a 2′-aminonucleotide at the cleavage site does not affect efficiency of the SER reaction, ruling out a requirement for the cleavage site 2′-hydroxyl to accept a hydrogen bond. Further analysis of the relative reactivity of substrates bearing a variety of 2′-substituents suggests that the cleavage site 2′-hydroxyl interacts with a water molecule, potentially reflecting an RNA–solvent interaction important for ribozyme activity. This study by Gordon and coworkers illustrates the utility and potential of parallel analysis of multiple 2′-substitutions, including 2′-amino, for investigating the various potential roles played by the 2′-hydroxyl in RNA structure and function.

6. Using 2′-Aminonucleotides to Investigate RNA Structure and Function: Case Studies

The remainder of the chapter will describe the use of 2′-aminonucleotides as probes of hydrogen bonding, metal ion coordination, and the electrostatic environment in the vicinity of a 2′-hydroxyl group. Whereas the examples provided focus on the use of 2′-aminonucleotides to investigate substrate binding and reactivity involved in ribozyme catalysis, analogous studies of the roles played by 2′-hydroxyl groups in RNA structure, folding, and RNA–RNA and RNA–protein interactions can be easily envisioned using folding or binding as the experimental readout. Each section will briefly describe the approach for assaying the specific type of interaction, the caveats and experimental requirements for each experiment, and the analysis of the effects of 2′-aminonucleotide substitution. A reference to a published report using the described experimental approach in the *Tetrahymena* group I ribozyme will serve as a practical example for each section.

6.1. Experiment: Investigating metal coordination to a 2′-hydroxyl group

6.1.1. Example: Metal coordination at the 2′-hydroxyl of the guanosine nucleophile in the *Tetrahymena* group I ribozyme (Shan and Herschlag, 1999)

Rescue of inhibition in the presence of a 2′-aminonucleotide by addition of soft metal ions provides evidence for metal coordination at the site of 2′-amino incorporation. However, enhancement of activity upon soft metal ion addition alone is insufficient to prove metal coordination at a given 2′-hydroxyl position, as nonspecific stimulation of ribozyme activity or stabilization of RNA structure in the presence of soft metal ions can manifest as rescue of the 2′-amino substitution (Persson *et al.*, 2003; Shan and Herschlag, 1999, 2000; Shan *et al.*, 1999b). To interpret metal rescue of 2′-aminonucleotide substitution as evidence for metal coordination to a specific 2′-hydroxyl group, the following conditions/criteria must be met:

1. *k_{rel} measurement.* The rate of the 2′-aminonucleotide-containing substrate must be measured relative to a substrate that will not preferentially coordinate to a soft metal ion with its 2′-substituent, such as the native 2′-hydroxyl or a 2′-deoxy substrate (if reactive). Measuring the relative effect of soft metal addition on substrate reactivity (k_{rel}) will cancel out nonspecific effects of soft metals, whether activating or inhibitory, yielding the specific effect of restoring metal coordination at the site of 2′-aminonucleotide substitution. In the case of the guanosine nucleophile in the *Tetrahymena* group I ribozyme reaction, the reactivity of 2′-amino-2′-deoxyguanosine was measured relative to guanosine as a function of added Mn^{2+} (Shan *et al.*, 1999b), Zn^{2+} (Shan *et al.*, 2001), or Cd^{2+} (Hougland *et al.*, 2005; Shan *et al.*, 2001). Nonspecific effects of added soft metal ions can also be suppressed through use of elevated background Mg^{2+} concentrations (Shan *et al.*, 1999b, 2001).
2. *Monitor the same reaction steps and proceed from the same ground state.* Over the entire course of a soft metal ion titration, the same reaction steps must be followed for both the 2′-amino substituted and the control substrate. This requirement is absolute to all methods that utilize relative reactivities or stabilities to elucidate the energetic effects of mutations. These steps may include binding, conformational changes, and the chemical step for a ribozyme reaction, depending on the system being studied; a well-defined kinetic framework for the reaction or pathway under investigation is extremely helpful for designing these experiments. Unless the same reaction steps are monitored for the 2′-amino substituted and control substrates, nonspecific suppression of a defect in binding or other reaction step caused by 2′-amino incorporation may

appear as a specific metal rescue. Such a nonspecific suppression effect has been observed in the *Tetrahymena* group I ribozyme, wherein careful study uncovered a metal ion that stabilizes docking but does not interact with any of the atoms involved in the ribozyme cleavage reaction (Shan and Herschlag, 2000). Furthermore, by careful choice of the starting state for the reactions (free enzyme or enzyme–substrate complex), metal ion rescue of interactions in either the ground state (substrate binding) or transition state (chemical step) can be interrogated (Shan and Herschlag, 1999).

3. *Effect of 2′-amino substitution on the affinity of the rescuing metal ion: bound versus unbound substrate.* To measure the apparent binding affinity of a rescuing metal ion for the enzyme, reactions must proceed from a starting state wherein the modified substrate is not bound to the enzyme. If the 2′-amino-modified substrate is bound to the enzyme in the starting state for the reaction, the measured binding affinity of the rescuing metal ion may be altered by the presence of the 2′-amino group (Shan and Herschlag, 1999). Alternatively, this perturbation of rescuing metal ion binding affinity can be used to demonstrate linkage between atoms in the ribozyme active site through coordination to the same metal ion (Hougland *et al.*, 2005).
4. pH dependence: 2′-NH_2 *versus* 2′-NH_3^+. Protonation of the 2′-amino group blocks its ability to coordinate a metal ion. To ensure that the 2′-amino group exists primarily in its deprotonated form, reactions must be performed at solution pH values above the pK_a of the 2′-amino group within the context of the RNA being studied. Whereas the pK_a of this group can shift from its value of ~6 in solution within the core of a structured RNA (Aurup *et al.*, 1994; Dai *et al.*, 2007; Shan and Herschlag, 1999; Shan *et al.*, 1999a), in general pH values above 7.5 are sufficient to ensure that the 2′-amino group remains deprotonated. Metal coordination can be further confirmed by loss of specific metal rescue at low pH values due to protonation of the 2′-amino group (Shan *et al.*, 1999a).

6.2. Experiment: Investigating hydrogen bonding involving the 2′-hydroxyl group

6.2.1. Example: Hydrogen bond donation by the cleavage site 2′-hydroxyl in the *Tetrahymena* group I ribozyme (Herschlag *et al.*, 1993b; Hougland *et al.*, 2004; Yoshida *et al.*, 1999)

As described above, comparison of changes in reactivity engendered by 2′-deoxy and 2′-amino modifications can provide qualitative evidence that a 2′-hydroxyl is involved in a functionally required hydrogen bond. The hydrogen bonding role of a 2′-hydroxyl can be explored further by AMC analysis, wherein the energetic effects from three atomic mutations ($2'$-OCH_3, $2'$-NH_2, and $2'$-$NHCH_3$) are measured relative to the ribonucleotide ($2'$-OH) (Fig. 6.1). Mutation of the 2′-hydroxyl to a 2′-methoxy group (left vertical) replaces the hydrogen atom of the hydroxyl group with a methyl group. A 2′-methoxynucleotide may affect activity through removal of the hydrogen atom ($\Delta G_{\mathrm{H\,removal}}$) and/or introduction of the bulky methyl group ($\Delta G_{\mathrm{CH_3\,installation}}$), which could cause steric clashes, loss of metal coordination, or other deleterious effects. To distinguish the effect of hydrogen atom removal from that of methyl group installation, the effect of the 2′-amino to 2′-methylamino mutation (right vertical) is measured, in which a methyl group replaces one of the hydrogen atoms on an amino group. The 2′-methylaminonucleotide imposes the consequences of the bulky methyl group, but, unlike the 2′-methoxynucleotide, also retains a heteroatom-bound hydrogen atom. Therefore, the energetic cost of the 2′-methylamino mutation relative to the 2′-amino mutation provides an independent measure of the effect of methyl group installation ($\Delta G_{\mathrm{CH_3\,installation}}$). Energetic differences between the vertical perturbations provide an operational estimate for $\Delta G_{\mathrm{H\,removal}}$, and thereby may implicate the 2′-hydroxyl as an important hydrogen bond donor. This analysis can be

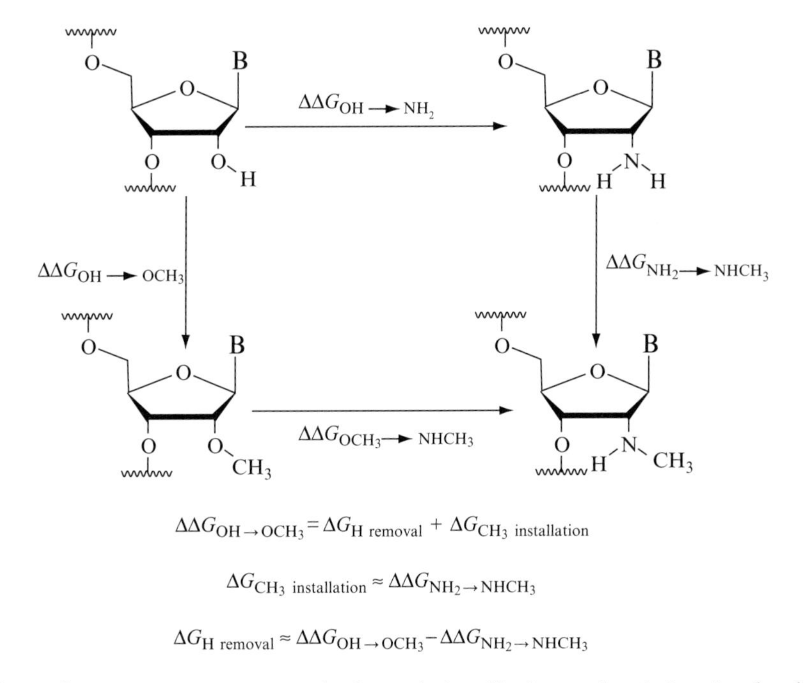

Figure 6.1 Atomic mutation cycle for analysis of hydrogen bond donation by the 2′-hydroxyl group. Adapted from Hougland *et al.* (2004).

applied to study the requirement for a hydrogen bond donor for both binding (Hougland *et al.*, 2008) and catalysis (Hougland *et al.*, 2004, 2008).

Similar to metal ion rescue experiments using 2′-amino-modified substrates, AMC analysis requires that the following set of conditions:

1. *Ground state and reaction steps.* Reactions for both wild-type and modified substrates must start from the same ground state and monitor the same reaction steps (e.g., binding or the chemical step), as described above.
2. *Protonation state of the 2′-amino group.* Analysis must be performed under conditions where the 2′-amino- and 2′-methylaminonucleotides are in the neutral state (as opposed to the protonated ammonium ion). If this requirement is not met, comparisons between the 2′-hydroxyl and 2′-methoxy substrates and the nitrogen-containing analogs not only entail the –O– to –NH– mutation, but also the introduction of a proton and the accompanying positive charge. These multiple changes would preclude assignment of mutation effects to either hydrogen removal or methyl group incorporation alone. In practice, this requirement may be addressed by determining cycles at different pH values. Care must also be taken to account for any changes in duplex stability or nucleotide binding that may occur as a result of 2′-aminonucleotide protonation that can alter the ground state of the reaction being studied.
3. *Potential complications of AMC analysis.* A significant caveat of AMC analysis is that the changes in ribose conformation and in hydrogen bonding strength may accompany the oxygen to nitrogen mutation in some contexts. For example, 2′-aminonucleosides in solution inhabit a more 2′-*endo* conformation than do 2′-ribonucleosides (Huang *et al.*, 1997; Uesugi *et al.*, 1979). This analysis assumes that methyl group incorporation does not significantly impact either of these attributes, and that effects caused by changing oxygen to nitrogen are cancelled out in the cycle. Although these assumptions appear to be appropriate in the systems already studied (Gordon *et al.*, 2004; Hougland *et al.*, 2004, 2008), they may not hold in other contexts.

6.3. Experiment: Using the 2′-aminonucleotide as a electrostatic probe

6.3.1. Example: Investigating the electrostatic environment of the guanosine 2′-hydroxyl in the group I ribozyme (Shan *et al.*, 1999a)

Protonation of the 2′-amino group can be used to probe the effects of introducing a positive charge within a structured RNA molecule. Analysis of experiments involving protonation of the 2′-amino group entails many of the same requirements as described above for analysis of metal rescue experiments, with the following changes:

1. k_{rel} *measurement.* The reaction rate or binding affinity of the 2′-aminonucleotide-containing substrate must be measured relative to a substrate that does not undergo protonation or deprotonation within the pH range examined. Measuring the relative binding affinities or reactivities of the control and 2′-amino containing substrates with the RNA/ribozyme under investigation as a function of pH will allow isolation of the functional effect (if any) of protonating the 2′-amino group while correcting for nonspecific pH effects on RNA structure or function.
2. *Metal ion titration.* In the event that the 2′-amino group lies in close proximity to a bound metal ion within RNA, protonation of this group will lead to electrostatic repulsion between the positively charged metal ion and the —NH_3^+ group. The functional signature of this interaction will be interdependence between metal concentration and the observed pK_a of the 2′-amino group when bound to the ribozyme. As the concentration of the metal ion concentration increases, the pK_a of the 2′-amino group will decrease due to competition between metal ion coordination to the deprotonated 2′-amino group and electrostatic interference between a bound metal ion and the protonated 2′-ammonium group (Shan *et al.*, 1999a).

7. Conclusions

Defining interactions involving 2′-hydroxyl groups is essential to understanding the complex relationship between RNA structure and function. 2′-Aminonucleotides are valuable analogs for investigating these contacts, presenting the opportunity to characterize local interactions within RNA both qualitatively and quantitatively. In investigations of hydrogen bonding, metal ion coordination, local electrostatics, and RNA structure and dynamics using covalently attached probes, studies employing 2′-aminonucleotides have provided a wealth of information regarding the roles of specific 2′-hydroxyl groups within multiple RNAs. This information has been crucial for defining the biophysical behavior of RNA and for developing an understanding of how these RNAs carry out their biological functions. The ready availability of 2′-aminonucleotides will hopefully encourage future research employing these nucleotide analogues that promises many future penetrating insights into RNA structure, folding, and RNA–RNA and RNA–protein interactions involved in both binding and catalysis.

Acknowledgments

The authors thank members of the Piccirilli and Fierke laboratories for helpful comments and editing. J. L. H. was supported by National Institutes of Health Postdoctoral Fellowship GM78894. J. A. P. is an Investigator of the Howard Hughes Medical Institute.

REFERENCES

Abramovitz, D. L., Friedman, R. A., and Pyle, A. M. (1996). Catalytic role of 2′-hydroxyl groups within a group II intron active site. *Science* **271,** 1410–1413.

Aurup, H., Williams, D. M., and Eckstein, F. (1992). 2′-Fluoro-2′-deoxynucleoside and 2′-amino-2′-deoxynucleoside 5′-triphosphates as substrates for T7 RNA polymerase. *Biochemistry* **31,** 9636–9641.

Aurup, H., Tuschl, T., Benseler, F., Ludwig, J., and Eckstein, F. (1994). Oligonucleotide duplexes containing 2′-amino-2′-deoxycytidines: Thermal stability and chemical reactivity. *Nucleic Acids Res.* **22,** 20–24.

Baum, D. A., and Silverman, S. K. (2008). Deoxyribozymes: Useful DNA catalysts *in vitro* and *in vivo*. *Cell. Mol. Life. Sci.* **65,** 2156–2174.

Beringer, M., and Rodnina, M. V. (2007). The ribosomal peptidyl transferase. *Mol. Cell* **26,** 311–321.

Brannvall, M., Kikovska, E., and Kirsebom, L. A. (2004). Cross talk between the +73/294 interaction and the cleavage site in RNase P RNA mediated cleavage. *Nucleic Acids Res.* **32,** 5418–5429.

Chamberlin, S., and Weeks, K. (2000). Mapping local nucleotide flexibility by selective acylation of 2′-amine substituted RNA. *J. Am. Chem. Soc.* **122,** 216–224.

Chamberlin, S. I., and Weeks, K. M. (2003). Differential helix stabilities and sites pre-organized for tertiary interactions revealed by monitoring local nucleotide flexibility in the bI5 group I intron RNA. *Biochemistry* **42,** 901–909.

Cochrane, J. C., and Strobel, S. A. (2008). Catalytic strategies of self-cleaving ribozymes. *Acc. Chem. Res.* **41,** 1027–1035.

Dai, Q., Deb, S. K., Hougland, J. L., and Piccirilli, J. A. (2006). Improved synthesis of 2′-amino-2′-deoxyguanosine and its phosphoramidite. *Bioorgan. Med. Chem.* **14,** 705–713.

Dai, Q., Lea, C. R., Lu, J., and Piccirilli, J. A. (2007). Syntheses of (2′)3′-*N*-15-amino-(2′) 3′-deoxyguanosine and determination of their pK(a) values by N-15 NMR spectroscopy. *Org. Lett.* **9,** 3057–3060.

Das, S. R., Fong, R., and Piccirilli, J. A. (2005). Nucleotide analogues to investigate RNA structure and function. *Curr. Opin. Chem. Biol.* **9,** 585–593.

Das, R., Kudaravalli, M., Jonikas, M., Laederach, A., Fong, R., Schwans, J. P., Baker, D., Piccirilli, J. A., Altman, R. B., and Herschlag, D. (2008). Structural inference of native and partially folded RNA by high-throughput contact mapping. *Proc. Natl. Acad. Sci. USA* **105,** 4144–4149.

Dertinger, D., Dale, T., and Uhlenbeck, O. C. (2001). Modifying the specificity of an RNA backbone contact. *J. Mol. Biol.* **314,** 649–654.

Earnshaw, D., and Gait, M. (1998). Modified oligoribonucleotides as site-specific probes of RNA structure and function. *Biopolymers* **48,** 39–55.

Edwards, T. E., and Sigurdsson, S. T. (2007). Site-specific incorporation of nitroxide spin-labels into 2′-positions of nucleic acids. *Nat. Protoc.* **2,** 1954–1962.

Edwards, T. E., Okonogi, T. M., and Sigurdsson, S. T. (2002). Investigation of RNA–protein and RNA–metal ion interactions by electron paramagnetic resonance spectroscopy: The HIV TAR-Tat motif. *Chem. Biol.* **9,** 699–706.

Egli, M., Portmann, S., and Usman, N. (1996). RNA hydration: A detailed look. *Biochemistry* **35,** 8489–8494.

Gordon, P. M., Sontheimer, E. J., and Piccirilli, J. A. (2000). Kinetic characterization of the second step of group II intron splicing: Role of metal ions and the cleavage site 2′-OH in catalysis. *Biochemistry* **39,** 12939–12952.

Gordon, P. M., Fong, R., Deb, S. K., Li, N. S., Schwans, J. P., Ye, J. D., and Piccirilli, J. A. (2004). New strategies for exploring RNA's 2′-OH expose the importance of solvent during group II intron catalysis. *Chem. Biol.* **11,** 237–246.

Gordon, P. M., Fong, R., and Piccirilli, J. A. (2007). A second divalent metal ion in the group II intron reaction center. *Chem. Biol.* **14,** 607–612.
Guillerez, J., Lopez, P. J., Proux, F., Launay, H., and Dreyfus, M. (2005). A mutation in T7 RNA polymerase that facilitates promoter clearance. *Proc. Natl. Acad. Sci. USA* **102,** 5958–5963.
Hermann, T., and Patel, D. J. (1999). Stitching together RNA tertiary architectures. *J. Mol. Biol.* **294,** 829–849.
Herschlag, D., Eckstein, F., and Cech, T. R. (1993a). Contributions of 2′-hydroxyl groups of the RNA substrate to binding and catalysis by the *Tetrahymena* ribozyme. An energetic picture of an active site composed of RNA. *Biochemistry* **32,** 8299–8311.
Herschlag, D., Eckstein, F., and Cech, T. R. (1993b). The importance of being ribose at the cleavage site in the *Tetrahymena* ribozyme reaction. *Biochemistry* **32,** 8312–8321.
Hou, Y. M., Zhang, X. L., Holland, J. A., and Davis, D. R. (2001). An important 2′-OH group for an RNA–protein interaction. *Nucleic Acids Res.* **29,** 976–985.
Hougland, J. L., Deb, S. K., Maric, D., and Piccirilli, J. A. (2004). An atomic mutation cycle for exploring RNA's 2′-hydroxyl group. *J. Am. Chem. Soc.* **126,** 13578–13579.
Hougland, J. L., Kravchuk, A. V., Herschlag, D., and Piccirilli, J. A. (2005). Functional identification of catalytic metal ion binding sites within RNA. *PLoS Biol.* **3,** 1536–1548.
Hougland, J. L., Sengupta, R. N., Dai, Q., Deb, S. K., and Piccirilli, J. A. (2008). The 2′-hydroxyl group of the guanosine nucleophile donates a functionally important hydrogen bond in the *Tetrahymena* ribozyme reaction. *Biochemistry* **47,** 7684–7694.
Huang, Y., Eckstein, F., Padilla, R., and Sousa, R. (1997). Mechanism of ribose 2′-group discrimination by an RNA polymerase. *Biochemistry* **36,** 8231–8242.
Huang, K. S., Carrasco, N., Pfund, E., and Strobel, S. A. (2008). Transition state chirality and role of the vicinal hydroxyl in the ribosomal peptidyl transferase reaction. *Biochemistry* **47,** 8822–8827.
John, D. M., and Weeks, K. M. (2000). Tagging DNA mismatches by selective 2′-amine acylation. *Chem. Biol.* **7,** 405–410.
John, D. M., and Weeks, K. M. (2002). Chemical interrogation of mismatches in DNA–DNA and DNA–RNA duplexes under nonstringent conditions by selective 2′-amine acylation. *Biochemistry* **41,** 6866–6874.
John, D. M., Merino, E. J., and Weeks, K. M. (2004). Mechanics of DNA flexibility visualized by selective 2′-amine acylation at nucleotide bulges. *J. Mol. Biol.* **337,** 611–619.
Juneau, K., Podell, E., Harrington, D. J., and Cech, T. R. (2001). Structural basis of the enhanced stability of a mutant ribozyme domain and a detailed view of RNA–solvent interactions. *Structure* **9,** 221–231.
Karpeisky, A., Sweedler, D., Haeberli, P., Read, J., Jarvis, K., and Beigelman, L. (2002). Scaleable and efficient synthesis of 2′-deoxy-2′-*N*-phthaloyl nucleoside phosphoramidites for oligonucleotide synthesis. *Bioorg. Med. Chem. Lett.* **12,** 3345–3347.
Klostermeier, D., and Millar, D. P. (2002). Energetics of hydrogen bond networks in RNA: Hydrogen bonds surrounding G+1 and U42 are the major determinants for the tertiary structure stability of the hairpin ribozyme. *Biochemistry* **41,** 14095–14102.
Knitt, D. S., Narlikar, G. J., and Herschlag, D. (1994). Dissection of the role of the conserved G*U pair in group-I RNA self-splicing. *Biochemistry* **33,** 13864–13879.
Kuo, L., Perera, N., and Tarpo, S. (2004). Metal ion coordination to 2′ functionality of guanosine mediates substrate-guanosine coupling in group I ribozymes: Implications for conserved role of metal ions and for variability in RNA folding in ribozyme catalysis. *Inorg. Chim. Acta* **357,** 3934–3942.
Moore, M. J., and Query, C. C. (2000). Joining of RNAs by splinted ligation. *Methods Enzymol.* **317,** 109–123.
Moore, M. J., and Sharp, P. A. (1992). Site-specific modification of pre-mRNA: The 2′-hydroxyl groups at the splice sites. *Science* **256,** 992–997.

Narlikar, G. J., Khosla, M., Usman, N., and Herschlag, D. (1997). Quantitating tertiary binding energies of 2′ OH groups on the P1 duplex of the *Tetrahymena* ribozyme: Intrinsic binding energy in an RNA enzyme. *Biochemistry* **36,** 2465–2477.

Padilla, R., and Sousa, R. (2002). A Y639F/H784A T7 RNA polymerase double mutant displays superior properties for synthesizing RNAs with non-canonical NTPs. *Nucleic Acids Res.* **30,** e138.

Persson, T., Cuzic, S., and Hartmann, R. K. (2003). Catalysis by RNase P RNA—Unique features and unprecedented active site plasticity. *J. Biol. Chem.* **278,** 43394–43401.

Pham, J. W., Radhakrishnan, I., and Sontheimer, E. J. (2004). Thermodynamic and structural characterization of 2′-nitrogen-modified RNA duplexes. *Nucleic Acids Res.* **32,** 3446–3455.

Raines, R. (1998). Ribonuclease A. *Chem. Rev.* **98,** 1045–1066.

Ryder, S. P., Ortoleva-Donnelly, L., Kosek, A. B., and Strobel, S. A. (2000). Chemical probing of RNA by nucleotide analog interference mapping. *Methods Enzymol.* **317,** 92–109.

Saenger, W. (1984). Principles of Nucleic Acid Structure. Springer-Verlag, New York, NY.

Schiemann, O., Piton, N., Plackmeyer, J., Bode, B. E., Prisner, T. F., and Engels, J. W. (2007). Spin labeling of oligonucleotides with the nitroxide TPA and use of PELDOR, a pulse EPR method, to measure intramolecular distances. *Nat. Protoc.* **2,** 904–923.

Schwans, J. P., Cortez, C. N., Olvera, J. M., and Piccirilli, J. A. (2003). 2′-Mercaptonucleotide interference reveals regions of close packing within folded RNA molecules. *J. Am. Chem. Soc.* **125,** 10012–10018.

Shan, S. O., and Herschlag, D. (1999). Probing the role of metal ions in RNA catalysis: Kinetic and thermodynamic characterization of a metal ion interaction with the 2′-moiety of the guanosine nucleophile in the *Tetrahymena* group I ribozyme. *Biochemistry* **38,** 10958–10975.

Shan, S. O., and Herschlag, D. (2000). An unconventional origin of metal-ion rescue and inhibition in the *Tetrahymena* group I ribozyme reaction. *RNA* **6,** 795–813.

Shan, S. O., Narlikar, G. J., and Herschlag, D. (1999a). Protonated 2′-aminoguanosine as a probe of the electrostatic environment of the active site of the *Tetrahymena* group I ribozyme. *Biochemistry* **38,** 10976–10988.

Shan, S. O., Yoshida, A., Sun, S. G., Piccirilli, J. A., and Herschlag, D. (1999b). Three metal ions at the active site of the *Tetrahymena* group I ribozyme. *Proc. Natl. Acad. Sci. USA* **96,** 12299–12304.

Shan, S., Kravchuk, A. V., Piccirilli, J. A., and Herschlag, D. (2001). Defining the catalytic metal ion interactions in the *Tetrahymena* ribozyme reaction. *Biochemistry* **40,** 5161–5171.

Silverman, S. K., and Cech, T. R. (1999a). Energetics and cooperativity of tertiary hydrogen bonds in RNA structure. *Biochemistry* **38,** 8691–8702.

Silverman, S. K., and Cech, T. R. (1999b). RNA tertiary folding monitored by fluorescence of covalently attached pyrene. *Biochemistry* **38,** 14224–14237.

Silverman, S. K., Deras, M. L., Woodson, S. A., Scaringe, S. A., and Cech, T. R. (2000). Multiple folding pathways for the P4–P6 RNA domain. *Biochemistry* **39,** 12465–12475.

Sjogren, A. S., Pettersson, E., Sjoberg, B. M., and Stromberg, R. (1997). Metal ion interaction with cosubstrate in self-splicing of group I introns. *Nucleic Acids Res.* **25,** 648–653.

Smalley, M. K., and Silverman, S. K. (2006). Fluorescence of covalently attached pyrene as a general RNA folding probe. *Nucleic Acids Res.* **34,** 152–166.

Smith, D., and Pace, N. R. (1993). Multiple magnesium ions in the ribonuclease P reaction mechanism. *Biochemistry* **32,** 5273–5281.

Stahley, M. R., and Strobel, S. A. (2006). RNA splicing: Group I intron crystal structures reveal the basis of splice site selection and metal ion catalysis. *Curr. Opin. Struct. Biol.* **16,** 319–326.

Steyaert, J. (1997). A decade of protein engineering on ribonuclease T1—Atomic dissection of the enzyme–substrate interactions. *Eur. J. Biochem.* **247,** 1–11.

Strobel, S. A., and Ortoleva-Donnelly, L. (1999). A hydrogen-bonding triad stabilizes the chemical transition state of a group I ribozyme. *Chem. Biol.* **6,** 153–165.

Tuschl, T., Ng, M. M., Pieken, W., Benseler, F., and Eckstein, F. (1993). Importance of exocyclic base functional groups of central core guanosines for hammerhead ribozyme activity. *Biochemistry* **32,** 11658–11668.

Uesugi, S., Miki, H., Ikehars, M., Iwahashi, H., and Kyogoku, Y. A. (1979). A linear relationship between electronegativity of 2′-substituents and conformation of adenine nucleosides. *Tetrahedron Lett.* **42,** 4073–4076.

Verma, S., and Eckstein, F. (1998). Modified oligonucleotides: Synthesis and strategy for users. *Annu. Rev. Biochem.* **67,** 99–134.

Weinger, J. S., Parnell, K. M., Dorner, S., Green, R., and Strobel, S. A. (2004). Substrate-assisted catalysis of peptide bond formation by the ribosome. *Nat. Struct. Mol. Biol.* **11,** 1101–1106.

Yoshida, A., Sun, S. G., and Piccirilli, J. A. (1999). A new metal ion interaction in the *Tetrahymena* ribozyme reaction revealed by double sulfur substitution. *Nat. Struct. Biol.* **6,** 318–321.

Yoshida, A., Shan, S., Herschlag, D., and Piccirilli, J. A. (2000). The role of the cleavage site 2′-hydroxyl in the *Tetrahymena* group I ribozyme reaction. *Chem. Biol.* **7,** 85–96.

Young, B. T., and Silverman, S. K. (2002). The GAAA tetraloop–receptor interaction contributes differentially to folding thermodynamics and kinetics for the P4–P6 RNA domain. *Biochemistry* **41,** 12271–12276.

CHAPTER SEVEN

RNA Crosslinking Methods

Michael E. Harris *and* Eric L. Christian

Contents

Abstract

RNA–RNA crosslinking provides a rapid means of obtaining evidence for the proximity of functional groups in structurally complex RNAs and ribonucleoproteins. Such evidence can be used to provide a physical context for interpreting structural information from other biochemical and biophysical methods and for the design of further experiments. The identification of crosslinks that accurately reflect the native conformation of the RNA of interest is strongly dependent on the position of the crosslinking agent, the conditions of the crosslinking reaction, and the method for mapping the crosslink position. Here, we provide an overview of protocols and experimental considerations for RNA–RNA crosslinking with the most commonly used long- and short-range photoaffinity reagents. Specifically, we describe the merits and strategies for random and site-specific incorporation of these reagents into RNA, the crosslinking reaction and isolation of crosslinked products, the mapping crosslinked sites, and assessment of the crosslinking data.

Center for RNA Molecular Biology, Department of Biochemistry, Case Western Reserve University School of Medicine, Cleveland, Ohio, USA

Methods in Enzymology, Volume 468
ISSN 0076-6879, DOI: 10.1016/S0076-6879(09)68007-1

1. Introduction

RNA–RNA crosslinking is a well-established and widely used method for obtaining secondary and tertiary structural information from structurally complex RNAs and ribonucleoproteins (RNPs) when classical methods of structural analysis such as X-ray crystallography and NMR are not practical (Branch *et al.*, 1985; Butcher and Burke, 1994; Datta and Weiner, 1992; Downs and Cech, 1990; Favre *et al.*, 1998; Harris *et al.*, 1994; Ofengand *et al.*, 1979; Sun *et al.*, 1998; Wassarman and Steitz, 1992; Yaniv *et al.*, 1969; Zwieb and Brimacombe, 1980; Zwieb *et al.*, 1978). Crosslinking is generally achieved using ultraviolet (UV) light to induce the formation of a covalent bond between unmodified RNAs or between RNA and a photoaffinity reagent incorporated randomly or at specific positions in the RNA structure (Elad, 1976). Due to the conformational flexibility and promiscuous chemical reactivity of most photoaffinity reagents, the conformational heterogeneity of RNAs in solution, the inherent ambiguities in mapping the site of crosslinking and other factors (see below), the resolution of this method can be highly variable (typically between 5 and 30 Å) (Huggins *et al.*, 2005; Sergiev *et al.*, 2001; Whirl-Carrillo *et al.*, 2002). While this variability generally precludes the use of crosslinking to demonstrate specific intra- or intermolecular interactions, this method nevertheless provides highly valuable evidence for the proximity of functional groups at the time of UV induction.

The power of crosslinking lies in the ability to rapidly identify potential regions of intra- and intermolecular interaction that can be tested by site-specific modification and subsequent kinetic and thermodynamic studies. In addition, crosslinking can generate a range of distance constraints useful for the development of three-dimensional models of complex RNA structures (Chen *et al.*, 1998; Harris *et al.*, 1994, 1997; Malhotra and Harvey, 1994; Mueller *et al.*, 1997; Pinard *et al.*, 2001). Moreover, the methods used for this approach are relatively straightforward, can be applied under a wide range of experimental conditions, and require relatively little material for analysis. Crosslinking, therefore, is not optimally used as a stand-alone approach but rather as a key analytical tool that can provide an explicit physical context for designing new structure–function experiments and for interpreting structural information when used in conjunction with other biochemical and biophysical methods.

Herein, we describe general experimental protocols for RNA–RNA crosslinking with the most common long- and short-range photoaffinity reagents, as well as strategies for random and site-specific incorporation of these reagents to initially survey and subsequently target regions of potential intra- or intermolecular interaction. We will initially describe the merits and incorporation of individual photoaffinity reagents into RNA (Section 2), followed by methods for the crosslinking reaction and isolation of crosslinked products (Section 3), mapping crosslinked sites (Section 4), and finally suggestions for assessing the validity of individual results (Section 5). The description is designed to be sufficiently general in order to be useful as a guideline for an experimentalist at the graduate level who is considering application of photocrosslinking of RNA in their research. However, a basic understanding of techniques for handling nucleic acids is assumed. Specific reaction conditions will be based on our studies with RNase P RNA; however, such conditions are similar to that in other *in vitro* studies of RNA biochemistry. Indeed, some or all of the methods described in this chapter have been successfully applied in analysis of structural and catalytic RNAs as well as the major cellular RNPs, including the ribosome, and the spliceosome (Christian *et al.*, 1998; Druzina and Cooperman, 2004; Harris *et al.*, 1994, 1997; Huggins *et al.*, 2005; Juzumiene and Wollenzien, 2001; Juzumiene *et al.*, 2001; Kim and Abelson, 1996; Leung and Koslowsky, 1999; Maroney *et al.*, 1996; Montpetit *et al.*, 1998; Newman *et al.*, 1995; Pinard *et al.*, 1999; Pisarev *et al.*, 2008; Podar *et al.*, 1998; Rinke-Appel *et al.*, 2002; Ryan *et al.*, 2004; Sergiev *et al.*, 1997; Wassarman and Steitz, 1992; Yu and Steitz, 1997). Crosslinking is thus likely to remain a very useful tool in deconstructing the structure and function of RNAs and RNPs for many years to come.

2. Synthesis of Modified RNA Crosslinking Substrates

In the absence of detailed structural information, a useful crosslinking strategy is to begin with a broad survey of potential regions of intra- and intermolecular contact followed by site-specific positioning of crosslinking agents, mutagenesis, and biochemical analysis to systematically derive structure–function relationships. Nonspecific crosslinking approaches using UV light or psoralin have been used to established distance constraints in a wide variety of experimental systems involving large structural RNAs and RNP complexes (Behlen *et al.*, 1992; Branch *et al.*, 1985; Brandt and Gualerzi, 1992; Downs and Cech, 1990; Noah *et al.*, 2000; Sawa and Abelson, 1992; Sun *et al.*, 1998; Wassarman and Steitz, 1992; Zwieb *et al.*, 1978). The relatively strict geometric constraints needed for bond formation by these simple and direct methods often produce more accurate structural information than other crosslinking methods when such data have been subsequently evaluated in light of crystal structures (Sergiev *et al.*, 2001; Whirl-Carrillo *et al.*, 2002). However, the number of crosslinks produced using UV light or psoralin is generally too small to be used to broadly survey potential regions of intra- and intermolecular interaction or proximity.

Photoaffinity reagents are generally much more sensitive to UV induction and have fewer geometric constraints for bond formation when incorporated into RNA. Thus, they will tend to produce a much larger number of crosslinks and consequently more information. There is, of course, the potential for a tradeoff between the number of crosslinks observed and accuracy since the increased ability to react may increase the probability of producing misleading crosslinking results, particularly for the longer range photoaffinity reagents (>9 Å). However, such a tradeoff is reasonable in an initial survey of an RNA structure, or when there is an insufficient level crosslinking from which to design experiments.

Long-range photoaffinity reagents placed site specifically at various positions throughout an RNA structure provide a practical means of generating a relatively large set of potential intra- or intermolecular interactions or distance constraints that can be subsequently refined using site-specific analysis of more geometrically restricted photoaffinity reagents or mutagenesis. Accordingly, we will initially describe methods for synthesis and incorporation of azidophenacyl (APA) derivatives, which are useful, highly reactive photoaffinity reagents for long-range RNA crosslinking. This section will be followed by methods for random and site-specific incorporation of the thionucleotide photoagents 6-thioguanosine (6sG) and 4-thiouridine (4sU), which are some of the most convenient and broadly applied short-range (~3 Å) photoaffinity reagents in RNA crosslinking studies.

2.1. Long-range photoaffinity crosslinking with azidophenacyl modifications

APA is an arylazide photoagent (Hixson and Hixson, 1975; Hixson *et al.*, 1980) that positions an azido moiety *ca.* 9 Å from its point of attachment to the RNA and can be used to identify general features of an RNA structure such as position and orientation of RNA helices (Chen *et al.*, 1998; Harris *et al.*, 1994, 1997; Nolan *et al.*, 1993). APA photoagents like their thionucleotide counterparts have the important chemical property of being reactive over a wide range of experimental conditions, but are inert until activated with UV light. Moreover, photocrosslinking with these reagents is very sensitive. Most applications use radiochemical detection, which requires only picomole or even femtomole amounts of material.

APA is conveniently attached to the 5′- or 3′-ends of RNA, and may also be placed at different positions in a molecule by developing a series of circularly permuted RNAs as described in the following section. Attachment of APA to the 5′-end of an RNA involves reaction with a 5′-guanosine monophosphorothioate (GMPS) (Burgin and Pace, 1990; Murray and Atkinson, 1968). Efficiencies of 5′-GMPS incorporation of 70–90% can be achieved with GMPS:GTP ratios between 10:1 and 40:1 in otherwise standard *in vitro* RNA transcription reactions. GMPS is available commercially (Amersham), but can also be made by chemical phosphorylation (Behrman, 2000). Gel-purified 5′-GMPS RNAs are subsequently reacted with azidophenacyl bromide (Fluka/Pierce) by resuspending the 5′-GMPS RNA in 100 μl of 40% methanol; 20 m*M* sodium bicarbonate, pH 9.0; 0.1% SDS; and 5 m*M* azidophenacyl bromide and incubating for 1 h at room temperature. Reactions are then brought to 200 μl by the addition of a solution containing 10 m*M* Tris–HCl, pH 8.0, and 1 m*M* EDTA. The residual uncoupled photoagent is then removed by phenol extraction and the RNA recovered by ethanol precipitation using standard methods.

Attachment of APA to the 3′-end of an RNA can be accomplished by chemically modifying the terminal ribose to contain a primary amine. The amine is subsequently modified with a bifunctional reagent containing both the azido group for crosslinking and a hydroxysuccinimidyl group for reacting with the amine (Oh and Pace, 1994). This modification is accomplished by initially oxidizing the 3′-*cis*-diol of 5–10 μg RNA in 50–100 μl of 3 m*M* sodium periodate, 100 m*M* sodium acetate, pH 5.4, for 1 h at room temperature to create the 3′-dialdehyde. The RNA is subsequently recovered by ethanol precipitation, resuspended in 100 μl of 20 m*M* imidazole, pH 8.0, 5 m*M* $NaCNBH_3$, 1 m*M* ethylene dimaine and incubated at 37 °C for 1 h. $NaBH_4$ is then added to 5 m*M* and the incubation continued for another 10 min. Following reaction with $NaBH_4$, RNAs are precipitated twice with ethanol and reacted with 10 m*M* photoagent [*N*-hydroxysuccinimidyl-4-azidobenzoate (Pierce)] in 50 m*M* HEPES/NaOH, pH 9.0, at room temperature for 1 h in the dark before being recovered by ethanol precipitation. Reactions from this point forward should all be performed in amber tubes and significant care should be taken to minimize exposure to ambient light to prevent preactivation of APA.

As noted above, APA photoaffinity reagents may be positioned at different specific sites within a folded structure through the use of circularly permutated RNAs (cpRNAs) that ideally reposition of the 5′- and 3′-ends of the molecule without significantly altering the primary sequence or the overall three-dimensional structure. The above criteria of maintaining the overall three-dimensional structure obviously limit the positions on an individual molecule where the rearranged 5′- and 3′-ends can be placed. That said, structural RNAs are often sufficiently stabilized by internal interactions to tolerate discontinuities at many positions in the polynucleotide backbone (Guerrier-Takada and Altman, 1992; Pan *et al.*, 1991; Reich *et al.*, 1986; van der Horst *et al.*, 1991; Waugh and Pace, 1993). Thus, with sufficient care cpRNAs can often be constructed to position APA throughout the molecule of interest. In cases where functional cpRNAs cannot be obtained, meaningful crosslinking analysis may still be possible both within or between fragments or independent folding domains of the RNA (Leonov *et al.*, 1999). Crosslinking has also been done from oligonucleotides annealed to a folded RNA target. However, this method necessarily disrupts local secondary and has often produced results in conflict with high-resolution structural data (Sergiev *et al.*, 2001).

Templates for circularly permuted RNAs are generated by PCR amplification of tandem genes that are generally separated by a sufficient length of added sequence to connect the original 5′- and 3′-ends of the molecule without perturbing overall folding or biological activity. Forward and reverse primers that define the 5′- and 3′-ends of the cpRNA and contain promoter sequences and restriction enzyme sites for cloning are directed in the appropriate direction in the upstream and downstream genes to yield a single gene product upon amplification. In choosing the position the 5′-end of the cpRNA it is important to take into account the sequence requirements of the polymerase used during *in vitro* transcription. T7 RNA polymerase optimally requires two consecutive G residues at the 5′-end; however, sufficient amounts of RNA can be obtained from transcripts beginning with GU, GC, and GA (Milligan and Uhlenbeck, 1989). SP6 polymerase, which has a less stringent sequence requirement for initiation than T7, can be used when the efficiency of T T7 RNA polymerase proves insufficient. Also note that since the 3′-end of the cpRNA is formed by runoff transcription, restriction sites that cut at a distance from their recognition site (e.g., *Fok*I or *Bbs*I) must be designed into the downstream PCR primer to avoid loss of RNA coding sequences.

Transcripts of cpRNAs are generated by standard *in vitro* transcription methods. In our bacterial RNase P studies, 2–4 μg of linearized template DNA was combined at room temperature with 40 units of phage T7 RNA polymerase in a 100 μl reaction containing 40 m*M* Tris–HCl, pH 8.0, 1 m*M* spermidine, 5 m*M* dithiothreitol, 0.1% Triton X-100, 20 m*M* magnesium chloride, and 1 m*M* NTPs and incubated for 5–12 h. RNAs are gel-purified by electrophoresis through 4% polyacrylamide/8 *M* urea gels, visualized by UV shadow and passively eluted into 0.3 *M* sodium acetate, 20 m*M* Tris–HCl, pH 8.0, 1 m*M* EDTA, 0.1% sodium dodecyl sulfate overnight. As noted above, efficient 5′-labeling with GMPS will require altering the above conditions with GMPS:GTP ratios between 10:1 and 40:1. Also, note for GMPS-containing RNAs that 1 m*M* dithiothreitol should be included in the elution buffer to inhibit formation of disulfide.

The importance of establishing that cpRNAs reflect the native structure cannot overstated since without such evidence crosslinking studies cannot be interpreted. For ribozymes, this can be achieved in the comparison of the kinetic parameters k_{cat} and K_m of cpRNAs with their native counterparts (Chen *et al.*, 1998; Harris *et al.*, 1994, 1997). Other structural RNAs can be compared on the basis of diagnostic metal ion cleavage, chemical modification, or their binding kinetics to other molecules. cpRNAs altered in catalytic function or some other diagnostic feature are likely to contain structural distortions that limit their experimental utility and thus must be excluded from subsequent structural studies.

2.2. Short-range crosslinking using thionucleosides

Although a wide variety of chemical and photoreagents are available for short-range (~3 Å) RNA–RNA crosslinking, the thionucleotides 6sG and 4sU are preferred due to their simple molecular structure, relative stability, and high reactivity (Christian *et al.*, 1998; Dubreuil *et al.*, 1991; Favre *et al.*, 1998; Sergiev *et al.*, 1997; Sontheimer, 1994). In thionucleotide photoagents, the crosslinking thio moiety is attached directly to the nucleotide base and can thus be used to refine the distance constraints between positions that appear to be proximal by long-range crosslinking or other phylogenetic or biochemical data (Favre *et al.*, 1998). Importantly, 6sG and 4sU differ from their corresponding "parent" nucleoside by only a single atomic substitution, the replacement of a nucleobase oxygen by sulfur, thus reducing the potential for significant perturbation of RNA structure. Exposure of these thionucleotides to UV light produces a sulfur radical that can react efficiently with functional groups that are in proximity. Crosslinking reactions involving these reagents can be very efficient, making it easier to isolate sufficient quantities of crosslinked species for mapping of crosslinked nucleotides and for assessing the extent to which the observed crosslinks reflect the biologically relevant structure.

6sG and 4sU can serve as effective structural probes by either random incorporation or site-specific substitution. Random incorporation of these analogs during *in vitro* transcription provides a rapid means of surveying potential contacts over an entire RNA transcript and can be used as an alternative, or preferably, a complementary strategy to the broad structural probing using APA and cpRNAs. While this approach can generate a large number of crosslinks, it can be difficult to separate individual crosslinked species on polyacrylamide gels and thus pinpoint the site of crosslinking. This difficulty can often be reduced or eliminated by a reduction in the concentration of added thionucleotide in the transcription mixture or by selective deletion or mutation of nucleotide positions contributing to the crosslinking signal (Christian and Harris, 1999).

Site-specific incorporation by transcription or RNA ligation (see below) eliminates both the ambiguity of the source of crosslinking and generally increases the experimental signal. This latter feature is due to the fact that essentially all RNAs in the reaction mixture will contain a photoagent agent at a single position rather than diluted throughout the molecule, allowing for the total crosslinking signal to be distributed among a smaller number of crosslinked species. Site-specific analysis, however, is inherently more labor intensive since each position must be examined independently and generally requires the synthesis of a distinct site-specifically modified RNA in sufficient yield.

Random incorporation of thionucleotides by transcription is achieved by supplementing standard reaction conditions with the thionucleotide triphosphate of interest. In practice, sufficient levels of thionucleotide incorporation often require a reduction in the concentration of the corresponding unmodified nucleotide triphosphate. The relative concentrations of thio- and unmodified nucleotide triphosphate required to produce optimal crosslink will of course vary for individual RNAs and must be determined empirically. Optimally, crosslinking substrates should contain no more than one photoagent per molecule. The shifted absorption maxima for 4sU (330 nm) and 6sG (~340 nm) are generally well separated from the average absorption maximum for RNA (~260 nm) and can be used to quantify levels of incorporation. In our studies, 4sU-containing pre-tRNA crosslinking substrates used to identify potential regions of intermolecular interaction with RNase P RNA are transcribed under the following conditions. Linearized plasmid template (2 μg) is diluted at room temperature into transcription buffer containing 40 m*M* Tris–HCl, pH 7.9, 6 m*M* $MgCl_2$, 2 m*M* spermidine, and 10 m*M* dithiothreitol, 1 m*M* each ATP, CTP, GTP, and 0.1 m*M* each UTP and s^4UTP (USB) in a total volume of 100 μl and incubated overnight at 37 °C. These reactions are terminated by the addition of 100 μl of 10 m*M* Tris–HCl, pH 8.0, and 1 m*M* EDTA, and purified on polyacrylamide gels as described above. Transcripts (20 pmol) are then often subsequently 5′-end-labeled for binding and crosslinking studies with 150 μCi of [γ-^{32}P] ATP (New England Nuclear) and T4 polynucleotide kinase (Life Sciences).

Site-specific incorporation of 6sG and 4sU can be achieved by priming transcription of circularly permuted constructs with the monophosphate form of s^6G (s^6GMP) (Christian *et al.*, 1998) or dinucleotide primers (e.g., s^4UpG) (Milligan and Uhlenbeck, 1989). The advantage of this approach is the ability to use the same set of cpRNAs to compare crosslinking reagents with different levels of geometric constraint. The main drawbacks of this approach are that s^6GMP is not commercially available and must be synthesized by chemical phosphorylation of s^6G (Behrman, 2000) and that priming with dinucleotide primers can be very inefficient. An alternative and more broadly used approach involves site-specific incorporation of 6sG or 4sU into a short oligonucleotide fragment by chemical synthesis and its subsequent joining to remaining sequences of a functional RNA by oligonucleotide-directed RNA ligation (Moore and Sharp, 1992). This approach allows the incorporation of a far greater range of modified nucleotides than can be incorporated by transcriptional initiation and avoids the potential structural complications of crosslinking adjacent to a gap in the polynucleotide backbone. The main drawback of this approach is that oligonucleotide-directed ligation can be inefficient, and thus it may be difficult to obtain sufficient material to map weaker crosslinks or to study biochemically.

Oligonucleotide-directed RNA ligation in our studies is done essentially as described by Moore and Sharp (Moore and Sharp, 1992). In general, synthetic RNA fragments containing the modified nucleotide of interest are joined to a single transcript containing the remaining sequences of the RNA being studied. In large molecules such as RNase P RNA, this is done using cpRNAs beginning immediately downstream of the 3′-end of the synthetic RNA fragment and ending immediately upstream of the fragment's 5′-end. In practice, we obtained significantly greater yields ligating only one of the junctions between the modified oligonucleotide and RNA transcript and that the remaining gap, when carefully positioned, rarely altered the kinetic features of the ribozyme. Ligation of two or more junctions is possible but often requires reoptimization of the reaction conditions. In our studies, synthetic RNA fragments (Dharmacon) are 5′-^{32}P-end-labeled with 150 μCi of [γ-^{32}P] ATP (New England Nuclear) and T4 polynucleotide kinase (Life Sciences) and purified on 22.5% polyacrylamide gels as described above. For the ligation reaction equal amounts (10 pmol) of the end-labeled oligonucleotide, RNA transcript, and a DNA oligonucleotide complementary to the synthetic RNA and the first 20 nucleotides of the transcript are combined in a total volume of 6.8 μl (in distilled water), heated to 60 °C for 2 min and then immediately frozen on dry ice for at least 2 min. Samples are subsequently thawed on wet ice and supplemented with 1 μl 10× T4 ligase buffer (New England Biolabs), 1 μl 50% 8000 g/mol polyethylene glycol (Fluka), 0.5 μl (200 units) RNasin (Promega), and 0.7 μl (280 units) T4 DNA ligase (New England Biolabs). Samples are then incubated for 2 h at 30 °C, combined with an equal volume (10 μl) of gel loading buffer, heated to 90 °C for 2 min, and then purified on polyacrylamide gels as described above. In our experience, 6sG- and 4sU-modified RNAs were more sensitive to preactivation by ambient light than those modified with APA, which again can be minimized by using dark amber tubes.

3. Generation of Crosslinked RNAs

3.1. General considerations of reaction conditions

The central difficulty in crosslinking studies is not the ability to generate crosslinked molecules, but rather the generation of crosslinks that reflect the native conformation or structure of interest. This difficulty stems from the strong tendency of RNA to adopt multiple conformations in solution as well as the sensitivity of RNA structure to mono- and divalent metal ion concentration and temperature (Brion and Westhof, 1997; Pyle and Green, 1995; Treiber and Williamson, 2001; Uhlenbeck, 1995). Non-native intra- or intermolecular interactions can be very stable, sometimes more stable than their biologically relevant counterparts, and can dominate a population of molecules even under optimal experimental conditions. And unfortunately, the efficient photocrosslinking agents so often necessary to generate sufficient crosslinking signal are sure to dutifully report the presence of both correctly folded and misfolded structures.

Ensuring that the presence of a mixed population of RNA structural isomers will not complicate the results is (or at least should be) built upon the often time consuming process of determining how to optimally fold the RNA prior to reaction. Such conditions are best derived from a detailed understanding of the influence of the individual parameters of the experimental system (e.g., mono- and divalent ion concentrations and identity, temperature, pH, and macromolecular concentration) on biological activity. In the course of establishing folding conditions, it is also useful to examine the effect of these parameters on the level of crosslinking observed in the reaction since gaining the highest efficiency possible is important for subsequent identification of crosslinked nucleotides and analysis of the retention of biological activity of the purified crosslinked species. The development of reaction conditions for crosslinking should thus be viewed as a reiterative process of testing and validation.

3.2. Crosslinking photoactivation

In contrast to the considerable experimental preparation note above, the crosslinking reaction itself is a quick and simple process. This procedure will be illustrated here using an example of pre-tRNA crosslinking to bacterial RNase P RNA. However, the conditions used are largely applicable as an outline for RNA crosslinking in general. Purified 5′-^{32}P-end-labeled pre-tRNA containing 4sU and unlabeled bacterial RNase P RNA transcripts are resuspended separately in reaction buffer, in this case (2 *M* ammonium acetate; 50 m*M* Tris–HCl, pH 8.0) for refolding. The RNA-containing solutions are heated to 90 °C for 1 min in a programmable heating block (MJ Research) and then cooled to room temperature using a standard water bath over a period of approximately 20 min. Divalent metal ions, in this example 25 m*M* $CaCl_2$, are then added and the RNAs incubated at 37 °C for 15–30 min to insure that as much of the RNA as possible has attained the native, folded form. Equal volumes of substrate and enzyme RNA are subsequently mixed and incubated for 2 min. In this instance Ca^{2+} is used to replace the optimal metal ion for the reaction, Mg^{2+}, in order to slow the rate of catalysis and permit the assessment of the binding affinity of the substrate (Smith *et al.*, 1992).

Intermolecular crosslinking reactions (100–200 μl) are done under conditions of enzyme excess ([E]/[S] = 10, [E] $\geq K_d$, often 100 nm photoagent-containing pre-tRNA and 1 μM RNase P ribozyme) in order to insure that the majority of the photoagent-modified substrate is bound to the ribozyme. For intermolecular interactions it is also import to demonstrate that the formation of crosslinks are dependent on the presence of the interacting RNA species (here the ribozyme) and occur in a concentration-dependent manner over a broad range of concentrations. The observation of the same crosslinked species at both high and low concentrations of the ribozyme provides evidence that the crosslinks are intermolecular in nature and reflect the structure of high affinity complexes.

We found that optimal crosslinking in our experimental system occurred at 4 °C, and thus the following crosslinking apparatus was assembled on a standard laboratory ice bucket. The most efficient levels of crosslinking were observed when 10–20 μl aliquots were transferred to a parafilm-covered aluminum block and irradiated as separated droplets rather than a single pool of liquid. Aluminum blocks can be obtained from a standard dry-bath incubator and are precooled in ice for at least 1 h prior to the experiment. Parafilm and samples should be placed on the block just before irradiation to minimize dilution or contamination by condensation. The top of the block should be set 2–3 cm below the surface of the ice bucket. A standard ~3 mm thick glass plate is set on top of the bucket over the samples to help filter out shorter wavelengths (<300 nm) of UV light that can damage the RNA sample and produce photoagent-independent crosslinks. In our studies samples were irradiated for 5–15 min at 366 nm at a distance of 3 cm using a model UVP Model UVGL-58 ultraviolet lamp from UVP, Upland, CA. Note, however, that while 6sG is also activated by 366 nm light, the APA-derived crosslinking agents are activated at 302 nm (UVP Model UVM-57). Aliquots are recovered from the block, diluted to 200 μl with 10 m*M* Tris–HCl, pH 8.0, 0.5 m*M* EDTA, 0.3 *M* sodium acetate, then extracted twice with 50%/50% phenol/chloroform and once with chloroform alone and precipitated by addition of three volumes of ethanol.

We observed that significantly greater crosslinking can be achieved by increasing time or lamp intensity. Such efforts, however, must be balanced against damage to the RNA structure and the understanding that the increase in crosslinking yield can also vary significantly from position to position. The conditions above were also chosen in part to slow the rate of reaction in studies of substrate binding to offset the relatively long periods of time, kinetically speaking, required to obtain sufficient levels of crosslinked material. More powerful lamps (e.g., UV Products, B-100AP, 365 nm) generally only reduce the time required by several fold.

Progress in increasing efficiency without damaging the RNA has come from the recent use of nanosecond pulse laser technology to form RNA–RNA crosslinks in bacterial ribosomes (Shapkina *et al.*, 2004). Crosslinks obtained with one (22 ns, 248 nm) pulse showed about fourfold greater yield than that obtained with transilluminator irradiation. Importantly, such crosslinking could be achieved with relatively low levels of UV-induced strand breakage. In contrast to low intensity UV light which leads to the absorption of single photons by bases to induce formation of the S_1 and T_1 excited states (Cadet and Vigny, 1990; Daniels, 1976; Fisher and Johns, 1976), the higher energy of the pulse laser is thought to allow the absorption of two photons to form the higher activation states of H_T and H_S (Budowsky *et al.*, 1986; Cadet and Vigny, 1990). Importantly, bases in H_T and H_S states crosslink more efficiently and may have expanded specificity relative to the excited states produced with low intensity UV (Budowsky *et al.*, 1986). Indeed, new crosslinks could be observed in a system that has been analyzed for decades by UV crosslinking (Shapkina *et al.*, 2004). Pulse laser crosslinking has not as yet been characterized using the photoagents above. However, it is not difficult to imagine the experimental possibilities and potential impact of site-specific photoagents being activated by pulse laser in a rapid quench or stop flow apparatus to monitor changes in secondary and tertiary structure as a function of time. For a group of methods that has not seen much change in recent years, this seems fertile ground for future work.

3.3. Isolation of crosslinked products

Identification and isolation of crosslinked RNAs is accomplished by taking advantage of the significantly reduced mobility of the crosslinked RNAs relative to uncrosslinked RNA on denaturing polyacrylamide gels. Crosslinked RNAs are thus purified on low percentage denaturing polyacrylamide gels (4%, 19:1 acrylamide, bis) and recovered as described above. Gels should contain an adjacent lane in which the photoagent is omitted from the reaction to demonstrate that the formation of the more slowly migrating species depends on presence of the crosslinking reagent and not from adventitious crosslinking due to UV light. An additional lane should contain samples that are not irradiated since crosslinking can occur during sample workup that may not necessarily reflect the functional structure. Intermolecularly crosslinked RNAs such that between RNase P RNA and its pre-tRNA substrate form Y-branched structures that usually migrate slower than lariat forms, and thus can be easily separated from a strong background level of intramolecular crosslinking. That said, intra- and intermolecular crosslinks formed independent of the photoagent should not necessarily be discarded as they may contain useful structural information.

Once a valid intra- or intermolecular crosslink has been identified, the next step is to isolate sufficient quantities of the individual crosslinked species for mapping. A general rule of thumb is that picomole amounts of material are optimal for primer extension mapping and functional studies used for validation. Thus it is necessary to scale the crosslinking reaction up accordingly. For primer extension analysis it is important that the radioactivity of the RNA crosslink be significantly less than the signal generated by primer extension sequencing. This can be achieved by the addition of a small quantity of radiolabeled RNA to the crosslinking reaction. In contrast, direct sequencing of the crosslinked RNA by alkaline hydrolysis (see below) is best done with RNAs with the highest specific activity possible. In this latter case, 5′-^{32}P-end-labeling can be done prior to the crosslinking reactions or to the purified crosslinked species itself. Note, however, that conditions required for the removal of the nonradioactive 5′-terminal phosphate prior to 5′-^{32}P-end-labeling can lead to nicks in the backbone of the RNA crosslink and reduce the quality of the sequencing data. Breaks introduced in the backbone of RNAs labeled prior to the crosslinking reaction, however, will generally not produce crosslinks that comigrate with their full-length counterparts and will be eliminated from the experimental background during purification.

Large-scale crosslinking reactions can be loaded into a continuous well across the entire top of a 35 × 45 cm × 0.4 mm thick sequencing gel and are typically run at 100 W and ~50 °C for sufficient time to separate adjacent crosslinked species by at least 1 cm to avoid cross-contamination when individual bands are physically excised from the gel. Electrophoresis conditions for preparative scale reactions often differ from those used for the initial identification of the crosslink species and may need to be reoptimized to achieve the best degree of separation between crosslinked bans or from uncrosslinked RNA. Crosslinked RNAs are purified from gels as described above with the exception that the scale in terms of both the volumes used for elution and subsequent extraction with phenol and chloroform are much larger. In practice, preparative scale phenol/chloroform extractions are more easily accomplished using 15 ml organic resistant polypropylene tubes (Falcon) and a standard clinical centrifuge. Following extraction we have found that addition of 0.01 μg/μl glycogen as a carrier greatly improves recovery of the relatively low concentration of crosslinked RNAs from larger volumes and does not interfere in subsequent primer extension mapping, alkaline hydrolysis sequencing, or monitoring the crosslinked species for biological activity. More detailed considerations for isolating large crosslinked RNAs (rRNA) has been reviewed by Wollenzien and coworkers (Juzumiene *et al.*, 2001).

4. Mapping of Crosslinked Nucleotides

4.1. Alkaline hydrolysis mapping

As noted above, sites of crosslinking are generally mapped by alkaline hydrolysis or primer extension mapping. In alkaline hydrolysis mapping 5′- or 3′-^{32}P-labeled crosslinked RNAs are treated with base to randomly cleave the phosphate backbone at a level of approximately one site per molecule. Reactions typically involve treating crosslinked RNAs with 0.1 *M* NaOH in 8 *M* urea at 80 °C for 1–10 min and terminating the reaction by the addition of a molar equivalent of HCl or an excess of buffer at neutral pH. When these RNA fragments are separated on polyacrylamide gels a gap is revealed in the sequencing ladder. The sequencing ladder below the gap accurately reflects the length of RNA fragments from the ^{32}P-labeled 5′- or 3′-end up to the point of the crosslink. Labeled RNAs containing a break in the backbone distal to the site of crosslinking, however, will be significantly larger in molecular weight due to covalent attachment of added sequence, and accordingly will migrate much more slowly in the sequencing gel. The first band below the gap thus reflects the break in the RNA backbone immediately 5′ to the site of crosslinking. The position of the crosslink can be identified by simply counting the number of bands from the labeled end of the molecule or by running the alkaline ladder adjacent to lanes of standard dideoxy sequencing. This approach is simple, quick and generally very accurate (Sergiev *et al.*, 2001; Whirl-Carrillo *et al.*, 2002). The main disadvantage is that in larger RNAs the site of crosslinking can be distant from the labeled end of the molecule (>150–200 nts) and thus more difficult to resolve on standard sequencing gels. Alternative, more complex, yet very accurate methods, however, have been developed in the mapping of crosslinks in ribosomal RNA of using RNase H and RNA fingerprinting (Mitchell *et al.*, 1990).

4.2. Primer extension mapping

In primer extension sequencing 5′-^{32}P-labeled primers are annealed to different positions within the crosslinked RNA and extended with reverse transcriptase. Because reverse transcriptase is unable to extend through the site of crosslinking, the last nucleotide incorporated by reverse transcriptase is interpreted as being immediately upstream of the site the crosslink itself. The site of crosslinking is thus derived from the terminal position of the longest primer extension product when compared alongside a standard sequencing reaction of unmodified RNA on polyacrylamide gels. The advantage of this approach is the ability to place primers throughout an RNA structure to optimize resolution of the position of blocked primer extension products. The disadvantage of this approach is that normal RNA structural features can block primer extension to yield misleading results. Indeed, the comparison of crosslinking methods with high-resolution structures has led to the observation that primer extension is the least reliable of the current mapping methods despite its widespread use (Sergiev *et al.*, 2001; Whirl-Carrillo *et al.*, 2002). That said, the majority of interactions or constraints predicted by primer extension are likely to be correct and thus effort should not be placed in avoiding this very useful approach but rather in optimizing reaction conditions and careful interpretation of the results. Indeed, the resolution of an individual distance constraint is often more dependent on how the crosslink is mapped than the photoagent itself (Sergiev *et al.*, 2001; Whirl-Carrillo *et al.*, 2002).

Conditions that have yielded consistently low background levels in primer extension sequencing in our experiments are as follows. Typically, 0.2 pmol of 5′-^{32}P end-labeled sequencing primer is annealed to 0.05–0.2 pmol of crosslinked material in a total volume of 5 μl at 65 °C for 3 min in 50 m*M* Tris–HCl, pH 8.3, 15 m*M* NaCl, and 10 m*M* DTT and set immediately on dry ice. Samples are then thawed on ice and $MgCl_2$ (1 μl) is added to a final concentration of 6 m*M*, followed by the addition of each of the four deoxynucleotides (dATP, dCTP, dGTP, dTTP) to a final concentration of 400 nm. Reactions (8 μl) are initiated by the addition of 2 units (2 μl) of AMV reverse transcriptase (Boehringer Mannheim) and then incubated at 47 °C for 5 min. Reactions are then quenched by the addition of an equal volume of 0.5 *M* NaCl, 20 m*M* EDTA, and 5 μg of glycogen (Boehringer Mannheim) and precipitated in 2.5 volumes of ethanol. Primer extension reactions are then resuspended in 2 μl dH_2O and denatured in the presence of an equal volume of gel loading buffer (95% formamide, 150 m*M* Tris–HCl, pH 8.0, 15 m*M* EDTA, 1 mg/ml each bromophenol blue, xylene cyanol FF) for 3 min at 95 °C. Samples are allowed to cool on ice before loading (2 μl) onto a 6% (19:1) polyacrylamide gel adjacent to standard dideoxy sequencing reaction of uncrosslinked RNA.

5. Assessing the Validity of Crosslinking Data

Determining the structural relevance of a crosslink should include consideration of the following criteria. First, the number and distribution of observed crosslinks should be consistent with that normally observed with a given crosslinking agent. Crosslinking with long-range structure probes such as APA usually involves several adjacent nucleotides in 1–4 distinct regions of the target RNA while the number of nucleotides and crosslinked regions of RNA is significantly reduced in short-range structural probes such as 6sG and 4sU. The general suspicion of a structurally heterogeneous population of RNA should thus be raised when the number or distribution of observed crosslinks exceeds the general guidelines noted above. This should be initially be addressed by a re-examination of renaturing conditions prior to the crosslinking reaction, followed changes in the placement of the crosslinking reagent itself.

Second, the efficiency of crosslinking provides correlative rather than direct evidence for structural proximity or conformational stability. The strong geometrical and chemical requirements for bond formation dictate that the relative proximity of two crosslinked sites is not strictly linked to the level of crosslinking that is actually observed. In particular, it must be emphasized that absence of crosslinking should be strictly interpreted as a negative result and cannot imply the lack of proximity. Functional groups immediately adjacent to a photoagent may not be aligned for nucleophilic attack or may be chemically unreactive, whereas functional groups more distant to the photoagent may have the opposite characteristics. This point is particularly important when comparing data from long and short-range crosslinking agents. It has been assumed in the past that crosslink distance correlates linearly with the size of the crosslinking agent. However, this correlation was not observed when long and short-range crosslinking studies were compared in the context of established crystallographic structures (Sergiev *et al.*, 2001; Whirl-Carrillo *et al.*, 2002). While the absence of correlation may be partially due to experimental error (e.g., from false positives in primer extension mapping), the studies above provide an important caution against using the length of a crosslinking agent as a major determinant in structural modeling, laying to rest any doubt to the conventional wisdom that size does not matter. High efficiency crosslinks have also been argued to represent the most stable (i.e., native) structure in the population. Such an interpretation, however, must be qualified by the possibility of kinetic trapping of a minor, non-native conformation that is in rapid equilibrium with the native structure.

Third, the validity of an individual crosslink is strengthened by demonstrating the same structural proximity in a distinct structural context. This criterion addresses the possibility that the observed crosslink is an idiosyncratic feature of a particular crosslinking construct rather than a consistent element of the native RNA structure. The most direct approach to addressing this concern is to determine whether the same nucleotides or regions of RNA structure become crosslinked regardless of which of the nucleotides or regions of RNA in question contains the crosslinking agent (Chen *et al.*, 1998; Harris *et al.*, 1997). The demonstration of reciprocal crosslinks from different photoagents or under different experimental conditions provides further support that the observed results are not due to the perturbation of the native structure. Generality of the crosslinking results can also be established by reproducing the crosslink in a homologous RNA (Chen *et al.*, 1998; Christian *et al.*, 1998; Harris *et al.*, 1994, 1997; Noah *et al.*, 2000). Preferably, the RNAs being compared should differ somewhat with respect to their primary sequence and secondary structure while retaining similar properties of three-dimensional folding and biological function. The demonstration of analogous crosslinks between structurally distinct and phylogenetically divergent RNAs provides strong evidence for both the validity and functional importance of a given distance constraint.

Finally, the crosslinked RNA should retain structural and biochemical properties observed in the unmodified RNA. Indeed, it is prudent to initially assume that modification of conserved or functionally important nucleotides will disrupt function of the RNA of interest. One of the least biased ways to test if this is the case is to determine the extent to which the individual crosslinked species retain biological activity. Since the function of

RNAs is tied directly to their structure, significant changes to structure are likely to be reflected in properties such as substrate binding or catalytic rate. Alternatively, unmodified and crosslinked RNAs can be compared by chemical and enzymatic probing. Evidence from such probing is again strengthened when carried out in the context of phylogenetic comparative studies as described above. Ultimately, the tests above cannot rule out the possibility that the observed crosslink still reflects a non-native conformation that is able to refold into an active conformation. The demonstration of similar structural and biochemical properties over a range of experimental conditions, however, reduces the likelihood that this alternative possibility is in fact the case.

REFERENCES

Behlen, L. S., Sampson, J. R., and Uhlenbeck, O. C. (1992). An ultraviolet light-induced crosslink in yeast tRNA(Phe). *Nucleic Acids Res.* **20,** 4055–4059.

Behrman, E. J. (2000). An improved synthesis of guanosine 5′-monothiophosphate. *J. Chem. Res.* **9,** 446–447.

Branch, A. D., Benenfeld, B. J., and Robertson, H. D. (1985). Ultraviolet light-induced crosslinking reveals a unique region of local tertiary structure in potato spindle tuber viroid and HeLa 5S RNA. *Proc. Natl. Acad. Sci. USA* **82,** 6590–6594.

Brandt, R., and Gualerzi, C. O. (1992). Ribosomal localization of the mRNA in the 30S initiation complex as revealed by UV crosslinking. *FEBS Lett.* **311,** 199–202.

Brion, P., and Westhof, E. (1997). Hierarchy and dynamics of RNA folding. *Annu. Rev. Biophys. Biomol. Struct.* **26,** 113–137.

Budowsky, E. I., Axentyeva, M. S., Abdurashidova, G. G., Simukova, N. A., and Rubin, L. B. (1986). Induction of polynucleotide-protein cross-linkages by ultraviolet irradiation. Peculiarities of the high-intensity laser pulse irradiation. *Eur. J. Biochem.* **159,** 95–101.

Burgin, A. B., and Pace, N. R. (1990). Mapping the active site of ribonuclease P RNA using a substrate containing a photoaffinity agent. *EMBO J.* **9,** 4111–4118.

Butcher, S. E., and Burke, J. M. (1994). A photo-cross-linkable tertiary structure motif found in functionally distinct RNA molecules is essential for catalytic function of the hairpin ribozyme. *Biochemistry* **33,** 992–999.

Cadet, J., and Vigny, P. (1990). The Photochemistry of Nucleic Acids. John Wiley & Sons, New York, NY.

Chen, J. L., Nolan, J. M., Harris, M. E., and Pace, N. R. (1998). Comparative photocross-linking analysis of the tertiary structures of *Escherichia coli* and *Bacillus subtilis* RNase P RNAs. *EMBO J.* **17,** 1515–1525.

Christian, E. L., and Harris, M. E. (1999). The track of the pre-tRNA 5′ leader in the ribonuclease P ribozyme–substrate complex. *Biochemistry* **38,** 12629–12638.

Christian, E. L., McPheeters, D. S., and Harris, M. E. (1998). Identification of individual nucleotides in the bacterial ribonuclease P ribozyme adjacent to the pre-tRNA cleavage site by short-range photo-cross-linking. *Biochemistry* **37,** 17618–17628.

Daniels, M. (1976). Excited States of the Nucleic Acids: Bases, Mononucleosides and Mononucleotides. Academic Press, New York, NY.

Datta, B., and Weiner, A. M. (1992). Cross-linking of U2 snRNA using nitrogen mustard. Evidence for higher order structure. *J. Biol. Chem.* **267,** 4497–4502.

Downs, W. D., and Cech, T. R. (1990). An ultraviolet-inducible adenosine-adenosine cross-link reflects the catalytic structure of the Tetrahymena ribozyme. *Biochemistry* **29,** 5605–5613.

Druzina, Z., and Cooperman, B. S. (2004). Photolabile anticodon stem-loop analogs of tRNAPhe as probes of ribosomal structure and structural fluctuation at the decoding center. *RNA* **10,** 1550–1562.

Dubreuil, Y. L., Expert-Bezancon, A., and Favre, A. (1991). Conformation and structural fluctuations of a 218 nucleotides long rRNA fragment: 4-thiouridine as an intrinsic photolabelling probe. *Nucleic Acids Res.* **19,** 3653–3660.

Elad, D. (1976). Photochemistry and Photobiology of Nucleic Acids, Vol. 1: Chemistry. Academic Press, New York, NY.

Favre, A., Saintome, C., Fourrey, J. L., Clivio, P., and Laugaa, P. (1998). Thionucleobases as intrinsic photoaffinity probes of nucleic acid structure and nucleic acid–protein interactions. *J. Photochem. Photobiol. B* **42,** 109–124.

Fisher, G. J., and Johns, H. E. (1976). Pyrimidine photodimers. Academic Press, New York, NY.

Guerrier-Takada, C., and Altman, S. (1992). Reconstitution of enzymatic activity from fragments of M1 RNA. *Proc. Natl. Acad. Sci. USA* **89,** 1266–1270.

Harris, M. E., Nolan, J. M., Malhotra, A., Brown, J. W., Harvey, S. C., and Pace, N. R. (1994). Use of photoaffinity crosslinking and molecular modeling to analyze the global architecture of ribonuclease P RNA. *EMBO J.* **13,** 3953–3963.

Harris, M. E., Kazantsev, A. V., Chen, J. L., and Pace, N. R. (1997). Analysis of the tertiary structure of the ribonuclease P ribozyme–substrate complex by site-specific photoaffinity crosslinking. *RNA* **3,** 561–576.

Hixson, S. H., and Hixson, S. S. (1975). *p*-Azidophenacyl bromide, a versatile photolabile bifunctional reagent. Reaction with glyceraldehyde-3-phosphate dehydrogenase. *Biochemistry* **14,** 4251–4254.

Hixson, S. H., Brownie, T. F., Chua, C. C., Crapster, B. B., Satlin, L. M., Hixson, S. S., Boyce, C. O., Ehrich, M., and Novak, E. K. (1980). Bifunctional aryl azides as probes of the active sites of enzymes. *Ann. N. Y. Acad. Sci.* **346,** 104–114.

Huggins, W., Ghosh, S. K., Nanda, K., and Wollenzien, P. (2005). Internucleotide movements during formation of 16 S rRNA–rRNA photocrosslinks and their connection to the 30 S subunit conformational dynamics. *J. Mol. Biol.* **354,** 358–374.

Juzumiene, D. I., and Wollenzien, P. (2001). Arrangement of the central pseudoknot region of 16S rRNA in the 30S ribosomal subunit determined by site-directed 4-thiouridine crosslinking. *RNA* **7,** 71–84.

Juzumiene, D., Shapkina, T., Kirillov, S., and Wollenzien, P. (2001). Short-range RNA–RNA crosslinking methods to determine rRNA structure and interactions. *Methods* **25,** 333–343.

Kim, C. H., and Abelson, J. (1996). Site-specific crosslinks of yeast U6 snRNA to the pre-mRNA near the 5′ splice site. *RNA* **2,** 995–1010.

Leonov, A. A., Sergiev, P. V., Dontsova, O. A., and Bogdanov, A. A. (1999). Directed introduction of photoaffinity reagents in internal segments of RNA. *Mol. Biol. (Mosk.)* **33,** 1063–1073.

Leung, S. S., and Koslowsky, D. J. (1999). Mapping contacts between gRNA and mRNA in trypanosome RNA editing. *Nucleic Acids Res.* **27,** 778–787.

Malhotra, A., and Harvey, S. C. (1994). A quantitative model of the *Escherichia coli* 16 S RNA in the 30 S ribosomal subunit. *J. Mol. Biol.* **240,** 308–340.

Maroney, P. A., Yu, Y. T., Jankowska, M., and Nilsen, T. W. (1996). Direct analysis of nematode *cis-* and *trans*-spliceosomes: A functional role for U5 snRNA in spliced leader addition *trans*-splicing and the identification of novel Sm snRNPs. *RNA* **2,** 735–745.

Milligan, J. F., and Uhlenbeck, O. C. (1989). Synthesis of small RNAs using T7 RNA polymerase. *Methods Enzymol.* **180,** 51–62.

Mitchell, P., Osswald, M., Schueler, D., and Brimacombe, R. (1990). Selective isolation and detailed analysis of intra-RNA cross-links induced in the large ribosomal subunit of *E. coli*: A model for the tertiary structure of the tRNA binding domain in 23S RNA. *Nucleic Acids Res.* **18,** 4325–4333.

Montpetit, A., Payant, C., Nolan, J. M., and Brakier-Gingras, L. (1998). Analysis of the conformation of the 3′ major domain of *Escherichia coli*16S ribosomal RNA using site-directed photoaffinity crosslinking. *RNA* **4,** 1455–1466.

Moore, M. J., and Sharp, P. A. (1992). Site-specific modification of pre-mRNA: The 2′-hydroxyl groups at the splice sites. *Science* **256,** 992–997.

Mueller, F., Stark, H., van Heel, M., Rinke-Appel, J., and Brimacombe, R. (1997). A new model for the three-dimensional folding of *Escherichia coli* 16 S ribosomal RNA. III. The topography of the functional centre. *J. Mol. Biol.* **271,** 566–587.

Murray, A. W., and Atkinson, M. R. (1968). Adenosine 5′-phosphorothioate. A nucleotide analog that is a substrate, competitive inhibitor, or regulator of some enzymes that interact with adenosine 5′-phosphate. *Biochemistry* **7,** 4023–4029.

Newman, A. J., Teigelkamp, S., and Beggs, J. D. (1995). snRNA interactions at 5′ and 3′ splice sites monitored by photoactivated crosslinking in yeast spliceosomes. *RNA* **1,** 968–980.

Noah, J. W., Shapkina, T., and Wollenzien, P. (2000). UV-induced crosslinks in the 16S rRNAs of *Escherichia coli, Bacillus subtilis* and *Thermus aquaticus* and their implications for ribosome structure and photochemistry. *Nucleic Acids Res.* **28,** 3785–3792.

Nolan, J. M., Burke, D. H., and Pace, N. R. (1993). Circularly permuted tRNAs as specific photoaffinity probes of ribonuclease P RNA structure. *Science* **261,** 762–765.

Ofengand, J., Liou, R., Kohut, J. III, Schwartz, I., and Zimmermann, R. A. (1979). Covalent cross-linking of transfer ribonucleic acid to the ribosomal P site. Mechanism and site of reaction in transfer ribonucleic acid. *Biochemistry* **18,** 4322–4332.

Oh, B. K., and Pace, N. R. (1994). Interaction of the 3′-end of tRNA with ribonuclease P RNA. *Nucleic Acids Res.* **22,** 4087–4094.

Pan, T., Gutell, R. R., and Uhlenbeck, O. C. (1991). Folding of circularly permuted transfer RNAs. *Science* **254,** 1361–1364.

Pinard, R., Heckman, J. E., and Burke, J. M. (1999). Alignment of the two domains of the hairpin ribozyme–substrate complex defined by interdomain photoaffinity crosslinking. *J. Mol. Biol.* **287,** 239–251.

Pinard, R., Lambert, D., Heckman, J. E., Esteban, J. A., Gundlach, C. W., Hampel, K. J., Glick, G. D., Walter, N. G., Major, F., and Burke, J. M. (2001). The hairpin ribozyme substrate binding-domain: A highly constrained D-shaped conformation. *J. Mol. Biol.* **307,** 51–65.

Pisarev, A. V., Kolupaeva, V. G., Yusupov, M. M., Hellen, C. U., and Pestova, T. V. (2008). Ribosomal position and contacts of mRNA in eukaryotic translation initiation complexes. *EMBO J.* **27,** 1609–1621.

Podar, M., Zhuo, J., Zhang, M., Franzen, J. S., Perlman, P. S., and Peebles, C. L. (1998). Domain 5 binds near a highly conserved dinucleotide in the joiner linking domains 2 and 3 of a group II intron. *RNA* **4,** 151–166.

Pyle, A. M., and Green, J. B. (1995). RNA folding. *Curr. Opin. Struct. Biol.* **5,** 303–310.

Reich, C., Gardiner, K. J., Olsen, G. J., Pace, B., Marsh, T. L., and Pace, N. R. (1986). The RNA component of the *Bacillus subtilis* RNase P. Sequence, activity, and partial secondary structure. *J. Biol. Chem.* **261,** 7888–7893.

Rinke-Appel, J., Osswald, M., von Knoblauch, K., Mueller, F., Brimacombe, R., Sergiev, P., Avdeeva, O., Bogdanov, A., and Dontsova, O. (2002). Crosslinking of

4.5S RNA to the *Escherichia coli* ribosome in the presence or absence of the protein Ffh. *RNA* **8,** 612–625.

Ryan, D. E., Kim, C. H., Murray, J. B., Adams, C. J., Stockley, P. G., and Abelson, J. (2004). New tertiary constraints between the RNA components of active yeast spliceosomes: A photo-crosslinking study. *RNA* **10,** 1251–1265.

Sawa, H., and Abelson, J. (1992). Evidence for a base-pairing interaction between U6 small nuclear RNA and 5′ splice site during the splicing reaction in yeast. *Proc. Natl. Acad. Sci. USA* **89,** 11269–11273.

Sergiev, P. V., Lavrik, I. N., Wlasoff, V. A., Dokudovskaya, S. S., Dontsova, O. A., Bogdanov, A. A., and Brimacombe, R. (1997). The path of mRNA through the bacterial ribosome: A site-directed crosslinking study using new photoreactive derivatives of guanosine and uridine. *RNA* **3,** 464–475.

Sergiev, P. V., Dontsova, O. A., and Bogdanov, A. A. (2001). Study of ribosome structure using the biochemical methods: Judgment day. *Mol. Biol. (Mosk.)* **35,** 559–583.

Shapkina, T., Lappi, S., Franzen, S., and Wollenzien, P. (2004). Efficiency and pattern of UV pulse laser-induced RNA–RNA cross-linking in the ribosome. *Nucleic Acids Res.* **32,** 1518–1526.

Smith, D., Burgin, A. B., Haas, E. S., and Pace, N. R. (1992). Influence of metal ions on the ribonuclease P reaction. Distinguishing substrate binding from catalysis. *J. Biol. Chem.* **267,** 2429–2436.

Sontheimer, E. J. (1994). Site-specific RNA crosslinking with 4-thiouridine. *Mol. Biol. Rep.* **20,** 35–44.

Sun, J. S., Valadkhan, S., and Manley, J. L. (1998). A UV-crosslinkable interaction in human U6 snRNA. *RNA* **4,** 489–497.

Treiber, D. K., and Williamson, J. R. (2001). Beyond kinetic traps in RNA folding. *Curr. Opin. Struct. Biol.* **11,** 309–314.

Uhlenbeck, O. C. (1995). Keeping RNA happy. *RNA* **1,** 4–6.

van der Horst, G., Christian, A., and Inoue, T. (1991). Reconstitution of a group I intron self-splicing reaction with an activator RNA. *Proc. Natl. Acad. Sci. USA* **88,** 184–188.

Wassarman, D. A., and Steitz, J. A. (1992). Interactions of small nuclear RNA's with precursor messenger RNA during *in vitro* splicing. *Science* **257,** 1918–1925.

Waugh, D. S., and Pace, N. R. (1993). Gap-scan deletion analysis of *Bacillus subtilis* RNase P RNA. *FASEB J.* **7,** 188–195

Whirl-Carrillo, M., Gabashvili, I. S., Bada, M., Banatao, D. R., and Altman, R. B. (2002). Mining biochemical information: Lessons taught by the ribosome. *RNA* **8,** 279–289.

Yaniv, M., Favre, A., and Barrell, B. G. (1969). Structure of transfer RNA. Evidence for interaction between two non-adjacent nucleotide residues in tRNA from *Escherichia coli*. *Nature* **223,** 1331–1333.

Yu, Y. T., and Steitz, J. A. (1997). Site-specific crosslinking of mammalian U11 and u6atac to the 5′ splice site of an AT–AC intron. *Proc. Natl. Acad. Sci. USA* **94,** 6030–6035.

Zwieb, C., and Brimacombe, R. (1980). Localisation of a series of intra-RNA cross-links in 16S RNA, induced by ultraviolet irradiation of *Escherichia coli* 30S ribosomal subunits. *Nucleic Acids Res.* **8,** 2397–2411.

Zwieb, C., Ross, A., Rinke, J., Meinke, M., and Brimacombe, R. (1978). Evidence for RNA–RNA cross-link formation in *Escherichia coli* ribosomes. *Nucleic Acids Res.* **5,** 2705–2720.

CHAPTER EIGHT

Chemical Probing of RNA and RNA/Protein Complexes

Zhili Xu *and* Gloria M. Culver

Contents

Abstract

Chemical probing is widely used as a rapid approach for assessing RNA structure, folding, and function. In this chapter, we outline procedures for handling and using chemicals commonly used to probe nucleic acids. Detailed experimental conditions and design for footprinting and modification interference are presented herein. Protocols for RNA extraction, normalization, primer extension, and data evaluation are also provided. The methods described are designed to aid in the study of large RNAs, but with slight modifications are applicable to smaller RNAs.

Department of Biology, University of Rochester, Rochester, New York, USA

Methods in Enzymology, Volume 468
ISSN 0076-6879, DOI: 10.1016/S0076-6879(09)68008-3

1. Introduction

RNA and ribonucleoprotein particles (RNPs) are involved in a large number of cellular processes for which intricate tertiary structures are often required. As a consequence of base pairing and higher order interactions, RNA molecules can form preferred secondary, tertiary, and quaternary structures. RNAs can also undergo conformational changes upon changes in solution conditions and binding of small molecules and proteins. Thus, both folding and function of RNA and RNPs are dynamic. Obtaining molecular details of RNA folding, RNA/protein interactions, and ultimately characterization of the biological functions requires detailed understanding of their structures. Although recent advances in NMR spectroscopy and X-ray crystallography have led to a substantial increase in atomic-level insights into RNA tertiary structure (see, e.g., Nagaswamy *et al.*, 2001; Yusupov *et al.*, 2001), less direct structural methods, such as chemical probing, remain essential for assessing RNA structure and dynamics.

Chemical probing is a rapid and powerful technique that can provide nucleotide-level resolution of almost each residue in an RNA molecule. By subjecting RNAs or RNPs to chemical modification of the nucleobases or to phosphodiester strand scission, differences in the accessibility of specific RNA residues to the probing molecule are revealed. Traditionally, such experiments have been used to map secondary structure within RNAs, to map RNA/protein interactions and to monitor conformational changes in Waston–Crick base pairing and/or backbone accessibility. More recently, chemical probing has been used, in combination with other methods to identify nucleotides that are involved in multiple biological processes including: ligand binding, folding pathways, and three-dimensional structures of increasingly large and complicated RNA molecules (see, e.g., Adilakshmi *et al.*, 2005; Moazed *et al.*, 1986; Shiraishi and Shimura, 1988).

The most commonly applied strategies for chemical probing can be grouped into three broad categories: footprinting, modification interference, and directed hydroxyl-radical probing. Footprinting can provide experimental information on the solvent accessibility of individual nucleotides in an RNA molecule. It is often used to define secondary structure, study RNA folding processes, or identify nucleotides that are protected from chemical attack in the presence of a ligand (Hingorani *et al.*, 1989; Shiraishi and Shimura, 1988; van Belkum *et al.*, 1988). Modification interference, on the other hand, is often used to identify nucleotides whose modification prevent ligand binding or function (see McGregor *et al.*, 1999). While the above mentioned two methods can provide information of direct contact or interaction sites, directed hydroxyl-radical probing informs on the immediate environment of the ligand which is the source of hydroxyl radicals. As detailed methods of directed hydroxyl-radical probing have been more recently described (Culver and Noller, 1998), footprinting and modification interference will be the major focus of this chapter. The focus herein is on larger RNAs. However, it should be noted that other methods, such as nucleotide analog interference mapping (NAIM) and selective 2′-hydroxyl acylation analyzed by primer extension (SHAPE), are ideal for study of relatively smaller RNAs (Waldsich, 2008; Wilkinson *et al.*, 2006). While these approaches are highly generalizable where applicable, we introduce specific examples to aid in illustrating these methods.

Detailed information of the common chemical probes for nucleic acid analyses and their target sites can be found in Brunel and Romby (2000). Of the chemical reagents used to probe RNA structure, dimethyl sulfate (DMS), 3-ethoxy-a-ketobutyraldehyde (kethoxal), 1-cyclohexyl-3-(2-morpholinoethyl)carbodiimide metho-*p*-toluene sulfonate (CMCT) and hydroxyl radicals will be discussed below. Chemical probes can be divided into two classes: base-specific probes and sugar-phosphate backbone probes. Base-specific probes are sensitive to secondary structure; they can react with the Waston–Crick face of the nucleobases and thus have much more greater reactivity with fully unpaired nucleotides of RNA than with base-paired residues. However, the dynamic nature of many RNA structures can lead to a "grayscale" of reactivity, thus, yielding additional information but also complicating experimental interpretation. DMS, kethoxal, and CMCT belong to the base-specific category. For the purposes of this chapter, we will only discuss RNA modifications that are readily detectable by primer extension, thus other modifications promoted by these reagents that can be detected by other methods (D'Alessio, 1982) will be omitted for clarity (Brunel and Romby, 2000). DMS modifies the N1 position of adenine and N3 position of cytosine; kethoxal modifies the N1 and N2 positions of guanine creating a cyclic adduct, whereas CMCT modifies the N3 position of uridine and N1 position of guanine. Hydroxyl radicals belong to the class of probes that cleave the sugar-phosphate backbone. They are unique among the mentioned probes because they are small enough to diffuse into structural regions that may not be accessible to base probes and they can act on the backbone of paired or unpaired nucleotides. Thus, this is a sequence and secondary structure-unbiased reagent. These reagents are highly informative but not necessarily benign. Thus, information regarding handling of these reagents is provided. In addition, we discuss how to optimize conditions for chemical probing, provide general guidance on how to set up a chemical probing experiment, primer extension and how to interpret data. Protocols using DMS, kethoxal, CMCT, and hydroxyl radical for identification of protein footprinting and similar protocols for modification interference will be separately detailed.

2. Materials

2.1. Reagents

The following reagents unless specified should be stored at room temperature:

- CMCT is purchased from Fluka analytical, Cat #29469.
- DMS is purchased from Sigma Aldrich, 99+% grade, Cat #D186309.
- Hydrogen peroxide 30% Ultrex II ultrapure solution is purchased from J.T. Baker, Cat #515501, store at 4 °C.
- Kethoxal is purchased from Research Organics, Cat #8005K, store at −20 °C.
- Ammonium Iron(II) sulfate hexahydrate is purchased from Sigma Aldrich, 99.996% grade, Cat #203505-256.
- Ascorbic acid is purchased from Fluka Biochemika, Cat #29469.
- Thiourea is purchased from Sigma Aldrich, ACS grade, Cat #T8656.
- Sephadex G-25 is purchased from Sigma Aldrich, DNA grade, Cat #S5897-25G.
- Sephacryl S200 is purchased from Sigma Aldrich, Cat #S200HR, store at 4 °C.
- 2-Mercaptoethanol is purchased from Sigma Aldrich, Cat #M6250, store at 4 °C.
- Phenol is purchased from Amresco, Cat #946, store at 4 °C.
- Chloroform is purchased from Fisher, ACS grade, Cat #295-500.
- [α-^{32}P]-dTTP is purchased from Perkin Elmer, 10 mCi/ml, Cat #BLU005A001MC, store at −20 °C.
- Reverse transcriptase, Avian Myeloblastosis Virus (AMV), is purchased from Seikagku American, Inc., aliquot and store at −20 °C.
- Urea is purchased from MP biomedicals, Cat #821530.
- Nikkol (Octaethyleneglycol Mono-*n*-dodecyl Ether) is purchased from Calbiochem, Cat #3055-98-9, store at 4 °C.
- dNTP set is purchased from ABgene, Cat #AB-0315/A, store at −20 °C.
- ddNTP set is purchased from GE Healthcare, Cat #27-2045-01, store at −20 °C.
- Glycogen is purchased from Sigma, Cat #G8751, store at 4 °C.

2.2. Solutions

- 500 m*M* potassium borate (pH 7.0), pH with potassium hydroxide pellets to 500 m*M* boric acid.
- *Precipitation mix.* 85% ethanol, 0.1 *M* NaOAC (pH 6.0), 25 μg/ml glycogen. Prepare just prior to use.

- *RNA extraction buffer.* 0.3 *M* NaOAC (pH 6.0), 0.5% SDS, 5 m*M* EDTA (pH 8.0).
- *CMCT modification buffer.* 50 m*M* potassium borate (pH 8.0), 10 m*M* $MgCl_2$, 100 m*M* NH_4Cl.
- *Primer extension precipitation mix.* 67% ethanol and 85 m*M* NaOAC (pH 6.5). Prepare just prior to use.
- *4.5× hybridization buffer.* 225 m*M* K^+-HEPES (pH 7.0) and 450 m*M* KC1, store at −80 °C.
- *ddNTP stocks.* Both 1.5 and 67 *μM* stocks each of all four individual nucleotides (pH 7.5), store at −80 °C.
- *10× extension buffer.* 1.3 *M* Tris–HC1 (pH 8.5), 100 m*M* $MgCl_2$, and 100 m*M* dithiothreitol, store at −80 °C.
- *dNTP(-T).* 110 m*M* dATP, 110 m*M* dCTP, 110 m*M* dGTP, 6 m*M* dTTP, store at −80 °C.
- *Chase.* 1 m*M* dATP, 1 m*M* dCTP, 1 m*M* dGTP, and 1 m*M* dTTP, store at −80 °C.
- *Primer extension loading buffer.* 7 *M* urea, 0.1× TBE, 0.03% xylene cyanol, and 0.03% bromphenol blue.

2.3. Equipment and supplies

- VWR 3MM blotting paper, Cat #28298-020
- Bio-Rad Bio-Spin disposable chromatography columns, Cat #732-6008
- Fisherbrand gel-loading aerosol-barrier tips, Cat #02-707-172
- Thermal Fisher Scientific Nanodrop 2000

2.4. General guidance

It is strongly recommended to wear appropriate protective equipment, including gloves, lab coat, and eye protection throughout the experiment. It is essential to use RNase-free equipment and reagents. Thus, all the glassware should be washed, rinsed with 70% ethanol and then covered and baked at 200 °C.

3. Handling of the Chemicals

DMS. DMS is extremely harmful due to its irreversible modification to nucleobases and absorbance through the skin. In addition, DMS is toxic through inhalation. Manipulation should be carried out in a fume hood until DMS is quenched (see below). DMS liquid waste along with microcentrifuge tubes and pipette tips coming into contact with DMS should be mixed with 5 N sodium hydroxide before disposal into a dedicated waste box.

The waste should then be treated/removed as instructed by your local environment health and safety department professionals. DMS should be stored away from oxidizing reagents in a well-ventilated, dry location and in a secondary container.

Kethoxal. Kethoxal is comparatively less mutagenic than DMS because the modifications generated by kethoxal are reversible. However, this necessitates that care must be taken to stabilize this adduct on RNA of interest (see below). Kethoxal is a viscous liquid with a light yellow color and has a unique odor. A 1:10 stock solution is made by diluting the kethoxal with 20% ethanol and is stored at -20 °C. Before each use, it is advisable to vortex the 1:10 stock solution and to check carefully that a uniform solution has been formed. While adding kethoxal to the sample, pipetting up and down or vortexing is suggested to obtain even modification.

CMCT. CMCT decomposes upon exposure to moisture and should be stored under Argon. Unlike the chemicals described above, which are not sensitive to pH, CMCT is only reactive in solutions with pH > 8.0. To limit base hydrolysis of RNA and to allow CMCT modification, reactions are carried out in CMCT modification buffer containing potassium borate and are adjusted to pH 8.0.

Hydroxyl radicals. Hydroxyl radicals can be generated by synchrotron radiolysis of water (Sclavi *et al.*, 1997), peroxynitrous acid disproportion (Chaulk *et al.*, 2000), or by Fenton chemistry which produces hydroxyl radicals from hydrogen peroxide which will be described in detail herein as it is the most accessible among the three (Tullius, 1991). Ammonium Iron(II) sulfate which should be stored dry and away from light is a common reagent used to form Fe (II)-EDTA. Ascorbic acid, used as a reducing reagent, is weakly light sensitive and subject to oxidation, and thus should also be stored away from light and in an air tight container. All of the solutions made by these reagents should be prepared freshly, prior to each modification experiment.

4. Optimization of the Chemical Probing Reactions

Modified sites in the RNAs are generally identified by primer extension analysis. The described modification of the nucleobases and strand scissions stall/stop the reverse transcriptase used in primer extension. As a control, untreated RNA should be included along with the modified RNA samples. Comparison of the unmodified with the modified RNA allows identification of regions where reverse transcription pauses at naturally modified nucleotides and in regions with particularly stable secondary structure elements and thus these can be distinguished from modification incorporated by the probing reagents. These natural stops will be evident in the sequencing gels even in control experiments that are untreated with chemical probes and thus will not be misinterpreted.

To facilitate evaluation and to limit secondary effects, the probing conditions are such that only a fraction of the RNA templates are modified. This allows read through to assess longer RNA templates in a single primer extension. Ideally one modification per molecule would be appropriate and one could calculate the number of possible modified nucleobases and the concentration of modification reagents necessary to achieve this. However, since the length, folding, structure, and complexity of RNAs and RNPs can greatly differ, the ratio of chemical to RNA needed to achieve appropriate modification is best determined empirically. This can be estimated by titrating modification reagent to a fixed concentration of RNA followed by primer extension analysis. Overmodification will result in the disappearance of longer extended products as a function of modification reagent concentration and levels of modification that are too low will become indistinguishable from the unmodified control. An appropriate "middle ground" is generally desired for most experiments.

Here, optimization experiments to determine the concentration of DMS needed to modify 16S rRNA for modification interference analysis is provided as an example. Preparation of 30S subunits, 16S rRNA, and protein is described in Xu *et al.* (2008). Similar experiments can be conducted for optimizing incubation temperature and time, for any given probe/RNA (RNP) ratio. A similar protocol may be followed to optimize conditions for other probes, such as kethoxal and CMCT.

In this experiment, 2 μl of serially diluted DMS (960, 800, 640, 480, 320, and 216 m*M*; made by diluting the 10.6 *M* stock into 95% ethanol) is added to 30 pmol of heat activated 16S rRNA (1 pmol/μl in 30 μl standard 30S subunit reconstitution buffer; see Culver and Noller, 2000). This series yields a final concentration of DMS ranging from 14.4 to 60 m*M*. Samples were mixed by pipetting up and down and incubated on ice for 5 min. The probing reactions are quenched by adding 1.6 μl of 14.4 *M* 2-mercaptoethanol (discussed in Section 5.2.2). After ethanol precipitation and RNA extraction (see Section 6), the RNA can be analyzed by primer extension analysis (see Section 8) to evaluate signals. Higher concentrations of DMS may show better signal levels of modification, but the function or structure of RNA may be significantly impaired by the chemicals. Thus, the amount of chemical probe added to the RNA sample needs to be adjusted to allow a compromise between signal intensity with the integrity of function and structure of the RNA and to limit secondary effects due to overmodification. In our case, we tested ability of modified 16S rRNA to assemble into functional 30S subunits as a parameter to determine an amount of DMS needed for these experiments. A final concentration of 50 m*M* DMS proved to be ideal for modification of naked 16S rRNA where half of the RNA molecules were still competent to assemble into functional 30S particles.

5. Procedure of Chemical Probing

Footprinting and modification interference are slightly different in terms of procedure. Generally, many of the steps are shared; however, modification interference has an extra step of separation of functional particles from nonfunctional particles, that is, a selection step prior to analysis. In addition, quenching is more vital for modification interference than for footprinting as emphasized in Section 5.2.2. A simplified flowchart (Fig. 8.1) to reveal the basic steps of footprinting and modification interference is provided.

5.1. Procedure of footprinting

Footprinting traditionally is used as a protection assay in which protein- or ligand-binding sites or conformational changes associated with protein/ligand binding or folding are identified (Schlegl *et al.*, 1997; von Ahsen and Noller, 1993). Here, we describe how to use footprinting to analyze the interaction of a protein with an RNA (16S rRNA) or RNP (30S subunits).

5.1.1. Protein/30S subunit complex formation

Choosing a proper molar ratio of RNA/RNP to protein is a critical step in this experiment. While a molar ratio of protein to RNA that is too low may not yield a good signal, molar ratio of protein to RNA which is too high increases the possibility of nonspecific binding and thus spurious results. The optimal ratio may be determined empirically with the aid of a titration experiment. This will provide simple means of assessing protein association with your RNA or RNP and allows a wide range of protein concentrations to be tested prior to probing. These same approaches can be used to optimize binding buffers for compatibility with probing conditions. Prior to probing, a complex of RNA and protein/ligand using various concentrations is formed and then the complex is separated from the unbound protein. Affinity chromatography or size exclusion chromatography, if your RNA has sufficiently larger molecular weight relative to the protein, can be used to separate the complex from free protein prior to assaying protein binding. This can be assayed by simply performing a slot blot where nucleic acid and associated proteins are retained on a filter but free protein is not. The blot can then be analyzed by western detection of protein. For many of our experiments, the size of 16S rRNA and 30S subunits allows rapid separation of bound/unbound protein using Sephacryl S-200 (see below).

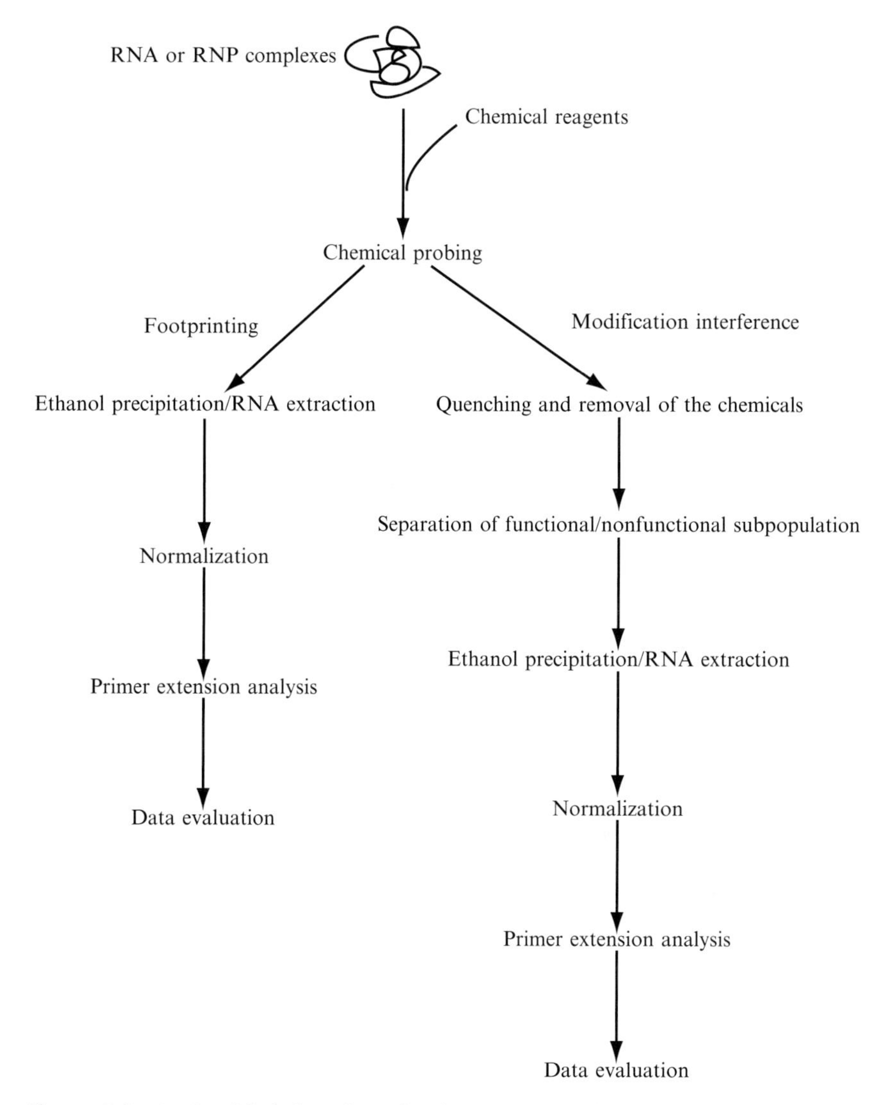

Figure 8.1 A simplified flowchart for footprinting and modification interference experiments.

5.1.2. Modification of the complex

It is important to emphasize that the protocols presented here are optimized for the study of large RNAs. Chemical modification conditions may need to be varied to accommodate smaller RNAs or less complex RNPs and thereby avoiding over modification (see above). We choose 0 °C as our probing temperature to minimize secondary effects, although this can sacrifice experimental speed. Increased temperature during chemical probing usually increases the speed at which modification is observed. An exception to this, hydroxyl-radical cleavage appears to occur very rapidly independent of temperature, but may also be the most likely to yield secondary effects as an initial cleavage of the backbone may then result in localized unfolding and subsequent "secondary" cleavage events.

DMS. To a 100 μl reaction mixture of protein/30S complex (40 pmol of 30S subunits and 80 pmol of KsgA, e.g., see Xu *et al.*, 2008), 3 μl of 867 m*M* DMS (made by mixing 10 μl stock 10.6 *M* DMS with 120 μl 95% ethanol) is added to achieve a final concentration of 34 m*M*. The whole reaction is incubated on ice for 1 h. The reaction is then terminated by adding 4 μl of 14.4 *M* 2-mercaptoethanol which is thoroughly mixed. This in essence quenches the DMS in the reaction, thereby, stopping the reaction and inactivating the DMS simultaneously. The reaction then proceeds as described in Fig. 8.1.

Kethoxal. To a 100 μl reaction mixture of protein/30S complex (40 pmol of 30S subunits and 80 pmol of KsgA, e.g., see Xu *et al.*, 2008), 4 μl of kethoxal (1:10 stock) is added to achieve final reaction condition of 0.4% kethoxal. The reaction is allowed to proceed on ice for 1 h. The reaction is then stopped and the modification of adduct on the RNA is stabilized by adding 5.5 μl 500 m*M* potassium borate to a final concentration of 25 m*M*. Since kethoxal modification is a reversible reaction, it is very important to keep 25 m*M* potassium borate present in all subsequent steps to stabilize the kethoxal adduct. Therefore, all solutions from here on contain 25 m*M* potassium borate in addition to its other components. For example, the precipitation mix for kethoxal modification contains 330 m*M* NaOAC, 10 mg/ml glycogen, and 25 m*M* potassium borate in 95% ethanol. The pellet recovered after RNA extraction should be resuspended in 25 m*M* potassium borate instead of water (see Section 6).

CMCT. CMCT reaches its optimal activity at pH 8.0 and 100 m*M* NH_4Cl. If the complex formation buffer has a different pH, and salt concentration, then, the complex needs to be equilibriated in CMCT modification buffer (50 m*M* potassium borate, pH 8.0, 10 m*M* $MgCl_2$, and 100 m*M* NH_4Cl) prior to modification. This can be achieved by a simple buffer exchange using a Sephadex G-25 sepharose spin column which is equilibrated with CMCT modification buffer. Sepharose G-25 is equilibrated in CMCT modification buffer as directed by the manufacturer (Sigma Aldrich) and made to approximately a 50% slurry. A 2 ml volume of sepharose G-25 is then packed in a Bio-Spin disposable column. The column void volume is then equilibrated with CMCT modification buffer by centrifugation in a tabletop Eppendorf 5702 centrifuge at $1000\times g$ for 3.5 min. Once the column is equilibrated, protein/30S subunit complex (40 pmol of 30S subunits and 80 pmol of KsgA, e.g., see Xu *et al.*, 2008) is loaded onto a prepared column and centrifuged at 1000×*g* for 3.5 min. Under these standardized conditions, small ions are retained in the column matrix while complexes are found in the void volume and thus are now in CMCT modification buffer. 100 μl of 100 m*M* CMCT in CMCT modification buffer is added to the treated protein/30S subunit complex. The reaction is carried out on ice for 1 h followed by ethanol precipitation which will terminate the reaction.

Hydroxyl radical. Two microliter of freshly made 100 m*M* Ammonium Iron(II) sulfate in water is mixed together with 2 μl of 200 m*M* EDTA (pH 8.0) to form Fe(II)-EDTA complexes such that the ratio is 1 Fe(II) to 2 EDTA. For a 100 μl protein/30S subunit complex (40 pmol of 30S subunits and 80 pmol of KsgA, e.g., see Xu *et al.*, 2008), 2 μl of 500 m*M* fresh ascorbic acid, 4 μl of the above Fe(II)-EDTA mixture, and 2 μl of 5% hydrogen peroxide are added to the side of the microcentrifuge tube sequentially. The final concentration of each chemical is 10 m*M*, 2 m*M*, and 0.1%, respectively. The reaction is then initiated by centrifugation to drive droplets of reactants from the side of the tube into the solution containing the RNA. Quenching is achieved by adding 50 μl of freshly prepared 0.1 *M* thiourea followed by a 10 min incubation on ice. All steps (such as centrifugation) should be performed at 4 °C, and incubations should be carried out on ice. Some chemicals, such as Tris, glycerol, sugars, and alcohols are shown to inhibit hydroxyl-radical-mediated cleavage (Tullius *et al.*, 1987). Thus, these chemicals should be avoided in a hydroxyl-radical probing reaction.

5.2. Procedure of modification interference

One critical requirement of modification interference is to physically separate the active and inactive subpopulations of complexes according to the study purpose. As the separation step may require RNA function, modification interference can be adapted to explore a number of biochemical features of RNA. For example, this technique has been used to identify residues in 30S subunits (von Ahsen and Noller, 1995) and 23S rRNA (Bocchetta *et al.*, 1998) important for tRNA binding. It has also been used to reveal nucleotides in 16S rRNA involved in codon–anticodon interaction (Yoshizawa *et al.*, 1999) and S15 binding (Batey and Williamson, 1996a,b). The experimental procedure described below is for a somewhat complicated modification interference scheme in which the target RNA is part of a large complex that contains many proteins and modification interference is performed at different stages of protein association. For example, 16S rRNA could be modified prior to reconstitution with ribosomal proteins to form functional 30S subunits. This would allow a means to assess 16S rRNA residues important for 30S subunit assembly. Subsequently, modified 16S rRNA that retains its ability to form 30S subunits and modified 16S rRNA that cannot assemble properly into 30S subunit can be physically separated through a tRNA-binding selection (von Ahsen and Noller, 1995). After RNA extraction, *t*otal population of *m*odified RNA (TM), bound modified RNA (bound), and unbound modified RNA (unselected) are normalized prior to primer extension analysis, and the relative intensity of the modified residues in each pool can then be compared.

5.2.1. Chemical probing of RNA modification interference

As described above, the probing condition should be such that it yields only a low level of modification per molecule and it needs to be adjusted according to the temperature, pH, buffer conditions, etc. Moreover, in modification interference generally a lower level of modification is tolerable; thus, conditions should again be determined empirically. As described earlier, a level of modification that allowed approximately half of the total RNA to retain its ability to assemble into a functional 30S subunit (as determined by tRNA binding) is deemed to be appropriate for such an experiment. For example, 240 pmol of 16S rRNA (1 pmol/μl) in 240 μl buffer is used to begin the experiment. Forty microliter of the 16S rRNA mixture is taken out as the unmodified control, and 2.9 μl of 3.5 *M* DMS (made by diluting 4 μl of 10.6 *M* stock into 8 μl 95% ethanol) is added to the mixture and incubated for 5 min on ice. Subsequently, DMS is quenched and removed together with quenching reagents (see below), prior to reconstitution of 30S subunits and selection by tRNA binding.

5.2.2. Quenching of the reaction and removal of chemical reagents

An effective quench is critical to keep constant levels of modification in the molecule and to limit modification throughout subsequent stages of binding/selection. However, the reagents used to terminate a probing reaction should not interfere with overall subsequent RNA–protein interactions or the overall functionality of RNA. For example, DMS activity can be quenched by either 2-mercaptoethanol or 5 N sodium hydroxide. The latter will induce hydrolysis of RNA backbone. Therefore, it should be excluded from use, and 2-mercaptoethanol should be used instead to terminate the reaction. Thus, 10 μl of 14.4 *M* 2-mercaptoethanol is added to 202.9 μl DMS containing reaction mix for final concentration of 709 m*M* 2-mercaptoethanol to quench the reactivity of DMS. The chemicals and quenching reagent can also be removed if required and this can be carried out in multiple ways: such as dialysis, gel filtration, or ultrafiltration. We have successfully used gel filtration, a size exclusion method, to eliminate the DMS and 2-mercaptoethanol from modified 16S rRNA. The 202.9 μl of reaction mixture is loaded onto two calibrated Sepharose G-25 columns (see above for preparation of the column), 101.5 μl of modified 16S rRNA mixture per column, and centrifuged at 1000×*g* for

3.5 min. Under these standardized conditions, DMS and 2-mercaptoethanol remain in the column matrix and 16S rRNA is recovered in the flow-through. The recovered 106 μl sample is collected and kept on ice until proceeding to the next step (see below).

5.2.3. Selection of functional population

As stated above, the selection step is fundamental to the entire approach and ideally should completely separate active species from inactive species. Consequently, this selection step defines the success of the modification interference study and requires considerable optimization. A variety of methods, such as affinity chromatography, can be used to separate the modified functional complexes from the modified nonfunctional complexes. A derivatized ligand for the functional complex is often a good means for affinity selection. However, many other possible approaches are also available. Batey *et al.* (Batey and Williamson, 1996a,b) used gel electrophoresis to separate RNA–S15 complexes from the free RNA. Also, one can use sedimentation velocity gradients to separate modified 30S subunits that can reassociate with 50S subunits to form 70S subunits from those cannot (Pulk *et al.*, 2006). As mentioned above, affinity chromatography can be used with a derivatized ligand as a selection scheme. This principle was demonstrated with a derivatized tRNA and a selection for functional 30S subunits (von Ahsen and Noller, 1995). Similar selection schemes can be devised to monitor assembly of rRNAs into functional particles. Again, these selection schemes are target/RNA dependent and need to be adjusted accordingly.

6. RNA Extraction

This step aims to separate RNA from protein or solution components that may affect reverse transcription during the primer extension analysis. For both modification interference and footprinting experiments after modification and quenching, RNA is recovered by addition of a 2.5-fold volume of precipitation mix. Samples are mixed by vortexing, and then quick frozen in a dry ice/ethanol bath for 10 min followed by centrifuging for 10 min at 16,200×g (these are the conditions for all subsequent centrifugation steps described herein) in a microcentrifuge. Ethanol is carefully decanted, taking care to pour so as not to disturb the pellet. The pellet is redissolved in 200 μl RNA extraction buffer. RNA extraction can be performed at room temperature until the final precipitation steps. Samples are shaken on an Eppendorf mixer for 5 min followed by addition of 200 μl water-saturated phenol (Tris-buffered, pH 7.5), and continued shaking for 5 min. Samples are centrifuged for 5 min and the phenol phase is carefully removed from the bottom of the tube by using a gel-loading tip. This step allows the extracted RNA to remain in a tube that has been "bathed" in phenol and thus aids in maintaining the integrity of the RNA. Phenol extraction is repeated two additional times in the same eppendorf tube. After the third phenol extraction, 200 μl chloroform is added followed by shaking for 3 min, and phases are separated by centrifugation for 5 min. Chloroform is removed from the bottom of the tube in an identical manner as phenol and the chloroform extraction is repeated. To precipitate the extracted RNA, a 2.5-fold volume of ice-cold 95% ethanol is added to samples, and mixed by gentle vortexing. The samples are then placed in a dry ice/ethanol bath for 10 min and then centrifuged as above. Ethanol is carefully and slowly poured out of the tube. Pellets are washed by addition of 1000 μl ice-cold 70% ethanol followed by centrifugation for 1 min at 4 °C. Ethanol is carefully decanted from the tube. The samples are pulse centrifuged and any residual ethanol is removed using a fine (gel-loading) pipette tip, taking care to avoid disturbing the pellet. Pellets are air dried in a hood with the tubes loosely covered with parafilm for 10 min and resuspended in 30 μl of water. The resulting RNA samples are then normalized to the same concentration (see below).

7. Normalization of the RNA Sample

From the beginning of the experiment to the final step of RNA extraction, the RNA samples have undergone multiple steps and therefore it is difficult to retain equal amounts of RNA in each reaction. Because modifications and changes therein are usually identified by comparison, it is critical to normalize the RNA concentration in each sample. Due to the limited volume of sample (usually only 30 μl) and the fragile nature of RNA, the best method is to use a microvolume spectrophotometer, such as Nanodrop (Thermo Scientific) to measure the OD_{260} of the RNA sample. Optical density readings with Nanodrop require only 1–2 μl of sample to yield an accurate reading. As an example, the samples achieved from modification interference after RNA extraction are diluted 1:20 in water to measure OD_{260}, and then the RNA concentration are obtained before normalization with water.

8. Primer Extension Analysis

Primer extension is an extremely sensitive technique for detecting cleavages and modifications in RNA. After annealing of synthetic DNA oligomers, which have a reverse complementary sequence to that of the target RNA, reverse transcriptase generates cDNA products. Generally, the cDNA products are internally radiolabeled or fluorescently end-labeled such that products can be visualized by autoradiography or fluorescent imaging after resolving on a sequencing gel. The fluorescently end-labeled products can also be analyzed by capillary gel electrophoresis (Mitra *et al.*, 2008). Recent developments in fluorescent labels make it possible to substitute hazardous and expensive radioisotope labeling for this "cleaner" technology (Ying *et al.*, 2007). For longer RNAs that require extension of multiple primers, internal incorporation of radioisotope or fluorescently labeled dNTP is preferable to yield longer reads than available from end-labeled primers. Since primers bind to the target RNA with different affinities, optimal primer concentrations must be determined empirically; these typically fall within the range from 0.01 to 0.5 μM final concentration. Suitable primer lengths are generally around 17 nucleotides. Primer stocks are prepared in water and stored at -20 °C. Dideoxy sequencing reactions of unmodified template RNA should accompany primer extension analysis of modified samples to allow specific nucleotides to be identified. Alternatively, a base hydrolysis or other cleavage ladder can accompany the extension if the products are end-labeled (Brunel and Romby, 2000).

The following protocol uses radioisotope to label the cDNA products which are resolved on a sequencing gel. Hybridization mixture is prepared by mixing equal volume of 4.5× hybridization buffer and primer (see above) such that 1 μl 4.5× hybrid buffer and 1 μl primer are prepared for a reaction. Assuming the number of samples to be analyzed is n, the number of reactions for which hybridization mixture should be prepared is $(n + 3)$ (2 "extra" for dideoxy sequencing reactions if only two sequencing lanes are used as is generally sufficient, the 1 extra allowing for pipetting inaccuracy). To each 2.5 μl sample of approximately 0.6 μM RNA, 2 μl of hybridization/primer mixture is added and mixed by brief vortexing and pulse centrifugation at room temperature. Samples are placed in a floating rack and incubated in a water bath at 92 °C for 60 s, the water bath is allowed to slowly cool for approximately 10 min to 47 °C. During hybridization, the extension mix should be prepared. For each sample, 2/3 μl 10× extension buffer, 1/3 μl dNTP(-T), 3/5 μl [α-^{32}P]dTTP, 2/5 μl water, and 1/10 μl undiluted reverse transcriptase are needed. Thus, the extension mix can be made by adding $(n + 3)$-fold volume of each reagent to a microcentrifuge tube and mixed by brief vortexing and pulse centrifugation at room temperature. Since this solution is now radioactive, care must be taken in handling, opening tubes, etc., throughout this procedure. The extension mix is kept on ice until needed. Once the temperature of the water bath has reached 47 °C, samples are centrifuged briefly at room temperature to collect any condensed sample. For the sequencing reactions, 1 μl of the appropriate 1.5 μM stock of ddNTP is added to the appropriate tube at this time. To each sample, 2 μl extension mix is added and mixed by multiple pipetting up and down. Samples are briefly pulse centrifuged at room temperature and incubated in a 42 °C dry-blot incubator for 30 min. Next, samples are centrifuged briefly at room temperature again before 1 μl chase is added to the side of each tube. For the sequencing reactions, a 1 μl of the appropriate 67 μ*M* stock of ddNTP is now added to the side of the appropriate tube to supplement the chase reaction. Samples are centrifuged at room temperature and then incubated at 42 °C for 15 min. To each sample, 120 μl primer extension precipitation mix is added and samples are vortexed thoroughly before incubation at room temperature for 10 min. Samples are centrifuged at room temperature for 10 min at 16,200×*g* and the supernatant is removed carefully using a fine gel-loading aerosol pipette tip. Removal of all ethanol at this stage is very important; however, the pellet will likely be invisible and must not be disturbed. Pellets are air dried for 5 min and resuspended in 20 μl primer extension loading buffer with extensive vortexing. Samples are heated to 92 °C for 2 min and quickly chilled on ice, and 1.5 μl is loaded on a 6% acrylamide, 6 *M* urea, 1× TBE sequencing gel (60 cm × 20 cm × 0.25 mm), which is run at 35 mA for 2 h with an aluminum heat-diffusing plate. The gel is transferred to VWR blotting paper and dried at 80 °C for 30 min. Autoradiography of the gel with autoradiography film or a phosphoimager is typically carried out overnight at room temperature. However, the length of time required for appropriate exposure varies for different primers, and intensifying screens may be used although this can lead to a less "crisp" autoradiogram.

9. Data Evaluation

For cDNA products analyzed by capillary electrophoresis, CAFA is available at https://simtk.org/home/cafa for evaluation of the data (Mitra *et al.*, 2008). Since many reactions are resolved on a sequencing gel, we discuss how to evaluate these data below. Figure 8.2 shows a representative autoradiogram of primer extension of the target RNA from a kethoxal footprinting experiment of protein/30S subunits. Modification of the RNA results in appearance of bands, or increased intensity of bands above background bands in the autoradiogram. These new bands correspond to abortive cDNA transcripts ending at the position of modification of the RNA template. The autoradiogram in Fig. 8.2 shows two sequencing lanes (A and G) followed by three experimental lanes. Lane 1 corresponds to unmodified 30S subunits, which was not subjected to chemical probing. Any bands appearing in lane 1 are not due to modification and are considered as the "background" for the other lanes. Such bands that are generally present in all lanes and an example of such is indicated as + in Fig. 8.2. Lanes 2 and 3 are kethoxal-modified 30S subunits and kethoxal-modified

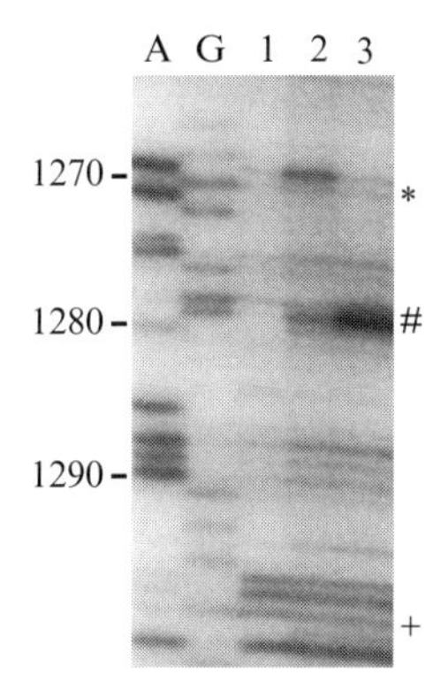

Figure 8.2 Kethoxal footprinting of 16S rRNA in 30S ribosome subunits and protein/30S complexes. The primer used in this extension is complementary to nucleotides 1376–1391 of *Escherichia coli* 16S rRNA. A and G are dideoxy sequencing lanes with 16S rRNA as the template. Lane 1, unmodified 30S subunits; lane 2, 30S subunits with kethoxal; lane 3, protein/30S complex with kethoxal. Band marked with "+" is considered as the "background" for the other lands. Band marked with "*" indicates that such a nucleotide is protected from chemical attack when the protein is bound to 30S subunits. Band marked with "#" indicates an example of enhancement as described in the text.

protein/30S complex, respectively. Additional bands that are not observed in lane 1 identify positions of modification by kethoxal and correspond to the positions of guanine residues in the sequencing lanes. Since reverse transcriptase cannot incorporate complementary nucleotide opposite the modified nucleotide (or at the cleavage site), the modified "band" appears one nucleotide 3′ of the actual modified site. Their relative intensities correspond to the extent of modification at each site in 16S rRNA as available in 30S subunits. Difference between the band intensities in lane 2 and lane 3 indicate the changes of nucleotides accessibility due to the protein/ligand association. Some nucleotides are protected from modification by protein association, for instance, the nucleotide marked as * in Fig. 8.2, whereas some nucleotides are more susceptible to chemical modification upon protein binding, that is, have enhanced reactivity upon protein binding. Enhanced reactivity is often interpreted as resulting from a conformational change in the RNA upon ligand binding. As an example, an enhancement is marked as # in Fig. 8.2. Either primer extension data can be annotated manually or programs such as SAFA (Laederach *et al.*, 2008) can also be used for automated analysis. Solution hydroxyl-radical probing yields a cleavage ladder where stretches of residues tend to be protected upon ligand binding. These cleavage data are generally interpreted very similarly to base-specific modification data, as discussed above.

10. Summary

Given the limitations and difficulties associated with crystallography and NMR, chemical probing offers a simple and precise way of analyzing RNA structure. Furthermore, chemical probing of RNA and RNA/protein complexes can yield large data sets and thus detailed information about RNA structure and interactions can be achieved by using the two described chemical probing strategies: footprinting and modification interference.

Acknowledgments

The authors are grateful to Deepika Calidas, Nathan Napper, Biswajoy Roy-Chaudhuri, Keith Connolly, Hiram Lyon, and Neha Gupta for their constant support and critical reading of this chapter. We also thank Dr. Laura Dutca for helpful discussions.

References

Adilakshmi, T., Ramaswamy, P., and Woodson, S. A. (2005). Protein-independent folding pathway of the 16S rRNA 5′ domain. *J. Mol. Biol.* **351,** 508–519.

Batey, R. T., and Williamson, J. R. (1996a). Interaction of the *Bacillus stearothermophilus* ribosomal protein S15 with 16 S rRNA. I. Defining the minimal RNA site. *J. Mol. Biol.* **261,** 536–549.

Batey, R. T., and Williamson, J. R. (1996b). Interaction of the *Bacillus stearothermophilus* ribosomal protein S15 with 16 S rRNA. II. Specificity determinants of RNA–protein recognition. *J. Mol. Biol.* **261,** 550–567.

Bocchetta, M., Xiong, L., and Mankin, A. S. (1998). 23S rRNA positions essential for tRNA binding in ribosomal functional sites. *Proc. Natl. Acad. Sci. USA* **95,** 3525–3530.

Brunel, C., and Romby, P. (2000). Probing RNA structure and RNA–ligand complexes with chemical probes. *Methods Enzymol.* **318,** 3–21.

Chaulk, S. G., Pezacki, J. P., and MacMillan, A. M. (2000). Studies of RNA cleavage by photolysis of *N*-hydroxypyridine-2(1*H*)-thione. A new photochemical footprinting method. *Biochemistry* **39,** 10448–10453.

Culver, G. M., and Noller, H. F. (1998). Directed hydroxyl radical probing of 16S ribosomal RNA in ribosomes containing Fe(II) tethered to ribosomal protein S20. *RNA* **4,** 1471–1480.

Culver, G. M., and Noller, H. F. (2000). *In vitro* reconstitution of 30S ribosomal subunits using complete set of recombinant proteins. *Methods Enzymol.* **318,** 446–460.

D'Alessio, J. M. (1982). RNA sequencing. *In* "Gel Electrophoresis of Nucleic Acids: A Practical Approach," (D. Rickwood and B. D. Hames, eds.), pp. 173–197. IRL Press, Oxford, England.

Hingorani, V. N., Chang, L. F., and Ho, Y. K. (1989). Chemical modification of bovine transducin: Probing the GTP-binding site with affinity analogues. *Biochemistry* **28,** 7424–7432.

Laederach, A., Das, R., Vicens, Q., Pearlman, S. M., Brenowitz, M., Herschlag, D., and Altman, R. B. (2008). Semiautomated and rapid quantification of nucleic acid footprinting and structure mapping experiments. *Nat. Protoc.* **3,** 1395–1401.

McGregor, A., Murray, J. B., Adams, C. J., Stockley, P. G., and Connolly, B. A. (1999). Secondary structure mapping of an RNA ligand that has high affinity for the MetJ repressor protein and interference modification analysis of the protein–RNA complex. *J. Biol. Chem.* **274,** 2255–2262.

Mitra, S., Shcherbakova, I. V., Altman, R. B., Brenowitz, M., and Laederach, A. (2008). High-throughput single-nucleotide structural mapping by capillary automated footprinting analysis. *Nucleic Acids Res.* **36,** e63.

Moazed, D., Stern, S., and Noller, H. F. (1986). Rapid chemical probing of conformation in 16 S ribosomal RNA and 30 S ribosomal subunits using primer extension. *J. Mol. Biol.* **187,** 399–416.

Nagaswamy, U., Gao, X., Martinis, S. A., and Fox, G. E. (2001). NMR structure of a ribosomal RNA hairpin containing a conserved CUCAA pentaloop. *Nucleic Acids Res.* **29,** 5129–5139.

Pulk, A., Maivali, U., and Remme, J. (2006). Identification of nucleotides in *E. coli* 16S rRNA essential for ribosome subunit association. *RNA* **12,** 790–796.

Schlegl, J., Gegout, V., Schlager, B., Hentze, M. W., Westhof, E., Ehresmann, C., Ehresmann, B., and Romby, P. (1997). Probing the structure of the regulatory region of human transferrin receptor messenger RNA and its interaction with iron regulatory protein-1. *RNA* **3,** 1159–1172.

Sclavi, B., Woodson, S., Sullivan, M., Chance, M. R., and Brenowitz, M. (1997). Time-resolved synchrotron X-ray "footprinting", a new approach to the study of nucleic acid structure and function: Application to protein–DNA interactions and RNA folding. *J. Mol. Biol.* **266,** 144–159.

Shiraishi, H., and Shimura, Y. (1988). Functional domains of the RNA component of ribonuclease P revealed by chemical probing of mutant RNAs. *EMBO J.* **7,** 3817–3821.

Tullius, T. D. (1991). DNA footprinting with the hydroxyl radical. *Free Radic. Res. Commun.* **12–13**(Pt. 2), 521–529.

Tullius, T. D., Dombroski, B. A., Churchill, M. E., and Kam, L. (1987). Hydroxyl radical footprinting: A high-resolution method for mapping protein–DNA contacts. *Methods Enzymol.* **155,** 537–558.

van Belkum, A., Verlaan, P., Kun, J. B., Pleij, C., and Bosch, L. (1988). Temperature dependent chemical and enzymatic probing of the tRNA-like structure of TYMV RNA. *Nucleic Acids Res.* **16,** 1931–1950.

von Ahsen, U., and Noller, H. F. (1993). Footprinting the sites of interaction of antibiotics with catalytic group I intron RNA. *Science* **260,** 1500–1503.

von Ahsen, U., and Noller, H. F. (1995). Identification of bases in 16S rRNA essential for tRNA binding at the 30S ribosomal P site. *Science* **267,** 234–237.

Waldsich, C. (2008). Dissecting RNA folding by nucleotide analog interference mapping (NAIM). *Nat. Protoc.* **3,** 811–823.

Wilkinson, K. A., Merino, E. J., and Weeks, K. M. (2006). Selective 2′-hydroxyl acylation analyzed by primer extension (SHAPE): Quantitative RNA structure analysis at single nucleotide resolution. *Nat. Protoc.* **1,** 1610–1616.

Xu, Z., O'Farrell, H. C., Rife, J. P., and Culver, G. M. (2008). A conserved rRNA methyltransferase regulates ribosome biogenesis. *Nat. Struct. Mol. Biol.* **15,** 534–536.

Ying, B. W., Fourmy, D., and Yoshizawa, S. (2007). Substitution of the use of radioactivity by fluorescence for biochemical studies of RNA. *RNA* **13,** 2042–2050.

Yoshizawa, S., Fourmy, D., and Puglisi, J. D. (1999). Recognition of the codon–anticodon helix by ribosomal RNA. *Science* **285,** 1722–1725.

Yusupov, M. M., Yusupova, G. Z., Baucom, A., Lieberman, K., Earnest, T. N., Cate, J. H., and Noller, H. F. (2001). Crystal structure of the ribosome at 5.5 Å resolution. *Science* **292,** 883–896.

CHAPTER NINE

RNA Folding During Transcription: Protocols and Studies

Terrence N. Wong *and* Tao Pan

Contents

Department of Biochemistry & Molecular Biology, University of Chicago, Chicago, Illinois, USA

Methods in Enzymology, Volume 468
ISSN 0076-6879, DOI: 10.1016/S0076-6879(09)68009-5

Abstract

RNA folds during transcription in the cell. Compared to most *in vitro* studies where the focus is generally on Mg^{2+}-initiated refolding of fully synthesized transcripts, cotranscriptional RNA folding studies better replicate how RNA folds in a cellular environment. Unique aspects of cotranscriptional folding include the 5′- to 3′-polarity of RNA, the transcriptional speed, pausing properties of the RNA polymerase, the effect of the transcriptional complex and associated factors, and the effect of RNA-binding proteins. Identifying strategic pause sites can reveal insights on the folding pathway of the nascent transcript. Structural mapping of the paused transcription complexes identifies important folding intermediates along these pathways. Oligohybridization assays and the appearance of the catalytic activity of a ribozyme either in *trans* or in *cis* can be used to monitor cotranscriptional folding under a wide range of conditions. In our laboratory, these methodologies have been applied to study the folding of three highly conserved RNAs (RNase P, SRP, and tmRNA), several circularly permuted forms of a bacterial RNase P RNA, a riboswitch (thiM), and an aptamer-activated ribozyme (glmS).

1. Introduction

The Mg^{2+}-initiated refolding of fully synthesized RNA transcripts has been the primary experimental method used to investigate RNA folding pathways (Draper *et al.*, 2005; Misra *et al.*, 2003; Sosnick and Pan, 2003; Treiber and Williamson, 2001; Woodson, 2002, 2005). These studies have shown that many RNAs fold through a rugged landscape kinetically dominated by the formation of and the subsequent escape from long-lived folding intermediates. Due to the small number of distinct nucleotide bases and the relative simplicity of Watson–Crick base pairing, RNA molecules exhibit a strong propensity to form non-native structures. Because of the thermodynamic strength of the Watson–Crick base pairs formed, these structures are often quite stable. The likelihood of forming long-lived folding intermediates grows as RNAs increase in size.

In vivo analyses of RNA folding have often revealed a different picture compared to *in vitro* studies performed on the same RNA molecule. The *Tetrahymena* group I intron is a widely studied model system for RNA folding (Cech and Bass, 1986; Doudna and Cech, 2002; Treiber and Williamson, 2001; Woodson, 2002). It is located in the gene for the large ribosomal subunit. *In vivo*, the half-life of this pre-rRNA has been measured at ~2 s (Brehm and Cech, 1983). However, when studied *in vitro*, this group I intron spliced at a rate ~20–50 times slower (Treiber and Williamson, 2001; Woodson, 2002). Presumably, when transcribed *in vivo*, this RNA manages to avoid the long-lived, nonfunctional structures which dominate its folding pathway *in vitro*.

Cotranscriptional RNA folding *in vivo* differs from Mg^{2+}-initiated refolding in several aspects. RNA has an inherent 5′- to 3′-polarity. During transcription, the 5′-portion of the nascent transcript emerges from the polymerase before its 3′-region. Therefore, the upstream region can begin folding at an earlier time. Factors influencing this timing window and, subsequently, the structural formation of the upstream regions include the transcriptional speed of the RNA polymerase and transcriptional pausing. Cotranscriptional folding studies replicate the 5′- to 3′-polarity of the RNA transcript and can account for the properties of the RNA polymerase and its various transcription factors (Pan and Sosnick, 2006). As such, they provide better models for the *in vivo* folding pathways of RNA (Al-Hashimi and Walter, 2008; Pan and Sosnick, 2006).

This chapter describes several techniques to study cotranscriptional folding of RNA. Some methods are useful only for particular RNAs while others are more widely applicable. Examples are provided in which these techniques have been used to investigate the folding of several noncoding RNAs during transcription. They include three highly conserved noncoding RNAs (RNase P, SRP, and tmRNA), several circularly permuted forms of a bacterial RNase P RNA, a riboswitch (thiM), and an aptamer-activated ribozyme (glmS). Through the study of cotranscriptional folding, it is shown how the process of transcription, the properties of the polymerase (transcriptional pausing in particular), and the presence of additional factors in the cellular environment can influence the RNA folding pathway.

2. Protocol 1: Determination of Transcriptional Pause Sites

RNA polymerases do not transcribe their nascent transcripts at a uniform rate. Rather they have been found to pause at specific sequence or structure-dependent locations (Artsimovitch and Landick, 2000). At these pause sites, the elongation rate of the polymerase slows dramatically. Transcriptional pausing can influence RNA folding by increasing the time window at strategic locations so that upstream regions of the transcript can form beneficial native or non-native interactions. Pausing may also facilitate interactions between the nascent RNA transcript and the RNA polymerase or its associated factors. These interactions can directly influence the RNA folding pathway. Studies have shown that strategically placed pause sites can alter the folding pathway of a nascent transcript and, in some cases, play an important role in allowing for efficient folding (Pan *et al.*, 1999; Wong *et al.*, 2005, 2007). Thus, the identification of these sites in a particular RNA can shed light on its cotranscriptional folding pathway.

Transcriptional pausing is a complex behavior affected by several factors including interactions between the RNA polymerase and the nascent transcript, the DNA template (i.e., the strength of the RNA–DNA hybrid and the nearby DNA sequence), and the nucleotide triphosphate in the polymerase's active site (Artsimovitch and Landick, 2000). The transcriptional pausing behavior of an RNA polymerase is also known to be species-specific (Artsimovitch *et al.*, 2000). Several different types of pause sites have been described (along with their distinguishing characteristics) (Artsimovitch and Landick, 2000). However, the reasoning behind polymerase pauses at many sites still remains mechanistically unexplained. These pause sites must be experimentally identified.

The following methods are variations from previously published protocols (Landick *et al.*, 1996). Of note, as they use single-turnover transcription, a stable ternary complex must be formed with the RNA polymerase before transcription is initiated. Therefore, they are well suited for determining the pause sites of bacterial RNA polymerases (e.g., *Escherichia coli* RNA polymerase) but not of phage RNA polymerases (e.g., T7 RNA polymerase).

2.1. Single-turnover transcription reactions

2.1.1. *E. coli* RNA polymerase

1. The template consists of either the *Bacillus subtilis* P RNA promoter or the T7A1 promoter followed by the RNA of interest. At least one of the four nucleotide bases must be absent from approximately the first 15 nucleotides in the RNA transcript sequence to form a stable ternary complex.
2. Ternary complexes are formed with 50 n*M* DNA template (obtained by PCR) in 20 m*M* Tris–HCl, pH 8.0, 20 m*M* NaCl, 14 m*M* $MgCl_2$, 14 m*M* 2-mercaptoethanol, 0.1 m*M* EDTA, 25 μg/ml BSA, 2.25% glycerol, and 0.09 U/μl RNA polymerase. The *E. coli* RNA polymerase contains the core subunits and sigma factor and can be purchased from USB (Cleveland, OH) or purified according to previously published protocol (Kuznedelov and Severinov, 2009; Landick *et al.*, 1990; Schmidt and Chamberlin, 1984).
3. The nascent RNA transcript is ^{32}P-labeled through the addition of either 200 μM of a 5′ ^{32}P-labeled dinucleotide representing the first

two nucleotides of the nascent transcript or 5–10 μCi [α-^{32}P]ATP or GTP. The reaction is incubated at 37 °C for 5 min.

4. The transcript is extended by at least 10 nucleotides through the addition of three NTPs ranging in concentrations from 0 to 45 μM. (If needed, we routinely use a promoter sequence containing a 5′-transcript extension 15 nucleotides long and lacking Uridine; Pan *et al.*, 1999.) Transcription halts at the first occurrence of the fourth nucleotide downstream of the start site, forming a stable complex. To prevent transcription through this site, only ultrapure NTPs (GE Healthcare, Piscataway, NJ) are used.
5. After a 10-min incubation at 37 °C, elongation is resumed with the addition of 100 μM–1 mM (final concentration) ATP, UTP, CTP, and GTP. The concentration of NTPs used is optimized depending on the pause site. Lower NTP concentrations amplify the strength and duration of pause sites (particularly useful with weaker pauses). Transcription is limited to a single round with the concurrent addition of rifampicin to a final concentration of 2 μM.

2.1.2. *B. subtilis* RNA polymerase

Single-round transcription is carried out in a manner similar to that of *E. coli* RNA polymerase. However, the ternary complex is formed with 50 nM DNA template in 25 mM Tris–HCl, pH 7.5, 10 mM $MgCl_2$, 50 mM KCl, 1 mM DTT, 0.2 μM RNA polymerase, and 2 μM recombinant *B. subtilis* σA protein. The polyhis-tagged RNA polymerase (Anthony *et al.*, 2000) and σA protein (Chang and Doi, 1990) are purified according to previously published protocol.

At designated time points, aliquots from the transcription reaction are removed and quenched with an excess of 100 mM EDTA and 9 M urea. The transcription products are separated by denaturing polyacrylamide gel electrophoresis, and the pauses are identified by comparing them to the products of a transcription reaction containing 3′-deoxy-NTPs (Landick *et al.*, 1996; Wong *et al.*, 2007).

The pausing profile of the transcription reaction can be analyzed qualitatively. Alternatively, the behavior of pause sites can be described by the following equation:

$$A(t) = A(0)e^{-kt}, \tag{9.1}$$

where $A(0)$ is the amount of the paused complex at an initial reference point and K is the rate at which the RNA polymerase leaves the pause site.

In this manner, the strength and duration of different pause sites can be compared with each other.

3. Protocol 2: Structural Mapping of Paused Complexes

Understanding how transcriptional pausing affects the folding of the nascent transcript requires the structural characterization of the RNA in the paused complex. Because these "truncated" RNAs lack significant portions of their full-length versions, the conformation of these nascent RNA transcripts most likely contain secondary and tertiary structures absent from the native conformation. Structural analyses of the paused complexes can shed light on these structures as intermediates in cotranscriptional folding. Additionally, as both the nascent transcript and the ternary complex are present in the paused complex, these analyses provide a closer representation of what is going on with RNA folding in the cell.

3.1. Isolation of paused complexes

1. Single-round transcription is performed as described in Protocol 1. The nascent RNA can be 5′-end or body-labeled with ^{32}P. The concentrations of ATP, CTP, GTP, and UTP used during transcription elongation will vary depending on the complex to be isolated. NTP concentrations used in the isolation of the P RNA, SRP, and tmRNA paused complexes ranged from 3.5 to 35 μM.
2. Upon the resumption of elongation, transcription is allowed to proceed at 37 °C for a defined period of time. For P RNA, SRP, and tmRNA, incubation times ranged from 45 to 60 s. Transcription products are identified via denaturing polyacrylamide gel electrophoresis, and the NTP concentrations and incubation time are adjusted to allow for the optimal enrichment of the paused complex.
3. After the incubation period, the reaction is transferred to ice.
4. Removal of the free NTPs is performed through use of a MicroSpin™ G-25 column (GE Healthcare, Piscataway, NJ), pre-equilibrated with the transcription buffer solution. Once the NTPs are removed, the enriched paused complex is stable for more than 1 h.

The difficulty in isolating the paused complex of interest is determined by (1) the strength and/or duration of the pause site, (2) the location of the pause site, and (3) the number (and strength) of competing pause sites. Complexes at pause sites of high duration and intensity and those closer to the 5′-end of the transcript are easier to isolate. On the other hand, an increasing number of competing pause sites may make it difficult to significantly enrich for the paused complex of interest. More difficult pause sites will require additional optimization of the NTP concentrations and incubation time to cleanly isolate their corresponding paused complexes.

3.2. Structural mapping of paused complexes

We use oligohybridization/RNase H and T1 ribonuclease structural mapping to characterize the isolated paused complexes. The following methods can also be adapted for hydroxyl-radical footprinting, chemical modification (i.e., DMS, kethoxal, DEPC, etc.), or partial digestion with different ribonucleases. However, as the interpretation from these methodologies depends on the degree of conformational homogeneity of the nascent RNA transcript, it is important that the paused complex isolation protocol be fully optimized.

3.3. Oligohybridization/RNase H

1. Probes are designed as 10-mer DNA oligonucleotides complementary to specific regions of the paused complex.
2. Aliquots of the isolated paused complex are mixed with an equal volume of an oligonucleotide–RNase H solution mixture consisting of 20 m*M* Tris–HCl, pH 7.5, 20 m*M* KCl, 26 m*M* $MgCl_2$, 0.1 m*M* EDTA, 0.1 m*M* DTT, 4 μM rifampicin, 1.2 m*M* $CaCl_2$, 0.3 $\mu g/\mu l$ DNase I (Sigma, St. Louis, MO), 200 μM DNA probe, and 0.4 U/μl RNase H (Epicentre, Madison, WI).
3. RNase H cleavage is allowed to proceed for 90 s at 37 °C before quenching with an equal volume of 100 m*M* EDTA and 2 mg/ml Proteinase K (Ambion, Austin, TX). The incubation time for the RNase H reaction may be altered depending on the accessibility of the complex to the oligoprobes.
4. The solution is incubated at 55 °C for 1 h to digest the remaining RNA polymerase proteins and the RNase H in the solution.
5. The cleaved and uncleaved RNA transcripts are separated on denaturing PAGE and identified by phosphorimaging.

3.4. Partial T1 ribonuclease cleavage

1. Aliquots of the isolated paused complex are mixed with an equal volume of an RNase T1 mixture consisting of 20 m*M* Tris–HCl, pH 7.5, 20 m*M* KCl, 14 m*M* $MgCl_2$, 0.1 m*M* EDTA, 0.1 m*M* DTT, 4 μM rifampicin, 1.2 m*M* $CaCl_2$, 0.3 $\mu g/\mu l$ DNase I, and 0.24 U/μl RNase T1 (Ambion, Austin, TX).
2. The reaction is allowed to proceed at 37 °C for 45–90 s. The incubation time is optimized depending on the complex's susceptibility toward T1 cleavage.
3. Aliquots are mixed with an equal volume of 2 mg/ml Proteinase K in 20 m*M* Tris–HCl, pH 8.0, 20 m*M* NaCl, 14 m*M* $MgCl_2$, 14 m*M* 2-mercaptoethanol, and 0.1 m*M* EDTA. The solution is incubated at

55 °C for 1 h. The amount of Proteinase K is sufficient to destroy the T1 nuclease on a shorter time scale when compared to the time allowed for the T1 nuclease reaction.

4. The cleavage products are separated on denaturing PAGE. They are identified and quantified by phosphorimaging.

4. Protocol 3: Cotranscriptional RNA Folding as Measured via Oligohybridization

Oligonucleotide hybridization with RNase H cleavage has been used to analyze the Mg^{2+}-initiated refolding pathways of the group I intron and RNase P (Treiber and Williamson, 2000; Zarrinkar and Williamson, 1994; Zarrinkar *et al.*, 1996). Briefly, as RNA folds into its native conformation, its secondary structures become more protected against hybridization to complementary DNA probes. The rates at which different regions of the RNA become protected can provide site-specific information on the folding pathway. This method has been adapted to the study of cotranscriptional RNA folding (Wong *et al.*, 2005, 2007).

1. Oligonucleotide probes are designed as 10-mers targeted against specific regions of the RNA transcript. These 10-mer oligoprobes have a rapid on-rate because of the high concentrations used in the cleavage reactions. Because of this, in a relatively short period of time (compared to the RNA folding rate), all accessible regions in the RNA are hybridized. This creates great reproducibility in terms of the absolute fraction of cleavage.
2. The regions targeted are typically areas/motifs expected to decrease in accessibility as the RNA folds into its native conformation. For example, probes typically target regions forming long-range secondary structures (e.g., pseudoknots and long-range helices) or tertiary structures in the native conformation. Regions forming short-range interactions in the final structure (e.g., hairpin loops) are generally avoided. Of note, certain regions can exhibit limited accessibility at early time points even though their native structures are not expected to form until later in the folding pathway. These regions are likely involved in the formation of non-native folding intermediates.
3. Transcription is allowed to proceed under single-turnover conditions as per Protocol 1. Multiple-turnover conditions may also be used. However, our experience has shown that the change in protection is greater (resulting in an easier analysis) under single-turnover conditions.
4. At designated time points, aliquots of the transcription reaction are taken and added to an equal volume of an oligonucleotide–RNase H solution consisting of 20 m*M* Tris–HCl, pH 7.5, 20 m*M* KCl, 14 m*M* $MgCl_2$,

0.1 mM EDTA, 0.1 mM DTT, 4 μM rifampicin, 200 μM DNA probe, and 0.4 U/μl RNase H.

5. The RNase H cleavage reaction is allowed to proceed for 30 s before quenching with an excess of 100 mM EDTA and 9 M urea.
6. The cleaved and uncleaved RNA transcripts are separated on denaturing PAGE and quantified by phosphorimaging.

We typically use the above oligohybridization protocol to search for qualitative changes in the folding pathway with different transcription conditions, RNA polymerases, RNA constructs, etc. We study the same RNA region under different conditions, searching for changes in the degree of protection. However, quantitative rates (k_f) can also be obtained from the changes in protection:

$$\frac{\text{Fraction RNA uncleaved}}{\text{Total RNA}} = B\left(1 - \frac{1 - e^{-k_f t}}{k_f t}\right) + A, \tag{9.2}$$

where A is the fraction protected initially and B is the gain in fraction protected upon folding.

This allows for the comparison of different oligoprobes with each other, further dissecting the folding pathway of the nascent transcript. For the majority of probes in cases of cotranscriptional folding, A is approximately equal to zero.

The above method can also be adapted to other mapping strategies (hydroxyl-radical footprinting, chemical modification, nuclease cleavage, etc.). These methodologies can provide additional structural information on the RNA folding pathway. However, two caveats should be noted. With oligohybridization, the structural motif being studied is already known. In contrast, other mapping methods likely result in multiple cleavages at unknown locations, making the interpretation of the data more difficult. To decrease the complexity of the analysis, the nascent transcript should be end-labeled with ^{32}P instead of being body-labeled. Also, oligohybridization benefits from the rapid on-rate of the oligos relative to the rate of RNA folding. This allows for great reproducibility in terms of the fraction of the RNA population cleaved. Many modifying agents react with RNA on a much slower time scale. Thus, the reproducibility of their results may be diminished.

5. Protocol 4: Cotranscriptional RNA Folding Measured via P RNA Catalytic Activity

RNase P is a highly conserved ribozyme in both prokaryotes and eukaryotes. Its primary function is to generate the mature 5′-end of tRNAs through an endonucleolytic cleavage reaction (Altman and Kirsebom, 1999;

Frank and Pace, 1998). In its native conformation, bacterial RNase P is catalytically active against precursor tRNAs. This can be used to investigate the cotranscriptional folding of this ribozyme. Of note, the following protocol may be modified for any *trans*-acting ribozyme as long as the ribozyme reaction time is significantly faster than the folding time.

5.1. Multiple-turnover transcription reactions

The folding of RNase P can be measured during transcription with either T7 RNA polymerase or bacterial RNA polymerases in multiple-turnover transcription reactions.

5.1.1. T7 RNA polymerase

1. The DNA template consists of the T7 promoter sequence followed by the RNA of interest.
2. Transcription with T7 RNA polymerase is performed in 40 m*M* Tris–HCl, pH 8.1, 1 m*M* spermidine, 50 μg/ml BSA, 5–10 μCi [α-^{32}P]ATP, 14 m*M* $MgCl_2$, 60 μg/ml plasmid DNA, and 40 μg/ml T7 RNA polymerase. The T7 RNA polymerase is purified from an overexpression clone as previously described (Studier *et al.*, 1990).
3. Transcription is initiated through the addition of ATP, CTP, GTP, and UTP. A default concentration is 2 m*M* of each NTP. However, varying NTP concentrations can be used to alter the speed of the RNA polymerase. The polymerase will transcribe at slower speeds as the NTP concentration is reduced.
4. Aliquots from the reaction mixture are removed at varying time points for analysis.

5.1.2. *E. coli* RNA polymerase

1. The DNA template consists of either the *B. subtilis* P RNA promoter (Reich *et al.*, 1986) or the T7A1 promoter (Dunn and Studier, 1983; Susa *et al.*, 2002) followed by the RNA of interest.
2. Transcription with *E. coli* RNA polymerase is carried out in 40 m*M* Tris–HCl, pH 7.9, 10 m*M* 2-mercaptoethanol, 4 m*M* spermidine, 5–10 μCi [α-^{32}P] ATP, 100 μg/ml BSA, 10 m*M* $MgCl_2$, 120 m*M* KCl, 0.05–0.1 μ*M* DNA template, and 0.07 U/μl *E. coli* RNA polymerase.
3. The reaction mixture is incubated at 37 °C for 5 min.
4. Transcription is initiated through the addition of ATP, CTP, GTP, and UTP. A default concentration is 1 m*M* of each NTP. However, as with the T7 RNA polymerase, varying NTP concentrations may be used to alter the speed of the RNA polymerase.

5. The transcription reaction is carried out for 8 min with the removal of aliquots at designated time points. During this time, transcription proceeds at a constant rate.
6. After transcription has proceeded for 8 min, rifampicin is added to a final concentration of 2 μM to halt transcription, and folding is allowed to proceed for up to 40 min with the removal of aliquots at various times.

5.1.3. *B. subtilis* RNA polymerase

Multiple-round transcription is carried out in a manner similar to that of *E. coli* RNA polymerase. However, transcription using *B. subtilis* RNA polymerase is carried out in 25 mM Tris–HCl, pH 7.5, 1 mM DTT, 1 mM each of ATP, GTP, CTP, UTP, 5–10 μCi [α-^{32}P]ATP, 100 μg/ml BSA, 10 mM $MgCl_2$, 50 mM KCl, 0.05–0.1 μM DNA template, 0.2 μM *B. subtilis* RNA polymerase, and 2 μM recombinant *B. subtilis* σA protein.

Bacterial RNase P consists of two components: a larger RNA subunit of ~300–400 nucleotides (denoted P RNA) and a smaller protein subunit of ~14 kDa (denoted P protein) (Altman, 1989; Pace and Brown, 1995). Under high ionic conditions, bacterial P RNA is capable of cleaving pre-tRNA without P protein present (Guerrier-Takada and Altman, 1984; Guerrier-Takada *et al.*, 1983). If pre-tRNA cleavage occurs under these conditions, it is possible to measure the cotranscriptional folding of P RNA in the absence of P protein. This protocol is described below. Of note, the RNase P holoenzyme consisting of P RNA and P protein is capable of cleaving precursor tRNA under lower ionic conditions (50 mM Tris–HCl, pH 8.0/10 mM $MgCl_2$/50 mM KCl) (Kurz *et al.*, 1998; Reich *et al.*, 1988). Thus, it is also possible to measure the formation of the full ribonucleoprotein complex by a similar method.

P RNA consists of two independently folding domains: a catalytic domain (C-domain) containing the catalytic site of pre-tRNA cleavage and a specificity domain (S-domain) responsible for pre-tRNA binding (Loria and Pan, 1996; Pan, 1995). Both domains must be in their native conformations for cleavage of a pre-tRNA substate. However, through *in vitro* evolution, a selected substrate has been developed which requires only a properly folded catalytic domain for its cleavage (Pan and Jakacka, 1996). Using these two substrates, folding of the P RNA ribozyme can be dissected at the domain level.

5.2. Renaturation of the pre-tRNA substrate

1. The precursor tRNA substrate is 5′ ^{32}P-labeled by standard methods using T4 polynucleotide kinase.
2. The 5′ ^{32}P-labeled pre-tRNA substrate in 250 mM Tris–HCl, pH 8.0, is incubated at 90 °C for 2 min. Afterwards, it is kept at room temperature for 3 min.

3. A solution of 1 *M* $MgCl_2$ is added to a concentration of 333 m*M* $MgCl_2$, and the resultant mixture is incubated at 37 °C for 5 min.
4. A solution of 3 *M* KCl is added to a concentration of 1.2 *M* KCl. The final conditions of the RNA solution are 100 m*M* Tris–HCl, pH 8.0/ 200 m*M* $MgCl_2$/1.2 *M* KCl.

5.3. Renaturation of the selected substrate

1. The selected substrate is 5′ ^{32}P-labeled by standard methods.
2. The 5′ ^{32}P-labeled selected substrate in 250 m*M* Tris–HCl, pH 8.1, is incubated at 90 °C for 2 min followed by incubation at room temperature for 3 min.
3. A solution of 1 *M* $MgCl_2$ is added to a concentration of 333 m*M* $MgCl_2$, and the resultant mixture is incubated at 37 °C for 5 min.
4. A solution of 5 m*M* spermine is added to a concentration of 2 m*M*. The final conditions of the RNA solution are 100 m*M* Tris–HCl, pH 8.1/ 200 m*M* $MgCl_2$/2 m*M* spermine.

5.4. Monitoring RNA folding via catalytic activity

At designated time points in the transcription reaction, aliquots of the transcription mixture and the ^{32}P-labeled substrate solution are mixed in equal volumes. Cleavage of the substrate by P RNA occurs at 37 °C. Aliquots from the cleavage reaction mixture are taken within 21 s (with a maximum of 7-s intervals) and quenched with an excess of 100 m*M* EDTA and 9 *M* urea. The time allowed for the cleavage reaction is short when compared to the folding time of the RNA transcript. Therefore, only a small fraction of the P RNA folds during the cleavage reaction. The relative amount of transcript synthesized ($S(t)$) is determined by detecting the incorporation of [α-^{32}P]ATP into the transcript. The cleaved substrate and product are then separated on denaturing polyacrylamide gels, and the fraction of pre-tRNA substrate cleaved is determined by phosphorimaging.

5.5. Conditions in which [E] > [S]

If the ribozyme is synthesized at a rate less than 10 n*M*/min (e.g., *E. coli* RNA polymerase), the amount of catalytically active P RNA ($A(t)$) is obtained by measuring the cleavage rate under single-turnover conditions in which the amount of P RNA transcript is in molar excess (~2–500 n*M* final concentration) of the substrate (~1 n*M* final concentration). In this scenario [E] ≫ [S], and the fraction of the pre-tRNA substrate cleaved versus time is fit to a single exponential equation to obtain the reaction rate. The rate of cleavage is proportional to the amount of folded ribozyme ($A(t)$). In the period during which the rate of transcription is constant (time < 10 min), the data before the addition of rifampicin are fit to the biphasic, Eq. (9.3), where k_1 and k_2 are the folding rates and f is the fraction of the population folding at rate k_1. This accounts for RNA synthesis. If the rate of RNA folding is monophasic, $f = 1$, and the second half of the equation drops out:

$$\frac{A(t)}{S(t)} = f\left(1 - \frac{1 - e^{-k_1 t}}{k_1 t}\right) + (1 - f)\left(1 - \frac{1 - e^{-k_2 t}}{k_2 t}\right). \quad (9.3)$$

After the addition of rifampicin, RNA synthesis ceases. Thus, the rate of RNA folding can be fit to a single exponential equation, Eq. (9.4), where f_1 (f_2) is the population still remaining to be folded at the same rate k_1 (k_2) as before the addition of rifampicin:

$$\frac{A(t)}{S(t)} = (1 - f_1 - f_2) + f_1(1 - e^{-k_1 t}) + f_2(1 - e^{-k_2 t}). \quad (9.4)$$

5.6. Conditions in which [E] < [S]

If the ribozyme is synthesized at rate greater than 20 n*M*/min (e.g., T7 RNA polymerase), the amount of the folded ribozyme can be measured with a molar excess of substrate (i.e., [S] > [E]). Under the high ionic conditions used for precursor tRNA cleavage, the bimolecular on-rate for the *B. subtilis* P RNA–substrate complex is $\sim 6 \times 10^6\ M^{-1}\ s^{-1}$, the rate of cleavage is $\sim 6\ s^{-1}$, and the disassociation rate of the tRNA product is $\sim 0.02\ s^{-1}$ (Kurz *et al.*, 1998). Therefore, at sufficiently high concentrations of substrate, tRNA (product) release is the rate-limiting step in the reaction. Subsequently, within the first 5–20 s of the reaction, each P RNA will have completed a single round of cleavage, and the amount of the cleaved tRNA product is equal to $A(t)$. The rate of ribozyme folding can then be determined as in the case when [E] > [S] using Eqs. (9.3) and (9.4). Of note, this analysis can be applied to any *trans*-acting ribozyme in which product release is the rate-limiting step in the reaction.

6. Protocol 5: The Folding of Self-Cleaving RNAs During Transcription

Many catalytic RNAs are characterized by their ability to self-cleave at specific locations upon folding into their native conformations (Been, 2006; Doherty and Doudna, 2000, 2001; Fedor, 2000; Fedor and Williamson,

2005; Long and Uhlenbeck, 1993). This fact can be exploited to study the cotranscriptional folding of these RNAs. *In vivo* folding during transcription studies have primarily focused on the group I intron (Emerick and Woodson, 1993; Hagen and Cech, 1999; Jackson *et al.*, 2006; Koduvayur and Woodson, 2004; Nikolcheva and Woodson, 1999; Zhang *et al.*, 1995). *In vitro* folding during transcription studies have also been performed on the hammerhead ribozyme (Monforte *et al.*, 1990; Tyagarajan *et al.*, 1991), the HDV ribozyme (Diegelman-Parente and Bevilacqua, 2002), and the hairpin ribozyme (Mahen *et al.*, 2005). Often, these *in vitro* studies use a protocol similar to one described below to study how these self-cleaving RNAs fold during transcription:

1. To obtain an accurate folding rate, the rate of RNA self-cleavage must be fast when compared to the rate of folding. In other words, during the *in vitro* transcription reaction, RNA folding, not the cleavage reaction is the rate-limiting step.
2. Multiple-turnover transcription is performed as described in Protocol 4. Either T7 RNA polymerase or a bacterial RNA polymerase is used to transcribe the RNA.
3. Any components needed for self-cleavage are included in the transcription mix. Many catalytic RNAs (e.g., the hammerhead ribozyme) will self-cleave under traditional transcription reaction conditions. Other RNAs (such as glmS) require a cofactor to be present for self-cleavage (Winkler *et al.*, 2004). In the case of glmS, varying amounts of glucosamine-6-phosphate (1–1000 μM) may be included in the transcription reaction.
4. At designated time points after the initiation of transcription, aliquots are taken and quenched with an excess of 100 mM EDTA and 9 M urea. The cleaved and full-length RNA fractions are separated by PAGE and quantified with phosphorimaging. During the period when the RNA transcription rate is constant (typically <10 min), the rate of folding (k_f) may be determined from the following equation:

$$\frac{\text{Cleaved RNA}}{\text{Total RNA}} = 1 - \frac{1 - e^{-k_f t}}{k_f t}. \tag{9.5}$$

This equation assumes that, given enough time, the entire RNA population will self-cleave. For smaller RNAs with natural sequences, this assumption typically holds. For RNAs of larger size and for mutational derivatives of natural RNAs, this may not hold true. A nonfolding population may exist, or two RNA populations may fold at different rates. In this case, the equation can be modified as in Eq. (9.3).

6.1. Cotranscriptional folding in the presence of a cofactor

Some self-cleaving ribozymes (e.g., glmS) require a cofactor to properly function. The dependence of these RNAs on their cofactors can be described by Eq. (9.6). This equation can also be used to describe RNAs requiring a cofactor to fold into a particular conformation (e.g., riboswitches):

$$\text{Folding rate} = \frac{k_{\max}[\mathrm{S}]}{K + [\mathrm{S}]}, \tag{9.6}$$

where $k_{\max}$ is the maximal rate of cleavage or folding and K is the cofactor concentration at 50% $k_{\max}$.

This equation holds true for RNAs that bind their cofactors in a 1:1 ratio (i.e., that do not exhibit binding cooperativity).

7. Additional Methodologies

Several techniques, not described in detail above, have also been used to investigate the cotranscriptional folding of RNA. These include native gel electrophoresis (Heilman-Miller and Woodson, 2003), temperature-gradient gel electrophoresis (Repsilber *et al.*, 1999), single-molecule techniques (Aleman *et al.*, 2008; Dalal *et al.*, 2006; Greenleaf *et al.*, 2008), and computational approaches (Geis *et al.*, 2008; Isambert and Siggia, 2000; Meyer and Miklos, 2004; Shapiro *et al.*, 2006; Xayaphoummine *et al.*, 2005, 2007). Additionally, the conformational changes of certain riboswitches from terminator to antiterminator structures have been used to monitor the cotranscriptional folding of these RNAs (Mandal *et al.*, 2004; Wickiser *et al.*, 2005).

8. Cotranscriptional Folding Studies from Our Laboratory

8.1. The folding of *B. subtilis* P RNA during transcription by T7 RNA polymerase (Pan *et al.*, 1999)

The appearance of catalytic activity against pre-tRNA was used to assay the folding of *B. subtilis* P RNA. The folding of this ribozyme when transcribed by T7 RNA polymerase was compared to its Mg^{2+}-initiated refolding. The cotranscriptional folding of *B. subtilis* P RNA entailed a pathway in which its C-domain folded approximately fourfold faster than its S-domain. This produced an experimentally observable folding intermediate that was catalytically active against the selected substrate but inactive against the pre-tRNA substrate. No such intermediate existed in its Mg^{2+}-initiated refolding pathway. Thus, the act of transcription can alter the folding pathway of a large noncoding RNA.

8.2. The effect of transcriptional pausing on the folding of a large circularly permuted RNA (Pan *et al.*, 1999; Wong *et al.*, 2005)

The folding during transcription of circularly permuted variants of *B. subtilis* P RNA was also studied with the appearance of their catalytic activity against a pre-tRNA substrate. In one construct, the entire C-domain was transcribed before the S-domain (i.e., it had the polarity 5′-C-domain → 3′-S-domain and was denoted CP240). The cotranscriptional folding of CP240 when transcribed by either T7 polymerase or *E. coli* RNA polymerase was similar to that of the native P RNA and involved its C-domain folding at a faster rate than its S-domain. The folding of this ribozyme was unaffected by changing the speed of transcription (as assayed by varying the NTP concentration). However, during transcription by *E. coli* RNA polymerase in the presence of the elongation factor, NusA, CP240 folded along a different pathway. In this pathway, the folding of the C-domain, not the S-domain, was the rate-limiting step. NusA greatly increased transcriptional pausing at the 225th nucleotide of CP240. When transcribed with a pausing-deficient *E. coli* RNA polymerase (β: P560S, T563I), CP240 folding was unaffected by the presence of NusA. This elongation factor also did not influence the folding of a CP240 mutant that eliminated this pause site. Site-specific transcriptional pausing at CP225 therefore altered the folding pathway of this ribozyme.

In CP240, the S-domain has not yet been transcribed when pausing occurs at the 225th position (the C-domain has 255 residues). However, pausing at this location significantly accelerated S-domain's folding. We hypothesized that NusA-induced pausing at this location prevented a misfold between a C-domain structure transcribed before the pause site and an S-domain structure transcribed after the site. The cotranscriptional folding of two other circular permutes of *B. subtilis* P RNA (CP1 = natural P RNA = 5′:$^{1/3}$C-domain → S-domain → $^{2/3}$C-domain; CP84 = 5′:S-domain → C-domain) was not affected by the pausing at the 225th position, indicating the importance of the order in which these structures are transcribed. Oligohybridization assays and mutational analyses were used to characterize the structures involved in the misfold. This misfold involved the S-domain region immediately 3′ to the C-domain (P7-J5.1/7) and the C-domain region that binds a portion of the pre-tRNA substrate (P15). We hypothesize that transcriptional pausing at the 225th position of CP240 sequesters P15 in a "locked down" conformation, preventing it from forming a stable misfold with P7-J5.1/7. These results indicate that transcriptional pausing can accelerate the folding of large RNAs through the avoidance of long-lived misfolded intermediates.

8.3. The facilitation of cotranscriptional RNA folding by pausing-induced non-native structures (Wong *et al.*, 2007)

The cotranscriptional folding of *E. coli* P RNA was monitored through the appearance of its catalytic activity against the pre-tRNA substrate. We discovered that this RNA folded more efficiently when transcribed by its cognate *E. coli* RNA polymerase versus by a noncognate *B. subtilis* RNA polymerase. Single-turnover transcription of this RNA carried out with *E. coli* RNA polymerase revealed a long-lived pause at residue C119. This pause site is phylogenetically conserved among many bacterial species, particularly among the γ-proteobacteria family to which *E. coli* belongs. In contrast, during transcription by *B. subtilis* RNA polymerase, this pause was reduced in both intensity and duration. Disruption of the C119 pause either through mutation of the RNA (G107C, C119G) or RNA polymerase (β: P560S, T563I) resulted in reduced P RNA folding efficiency during transcription by *E. coli* RNA polymerase. We concluded that pausing at C119 by *E. coli* RNA polymerase plays an important role in the cotranscriptional folding of this P RNA ribozyme.

The C119 paused complex was isolated and structural mapping was performed with both T1 nuclease and oligohybridization/RNase H methods. Based on the structural mapping data and phylogenetic analysis, we proposed that this paused complex formed a labile intermediate comprised of non-native interactions. This intermediate possessed a stable core with a more labile periphery. Mutations that altered the stability of this intermediate resulted in reduced P RNA folding efficiency. Interestingly, this intermediate effectively sequestered the 5′-portions of P RNA's six long-range helices. We propose that this structural intermediate helps guide the formation of *E. coli* P RNA's native long-range helices by preventing the formation of long-lived unproductive species. This is accomplished by sequestering the upstream strands of the long-range helices until their downstream partners have been transcribed.

We hypothesized that the sequestration of the upstream strands of long-range helices through pausing at strategic locations could be a general strategy for efficient cotranscriptional folding. Therefore, we analyzed the folding of *E. coli* SRP RNA and tmRNA. Like P RNA, these two noncoding RNAs are characterized by the presence of long-range helices. The folding of these two RNAs during transcription by *E. coli* RNA polymerase was analyzed by oligohybridization. Like P RNA, the efficient cotranscriptional folding of SRP RNA and tmRNA was dependent on site-specific pausing at strategic positions. Disruption of these specific pause sites resulted in reduced folding efficiency during transcription. As with P RNA, the paused complexes of these two RNAs were isolated and mapped with T1 nuclease and oligonucleotide hybridization. With both RNAs, we found that the 5′-portions of the long-range helices were sequestered in phylogenetically conserved non-native intermediates.

These results show that site-specific transcriptional pausing is a general strategy to allow for the efficient cotranscriptional folding of noncoding RNAs. This applies particularly to those RNAs containing long-range helices. In these RNAs, the upstream strands of the helices can be sequestered in non-native structural intermediates. Coevolution likely has occurred between the sequences of these RNAs and the pausing properties of their cognate polymerases to allow for efficient folding during transcription. Efficient cotranscriptional folding may be an evolutionarily selected attribute in RNA sequence determination.

8.4. The influence of pausing and small molecule cofactors on the folding of riboswitches

In *E. coli*, the thiM riboswitch is located in the 5′-UTR of the mRNA coding for hydroxyethylthiazole kinase (Winkler *et al.*, 2002a). This riboswitch is highly conserved among prokaryotes with almost 500 sequences currently deposited in the Rfam database (http://www.sanger.ac.uk/Software/Rfam/). A crystal structure of the TPP-bound aptamer domain has been solved (Edwards and Ferre-D'Amare, 2006; Serganov *et al.*, 2006). In-line probing data and chemical/enzymatic mapping have shown that thiM undergoes significant conformational changes (Fig. 9.1B) upon its binding to thiamine pyrophosphate (TPP) (Rentmeister *et al.*, 2008; Winkler *et al.*, 2002b). In the presence of TPP, these conformational changes prevent the expression of its downstream gene by reducing the ribosome accessibility of the gene's ribosomal binding site (RBS) (Ontiveros-Palacios *et al.*, 2008).

As riboswitches regulate gene expression in *cis*, the kinetics of their cotranscriptional folding may significantly impact how effectively they function. We performed single-round pausing assays on the thiM riboswitch to determine if pausing could play a role in the cotranscriptional folding of this RNA (Fig. 9.1A). In the absence of TPP, the major transcriptional pause sites were located from nucleotides ~102 to 125. These pauses are located just 3′ to the end of the TPP aptamer domain (Fig. 9.1B). Transcriptional pausing in this region could therefore allow thiM to bind TPP but prevent its TPP-unbound conformation from being formed. It has been shown using refolded RNA transcripts that both full-length thiM and its TPP-binding aptamer domain bind TPP significantly more efficiently than constructs intermediate in length (Lang *et al.*, 2007). Thus, pausing

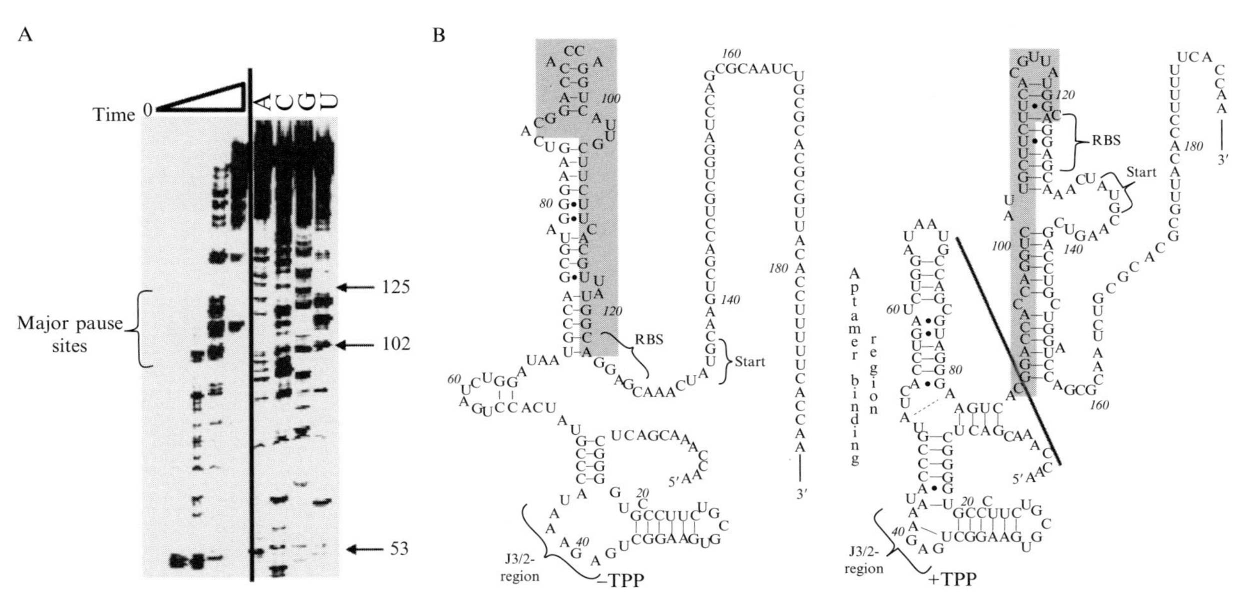

Figure 9.1 Transcriptional pausing of the thiM riboswitch. (A) Pausing of thiM when transcribed by *E. coli* RNA polymerase (in the absence of TPP). The addition of TPP did not affect the pausing profile of this riboswitch (data not shown). (B) Proposed secondary structures of the thiM construct used in this study in the absence and presence of TPP. The region shaded in gray includes the major pause sites (residues 102–125) plus the preceding 13 nucleotides which reside in the exit channel of the RNA polymerase during pausing. RBS, ribosome-binding site.

could allow thiM to fold into its TPP-bound conformer without first folding into its TPP-unbound state.

Riboswitches are often transcribed in the presence of the metabolite to which they bind. How this affects the folding of these RNAs is an open question. To investigate the role that transcription and TPP play in the conformational changes of the thiM riboswitch, we monitored thiM's folding to its metabolite-bound conformation under two different conditions: (1) the metabolite was added after transcription was already complete or (2) the metabolite was present during the transcription of the RNA.

To monitor the conformational changes of thiM upon the addition of TPP, we used the same oligohybridization/RNase H assay we applied for cotranscriptional folding studies of noncoding RNAs. The crystal structures of the TPP-bound thiM riboswitch and structural mapping data have identified several interactions that form upon TPP binding. In particular, J3/2 forms direct contacts with the 4-amino-5-hydroxymethyl-2-methylpyrimidine (HMP) ring of TPP. Upon thiM's binding of TPP, this region became approximately ninefold more protected against the hybridization of a complementary oligoprobe (Fig. 9.2A). Therefore, we elected to use an oligoprobe targeting J3/2 (probe 35) as a monitor for RNA conformational change. Of note, other regions also became more protected against oligohybridization upon thiM's binding to TPP (Fig. 9.2A). In future studies, these may be used to further elucidate the folding pathway of this riboswitch.

To measure the kinetics by which thiM changes conformation if TPP is added after transcription is complete, single-round transcription was allowed to proceed for 15 min before the addition of the metabolite. TPP was then added to a final concentration of 0.1–100 μM. Under this condition, the rate at which thiM folded into its metabolite-bound conformation first increased linearly with the TPP concentration and then leveled off at higher TPP concentration at $\sim 0.1\ s^{-1}$ (Fig. 9.2B). Therefore, at high enough TPP concentrations, a TPP-independent step was rate-limiting. The J3/2 probe targeted the metabolite-binding domain of thiM, which should show a decrease in accessibility soon after TPP binding. Additionally, when Lang *et al.* investigated the response of the thiM aptamer domain to TPP binding using full-length transcripts (Lang *et al.*, 2007), they found that tertiary contacts between L5 and P3 formed quickly and were soon followed by contacts between TPP and both J3/2 and P4/P5. In contrast, formation of the three-way junction and P1 proceeded significantly more slowly. These data suggest that the TPP-independent step occurs before TPP has bound. A number of Watson–Crick base pairs must be broken in the TPP-unbound conformer for the TPP-bound structure to be formed (Fig. 9.1B). We hypothesize that the breaking of these base pairs is the TPP-independent rate-limiting step.

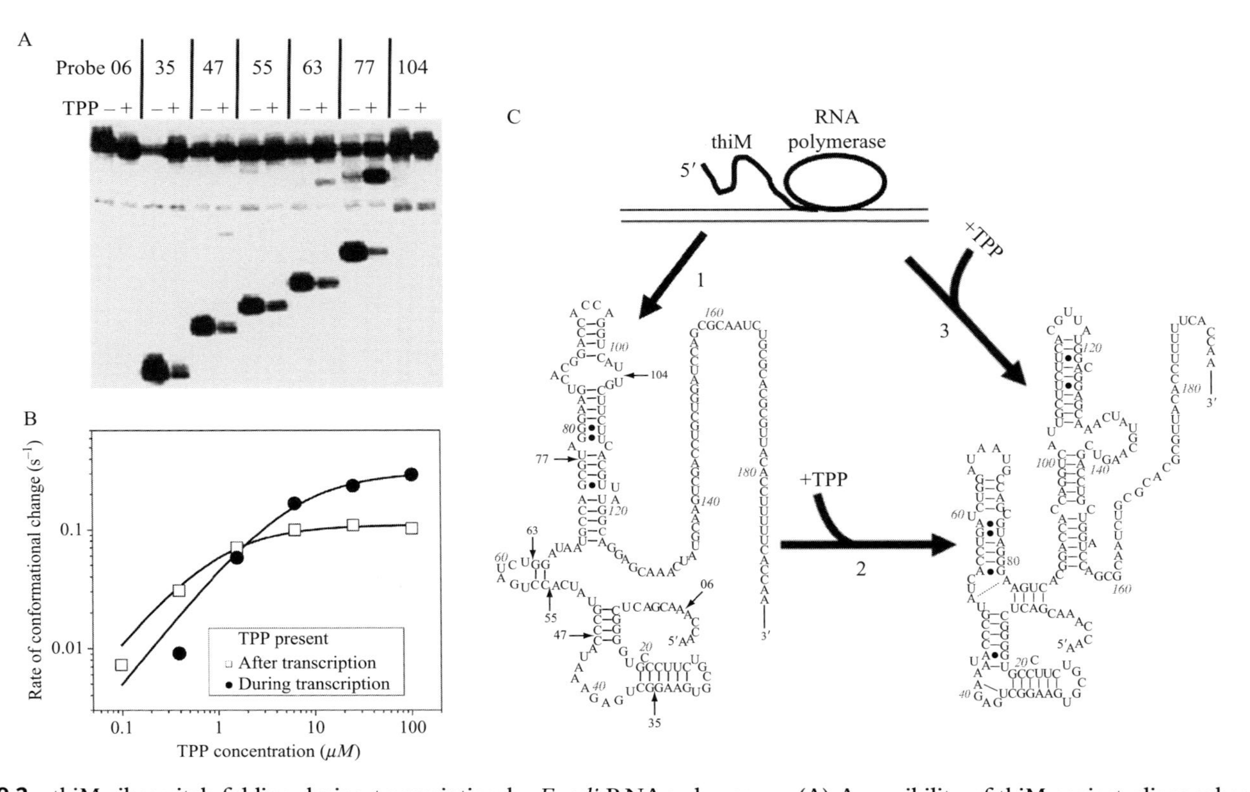

Figure 9.2 thiM riboswitch folding during transcription by *E. coli* RNA polymerase. (A) Accessibility of thiM against oligoprobes in the absence and presence of TPP. Probe 35 targeted the J3/2 region and was used in (B). (B) Rate of thiM conformational change upon the addition of TPP. When TPP was added after transcription was complete, $k_{\max} = 0.11 \pm 0.01\ s^{-1}$ and [TPP] at 50% $k_{\max} = 0.9 \pm 0.1\ \mu M$;

To measure the kinetics by which thiM changes conformation if TPP is present during transcription, the metabolite was added at the same time as the NTPs used to begin the transcription reaction. Under this condition, the thiM riboswitch exhibited a quite different folding behavior. It required sixfold higher concentrations of TPP to fold rapidly into its TPP-bound conformation, but its maximal rate of folding was three fold greater compared to when TPP was added after transcription was complete (Fig. 9.2B). With TPP present during transcription, thiM folded into its TPP-bound state along a different pathway. This pathway involved a faster rate-limiting step at high TPP concentrations.

From these data, we propose the following model for the folding of thiM during transcription (Fig. 9.2C). When TPP is absent, the TPP-unbound conformer is formed (step 1). Upon the later addition of TPP, thiM folds into its TPP-bound conformer (step 2). At high enough TPP concentrations, the rate-limiting step of this transition is to escape from the unbound conformer. With TPP present during transcription, the TPP-unbound conformer is not formed, and the rate-limiting step of escaping from the TPP-unbound conformer is avoided (step 3). Transcriptional pausing after the aptamer domain may allow thiM to fold directly into its TPP-bound state, bypassing the unbound conformer. Thus, both pausing and the presence of the small molecule cofactor can influence the cotranscriptional folding of this riboswitch.

The *B. subtilis* glmS ribozyme responds to the binding of glucosamine-6-phosphate (G6P, Fig. 9.3A; Winkler *et al.*, 2004). It is located in the 5′-UTR of the glucosamine-6-phosphate synthase mRNA. Upon the binding of G6P, *glmS* performs an endonucleolytic cleavage at a specific location in the 5′-UTR, resulting in the downregulation of this gene. Unlike the thiM riboswitch, the glmS ribozyme has the same conformation whether G6P is bound or not (Cochrane *et al.*, 2007; Hampel and Tinsley, 2006; Klein and Ferre-D'Amare, 2006; Tinsley *et al.*, 2007). Thus, there is no G6P-unbound conformer, and cofactor binding may play a different role in glmS's cotranscriptional folding pathway than thiM's. We investigated the folding of glmS when transcribed by *E. coli* RNA polymerase. The self-cleavage of glmS can be used to monitor its folding to the native state (with G6P present). At lower concentrations of G6P, the cleavage rate increased linearly with G6P concentration. The rate leveled off when

---

when TPP was present during transcription, $k_{max} = 0.31 \pm 0.01\ s^{-1}$ and [TPP] at 50% $k_{max} = 6 \pm 1\ \mu M$. (C) Model of thiM folding during transcription. In the absence of TPP, thiM folds into its −TPP form first (step 1) which switches its conformation upon TPP binding at $k_{max} \sim 0.1\ s^{-1}$ (step 2). When TPP is present during transcription, thiM directly folds into its +TPP form at $k_{max} \sim 0.3\ s^{-1}$. Arrows mark the 5′-end of the 10-mer regions targeted by the oligoprobes in (A).

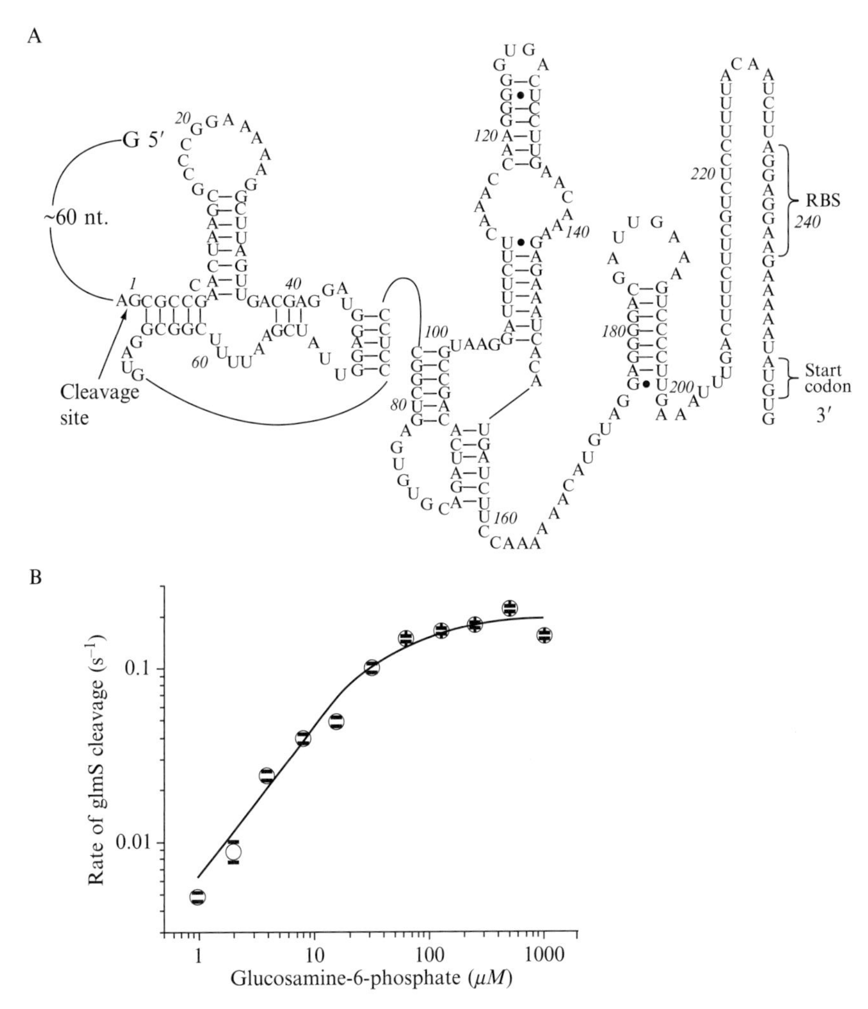

Figure 9.3 *glmS riboswitch folding during transcription by E. coli RNA polymerase.* (A) Proposed secondary structure of the glmS construct used in this study. RBS, ribosome-binding site. (B) Rate of glmS self-cleavage in the presence of glucosamine-6-phosphate. With G6P present during transcription, $k_{max} = 0.20 \pm 0.01$ s^{-1} and [G6P] at 50% $k_{max} = 30 \pm 1$ μM.

[G6P] > 1 m*M*. Overall, glmS had a $K_{1/2}$ of 30 μM with a maximal cleavage rate of 0.2 s^{-1} (Fig. 9.3B). Winkler *et al.* (2004) investigated the responsiveness of glmS to G6P after refolding of the RNA transcript; they determined a $K_{1/2}$ of 200 μM with a maximal cleavage rate of 0.05 s^{-1} at 50 m*M* Tris–HCl, pH 7.5, 10 m*M* $MgCl_2$, 200 m*M* KCl, and 23 °C. The lower $K_{1/2}$ observed in our experiment may be due to altered reaction conditions (e.g., 37 °C). Alternatively, the rate-limiting steps in the two experiments may be different (cotranscriptional vs refolding) and have differing degrees of responsiveness to the presence of G6P. It will be interesting to determine if cotranscriptional cleavage of glmS is folding or catalysis-limited and if the presence of G6P during transcription alters its dependence on G6P.

REFERENCES

Al-Hashimi, H. M., and Walter, N. G. (2008). RNA dynamics: It is about time. *Curr. Opin. Struct. Biol.* **18,** 321–329.

Aleman, E. A., Lamichhane, R., and Rueda, D. (2008). Exploring RNA folding one molecule at a time. *Curr. Opin. Chem. Biol.* **12,** 647–654.

Altman, S. (1989). Ribonuclease P: An enzyme with a catalytic RNA subunit. *Adv. Enzymol. Relat. Areas Mol. Biol.* **62,** 1–36.

Altman, S., and Kirsebom, L. (1999). Ribonuclease P. *In* "The RNA World," (R. Gesteland, T. Cech, and J. Atkins, eds.), 2nd ed. pp. 351–380. Cold Spring Harbor Laboratory Press, Cold Spring Harbor, NY.

Anthony, L. C., Artsimovitch, I., Svetlov, V., Landick, R., and Burgess, R. R. (2000). Rapid purification of His(6)-tagged *Bacillus subtilis* core RNA polymerase. *Protein Expr. Purif.* **19,** 350–354.

Artsimovitch, I., and Landick, R. (2000). Pausing by bacterial RNA polymerase is mediated by mechanistically distinct classes of signals. *Proc. Natl. Acad. Sci. USA* **97,** 7090–7095.

Artsimovitch, I., Svetlov, V., Anthony, L., Burgess, R. R., and Landick, R. (2000). RNA polymerases from *Bacillus subtilis* and *Escherichia coli* differ in recognition of regulatory signals *in vitro*. *J. Bacteriol.* **182,** 6027–6035.

Been, M. D. (2006). HDV ribozymes. *Curr. Top. Microbiol. Immunol.* **307,** 47–65.

Brehm, S. L., and Cech, T. R. (1983). Fate of an intervening sequence ribonucleic acid: Excision and cyclization of the Tetrahymena ribosomal ribonucleic acid intervening sequence *in vivo*. *Biochemistry* **22,** 2390–2397.

Cech, T. R., and Bass, B. L. (1986). Biological catalysis by RNA. *Annu. Rev. Biochem.* **55,** 599–629.

Chang, B. Y., and Doi, R. H. (1990). Overproduction, purification, and characterization of *Bacillus subtilis* RNA polymerase sigma A factor. *J. Bacteriol.* **172,** 3257–3263.

Cochrane, J. C., Lipchock, S. V., and Strobel, S. A. (2007). Structural investigation of the GlmS ribozyme bound to its catalytic cofactor. *Chem. Biol.* **14,** 97–105.

Dalal, R. V., Larson, M. H., Neuman, K. C., Gelles, J., Landick, R., and Block, S. M. (2006). Pulling on the nascent RNA during transcription does not alter kinetics of elongation or ubiquitous pausing. *Mol. Cell* **23,** 231–239.

Diegelman-Parente, A., and Bevilacqua, P. C. (2002). A mechanistic framework for co-transcriptional folding of the HDV genomic ribozyme in the presence of downstream sequence. *J. Mol. Biol.* **324,** 1–16.

Doherty, E. A., and Doudna, J. A. (2000). Ribozyme structures and mechanisms. *Annu. Rev. Biochem.* **69,** 597–615.

Doherty, E. A., and Doudna, J. A. (2001). Ribozyme structures and mechanisms. *Annu. Rev. Biophys. Biomol. Struct.* **30,** 457–475.

Doudna, J. A., and Cech, T. R. (2002). The chemical repertoire of natural ribozymes. *Nature* **418,** 222–228.

Draper, D. E., Grilley, D., and Soto, A. M. (2005). Ions and RNA folding. *Annu. Rev. Biophys. Biomol. Struct.* **34,** 221–243.

Dunn, J. J., and Studier, F. W. (1983). Complete nucleotide sequence of bacteriophage T7 DNA and the locations of T7 genetic elements. *J. Mol. Biol.* **166,** 477–535.

Edwards, T. E., and Ferre-D'Amare, A. R. (2006). Crystal structures of the thi-box riboswitch bound to thiamine pyrophosphate analogs reveal adaptive RNA-small molecule recognition. *Structure* **14,** 1459–1468.

Emerick, V. L., and Woodson, S. A. (1993). Self-splicing of the Tetrahymena pre-rRNA is decreased by misfolding during transcription. *Biochemistry* **32,** 14062–14067.

Fedor, M. J. (2000). Structure and function of the hairpin ribozyme. *J. Mol. Biol.* **297,** 269–291.

Fedor, M. J., and Williamson, J. R. (2005). The catalytic diversity of RNAs. *Nat. Rev. Mol. Cell Biol.* **6,** 399–412.

Frank, D. N., and Pace, N. R. (1998). Ribonuclease P: Unity and diversity in a tRNA processing ribozyme. *Annu. Rev. Biochem.* **67,** 153–180.

Geis, M., Flamm, C., Wolfinger, M. T., Tanzer, A., Hofacker, I. L., Middendorf, M., Mandl, C., Stadler, P. F., and Thurner, C. (2008). Folding kinetics of large RNAs. *J. Mol. Biol.* **379,** 160–173.

Greenleaf, W. J., Frieda, K. L., Foster, D. A., Woodside, M. T., and Block, S. M. (2008). Direct observation of hierarchical folding in single riboswitch aptamers. *Science* **319,** 630–633.

Guerrier-Takada, C., and Altman, S. (1984). Catalytic activity of an RNA molecule prepared by transcription *in vitro*. *Science* **223,** 285–286.

Guerrier-Takada, C., Gardiner, K., Marsh, T., Pace, N., and Altman, S. (1983). The RNA moiety of ribonuclease P is the catalytic subunit of the enzyme. *Cell* **35,** 849–857.

Hagen, M., and Cech, T. R. (1999). Self-splicing of the Tetrahymena intron from mRNA in mammalian cells. *EMBO J.* **18,** 6491–6500.

Hampel, K. J., and Tinsley, M. M. (2006). Evidence for preorganization of the glmS ribozyme ligand binding pocket. *Biochemistry* **45,** 7861–7871.

Heilman-Miller, S. L., and Woodson, S. A. (2003). Effect of transcription on folding of the Tetrahymena ribozyme. *RNA* **9,** 722–733.

Isambert, H., and Siggia, E. D. (2000). Modeling RNA folding paths with pseudoknots: Application to hepatitis delta virus ribozyme. *Proc. Natl. Acad. Sci. USA* **97,** 6515–6520.

Jackson, S. A., Koduvayur, S., and Woodson, S. A. (2006). Self-splicing of a group I intron reveals partitioning of native and misfolded RNA populations in yeast. *RNA* **12,** 2149–2159.

Klein, D. J., and Ferre-D'Amare, A. R. (2006). Structural basis of glmS ribozyme activation by glucosamine-6-phosphate. *Science* **313,** 1752–1756.

Koduvayur, S. P., and Woodson, S. A. (2004). Intracellular folding of the Tetrahymena group I intron depends on exon sequence and promoter choice. *RNA* **10,** 1526–1532.

Kurz, J. C., Niranjanakumari, S., and Fierke, C. A. (1998). Protein component of *Bacillus subtilis* RNase P specifically enhances the affinity for precursor-tRNAAsp. *Biochemistry* **37,** 2393–2400.

Kuznedelov, K., and Severinov, K. (2009). Recombinant bacterial RNA polymerase: Preparation and applications. *Methods* **47,** 44–52.

Landick, R., Stewart, J., and Lee, D. N. (1990). Amino acid changes in conserved regions of the beta-subunit of *Escherichia coli* RNA polymerase alter transcription pausing and termination. *Genes Dev.* **4,** 1623–1636.

Landick, R., Wang, D., and Chan, C. L. (1996). Quantitative analysis of transcriptional pausing by *Escherichia coli* RNA polymerase: His leader pause site as paradigm. *Methods Enzymol.* **274,** 334–353.

Lang, K., Rieder, R., and Micura, R. (2007). Ligand-induced folding of the thiM TPP riboswitch investigated by a structure-based fluorescence spectroscopic approach. *Nucleic Acids Res.* **35,** 5370–5378.

Long, D. M., and Uhlenbeck, O. C. (1993). Self-cleaving catalytic RNA. *FASEB J.* **7,** 25–30.

Loria, A., and Pan, T. (1996). Domain structure of the ribozyme from eubacterial ribonuclease P. *RNA* **2,** 551–563.

Mahen, E. M., Harger, J. W., Calderon, E. M., and Fedor, M. J. (2005). Kinetics and thermodynamics make different contributions to RNA folding *in vitro* and in yeast. *Mol. Cell* **19,** 27–37.

Mandal, M., Lee, M., Barrick, J. E., Weinberg, Z., Emilsson, G. M., Ruzzo, W. L., and Breaker, R. R. (2004). A glycine-dependent riboswitch that uses cooperative binding to control gene expression. *Science* **306,** 275–279.

Meyer, I. M., and Miklos, I. (2004). Co-transcriptional folding is encoded within RNA genes. *BMC Mol. Biol.* **5,** 10.

Misra, V. K., Shiman, R., and Draper, D. E. (2003). A thermodynamic framework for the magnesium-dependent folding of RNA. *Biopolymers* **69,** 118–136.

Monforte, J. A., Kahn, J. D., and Hearst, J. E. (1990). RNA folding during transcription by *Escherichia coli* RNA polymerase analyzed by RNA self-cleavage. *Biochemistry* **29,** 7882–7890.

Nikolcheva, T., and Woodson, S. A. (1999). Facilitation of group I splicing *in vivo*: Misfolding of the Tetrahymena IVS and the role of ribosomal RNA exons. *J. Mol. Biol.* **292,** 557–567.

Ontiveros-Palacios, N., Smith, A. M., Grundy, F. J., Soberon, M., Henkin, T. M., and Miranda-Rios, J. (2008). Molecular basis of gene regulation by the THI-box riboswitch. *Mol. Microbiol.* **67,** 793–803.

Pace, N. R., and Brown, J. W. (1995). Evolutionary perspective on the structure and function of ribonuclease P, a ribozyme. *J. Bacteriol.* **177,** 1919–1928.

Pan, T. (1995). Higher order folding and domain analysis of the ribozyme from *Bacillus subtilis* ribonuclease P. *Biochemistry* **34,** 902–909.

Pan, T., and Jakacka, M. (1996). Multiple substrate binding sites in the ribozyme from *Bacillus subtilis* RNase P. *EMBO J.* **15,** 2249–2255.

Pan, T., and Sosnick, T. (2006). RNA folding during transcription. *Annu. Rev. Biophys. Biomol. Struct.* **35,** 161–175.

Pan, T., Artsimovitch, I., Fang, X. W., Landick, R., and Sosnick, T. R. (1999). Folding of a large ribozyme during transcription and the effect of the elongation factor NusA. *Proc. Natl. Acad. Sci. USA* **96,** 9545–9550.

Reich, C., Gardiner, K. J., Olsen, G. J., Pace, B., Marsh, T. L., and Pace, N. R. (1986). The RNA component of the *Bacillus subtilis* RNase P. Sequence, activity, and partial secondary structure. *J. Biol. Chem.* **261,** 7888–7893.

Reich, C., Olsen, G. J., Pace, B., and Pace, N. R. (1988). Role of the protein moiety of ribonuclease P, a ribonucleoprotein enzyme. *Science* **239,** 178–181.

Rentmeister, A., Mayer, G., Kuhn, N., and Famulok, M. (2008). Secondary structures and functional requirements for thiM riboswitches from *Desulfovibrio vulgaris, Erwinia carotovora* and *Rhodobacter spheroides. Biol. Chem.* **389,** 127–134.

Repsilber, D., Wiese, S., Rachen, M., Schroder, A. W., Riesner, D., and Steger, G. (1999). Formation of metastable RNA structures by sequential folding during transcription: Time-resolved structural analysis of potato spindle tuber viroid (−)-stranded RNA by temperature-gradient gel electrophoresis. *RNA* **5,** 574–584.

Schmidt, M. C., and Chamberlin, M. J. (1984). Amplification and isolation of *Escherichia coli* nusA protein and studies of its effects on *in vitro* RNA chain elongation. *Biochemistry* **23,** 197–203.

Serganov, A., Polonskaia, A., Phan, A. T., Breaker, R. R., and Patel, D. J. (2006). Structural basis for gene regulation by a thiamine pyrophosphate-sensing riboswitch. *Nature* **441,** 1167–1171.

Shapiro, B. A., Kasprzak, W., Grunewald, C., and Aman, J. (2006). Graphical exploratory data analysis of RNA secondary structure dynamics predicted by the massively parallel genetic algorithm. *J. Mol. Graph. Model.* **25,** 514–531.

Sosnick, T. R., and Pan, T. (2003). RNA folding: Models and perspectives. *Curr. Opin. Struct. Biol.* **13,** 309–316.

Studier, F. W., Rosenberg, A. H., Dunn, J. J., and Dubendorff, J. W. (1990). Use of T7 RNA polymerase to direct expression of cloned genes. *Methods Enzymol.* **185,** 60–89.

Susa, M., Sen, R., and Shimamoto, N. (2002). Generality of the branched pathway in transcription initiation by *Escherichia coli* RNA polymerase. *J. Biol. Chem.* **277,** 15407–15412.

Tinsley, R. A., Furchak, J. R., and Walter, N. G. (2007). *Trans*-acting glmS catalytic riboswitch: Locked and loaded. *RNA* **13,** 468–477.

Treiber, D. K., and Williamson, J. R. (2000). Kinetic oligonucleotide hybridization for monitoring kinetic folding of large RNAs. *Methods Enzymol.* **317,** 330–353.

Treiber, D. K., and Williamson, J. R. (2001). Beyond kinetic traps in RNA folding. *Curr. Opin. Struct. Biol.* **11,** 309–314.

Tyagarajan, K., Monforte, J. A., and Hearst, J. E. (1991). RNA folding during transcription by T7 RNA polymerase analyzed using the self-cleaving transcript assay. *Biochemistry* **30,** 10920–10924.

Wickiser, J. K., Winkler, W. C., Breaker, R. R., and Crothers, D. M. (2005). The speed of RNA transcription and metabolite binding kinetics operate an FMN riboswitch. *Mol. Cell* **18,** 49–60.

Winkler, W., Nahvi, A., and Breaker, R. R. (2002a). Thiamine derivatives bind messenger RNAs directly to regulate bacterial gene expression. *Nature* **419,** 952–956.

Winkler, W. C., Cohen-Chalamish, S., and Breaker, R. R. (2002b). An mRNA structure that controls gene expression by binding FMN. *Proc. Natl. Acad. Sci. USA* **99,** 15908–15913.

Winkler, W. C., Nahvi, A., Roth, A., Collins, J. A., and Breaker, R. R. (2004). Control of gene expression by a natural metabolite-responsive ribozyme. *Nature* **428,** 281–286.

Wong, T., Sosnick, T. R., and Pan, T. (2005). Mechanistic insights on the folding of a large ribozyme during transcription. *Biochemistry* **44,** 7535–7542.

Wong, T. N., Sosnick, T. R., and Pan, T. (2007). Folding of noncoding RNAs during transcription facilitated by pausing-induced nonnative structures. *Proc. Natl. Acad. Sci. USA* **104,** 17995–18000.

Woodson, S. A. (2002). Folding mechanisms of group I ribozymes: Role of stability and contact order. *Biochem. Soc. Trans.* **30,** 1166–1169.

Woodson, S. A. (2005). Metal ions and RNA folding: A highly charged topic with a dynamic future. *Curr. Opin. Chem. Biol.* **9,** 104–109.

Xayaphoummine, A., Bucher, T., and Isambert, H. (2005). Kinefold web server for RNA/DNA folding path and structure prediction including pseudoknots and knots. *Nucleic Acids Res.* **33,** W605–W610.

Xayaphoummine, A., Viasnoff, V., Harlepp, S., and Isambert, H. (2007). Encoding folding paths of RNA switches. *Nucleic Acids Res.* **35,** 614–622.

Zarrinkar, P. P., and Williamson, J. R. (1994). Kinetic intermediates in RNA folding. *Science* **265,** 918–924.

Zarrinkar, P. P., Wang, J., and Williamson, J. R. (1996). Slow folding kinetics of RNase P RNA. *RNA* **2,** 564–573.

Zhang, F., Ramsay, E. S., and Woodson, S. A. (1995). *In vivo* facilitation of Tetrahymena group I intron splicing in *Escherichia coli* pre-ribosomal RNA. *RNA* **1,** 284–292.

CHAPTER TEN

Catalytic Activity as a Probe of Native RNA Folding

Yaqi Wan, David Mitchell III, *and* Rick Russell

Contents

Abstract

As RNAs fold to functional structures, they traverse complex energy landscapes that include many partially folded and misfolded intermediates. For structured RNAs that possess catalytic activity, this activity can provide a powerful means of monitoring folding that is complementary to biophysical approaches. RNA catalysis can be used to track accumulation of the native RNA specifically and quantitatively, readily distinguishing the native structure from intermediates that resemble it and may not be differentiated by other approaches. Here, we outline how to

Department of Chemistry and Biochemistry, Institute for Cellular and Molecular Biology, University of Texas at Austin, Austin, Texas, USA

Methods in Enzymology, Volume 468
ISSN 0076-6879, DOI: 10.1016/S0076-6879(09)68010-1

design and interpret experiments using catalytic activity to monitor RNA folding, and we summarize adaptations of the method that have been used to probe aspects of folding well beyond determination of the folding rates.

1. Introduction

The discovery in 1982 by Cech and coworkers that a group I intron from *Tetrahymena thermophila* folds into a specific tertiary structure and catalyzes the chemical reactions necessary to excise itself from a precursor rRNA (Kruger *et al.*, 1982) ushered in a new age of interest in understanding how RNAs fold to functional structures. Since then, interest in RNA folding has continued to grow as structured RNAs have been implicated in a dizzying array of cellular functions. In addition to performing self-splicing and cleavage reactions in a wide variety of contexts (Fedor, 2009), RNAs process and translate messages as core components of the spliceosome and ribosome (Moore *et al.*, 2002; Noller, 2005; Wahl *et al.*, 2009), process tRNA precursors (Evans *et al.*, 2006), protect chromosomal ends (Collins, 2006; Theimer and Feigon, 2006), mediate protein trafficking events (Egea *et al.*, 2005), and regulate gene expression by a variety of mechanisms (Bartel, 2004; Carthew and Sontheimer, 2009).

In parallel, the known catalytic RNAs and the corresponding catalytic repertoire have increased dramatically. The natural catalytic RNAs include group I and group II introns, RNase P, and self-cleaving RNA elements including the hairpin, hammerhead, Varkud satellite, and hepatitis delta virus (HDV) ribozymes, and most recently the glmS riboswitch (Cech, 2002; Fedor, 2009). Self-cleaving elements have been found in the genomes of humans and other mammals (Martick *et al.*, 2008; Salehi-Ashtiani *et al.*, 2006; Teixeira *et al.*, 2004), and it seems likely that many more remain undiscovered (Carninci *et al.*, 2005; Katayama *et al.*, 2005; Salehi-Ashtiani *et al.*, 2006). The catalytic capabilities of RNA have even been extended by *in vitro* selection methods beyond the range identified in nature to include phosphorylation and ligation reactions (Lincoln and Joyce, 2009; Lorsch and Szostak, 1994), and even more diverse reactions such as alkylations and condensations (Chen *et al.*, 2007).

As described throughout this volume, a powerful arsenal of biophysical approaches has been developed and applied to study the formation of specific structures by RNA. These approaches have been used to probe questions such as what provides the energetic driving force for formation of compact RNA structures? What features of the RNAs provide specificity to stabilize the functional structures relative to others? How fast do RNAs form specific structures, and what physical steps limit the folding rates? More generally, what do the energy landscapes of RNA folding look like?

Beginning in the early 1990s, catalytic RNAs became prominent in studies of RNA folding, and work using these RNAs has fundamentally advanced our understanding of folding processes and landscapes. Two powerful time-resolved footprinting methods were developed and first applied to the *Tetrahymena* group I ribozyme. One of them followed accessibility of the RNA to complementary oligonucleotide probes (Zarrinkar and Williamson, 1994, 1996), and the other used hydroxyl radicals to monitor backbone exposure of RNA within milliseconds after initiation of folding by addition of Mg^{2+} ions (Sclavi *et al.*, 1997, 1998).

The most general conclusion from these studies was that RNA folding involves multiple steps, with highly structured intermediates being populated along the way. Since then, the presence of intermediates has been shown to be a general feature of RNA folding (Chadalavada *et al.*, 2002; Chauhan and Woodson, 2008; Pan and Sosnick, 1997; Pan and Woodson, 1998; Russell and Herschlag, 1999; Treiber *et al.*, 1998). The extensive formation of intermediates creates a significant challenge for studying RNA folding. It is critical to know whether the RNA is completely folded to the native, functional structure or instead is present as a folding intermediate, but this can be exceedingly difficult for folding intermediates that are extensively structured and bear a strong resemblance to the native structure.

Catalytic RNAs provide a unique window to view the folding process, because the onset of catalytic activity can provide a powerful and specific readout for formation of the native structure. This approach allows the native state to be distinguished from alternative long-lived intermediates, regardless of their physical similarities, as only the native state is capable of performing the catalytic reaction. In the most favorable cases, this method can provide an unambiguous measure of the amount of native ribozyme as a function of time, which can then be integrated with data from other approaches. It was used in early studies of ribosome assembly (Held and Nomura, 1973; Traub and Nomura, 1969), and since then has been applied to a broad range of RNAs (e.g., Brooks and Hampel, 2009; Chadalavada *et al.*, 2002; Emerick *et al.*, 1996; Pan and Sosnick, 1997; Russell and Herschlag, 1999; Su *et al.*, 2003; Woodson and Cech, 1991; Xiao *et al.*, 2005). Catalytic activity has also been used to follow protein folding (Kiefhaber, 1995; Kiefhaber *et al.*, 1990), but it is uniquely well suited to RNA because the propensity of RNA to form kinetically trapped intermediates causes folding of many RNAs to be slower than their catalytic reactions (Treiber and Williamson, 1999).

Nevertheless, there are also limitations to the activity approach. Of course, it is limited to RNAs that possess catalytic activity, and further, it can only be used for RNAs whose catalytic activity is relatively robust.

Also, the method provides limited information on steps in folding and the structures of intermediates.

There are two general classes of catalytic activity assay, termed continuous and discontinuous assays (Fig. 10.1). In the continuous assay, folding and catalysis are monitored together, and the rate constant observed by monitoring the catalytic reaction reflects the folding process if folding is rate-limiting. In the discontinuous assay, folding is separated from the chemical reaction by performing the reaction in two stages, where the first stage allows folding but not catalysis, and the second stage allows catalysis but not further formation of native RNA.

In the sections below, we describe how to perform and interpret RNA folding experiments using each of these assays. We and others have used catalytic activity extensively to study ribozymes that cleave RNA, particularly group I introns (Pan and Woodson, 1998; Russell and Herschlag, 1999, 2001; Russell *et al.*, 2007; Zarrinkar and Williamson, 1994). For simplicity, we focus on these ribozymes in our descriptions, but the same methods are applicable to ribozymes that perform other types of reactions. Also for simplicity, we assume in the sections below that the reaction substrate is radiolabeled and is separated from the reaction product using polyacrylamide gel electrophoresis, but any method that separates reaction substrates and products quantitatively can be used. Our goal in the sections below is not to list detailed protocols, as the details will differ for different RNAs and different methods of labeling and detecting substrates, but rather to outline general strategies of experimental design and interpretation that allow catalytic activity to be an effective monitor of RNA folding.

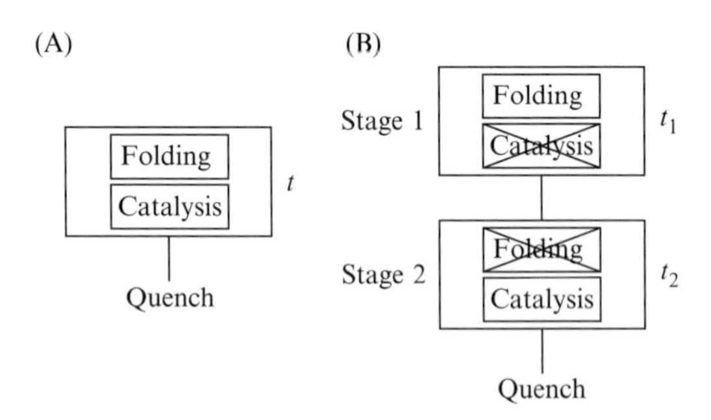

Figure 10.1 Schematics of continuous and discontinuous catalytic activity assays for folding. (A) Continuous assay. The reaction is performed in a single stage, in which both folding and catalysis occur. Aliquots are quenched at various times t. (B) Discontinuous assay. In the first stage, folding is allowed but the catalytic reaction is blocked. At various times t_1, aliquots are transferred to stage 2. Here, the conditions allow robust catalysis while preventing further accumulation of native ribozyme, allowing determination of the fraction of native ribozyme at time t_1.

2. Preliminary Measurements of Catalytic Reaction

The only absolute requirement for using catalytic activity to monitor folding is that conditions exist under which the overall rate constant for the catalytic reaction, including all steps leading up to detectable product formation, exceeds the rate constant for folding. The degree to which catalysis must be faster than folding depends on a variety of factors, but as a general rule the difference must be at least three- to fivefold. As a first step, it is important to measure the catalytic reaction, independently from folding, and to survey conditions to maximize the fraction of active ribozyme and the catalytic rate constant. In practice, these measurements may be performed in parallel with trial folding measurements using the continuous assay described in Section 3.

2.1. How to measure the catalytic reaction and establish prefolding conditions

The basic experiment for monitoring the catalytic reaction is described in Fig. 10.2A. First, the RNA is prefolded in the presence of Mg^{2+}, typically at elevated temperature (10–50 m*M* Mg^{2+}, 37–60 °C, 10–60 min) and in the absence of a substrate or cofactor. In practice, the prefolding conditions will not yet be established for a new ribozyme, so the time and conditions of this incubation should be varied. Next, samples are adjusted to a constant set of conditions to measure the catalytic reaction rate, and the substrate is added. In an initial experiment, the substrate should be present in modest excess of the ribozyme (two- to fivefold) so that the first round of the catalytic reaction can be observed and distinguished from subsequent rounds. Aliquots are quenched at various times by adding formamide (50–90% final concentration) and EDTA ($\geq$twofold in excess of Mg^{2+}), and products are separated from substrate, typically by denaturing polyacrylamide gel electrophoresis (Herschlag and Cech, 1990; Zaug *et al.*, 1988). The fraction of substrate converted to product is normalized by the substrate and the ribozyme concentrations (as determined by UV absorbance) and plotted against reaction time.

A set of preincubation conditions should be selected that maximizes the fraction of active ribozyme, as determined from the amount of product in a burst or from the steady-state rate (see below). Once optimal preincubation conditions are established, this step is kept constant while the reaction conditions are surveyed (Section 2.3).

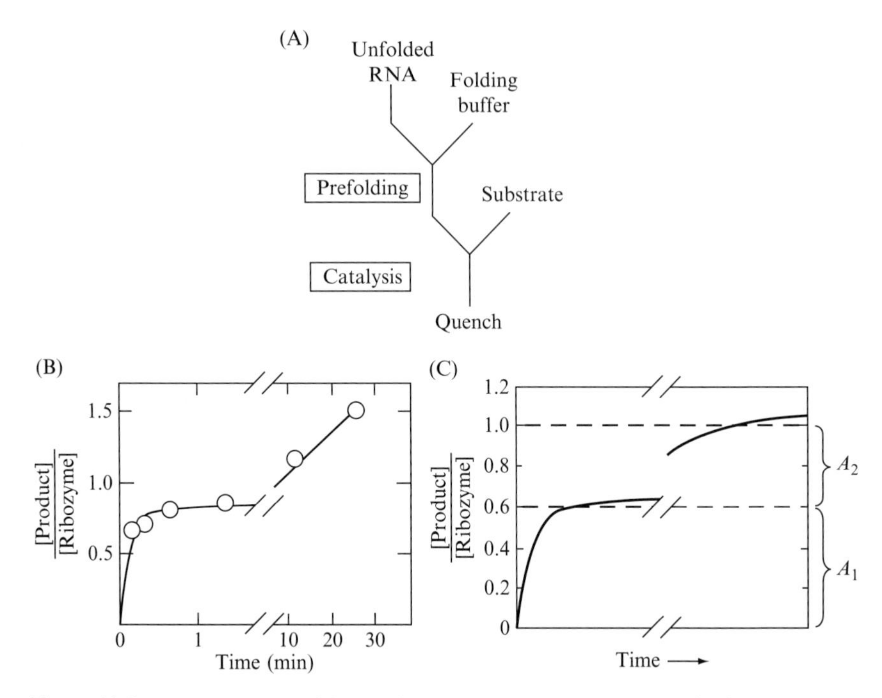

Figure 10.2 Measurement of the catalytic reaction. (A) Experimental scheme. RNA is prefolded (typically 10–50 m*M* Mg^{2+}, 37–60 °C, 10–60 min) and then mixed with substrate. (B) Measurement of substrate cleavage by the *Tetrahymena* ribozyme (Russell and Herschlag, 1999). The reaction included 500 n*M* oligonucleotide substrate (with trace 5′-radiolabeled substrate) and 200 n*M* prefolded ribozyme and was performed at 37 °C, pH 7.0, 10 m*M* Mg^{2+}, with 1 m*M* guanosine. The cleavage reaction gave a burst of 0.8 product/ribozyme, indicating that most of the ribozyme was catalytically active. The rate constant was ≥ 8 min^{-1}, faster than folding under these conditions. (C) Simulated data showing biphasic kinetics from an internal equilibrium (K_{Int}) between substrate cleavage and ligation. The approach to equilibrium occurs rapidly (lower dashed line at 0.6), and then dissociation of one of the products drives the reaction forward to a [product]/[ribozyme] value of 1 (upper dashed line, amplitude of 0.4 for the slow phase). From Eqs. (10.3) and (10.4), $K_{\mathrm{Int}} = 1.5$ (0.6/0.4) and the correction factor $C_{\mathrm{F}} = 0.6$. Panel (B) reprinted from Russell and Herschlag (1999) with permission from Elsevier.

2.2. Interpreting results from catalytic rate measurements

The method used to analyze the progress curve depends on the shape of the curve, which in turn depends on whether the reaction products are released faster or slower than they are produced. For a ribozyme that cleaves a substrate to generate two products, there are three regimes (Table 10.1; corresponding regimes for ribozymes that self-cleave or produce only one reaction product are also shown). If both products are released substantially

Table 10.1 Effects of product release rates on reaction and folding kinetics

Relative product release rates[a]	Intramolecular cleavage	Intermolecular reaction 1 product	Intermolecular reaction 2 products (e.g., substrate cleavage)
Fast	Burst kinetics, amplitude corresponds to active ribozyme	Linear product accumulation, no burst. Rate-limiting folding detectable by lag kinetics	Linear product accumulation, no burst. Rate-limiting folding detectable by lag kinetics
One product fast, one product slow	NA	NA	Burst kinetics, amplitude corresponds to active ribozyme
Slow	Burst kinetics, amplitude may be reduced by reverse reaction	Burst kinetics, amplitude may be reduced by reverse reaction	Burst kinetics, amplitude may be reduced by reverse reaction

[a] Rate constants for product release are considered relative to those for earlier steps of the catalytic reaction for reactions with prefolded ribozyme, and relative to the folding rate constant for experiments using the continuous activity assay (Section 3).

faster than the substrate is cleaved ($\geq$three- to fivefold), a linear accumulation of product will be observed. Changes in the fraction of native ribozyme from different prefolding conditions are reflected in corresponding changes in the reaction rate. If dissociation of one product is faster than cleavage and the other is slower, a rapid burst of product will be generated, which gives values both for the cleavage rate and for the amount of active ribozyme. If both products are released slower, a burst will be present, but the amplitude of the burst may correspond to less than the total amount of active ribozyme (see below).

If the progress curve shows a burst of product formation, the normalized data are fit by Eq. (10.1), in which A is the amplitude of the product burst, which reflects the amount of active ribozyme. Prefolding conditions should be chosen to maximize this value. k_{obs} is the rate constant for the catalytic reaction for steps leading up to product formation, s is the rate constant for the slower steady-state reaction, and t is the reaction time:

$$\frac{[\text{Product}]}{[\text{Ribozyme}]} = A(1 - \mathrm{e}^{-k_{obs}t}) + st. \tag{10.1}$$

A cleavage reaction of the *Tetrahymena* ribozyme, for which one product is released rapidly and the other slowly (middle regime in Table 10.1; Herschlag and Cech, 1990), is shown in Fig. 10.2B. After folding (30 min, 50 °C, 10 mM Mg^{2+}), the ribozyme rapidly performs a single round of oligonucleotide substrate cleavage at 37 °C and 10 mM Mg^{2+} (Herschlag and Cech, 1990; Russell and Herschlag, 1999; Zarrinkar and Williamson, 1994). The burst amplitude indicates that nearly all of the ribozyme is active. One of the products is then released slowly, as is often observed for RNA products that bind by base-pairing interactions, producing a slow steady-state phase.

If both products are released slowly, the burst amplitude may correspond to less than the total amount of active ribozyme because the cleavage products can religate on the ribozyme (Scheme 10.1), reaching an internal equilibrium (Eq. 10.2) (Buzayan *et al.*, 1986; Golden and Cech, 1996; Hegg and Fedor, 1995; Karbstein *et al.*, 2002):

$$K_{\mathrm{Int}} = \frac{k_{\mathrm{forward}}}{k_{\mathrm{reverse}}}. \tag{10.2}$$

However, a small amplitude may indicate instead that the ribozyme is not fully folded to the native state or that some of the ribozyme is damaged and therefore inactive.

Although it is not trivial to distinguish between these possibilities, a strong indicator can come from the shape of the reaction progress curve. If the reaction is reversible and the products dissociate with distinct rate constants, the progress curve will give a double exponential within the first turnover, where the first phase, with amplitude A_1, represents the approach to K_{Int} and the second phase, with amplitude A_2, represents dissociation of the first product (Fig. 10.2C). In contrast, the presence of inactive or misfolded ribozyme will lead to a single exponential followed by a linear phase. The amplitudes of the two exponential phases allow calculation of K_{Int} (Eq. 10.3) and a correction factor, C_{F} (Eq. 10.4). The measured burst amplitude is divided by C_{F} to obtain the fraction of native ribozyme:

$$K_{\mathrm{Int}} = \frac{A_1}{A_2}, \tag{10.3}$$

$$C_{\mathrm{F}} = \frac{A_1}{A_1 + A_2}. \tag{10.4}$$

$$\mathrm{E{\bullet}S} \rightleftharpoons \mathrm{E{\bullet}P_1{\bullet}P_2} \xrightarrow[\mathrm{P_1}]{} \mathrm{E{\bullet}P_2} \xrightarrow[\mathrm{P_2}]{} \mathrm{E}$$

Scheme 10.1

It should also be confirmed that one of the products is released with the rate constant of the slower phase by measuring its dissociation directly using pulse-chase procedures (Herschlag and Cech, 1990; Rose *et al.*, 1974). To confirm the conclusion of an internal equilibrium, the reverse reaction should also be probed directly by monitoring reaction beginning from products (Hertel and Uhlenbeck, 1995; Karbstein *et al.*, 2002).

2.3. Surveying conditions for use in folding experiments

After preincubation conditions are established, it is important to measure the catalytic rate under various conditions. Conditions that should be varied include monovalent and divalent cation concentrations, pH, and temperature. Substrate concentration should also be varied to establish saturation. The main goals are to identify conditions under which the catalytic rate exceeds the folding rate and to maximize this difference. Increases in pH and Mg^{2+} concentration increase the catalytic rates of many RNAs (Herschlag and Khosla, 1994; Hertel and Uhlenbeck, 1995; Narlikar and Herschlag, 1997), and a decrease in temperature may be advantageous because it may give a smaller decrease in the catalytic rate than the folding rate (Russell *et al.*, 2007).

For all reaction conditions, the ribozyme should be first prefolded under the optimal set of conditions established above. This allows changes in observed reaction rate to be interpreted in terms of effects on the catalytic steps, not on the fraction of active ribozyme (assuming that the reaction conditions do not give loss of native ribozyme). In practice, reactions in which the ribozyme is not prefolded are conveniently performed in parallel (Section 3). This allows scanning of conditions to find those under which the catalytic rate exceeds the folding rate, allowing the continuous assay to be used. These conditions can also be used as the second stage of a discontinuous assay (Section 4), allowing folding to be measured even under first-stage conditions that do not support robust activity, as well as allowing a broad set of applications (Section 5).

3. Following RNA Folding by Continuous Activity Assay

This section describes a simple assay for measuring RNA folding in which the adoption of the native state is determined by onset of catalytic activity. RNA folding is initiated at the same time as the catalytic reaction and, provided that folding is the rate-limiting step, the observed rate constant for the catalytic activity reflects the rate constant for folding. The continuous assay is very simple, so it is a good first choice for ribozymes whose catalytic reactions are rapid. It is also useful for studying RNAs for which folding and cleavage steps cannot easily be separated, such as self-processing RNAs that do not require cofactors. The main limitations are that it cannot be used to follow folding under conditions where folding is faster than the catalytic reaction, and it cannot generally be used to track unfolding processes, with the fraction of native ribozyme decreasing over time.

3.1. How to perform experiments using the continuous assay

The overall technique for measuring the rate of RNA folding is very similar to that for measuring the rate of substrate cleavage. The only difference is that the substrate, in small excess relative to the ribozyme, is added as folding is initiated, typically by Mg^{2+} addition (Fig. 10.3A). As in Section 2, aliquots are removed at various times and quenched.

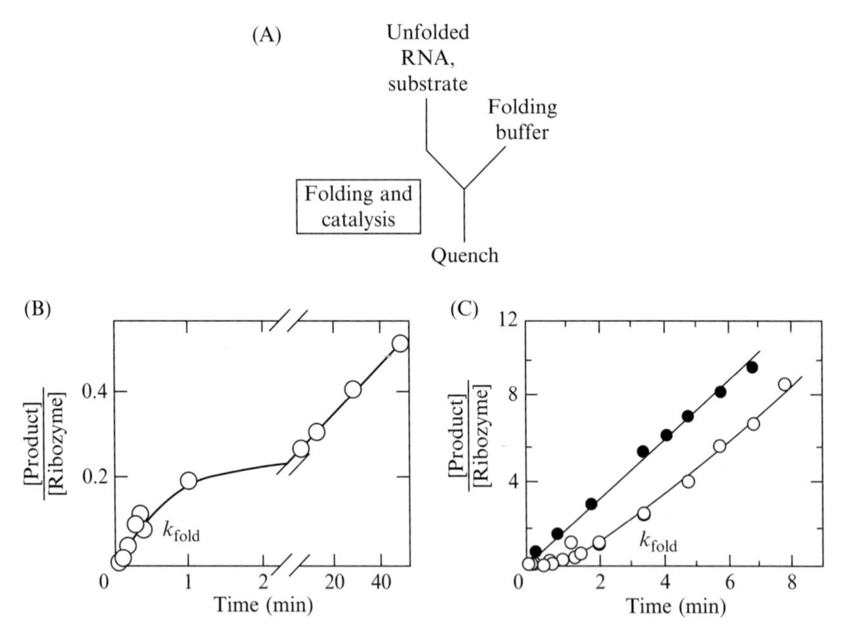

Figure 10.3 Monitoring RNA folding using the continuous assay. (A) Experimental scheme. Folding is initiated in the presence of substrate, and folding and catalysis are quenched with formamide and EDTA. (B) Example of burst kinetics for the wild-type *Tetrahymena* ribozyme (Russell and Herschlag, 1999). The observed rate constant reflects folding, and the slow, linear phase is limited by product release. The burst amplitude (0.27) is lower than in an equivalent reaction of prefolded ribozyme (see Fig. 10.2B), indicating that most of the ribozyme did not reach the native state on this time scale. (C) Lag kinetics for the $E^{\Delta P5abc}$ *Tetrahymena* variant ribozyme. The steady-state rate was the same as that for prefolded ribozyme (see text). Panels (B) and (C) adapted from Russell and Herschlag (1999) with permission from Elsevier.

3.2. Interpreting results from the continuous assay

The results are interpreted in conjunction with catalytic rate measurements described in Section 2. If the progress curves are the same with or without prefolding of the ribozyme, folding is more rapid than cleavage and only a lower limit on the folding rate is obtained. On the other hand, if the reaction with coinitiation of folding and cleavage gives slower or less product formation, folding is limiting for at least a fraction of the population and further interpretation of the data is warranted.

There are two general types of behavior that may be observed. If the rate constant for release of at least one product is slower than folding and the catalytic reaction, product will be formed in an initial burst. In this case, the data are normalized for substrate and ribozyme concentration as in Section 2 and fit by Eq. (10.1). Here, k_{obs} reflects the rate constant for folding (k_{fold}), and the amplitude gives the fraction of ribozyme that folds to the native state in this transition. If the burst amplitude indicates less than stoichiometric formation of product, it suggests that some of the ribozyme does not reach the native state on this time scale (or that the reaction reaches an internal equilibrium before product dissociation, see Section 2.2). To rule out the possibility that the low value results from damaged ribozyme, the amplitude should be compared to that from a reaction with prefolded ribozyme. Using this assay, it was shown that a small fraction of the *Tetrahymena* ribozyme reaches the native state with a rate constant of 1 min^{-1} (Fig. 10.3B), whereas the rest misfolds and remains non-native on the time scale of the experiment (Russell and Herschlag, 1999, 2001). In principle, the continuous assay can also reveal multiple folding pathways by the presence of additional exponential phases. However, in practice it can be difficult to distinguish these phases from continued cleavage by native ribozyme formed in the fastest phase of folding.

The second possible behavior is a steady-state accumulation of product without a burst, which may be preceded by a lag. A lag appears if the folding process is slower than the entire catalytic cycle, because the steady-state rate of product formation depends on the concentration of active ribozyme, and this concentration increases slowly as the native state accumulates in folding. An example of lag kinetics is shown in Fig. 10.3C (Russell and Herschlag, 1999). In this case, the data are fit by Eq. (10.5), in which s represents the rate constant for the steady-state phase and k_{obs} is the rate constant for the lag, which is equal to the folding rate constant k_{fold}. To maximize the signal from a lag, a large excess of substrate should be used relative to the ribozyme ($\geq$tenfold) to increase the length and linearity of the steady-state phase:

$$\frac{[\text{Product}]}{[\text{Ribozyme}]} = s\left[t - \frac{1}{k_{obs}} + \frac{1}{k_{obs}}\left(e^{-k_{obs}t}\right)\right]. \tag{10.5}$$

To obtain a relative measure of the fraction of ribozyme present in the native state, the steady-state rate is compared to that for the prefolded RNA. If the rate is lower without prefolding, not all of the ribozyme has folded to the native state, whereas if they are the same, there is as much native ribozyme as in the prefolded control. However, it is important that this comparison does not give an absolute measure of the fraction of native ribozyme, only a relative one. For example, a *Tetrahymena* ribozyme variant that lacks the P5abc peripheral element ($E^{\Delta P5abc}$) gives the same steady-state rate with or without prefolding (Fig. 10.3C; Russell and Herschlag, 1999). Nevertheless, this ribozyme equilibrates between the native and misfolded conformers, with the native state populated only by about 60% of the ribozyme even after prefolding (Johnson *et al.*, 2005). This incomplete occupancy was not detectable using the continuous assay.

4. Following RNA Folding by Discontinuous Activity Assay

The essence of the discontinuous assay is that the reaction is performed in two stages (see Fig. 10.1). In the first stage, folding is allowed to proceed but the catalytic reaction is blocked. In the second stage, the conditions are changed such that folding is arrested and the catalytic reaction is initiated, allowing determination of the fraction of native ribozyme that accumulated during the first stage. The discontinuous assay has several advantages over the continuous assay. By separating folding from the catalytic reaction, conditions can be chosen for stage 2 that maximize the catalytic rate, increasing the robustness of the signal. Indeed, using this assay it is even possible to measure the folding rate under conditions in stage 1 that do not give rapid catalysis, such as low Mg^{2+} concentration. Further, decreases in the fraction of native ribozyme can be determined, allowing unfolding processes to be followed.

There are also disadvantages and limitations to the discontinuous assay. It can only be used if conditions exist that arrest folding. Minimally, folding must become much slower than catalysis, and in practice it is valuable to establish conditions that arrest folding on the experimental time scale (minutes for experiments performed by hand). Another limitation is that the discontinuous assay requires more work than the continuous assay, because each folding time course consists of a series of reaction time courses. This disadvantage can be minimized by collecting only a small number of points during each cleavage reaction, which can be sufficient to measure reliably the fraction of native ribozyme if the reaction kinetics are well understood.

4.1. How to design and perform experiments using the discontinuous assay

The key to using the discontinuous assay is to establish conditions for the two reaction stages (Fig. 10.4A). In the first stage, the catalytic reaction must be inhibited, most commonly by omitting the substrate or a necessary cofactor. In the second stage, folding must be arrested, or at least slowed to a lower rate than the catalytic reaction. This goal has been achieved by using high Mg^{2+}

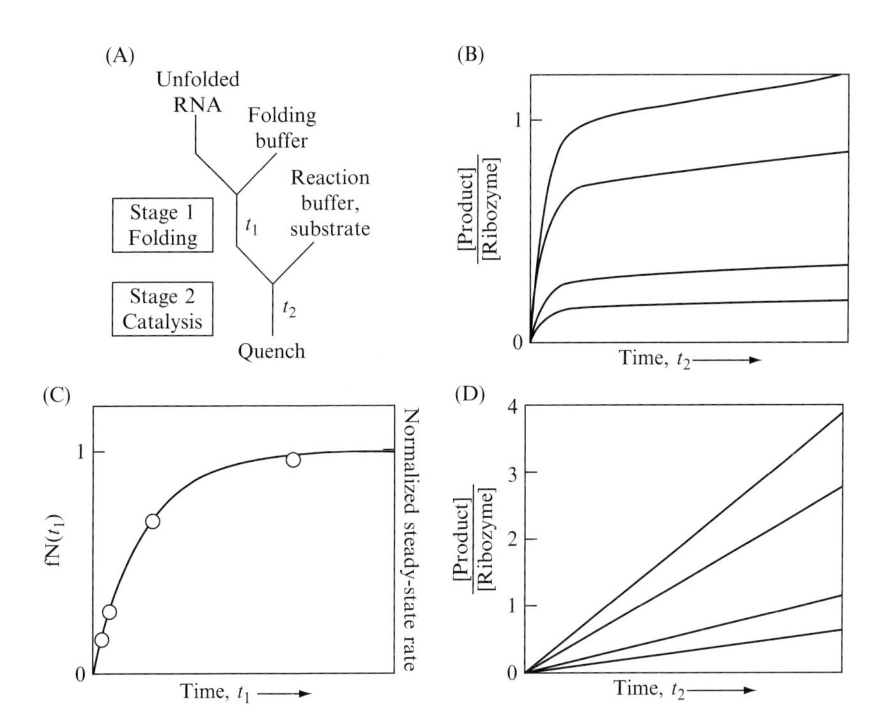

Figure 10.4 Monitoring RNA folding by the discontinuous activity assay. (A) Experimental design. Unfolded RNA is first mixed with Mg^{2+} to initiate folding in stage 1. After various times t_1, aliquots are transferred to stage 2, blocking further native folding, and are mixed with substrate. Aliquots from stage 2 are quenched at various times t_2 by formamide and EDTA. (B) Simulated data for a ribozyme that gives burst kinetics. The four curves represent stage 2 reactions after different folding times t_1. Each curve has the same rate constant, which reflects the catalytic reaction, followed by a steady-state increase that reflects subsequent turnovers. (C) Simulated folding curve. Plotting the burst amplitudes from (B) against folding time t_1 gives an exponential curve with rate constant k_{fold}. (D) Simulated product accumulation for four stage 2 reactions with a ribozyme that rapidly releases products. The slopes are also plotted in (C) and give the same folding curve as the burst amplitudes from (B), except that here the amounts of native ribozyme are on a relative scale (alternate y-axis labels at right).

concentration (50–100 m*M*) and/or low temperature (Pan *et al.*, 1999; Russell and Herschlag, 2001; Russell *et al.*, 2006, 2007) (see Section 2.3).

When conditions for the two stages are established, folding experiments can be performed. Folding is typically initiated in stage 1 by adding divalent ions (Mg^{2+}). Aliquots are removed from the mixture at different times t_1 and transferred to conditions for stage 2. This can be done in a single step or, if the folding reaction is sufficiently slow under stage 2 conditions, it can be performed in multiple steps. For example, the aliquot can first be transferred to low temperature and/or high Mg^{2+} conditions, and then substrate can be added. Because aliquots must be quenched at relatively defined times after substrate addition for a good determination of the burst amplitude or reaction rate, it is helpful to "unlink" this reaction from the stage 1 reaction by adding substrate separately. Quenched aliquots are processed as above.

4.2. Interpreting results from the discontinuous assay

For each folding time t_1, the fraction of cleaved substrate is normalized by the substrate and ribozyme concentrations and plotted against t_2. Most commonly, the progress curve can be fit to a single exponential phase of product formation followed by a linear increase (Fig. 10.4B, Eq. 10.1). Here, the rate constant k_{obs} reflects the catalytic reaction, as in Section 2. The burst amplitude A reflects the fraction of native ribozyme at the time the aliquot was transferred from stage 1 to stage 2 ($fN(t_1)$).

To obtain the folding rate, $fN(t_1)$ is plotted against folding time t_1 (Fig. 10.4C). If a simple pathway is traversed, this curve will give a single exponential curve in which the rate constant reflects the observed folding process and the amplitude reflects the fraction of native ribozyme present at the endpoint. The amplitude value will be less than 1 if the ribozyme is not completely folded at equilibrium, if there are additional folding pathways that do not give native ribozyme on the time scale observed, or if some of the ribozyme is damaged. It may be necessary to include a constant term if additional folding pathways give rapid, unresolved phases of native ribozyme formation or to include additional exponential terms if there are multiple pathways that produce native ribozyme on the observed time scale (Russell *et al.*, 2007).

If, under the conditions of the stage 2 catalytic reaction, the first turnover is not faster than subsequent turnovers, the product will not accumulate in a burst, but will instead accumulate linearly (Fig. 10.4D). This rate can be used to provide a relative measure of the fraction of native ribozyme, as it will be proportional to the fraction of active ribozyme. Therefore, the rate can be plotted against folding time t_1 to monitor the progress of folding. However, in this case the fraction of native ribozyme cannot be determined absolutely, only relative to a prefolded control. This situation is analogous to that of lag kinetics in the continuous assay (Section 3.2).

For certain applications of the discontinuous assay, the substrate cleavage reaction has been performed under single-turnover conditions, with ribozyme in excess of substrate (Pan *et al.*, 1999; Russell and Herschlag, 2001; Russell *et al.*, 2006, 2007; see Figs. 10.6 and 10.8). Here, trace substrate is added in stage 2, and either the observed rate constant (Pan *et al.*, 1999) or the fraction of the substrate cleaved rapidly (Russell and Herschlag, 2001) is taken as a measure of the fraction of native ribozyme. This method has the advantage that the burst amplitude does not depend on substrate concentration, or on ribozyme concentration in the latter case, which can increase precision and reproducibility. However, the single-turnover assay has more extensive and specific requirements for the relative rate constants of substrate binding, release, and catalysis, and it should only be used under well-defined circumstances.

5. Other Applications of Catalytic Activity as a Probe of Folding

The preceding sections describe how to use catalytic activity to follow folding of structured RNAs. The same basic approach has been used in creative ways to gain additional information about RNA folding processes. As described below, the continuous assay has been applied to follow assembly of RNA–protein complexes. The discontinuous assay has been used to measure the equilibrium of native RNA formation and to follow folding cotranscriptionally or in the presence of chaperones.

5.1. Assembly of RNA–protein complexes

In addition to monitoring intramolecular folding events, the continuous activity assay has been used to follow assembly of a catalytic RNA with protein (Webb *et al.*, 2001) or an activating RNA (Russell and Herschlag, 1999). The most general requirement here is that the catalytic reaction must be faster than the assembly process, which can be modulated by changing the concentrations of the interacting molecules.

Weeks and colleagues used this method to measure association and assembly of a group I intron RNA with its cofactor protein Cbp2p (Webb *et al.*, 2001). They first preincubated the intron to allow formation of a near-native collapsed state and then added Cbp2p to allow assembly and subsequent splicing (Fig. 10.5A). The observed rate constant was compared to that from a reaction in which Cbp2p and the intron were preincubated to allow complex formation, and then GMP was added to initiated splicing (analogous to prefolded reactions of Section 2). The observed rate constant was smaller for

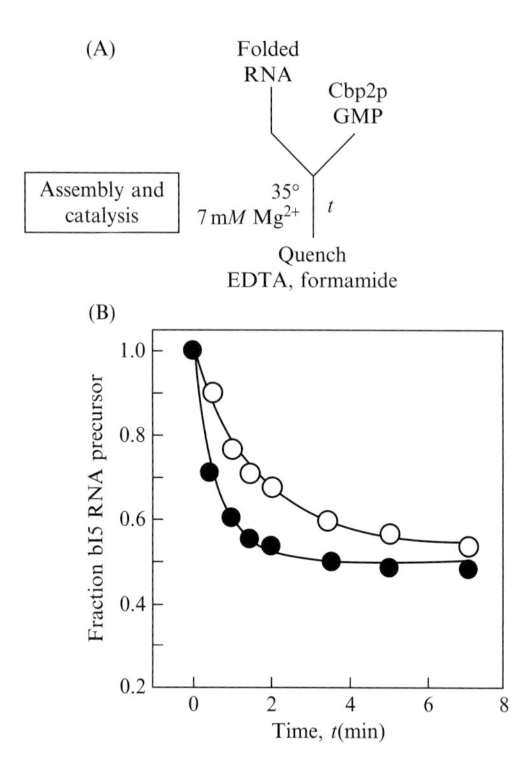

Figure 10.5 Continuous activity assay to follow RNA–protein complex assembly. (A) Experimental design. Radiolabeled precursor of the bI5 group I intron was mixed with protein Cbp2p and cofactor GMP to initiate the folding and cleavage assay. (B) This continuous assay gave a rate constant of 0.5 min^{-1} (○). A control reaction in which the RNA was preincubated with Cbp2p to allow complex formation and then mixed with GMP to initiate cleavage gave a larger rate constant, 1.6 min^{-1} (●). Panel (B) reprinted from Webb *et al.* (2001) with permission from Elsevier.

the reaction initiated with Cbp2p, indicating that complex assembly was rate-limiting for splicing under the experimental conditions (Fig. 10.5B).

5.2. Native folding at equilibrium

The discontinuous assay has been used to measure the Mg^{2+}-dependent equilibrium for native RNA folding of the *Tetrahymena* $E^{\Delta P5abc}$ ribozyme (Russell *et al.*, 2007). In stage 1, the ribozyme was incubated at various Mg^{2+} concentrations to allow equilibration between intermediates and the native state, and then the fraction of native ribozyme was determined in stage 2 by adding 50 m*M* Mg^{2+} and P5abc RNA and then substrate. These additions allowed catalytic activity, while trapping most of the ribozyme that was non-native in stage 1 as a long-lived misfolded intermediate

(Fig. 10.6A). A small fraction of the ribozyme did fold to the native state during stage 2 prior to substrate addition (10%, Fig. 10.6B), as expected from earlier results (Russell and Herschlag, 1999). Because this fraction was small and expected to be constant, it could be subtracted to calculate the fraction of native ribozyme in the stage 1 incubation (not shown).

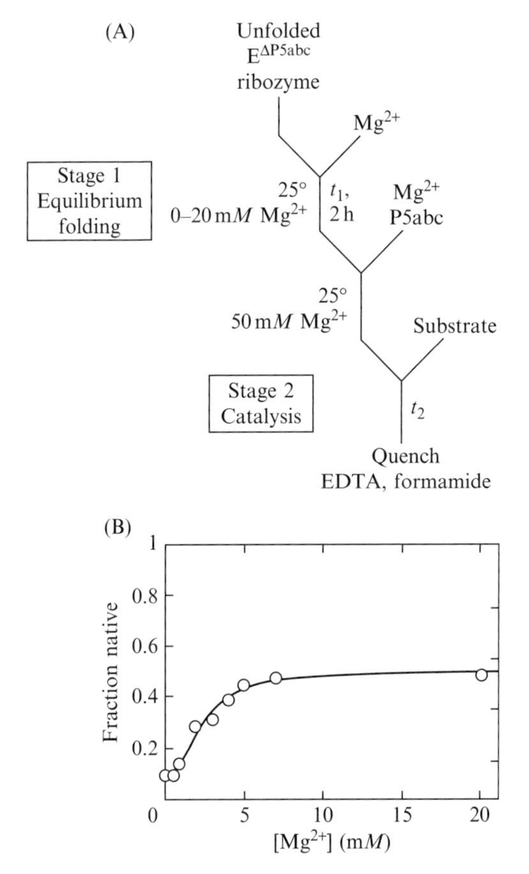

Figure 10.6 Mg^{2+} concentration dependence of native ribozyme formation. (A) Reaction scheme. The $E^{\Delta P5abc}$ *Tetrahymena* ribozyme was incubated in stage 1 with 0–20 m*M* Mg^{2+}, allowing equilibration of the native RNA and folding intermediates. Additional Mg^{2+} (50 m*M* final) was then added, which induces folding predominantly to the long-lived misfolded form, and substrate was added to initiate the catalytic reaction in stage 2. (B) Burst amplitude versus Mg^{2+} concentration in stage 1. The amplitude does not reach 1 because this RNA equilibrates between the native and misfolded conformations. Panel (B) adapted with permission from Russell *et al.* (2007), copyright 2007 American Chemical Society.

5.3. Cotranscriptional folding

A discontinuous activity assay has been used by Pan, Sosnick and colleagues to follow folding of the RNase P RNA during *in vitro* transcription (Pan *et al.*, 1999; Wong *et al.*, 2005, 2007). In this work, stage 1 included a DNA template, T7 RNA polymerase, and NTPs to allow transcription. At various times, aliquots were transferred to a stage 2 reaction that included a tRNA precursor substrate and high ionic strength conditions (100 mM $MgCl_2$ and 0.06–0.6 M KCl), permitting cleavage of the substrate but blocking transcription initiation (Fig. 10.7). The activity assay was used to monitor active RNase P RNA formed over time in the stage 1 reaction, and the amount of RNA synthesized was determined from incorporation of radiolabeled nucleotide. The lag between the accumulation of large RNA and the appearance of cleavage activity provided evidence for slow cotranscriptional folding of the RNase P RNA, and further work showed that pausing of RNA polymerase during transcription accelerated folding.

The use of RNA catalysis to follow cotranscriptional folding was taken a step further by Fedor and colleagues, who studied the hairpin self-cleaving RNA *in vivo* using a continuous assay (Mahen *et al.*, 2005). Starting from a steady state of precursors and RNA cleavage products, transcription was shut off using a glucose-sensitive promoter, and the rate of the subsequent loss of precursor RNA was used to determine the efficiency of folding to the native state versus alternative structures.

5.4. RNA chaperone-mediated folding and unfolding

Certain DEAD-box "RNA helicase" proteins possess RNA chaperone activity and function during folding of several group I introns (Huang *et al.*, 2005; Mohr *et al.*, 2002). Our group has applied the discontinuous assay to study refolding of the *Tetrahymena* ribozyme in the presence of the CYT-19 DEAD-box protein (Bhaskaran and Russell, 2007; Tijerina *et al.*, 2006). The ribozyme was first misfolded by incubation with Mg^{2+} (Fig. 10.8A). Then, CYT-19 and ATP were added in stage 1. In stage 2, Proteinase K was added to inactivate CYT-19, and the Mg^{2+} concentration was increased to 50 mM to block further refolding while allowing robust catalytic activity. By this assay, CYT-19 was shown to accelerate refolding in a concentration-dependent manner (Fig. 10.8B).

The discontinuous assay was also used to determine whether CYT-19 mediates unfolding of the native ribozyme (Bhaskaran and Russell, 2007). The motivation was to test whether CYT-19 acts as a general chaperone without recognizing structural features of misfolded RNA. Here, the ribozyme was first prefolded to the native state, and the stage 1 reaction was designed to destabilize the native structure, either with low Mg^{2+}

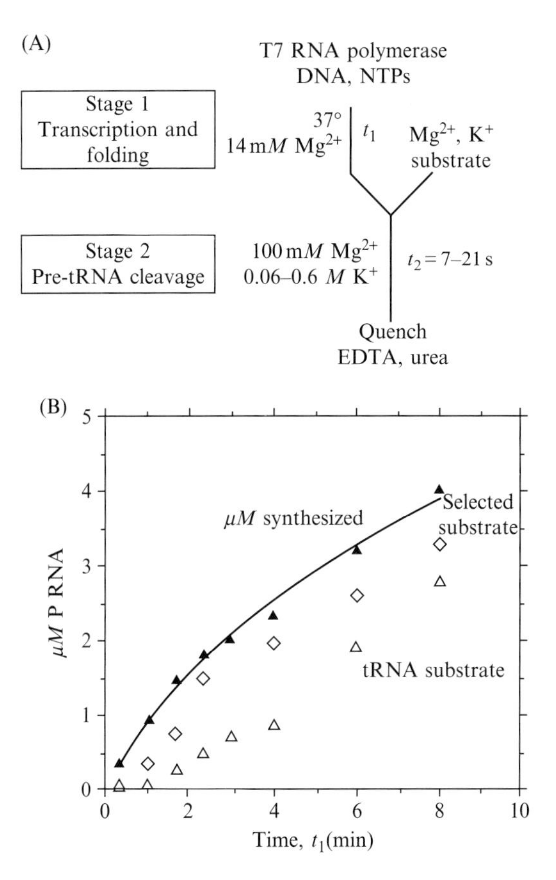

Figure 10.7 Cotranscriptional folding of RNase P RNA monitored by a discontinuous activity assay. (A) Reaction scheme. In stage 1, T7 RNA polymerase was incubated with template DNA and NTPs (including [α-^{32}P]CTP) at 37 °C and pH 8.1 to initiate transcription and concurrent folding. At various times t_1, transcription was arrested by transferring aliquots to high ionic strength conditions (stage 2), and the catalytic reaction was initiated by adding substrate, either a pre-tRNA or an oligonucleotide substrate obtained by *in vitro* selection that is cleaved efficiently by the catalytic domain of RNase P RNA (Pan and Jakacka, 1996). Stage 2 reactions were quenched after 7–21 s, allowing determination of the relative amount of native RNase P RNA. (B) The amount of RNA synthesized (▲) was measured by incorporation of labeled CTP, while native folding was monitored by cleavage of the tRNA precursor substrate (△) or the selected substrate for folding of the catalytic domain (◇). Panel (B) reprinted with permission from Pan *et al.* (1999), copyright 1999 National Academy of Sciences, USA.

concentration or with destabilizing mutations in the RNA. In stage 2, Proteinase K and 50 m*M* Mg^{2+} were added and the assay was performed exactly as described above.

Using this assay, cycles of folding and unfolding of the ribozyme were observed (Fig. 10.8C). Initial folding gave native ribozyme formation, and

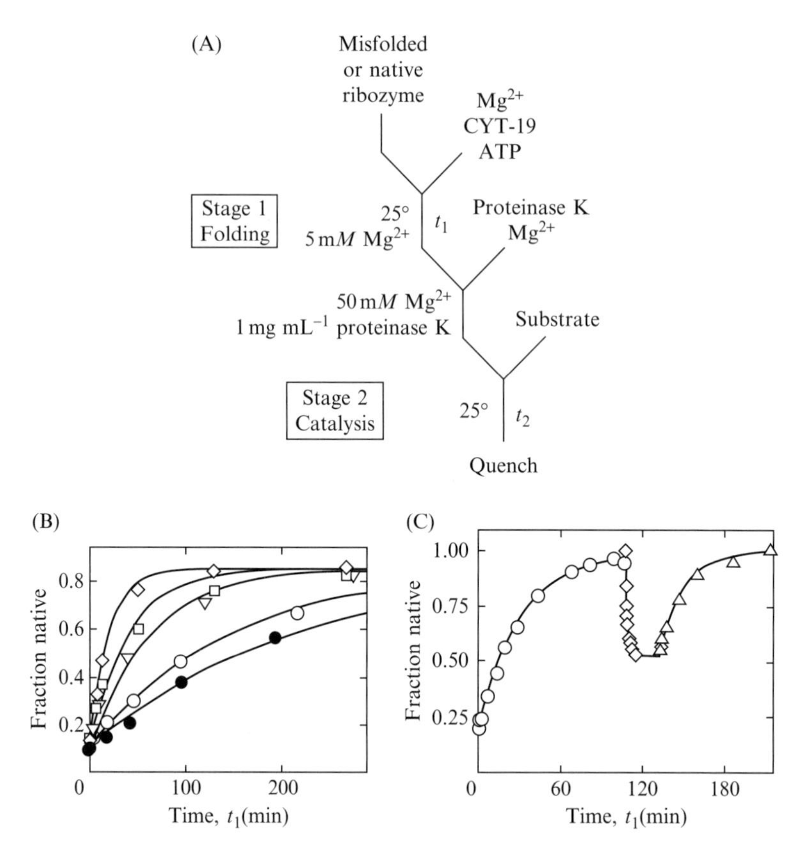

Figure 10.8 Chaperone-mediated RNA folding monitored by a discontinuous activity assay. (A) Reaction scheme. In stage 1, CYT-19 protein was added with ATP to populations of native or misfolded *Tetrahymena* ribozyme. At various times, further transitions between the native and misfolded conformations were blocked by Proteinase K and 50 mM Mg^{2+}, and the fraction of native ribozyme was determined by adding substrate for 1 min (stage 2). (B) Progress curves without CYT-19 (○), with 500 nM CYT-19 without ATP (●), or with ATP and 100 nM (▽), 200 nM (□), or 500 nM (◇) CYT-19. (C) Unfolding of native ribozyme by CYT-19. Using a tertiary contact mutant, the experiment monitored initial folding to the native state (○), unfolding of native ribozyme by CYT-19 (◇), and subsequent refolding after Proteinase K digestion of CYT-19 in stage 1 (△). Y-axis values are larger in (C) because data were normalized for 10–20% damaged ribozyme. Panel (B) adapted with permission from Tijerina *et al.* (2006), copyright 2006 National Academy of Sciences, USA. Panel (C) reprinted from Bhaskaran and Russell (2007) with permission from Macmillan Publishers Ltd: Nature, copyright 2007.

then after addition of CYT-19, the cleavage burst amplitude decreased, indicating loss of the native ribozyme. After CYT-19 proteolysis, the ribozyme returned to the native state. Note that it would not be possible to use the continuous assay to monitor the unfolding transition because the substrate would already be cleaved prior to CYT-19 addition. However, it is straightforward with the discontinuous assay, requiring only that unfolding be faster than folding in stage 1, so that net unfolding occurs, and that the stage 2 conditions permit catalytic activity and block refolding of RNA that unfolded in stage 1.

ACKNOWLEDGMENTS

We thank members of the Russell lab for comments on the chapter. Research in the Russell lab is funded by grants from the NIH (GM 070456), the Welch Foundation (F-1563), and the Norman Hackerman Advanced Research Program (003658-0242-2007).

REFERENCES

Bartel, D. P. (2004). MicroRNAs: Genomics, biogenesis, mechanism, and function. *Cell* **116,** 281–297.

Bhaskaran, H., and Russell, R. (2007). Kinetic redistribution of native and misfolded RNAs by a DEAD-box chaperone. *Nature* **449,** 1014–1018.

Brooks, K. M., and Hampel, K. J. (2009). A rate-limiting conformational step in the catalytic pathway of the glmS ribozyme. *Biochemistry* **48,** 5669–5678.

Buzayan, J. M., Gerlach, W. L., and Bruening, G. (1986). Non-enzymatic cleavage and ligation of RNAs complementary to a plant virus satellite RNA. *Nature* **323,** 349–353.

Carninci, P., Kasukawa, T., Katayama, S., Gough, J., Frith, M. C., Maeda, N., Oyama, R., Ravasi, T., Lenhard, B., Wells, C., Kodzius, R., Shimokawa, K., *et al.* (2005). The transcriptional landscape of the mammalian genome. *Science* **309,** 1559–1563.

Carthew, R. W., and Sontheimer, E. J. (2009). Origins and mechanisms of miRNAs and siRNAs. *Cell* **136,** 642–655.

Cech, T. R. (2002). Ribozymes, the first 20 years. *Biochem. Soc. Trans.* **30,** 1162–1166.

Chadalavada, D. M., Senchak, S. E., and Bevilacqua, P. C. (2002). The folding pathway of the genomic hepatitis delta virus ribozyme is dominated by slow folding of the pseudoknots. *J. Mol. Biol.* **317,** 559–575.

Chauhan, S., and Woodson, S. A. (2008). Tertiary interactions determine the accuracy of RNA folding. *J. Am. Chem. Soc.* **130,** 1296–1303.

Chen, X., Li, N., and Ellington, A. D. (2007). Ribozyme catalysis of metabolism in the RNA world. *Chem. Biodivers.* **4,** 633–655.

Collins, K. (2006). The biogenesis and regulation of telomerase holoenzymes. *Nat. Rev. Mol. Cell Biol.* **7,** 484–494.

Egea, P. F., Stroud, R. M., and Walter, P. (2005). Targeting proteins to membranes: Structure of the signal recognition particle. *Curr. Opin. Struct. Biol.* **15,** 213–220.

Emerick, V. L., Pan, J., and Woodson, S. A. (1996). Analysis of rate-determining conformational changes during self-splicing of the *Tetrahymena* intron. *Biochemistry* **35,** 13469–13477.

Evans, D., Marquez, S. M., and Pace, N. R. (2006). RNase P: Interface of the RNA and protein worlds. *Trends Biochem. Sci.* **31,** 333–341.

Fedor, M. J. (2009). Comparative enzymology and structural biology of RNA self-cleavage. *Annu. Rev. Biophys.* **38,** 271–299.

Golden, B. L., and Cech, T. R. (1996). Conformational switches involved in orchestrating the successive steps of group I RNA splicing. *Biochemistry* **35,** 3754–3763.

Hegg, L. A., and Fedor, M. J. (1995). Kinetics and thermodynamics of intermolecular catalysis by hairpin ribozymes. *Biochemistry* **34,** 15813–15828.

Held, W. A., and Nomura, M. (1973). Rate determining step in the reconstitution of *Escherichia coli* 30S ribosomal subunits. *Biochemistry* **12,** 3273–3281.

Herschlag, D., and Cech, T. R. (1990). Catalysis of RNA cleavage by the *Tetrahymena thermophila* ribozyme. 1. Kinetic description of the reaction of an RNA substrate complementary to the active site. *Biochemistry* **29,** 10159–10171.

Herschlag, D., and Khosla, M. (1994). Comparison of pH dependencies of the *Tetrahymena* ribozyme reactions with RNA 2′-substituted and phosphorothioate substrates reveals a rate-limiting conformational step. *Biochemistry* **33,** 5291–5297.

Hertel, K. J., and Uhlenbeck, O. C. (1995). The internal equilibrium of the hammerhead ribozyme reaction. *Biochemistry* **34,** 1744–1749.

Huang, H. R., Rowe, C. E., Mohr, S., Jiang, Y., Lambowitz, A. M., and Perlman, P. S. (2005). The splicing of yeast mitochondrial group I and group II introns requires a DEAD-box protein with RNA chaperone function. *Proc. Natl. Acad. Sci. USA* **102,** 163–168.

Johnson, T. H., Tijerina, P., Chadee, A. B., Herschlag, D., and Russell, R. (2005). Structural specificity conferred by a group I RNA peripheral element. *Proc. Natl. Acad. Sci. USA* **102,** 10176–10181.

Karbstein, K., Carroll, K. S., and Herschlag, D. (2002). Probing the *Tetrahymena* group I ribozyme reaction in both directions. *Biochemistry* **41,** 11171–11183.

Katayama, S., Tomaru, Y., Kasukawa, T., Waki, K., Nakanishi, M., Nakamura, M., Nishida, H., Yap, C. C., Suzuki, M., Kawai, J., Suzuki, H., Carninci, P., *et al.* (2005). Antisense transcription in the mammalian transcriptome. *Science* **309,** 1564–1566.

Kiefhaber, T. (1995). Protein folding kinetics. *Methods Mol. Biol.* **40,** 313–341.

Kiefhaber, T., Quaas, R., Hahn, U., and Schmid, F. X. (1990). Folding of ribonuclease T1. 1. Existence of multiple unfolded states created by proline isomerization. *Biochemistry* **29,** 3053–3061.

Kruger, K., Grabowski, P. J., Zaug, A. J., Sands, J., Gottschling, D. E., and Cech, T. R. (1982). Self-splicing RNA: Autoexcision and autocyclization of the ribosomal RNA intervening sequence of *Tetrahymena. Cell* **31,** 147–157.

Lincoln, T. A., and Joyce, G. F. (2009). Self-sustained replication of an RNA enzyme. *Science* **323,** 1229–1232.

Lorsch, J. R., and Szostak, J. W. (1994). *In vitro* evolution of new ribozymes with polynucleotide kinase activity. *Nature* **371,** 31–36.

Mahen, E. M., Harger, J. W., Calderon, E. M., and Fedor, M. J. (2005). Kinetics and thermodynamics make different contributions to RNA folding *in vitro* and in yeast. *Mol. Cell* **19,** 27–37.

Martick, M., Horan, L. H., Noller, H. F., and Scott, W. G. (2008). A discontinuous hammerhead ribozyme embedded in a mammalian messenger RNA. *Nature* **454,** 899–902.

Mohr, S., Stryker, J. M., and Lambowitz, A. M. (2002). A DEAD-box protein functions as an ATP-dependent RNA chaperone in group I intron splicing. *Cell* **109,** 769–779.

Moore, P. B., Steitz, T. A., Nissen, P., Hansen, J., and Ban, N. (2002). The involvement of RNA in ribosome function. *Nature* **418,** 229–235.

Narlikar, G. J., and Herschlag, D. (1997). Mechanistic aspects of enzymatic catalysis: Lessons from comparison of RNA and protein enzymes. *Annu. Rev. Biochem.* **66,** 19–59.

Noller, H. F. (2005). RNA structure: Reading the ribosome. *Science* **309,** 1508–1514.

Pan, T., and Jakacka, M. (1996). Multiple substrate binding sites in the ribozyme from *Bacillus subtilis* RNase P. *EMBO J.* **15,** 2249–2255.

Pan, T., and Sosnick, T. R. (1997). Intermediates and kinetic traps in the folding of a large ribozyme revealed by circular dichroism and UV absorbance spectroscopies and catalytic activity. *Nat. Struct. Biol.* **4,** 931–938.

Pan, J., and Woodson, S. A. (1998). Folding intermediates of a self-splicing RNA: Mispairing of the catalytic core. *J. Mol. Biol.* **280,** 597–609.

Pan, T., Artsimovitch, I., Fang, X. W., Landick, R., and Sosnick, T. R. (1999). Folding of a large ribozyme during transcription and the effect of the elongation factor NusA. *Proc. Natl. Acad. Sci. USA* **96,** 9545–9550.

Rose, I. A., O'Connell, E. L., and Litwin, S. (1974). Determination of the rate of hexokinase–glucose dissociation by the isotope-trapping methods. *J. Biol. Chem.* **249,** 5163–5168.

Russell, R., and Herschlag, D. (1999). New pathways in folding of the *Tetrahymena* group I RNA enzyme. *J. Mol. Biol.* **291,** 1155–1167.

Russell, R., and Herschlag, D. (2001). Probing the folding landscape of the *Tetrahymena* ribozyme: Commitment to form the native conformation is late in the folding pathway. *J. Mol. Biol.* **308,** 839–851.

Russell, R., Das, R., Suh, H., Travers, K., Laederach, A., Engelhardt, M., and Herschlag, D. (2006). The paradoxical behavior of a highly structured misfolded intermediate in RNA folding. *J. Mol. Biol.* **363,** 531–544.

Russell, R., Tijerina, P., Chadee, A. B., and Bhaskaran, H. (2007). Deletion of the P5abc peripheral element accelerates early and late folding steps of the *Tetrahymena* group I ribozyme. *Biochemistry* **46,** 4951–4961.

Salehi-Ashtiani, K., Luptak, A., Litovchick, A., and Szostak, J. W. (2006). A genomewide search for ribozymes reveals an HDV-like sequence in the human CPEB3 gene. *Science* **313,** 1788–1792.

Sclavi, B., Woodson, S., Sullivan, M., Chance, M. R., and Brenowitz, M. (1997). Time-resolved synchrotron X-ray "footprinting", a new approach to the study of nucleic acid structure and function: Application to protein–DNA interactions and RNA folding. *J. Mol. Biol.* **266,** 144–159.

Sclavi, B., Sullivan, M., Chance, M. R., Brenowitz, M., and Woodson, S. A. (1998). RNA folding at millisecond intervals by synchrotron hydroxyl radical footprinting. *Science* **279,** 1940–1943.

Su, L. J., Brenowitz, M., and Pyle, A. M. (2003). An alternative route for the folding of large RNAs: Apparent two-state folding by a group II intron ribozyme. *J. Mol. Biol.* **334,** 639–652.

Teixeira, A., Tahiri-Alaoui, A., West, S., Thomas, B., Ramadass, A., Martianov, I., Dye, M., James, W., Proudfoot, N. J., and Akoulitchev, A. (2004). Autocatalytic RNA cleavage in the human beta-globin pre-mRNA promotes transcription termination. *Nature* **432,** 526–530.

Theimer, C. A., and Feigon, J. (2006). Structure and function of telomerase RNA. *Curr. Opin. Struct. Biol.* **16,** 307–318.

Tijerina, P., Bhaskaran, H., and Russell, R. (2006). Non-specific binding to structured RNA and preferential unwinding of an exposed helix by the CYT-19 protein, a DEAD-box RNA chaperone. *Proc. Natl. Acad. Sci. USA* **103,** 16698–16703.

Traub, P., and Nomura, M. (1969). Structure and function of *Escherichia coli* ribosomes. VI. Mechanism of assembly of 30 s ribosomes studied *in vitro*. *J. Mol. Biol.* **40,** 391–413.

Treiber, D. K., and Williamson, J. R. (1999). Exposing the kinetic traps in RNA folding. *Curr. Opin. Struct. Biol.* **9,** 339–345.

Treiber, D. K., Rook, M. S., Zarrinkar, P. P., and Williamson, J. R. (1998). Kinetic intermediates trapped by native interactions in RNA folding. *Science* **279,** 1943–1946.

Wahl, M. C., Will, C. L., and Luhrmann, R. (2009). The spliceosome: Design principles of a dynamic RNP machine. *Cell* **136,** 701–718.

Webb, A. E., Rose, M. A., Westhof, E., and Weeks, K. M. (2001). Protein-dependent transition states for ribonucleoprotein assembly. *J. Mol. Biol.* **309,** 1087–1100.

Wong, T., Sosnick, T. R., and Pan, T. (2005). Mechanistic insights on the folding of a large ribozyme during transcription. *Biochemistry* **44,** 7535–7542.

Wong, T. N., Sosnick, T. R., and Pan, T. (2007). Folding of noncoding RNAs during transcription facilitated by pausing-induced nonnative structures. *Proc. Natl. Acad. Sci. USA* **104,** 17995–18000.

Woodson, S. A., and Cech, T. R. (1991). Alternative secondary structures in the 5′ exon affect both forward and reverse self-splicing of the *Tetrahymena* intervening sequence RNA. *Biochemistry* **30,** 2042–2050.

Xiao, M., Li, T., Yuan, X., Shang, Y., Wang, F., Chen, S., and Zhang, Y. (2005). A peripheral element assembles the compact core structure essential for group I intron self-splicing. *Nucleic Acids Res.* **33,** 4602–4611.

Zarrinkar, P. P., and Williamson, J. R. (1994). Kinetic intermediates in RNA folding. *Science* **265,** 918–924.

Zarrinkar, P. P., and Williamson, J. R. (1996). The kinetic folding pathway of the *Tetrahymena* ribozyme reveals possible similarities between RNA and protein folding. *Nat. Struct. Biol.* **3,** 432–438.

Zaug, A. J., Grosshans, C. A., and Cech, T. R. (1988). Sequence-specific endoribonuclease activity of the *Tetrahymena* ribozyme: Enhanced cleavage of certain oligonucleotide substrates that form mismatched ribozyme–substrate complexes. *Biochemistry* **27,** 8924–8931.

CHAPTER ELEVEN

Probing RNA Structure Within Living Cells

Andreas Liebeg *and* Christina Waldsich

Contents

Abstract

RNA folding is the most fundamental process underlying RNA function. RNA structure and associated folding paradigms have been intensively studied *in vitro*. However, *in vivo* RNA structure formation has only been explored to a limited extent. To determine the influence of the cellular environment, which differs significantly from the *in vitro* refolding conditions, on RNA architecture, we have applied a chemical probing technique to assess the structure of catalytic RNAs in living cells. This method is based on the fact that chemicals like dimethyl sulfate readily penetrate cells and modify specific atoms of RNA bases (N1-A, N3-C), provided that these positions are solvent accessible. By mapping the modified residues, one gains substantial information on the architecture of the target RNA on the secondary and tertiary structure level. This method also allows exploration of interactions of the target RNA with ligands such as proteins, metabolites, or other RNA molecules and associated

Max F. Perutz Laboratories, Department of Biochemistry and Cell Biology, University of Vienna, Vienna, Austria

Methods in Enzymology, Volume 468
ISSN 0076-6879, DOI: 10.1016/S0076-6879(09)68011-3

conformational changes. In brief, *in vivo* chemical probing is a powerful tool to investigate RNA structure in its natural environment and can be easily adapted to study RNAs in different cell types.

1. Introduction

Most RNA molecules have to adopt a specific 3D structure to be functional. Unraveling the architecture of RNA molecules therefore provides profound insights into the basic mechanism of RNA-dependent cellular processes. Several paradigms for how RNA molecules undergo the transition from the unfolded, disordered state to the native, productive conformation have so far been described *in vitro* (Herschlag, 1995; Pyle *et al.*, 2007; Schroeder *et al.*, 2002; Thirumalai, 1998; Thirumalai *et al.*, 2001; Treiber and Williamson, 1999, 2001; Woodson, 2000a,b, 2005). However, the cellular environment differs significantly the *in vitro* refolding conditions, which typically involve nonphysiological metal ion concentrations and elevated temperature. In the cell the directionality and velocity of transcription and translation as well as *trans*-acting factors such as proteins, small ligands, or other RNA molecules are likely to influence RNA folding (reviewed in Schroeder *et al.*, 2002). Hitherto little is known about intracellular RNA structure formation, which is in part due to the complexity of the cellular environment, but also to the limited availability of suitable techniques to explore RNA architecture in the cell. A first step toward understanding RNA folding *in vivo* is to determine the structure of RNAs in the living cells.

A powerful tool for structure mapping, which has been widely used to investigate RNA architecture *in vitro*, is chemical probing (e.g., Brunel and Romby, 2000; Brunel *et al.*, 1991; Ehresmann *et al.*, 1987; Moazed and Noller, 1986, 1989). This technique is a well-suited, readily adaptable and rather inexpensive approach to explore RNA structure, folding, and ligand-induced conformational changes in living cells as well. Thus, chemical probing *in vivo* has mostly been the method of choice in the few studies attempting to investigate intracellular structure of RNA molecules (Ares and Igel, 1990; Balzer and Wagner, 1998; Climie and Friesen, 1988; Doktycz *et al.*, 1998; Harris *et al.*, 1995; Mereau *et al.*, 1997; Moazed *et al.*, 1988; Senecoff and Meagher, 1992; Waldsich *et al.*, 2002a,b; Wells and Ares, 1994; Zaug and Cech, 1995). Notably, there are also a few studies reporting the *in vivo* application of crosslinking (Gallouzi *et al.*, 2000; Pinol-Roma *et al.*, 1989), metal-induced cleavage assay (Lindell *et al.*, 2002, 2005) and, most recently, of nucleotide analog interference mapping (NAIM; Szewczak, 2008) to map RNA structure or RNA–protein interactions within cells.

Here, we present a detailed description of how to perform *in vivo* chemical probing with DMS. The power of this method is highlighted by applying it to monitor the intracellular structure of catalytic RNAs in yeast (A. Liebeg and C. Waldsich, unpublished data) and in *Escherichia coli* (Waldsich *et al.*, 2002b) as well as of protein-induced conformational changes within group I introns (Waldsich *et al.*, 2002a).

2. Experimental Procedure

2.1. Theory

Of all chemical probing reagents (Brunel and Romby, 2000) dimethyl sulfate (DMS) has been most widely used to analyze RNA structure *in vivo*. DMS is a base-specific probe capable of methylating the Watson–Crick positions N1 of adenines and N3 of cytidines, in addition to the Hoogsteen position N7 of guanines (Fig. 11.1A; Ehresmann *et al.*, 1987). Notably, certain uridines and guanines are occasionally stabilized in an enol-tautomer due to a specific local environment, and are therefore reactive to DMS at their N3 or N1 position (Ehresmann *et al.*, 1987). Importantly, DMS can only donate a methyl group to proton accepting ring nitrogens, if these atoms are not engaged in hydrogen bonding (e.g., Watson–Crick base pairing or sheared AA base pairs; Leontis and Westhof, 2001; Leontis *et al.*, 2006) and if they are solvent accessible, while a reduced solvent exposure or binding of a protein ligand results in protection from DMS modification (Fig. 11.1B). Accordingly, DMS modification of RNA allows gaining insights into secondary and tertiary structure elements formed within an RNA as well as assessing the binding sites of ligands and associated structural changes. The main advantage of DMS is that the chemical easily and rapidly penetrates cells and all of their compartments. Thus, this method has been successfully applied to a variety of organisms including bacteria (Balzer and Wagner, 1998; Climie and Friesen, 1988; Moazed *et al.*, 1988; Waldsich *et al.*, 2002a,b), protozoa (Harris *et al.*, 1995; Zaug and Cech, 1995), yeast (Ares and Igel, 1990), and plants (Senecoff and Meagher, 1992).

The sites of modification are detected by reverse transcription (Fig. 11.1C and D) and subsequently compiled into a pattern profile of nucleotides protected from or accessible to DMS, which can be plotted onto the secondary structure map of the RNA of interest. The observed modification pattern reveals residues that do not participate in any intra- or intermolecular contacts—at least not through their N1 or N3 atoms—and are thus accessible to DMS methylation. In contrast, a protection from DMS modification results either from base pairing or from an interaction with a protein (or other ligand).

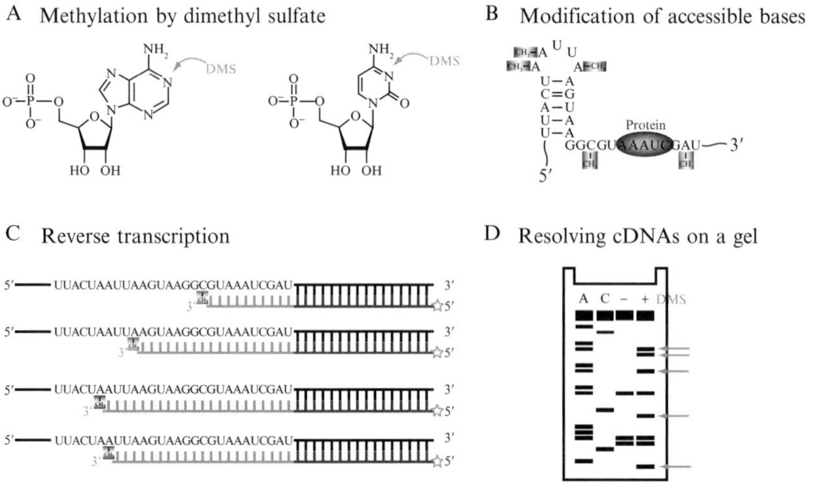

Figure 11.1 *In vivo* chemical probing assay using dimethyl sulfate (DMS). (A) DMS methylates the N1 atom of adenosines and the N3 atom of cytosines. Both of these positions are located on the Watson–Crick face of the respective nucleotides. (B) Scheme of a structured RNA interacting with a protein ligand to exemplify residues accessible to DMS modification. (C) Mapping the modified positions by reverse transcription. The enzyme terminates the extension due to steric hindrance by the bulky methyl group at the Watson–Crick face of A and Cs. (D) The pool of cDNAs is separated on a standard denaturing gel. A and C denote the sequencing lanes, which are essential to map the modified positions. In addition, an RT control, which is prepared from RNA extracted from untreated cells, is essential to determine any natural stops of the RT (lane −). Comparison of the RT control lane with the cDNA pool created from DMS-modified RNA (lane +) reveals sites of modification. Note that the band resulting from a DMS modification migrates one nucleotide faster than the corresponding band in the sequencing lane, because the ddNTP is incorporated into the nascent cDNA, while the methyl group results in a stop one residue prior to the modification.

2.2. General setup

DMS probing allows taking a snapshot of how structured an RNA is in living cells, as DMS rapidly enters cells without requiring prior permeabilization treatment of cells. After the modification, it is particularly critical that the reaction is stopped efficiently with β-mercaptoethanol to avoid continuous DMS modification during RNA preparation. Thus, high concentrations of β-mercaptoethanol are used to quench excess soluble DMS, while aqueous insoluble DMS that may contaminate the cell pellet can be removed by extraction of cells with water-saturated isoamyl alcohol prior to centrifugation. Both the DMS modification and quenching reactions have to be optimized for each novel target RNA and/or organism. DMS remnants do significantly reduce the recovery of RNA.

During preparation the entire RNA pool of a cell, which mostly consists of ribosomal RNAs and tRNAs, is isolated. Mapping the methylated sites within the RNA of interest is accomplished by using target molecule-specific DNA oligos as primers for reverse transcription. Importantly, the methyl group at the Watson–Crick face of either A or C residues poses a steric block and consequently causes the Reverse Transcriptase to stop the extension of the primer. Detecting G-N7 modifications necessitates treatment of the RNA pool with aniline and borohydride (Brunel and Romby, 2000), which induce cleavage of the RNA backbone at methylated guanosines, prior to reverse transcription. Also, it has to be kept in mind that naturally modified nucleotides like 7mG in rRNAs could occur in the target RNA. Therefore, it is crucial to determine any natural stops of the Reverse Transcriptase along the target RNA, which are the result of natural modifications, sequence context (e.g., stretch of G), RNase contamination or intrinsic breaks occurring at sites of a strained RNA backbone. This is generally accomplished by reverse transcribing RNAs extracted from cells which were not treated with DMS. Comparing the band pattern of RNAs from untreated and treated cells reveals residues accessible to DMS modification, eventually elucidating the intracellular RNA structure and its association with proteins or other ligands.

2.3. Optimizing the individual steps of the *in vivo* chemical probing assay

The most crucial aspect is to optimize the DMS modification step for the RNA of interest. Specifically, determining the optimal DMS concentration as well as incubation temperature and time is most integral to successfully employing *in vivo* structural probing (Fig. 11.2). To avoid secondary effects caused by excessive DMS modification, a DMS concentration has to be chosen, at which a maximum of one methylation event per RNA molecule occurs. This is most easily evaluated by analyzing the amount of full-length cDNA produced from RNA isolated from untreated versus DMS-treated cells. Comparing lane 3 with lanes 4–7 [RNA from cells which were treated with increasing (DMS)] in Fig. 11.2 reveals that a final concentration of 50 m*M* DMS (lane 5) is sufficient to obtain a modification pattern with good signal-to-noise ratio and at the same time the amount of full-length cDNA is comparable to the untreated control sample (in contrast to lane 7). Also, it is essential to ensure that the reaction is stopped efficiently with β-mercaptoethanol to prevent continuous modification during RNA preparation, which may result in misleading data. In the stop control (Fig. 11.2, lane 8), in which we first add β-mercaptoethanol and subsequently DMS, we did not detect any methylation sites, implying that the amount of β-mercaptoethanol is sufficient to block DMS modification.

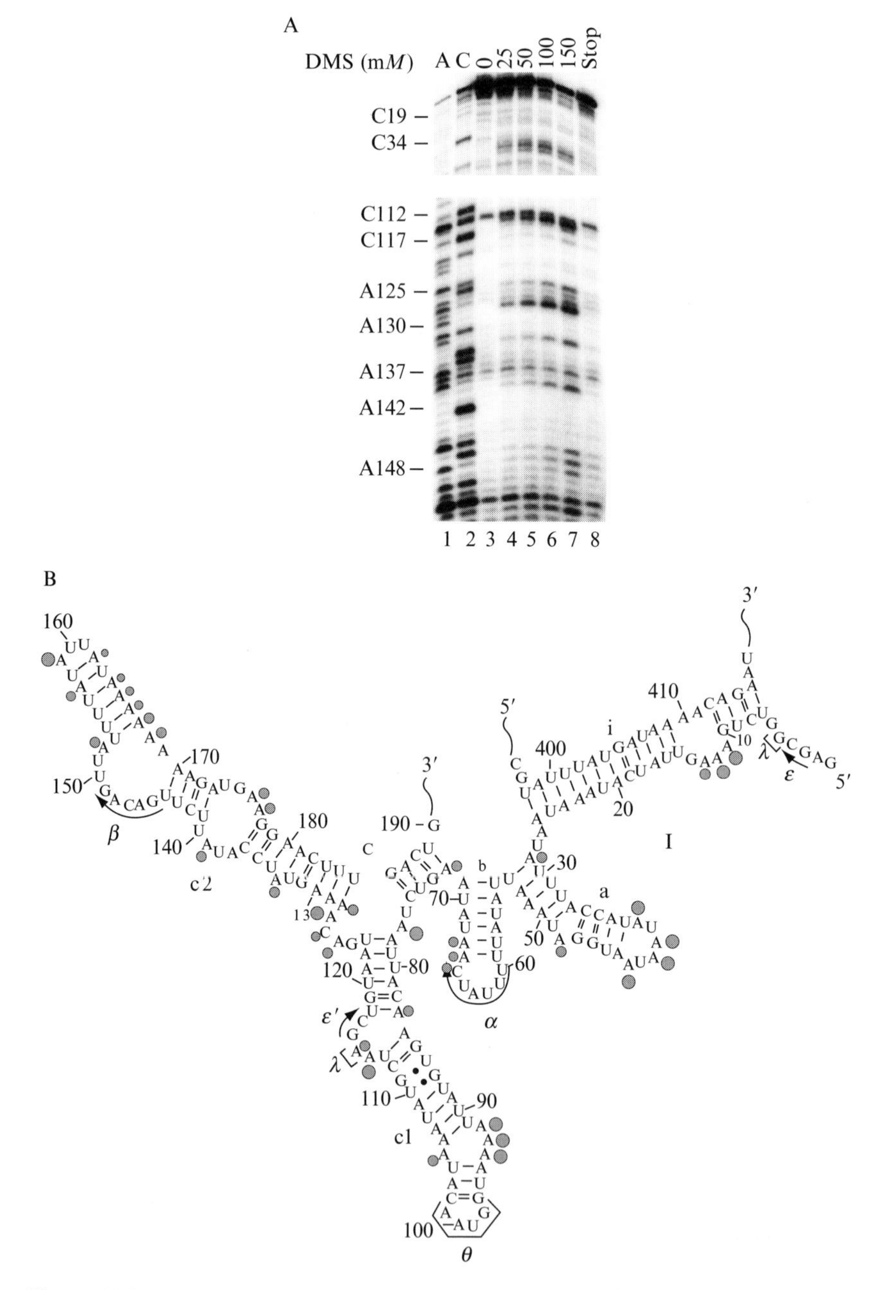

Figure 11.2 Optimizing the modification step to explore the architecture of a group II intron in yeast mitochondria. (A) DMS concentration series. A and C denote the sequencing lanes. In lane "0" the RT control that reveals any natural stops of the RT enzyme is shown (reverse transcription of nonmodified RNA). Lanes "25" to "150"

Apart from the modification step, it is also essential to find an RNA isolation procedure that yields pure RNA of good quality and in large quantities. We compared the yield and quality of several RNA preparation methods, including "home-made" protocols as well as kits from different companies (Table 11.1). In the end, we chose to use the RiboPure Yeast Kit (Ambion) to extract RNA from yeast cells following the instructions of the manufacturer. While the total RNA yield was lower compared to other methods, the extracted RNA was of excellent quality. To assess the quality of the RNA, we loaded a small aliquot of the total RNA on an agarose gel to detect the ribosomal RNAs: the sharper and more defined the bands are and the less "smear" occurs along the lane, the better is the RNA quality; this was further confirmed by reverse transcription of the RNA revealing that the amount of "natural" breaks varied significantly between the different RNA isolation protocols. In addition, the RiboPure Yeast kit (Ambion) was the most efficient in lysis of yeast mitochondria containing our RNA of interest. This was based on the comparison of the amount of full-length cDNAs reverse transcribed from RNA prepared with different methods (data not shown). Nevertheless, we considered providing a "home-made" protocol for isolating total yeast RNA to be useful, because it yielded the highest amount of RNA (Table 11.1). Notably, to remove genomic DNA entirely repeated DNAse I treatments may be necessary (data not shown). In general, removal of any DNA contaminants is of great importance and its efficiency varied significantly for the various protocols (data not shown).

Working with endogenous levels of RNA required some optimization of the reverse transcription protocol. The amount of total RNA necessary to obtain a DMS band pattern with good signal-to-noise ratio varies with different target RNAs and it has to be determined individually. Also, it may be necessary to use a RT enzyme other than the commonly used AMV Reverse Transcriptase (Promega) in case the target RNA is highly structured and GC-rich in sequence. For example, we had to use ThermoScript (Invitrogen) or Transcriptor Reverse Transcriptase (Roche) to map a GC-rich hairpin within the group II intron. In addition, primer design and specificity are crucial to detect the favorite RNA within the pool of total RNA.

show the DMS-accessible residues within the group II intron at increasing DMS concentrations (25–150 m*M*). In lane "stop" we show the "stop control," in which β-mercaptoethanol was added prior to DMS in order to determine the efficiency of quenching. The gel part displaying residues 46–111 is not shown. (B) The DMS modification pattern of lane "50" is plotted onto the secondary structure map of the *Sc.* ai5γ group II intron. This region corresponds to the c–c1–c2 three-way junction in the scaffolding domain D1 (A. Liebeg and C. Waldsich, unpublished data). The size of the gray filled circles corresponds to the relative modification intensity.

Table 11.1 Comparing different RNA isolation procedures

Method	RNA (μg) per 10^7 cells	DNA contamination	RNA quality	Efficiency of mitochondrial lysis
Hot acid phenol I[a]	10	ND	ND	ND
Hot acid phenol II[b]	20	+++	++	+
Glass bead I[c]	10	++	++	+
Glass bead II[d]	6	ND	ND	ND
NucleoSpin (Macherey-Nagel)[e]	3.5	+	+	++
RiboPure Yeast (Ambion)[e]	7	+	+++	+++

[a] This method was performed as described herein.
[b] This method was performed as described herein with the notable exception of including a douncing step at the beginning of the procedure attempting to improve lysis of yeast cells.
[c] Cell lysis was achieved with glass beads followed by removal of DNA with DNAse I (RNAse-free) and phenol extraction to remove proteins and cell debris (Collart and Oliviero, 2001).
[d] In addition to breaking the cells with the glass beads, a freeze–thaw step (liquid nitrogen) was incorporated into the "glass bead" method (Collart and Oliviero, 2001).
[e] RNA was isolated following the manufacturer's instruction.
ND, not determined.

To prevent fundamental pitfalls we recommend analyzing whether the modified RNA remains functional. Thus, it is advisable to incorporate the *in vivo* DMS modification step into a well-established experimental procedure that in the end allows testing the activity of the target RNA. For example, we incorporated the modification step in our *in vivo* splicing analysis procedure. As the splicing efficiency and RNA levels were not affected by DMS treatment of cells (data not shown), we were confident that the RNA we were analyzing was in good condition (Waldsich *et al.*, 2002a,b). To this end, as we demonstrate the power of this technique by presenting structural aspects of catalytic RNAs from different organisms, we also provide modification protocols for monitoring the intracellular structure of RNAs in *Escherichia coli* and *Saccharomyces cerevisiae*.

2.4. The *in vivo* DMS modification

2.4.1. DMS modification of *E. coli* RNA

This protocol was originally adapted from Zaug and Cech (1995):

1. Grow an overnight culture of the desired *E. coli* strain at 37 °C.
2. Inoculate 100 ml growth medium (e.g., LB) with 1 ml of the overnight culture and incubate at 37 °C. Note add antibiotics or IPTG if required.

3. Harvest the cells (2× 50 ml) at an $OD_{600\ nm}$ of 0.2–0.3 by centrifugation at 4000×*g* at 4 °C for 5 min.
4. Discard the supernatant and resuspend the cell pellet in 1 ml of ice-cold TM buffer (10 m*M* Tris pH 7.5, 10 m*M* $MgCl_2$).
5. Add DMS to a final concentration of 150 m*M* to one of the two samples (vortex briefly). Incubate the cells at room temperature for 1 min.
6. Afterwards, add β-mercaptoethanol to a final concentration of 0.7 *M* in order to quench DMS. Vortex strongly to avoid pelleting of insoluble DMS with the cells.
7. Centrifuge the tubes immediately at 6000×*g* for 2 min. After discarding the supernatant freeze the pellet at −80 °C.

2.4.2. DMS modification of yeast RNA

1. Grow an overnight culture of the yeast strain of interest at 30 °C.
2. Inoculate 100 ml growth medium (e.g., YPD) with 5 ml of the overnight culture. Grow the culture at 30 °C to an $OD_{600\ nm}$ of 1.0. Harvest 2× 30 ml of the culture (spin with 4000×*g* at 20 °C for 5 min).
3. Discard the supernatant and resuspend the cell pellet in 1 ml of prewarmed growth medium (e.g., YPD).
4. Add DMS to a final concentration of 50 m*M* to one of the two samples (vortex briefly). Incubate the cells at 30 °C for 2 min.
5. Afterwards, stop the DMS reaction by adding β-mercaptoethanol (final concentration of 0.7 *M*) and 50 μl isoamyl alcohol. Vortex strongly and centrifuge at 6000×*g* for 2 min.
6. Carefully remove the supernatant and resuspend the cells in 1 ml of TM buffer (10 m*M* Tris pH 7.5, 10 m*M* $MgCl_2$) and add another 50 μl β-mercaptoethanol (vortex).
7. Repeat the centrifugation step and freeze the cell pellet at −80 °C (for at least 15 min).

2.5. RNA preparation

2.5.1. RNA preparation from *E. coli*

1. Resuspend the frozen cell pellet carefully in 157 μl Solution A (vortex). This solution consists of 150 μl TE (10 m*M* Tris pH 8.0, 1 m*M* EDTA pH 8.0), 1.5 μl 1 *M* DTT, 0.75 μl RNase inhibitor (40 U/μl), 4 μl lysozyme (10 mg/ml), 0.75 μl ddH_2O.
2. The cell suspension is frozen in liquid nitrogen and then thawed in a room-temperature water bath. These steps (freeze and thaw) have to be performed three times.

3. Add 20 μl Solution B [4 μl 1 *M* $MgOAc_2$, 3.5 μl DNAse I (RNase-free) (10 U/μl), 0.1 μl RNase inhibitor (40 U/μl), 12.4 μl ddH_2O], mix gently and incubate the samples on ice for 1 h.
4. Add 20 μl Solution C (10 μl 0.2 *M* acetic acid, 10 μl 10% SDS), mix gently and incubate the samples at room temperature for 5 min.
5. *Phenol extraction.* Add 200 μl phenol (pH < 6.0), vortex and centrifuge at 18,000×*g* for 5 min. Collect the upper (aqueous) phase and mix it with 200 μl PCI (phenol/chloroform/isoamyl alcohol = 50:48:2), vortex and centrifuge at 18,000×*g* for 5 min. Transfer the upper phase into a new tube and add 200 μl CI (chloroform/isoamyl alcohol = 24:1), vortex and centrifuge at 18,000×*g* for 3 min. Take the aqueous phase and transfer it into a new 1.5 ml reaction tube.
6. Precipitate the RNA with 2 μl 0.5 *M* EDTA pH 8.0, 600 μl ethanol/0.3 *M* NaOAc pH 5.0.
7. Freeze the samples at −20 °C for 1 h and then centrifuge at 18,000×*g* at 4 °C for 30 min. Discard the supernatant, wash the RNA pellet with 70% ethanol, and dry it carefully but briefly (5 min at room temperature). Resuspend the pellet in about 30 μl ddH_2O. Store RNA at −80 °C.

2.5.2. Isolating RNA from yeast

Alternatively, we recommend using the RiboPure Yeast Kit (Ambion):

1. Resuspend frozen cells in 500 μl AE buffer (50 m*M* NaOAc pH 5.0, 10 m*M* EDTA pH 8.0) and 100 μl of 10% SDS.
2. Add cells to 700 μl of hot (65 °C) phenol in a 1.5 ml reaction tube. Vortex the mixture strongly.
3. Chill the sample rapidly in liquid nitrogen. Thaw the samples by shaking at 65 °C for ~1 min. Vortex the samples strongly and repeat the freeze–thaw step.
4. Shake tubes at 65 °C for 4 min. Centrifuge at 18,000×*g* for 5 min.
5. Add the aqueous phase to 700 μl PCI (phenol/chloroform/isoamyl alcohol = 50:48:2). After vortexing strongly, spin the sample at 18,000×*g* for 5 min.
6. Take the upper phase and mix it with to 700 μl CI (chloroform/isoamyl alcohol = 24:1), vortex intensively and then centrifuge at 18,000×*g* for 3 min.
7. Precipitate RNA with 1.4 ml ethanol/0.3 *M* NaOAc pH 5.0 at −20 °C overnight. Centrifuge at 18,000×*g* for 30 min (4 °C). Remove supernatant, wash pellet by adding 100 μl 70% EtOH. Remove supernatant, dry pellets at room temperature (~5 min); do not overdry. Add 180 μl ddH_2O to resuspend the pellet (containing both RNA and remaining DNA).
8. *DNAse I treatment.* Add 20 μl of a solution consisting of 4 μl 1 *M* $MgOAc_2$, 3.5 μl DNAse I (RNase-free) (10 U/μl), 0.1 μl RNase

inhibitor (40 U/μl), 12.4 μl ddH_2O to the sample. Mix gently and incubate the samples on ice for 1 h.

9. Repeat steps 5 and 6 using 200 μl PCI or CI, respectively.
10. Precipitate the RNA with 3× volume ethanol/0.3 *M* NaOAc pH 5.0. Eventually, resuspend the RNA pellet in 10–30 μl ddH_2O. Store RNA at −80 °C.

2.6. Reverse transcription

2.6.1. 5′-end-labeling of primer

1. Set up the labeling reaction by mixing 10 pmol DNA oligo, 10 pmol γ-P^{32}-ATP (6000 Ci/mmol), 2 μl 10× PNK buffer; and add ddH_2O to a final volume of 19 μl. Add 1 μl T4-polynucleotide kinase (10 U/μl) and incubate the sample at 37 °C for 30 min.
2. Stop the reaction by adding 1 μl 0.5 *M* EDTA pH 8.0.
3. Place the sample on a 95 °C heat block for 1 min, then immediately place the tubes on ice for 2 min.
4. To remove unused, excess radioactive nucleotides, precipitate the stopped labeling reaction by adding 1 μl glycogen (10 mg/ml) and 2.5×volume ethanol/0.3 *M* NaOAc mix. Freeze at −20 °C for 30 min; centrifuge at 4 °C for 30 min (18,000×*g*). Remove the supernatant and air dry the pellet.
5. Resuspend the 5′-end-labeled oligo in 50 μl ddH_2O. Store at −20 °C.

2.6.2. Primer extension

1. Set up the annealing reaction by mixing 2.5 μl RNA (40 μg), 1 μl P^{32}-labeled primer (0.2 μ*M*), and 1 μl 4.5× hybridization buffer (225 m*M* Hepes pH 7.0; 450 m*M* KCl).
2. Incubate the sample at 95 °C for 1 min.
3. Cool the samples slowly to 42 °C.
4. Add 2.2 μl extension mix consisting of 0.6 μl 6.7× extension buffer (1.3 *M* Tris–HCl pH 8.0, 100 m*M* $MgCl_2$, 100 m*M* DTT), 0.3 μl 2.5 m*M* NTP mix, 1.0 μl ddH_2O, 0.3 μl AMV Reverse Transcriptase (10 U/μl) and incubate the samples at 42 °C for 1 h. *Note.* To generate sequencing lanes (Figs. 11.1–11.4, lanes denoted A and C) add 1 μl of a 1 m*M* ddTTP or ddGTP solution to the annealing reaction in addition to the extension mix.
5. Degrade the RNA using 1.5 μl 1 *M* NaOH. Incubate the sample at 42 °C for 30–60 min.
6. Neutralize the pH by adding 1.5 μl 1 *M* HCl.

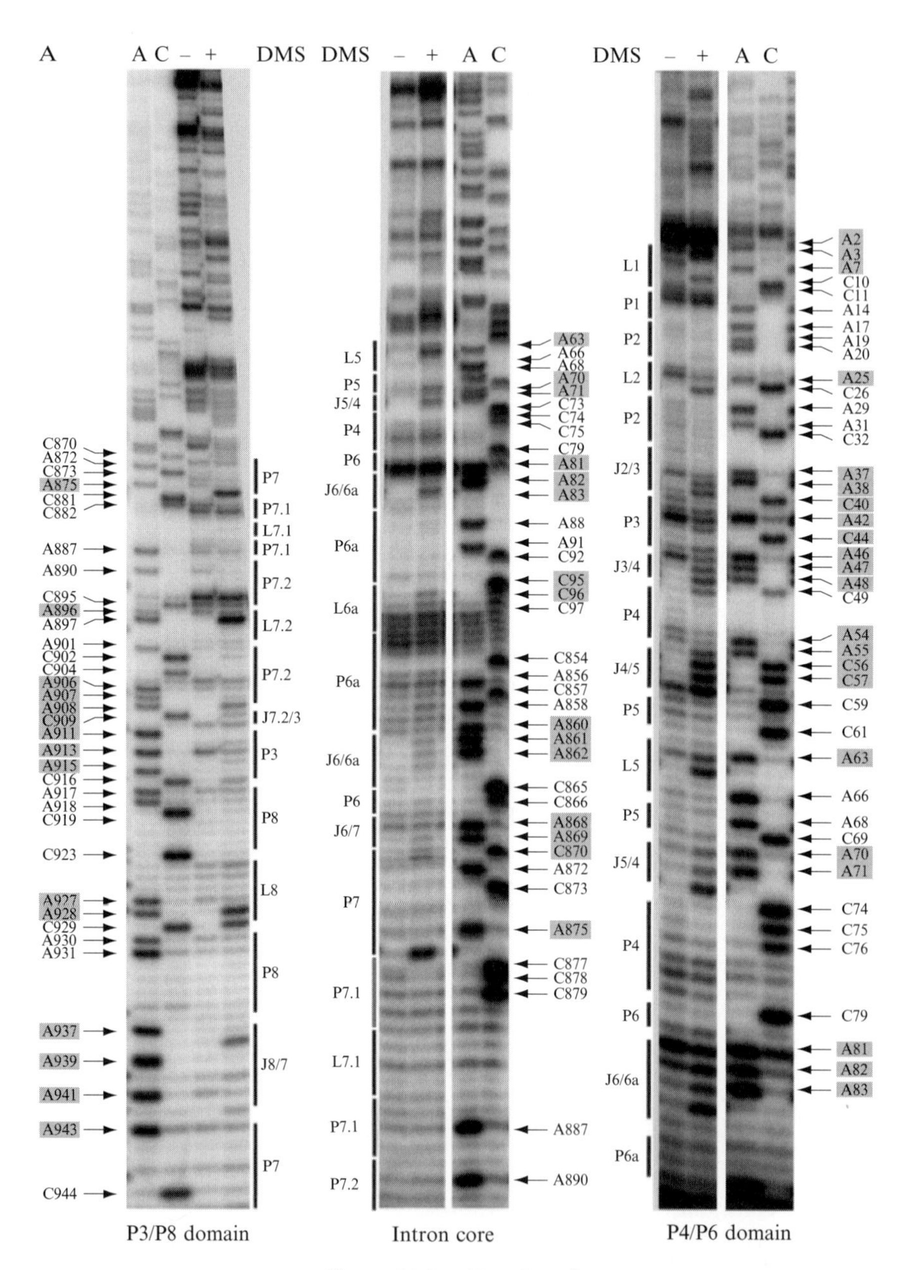

Figure 11.3 (Continued)

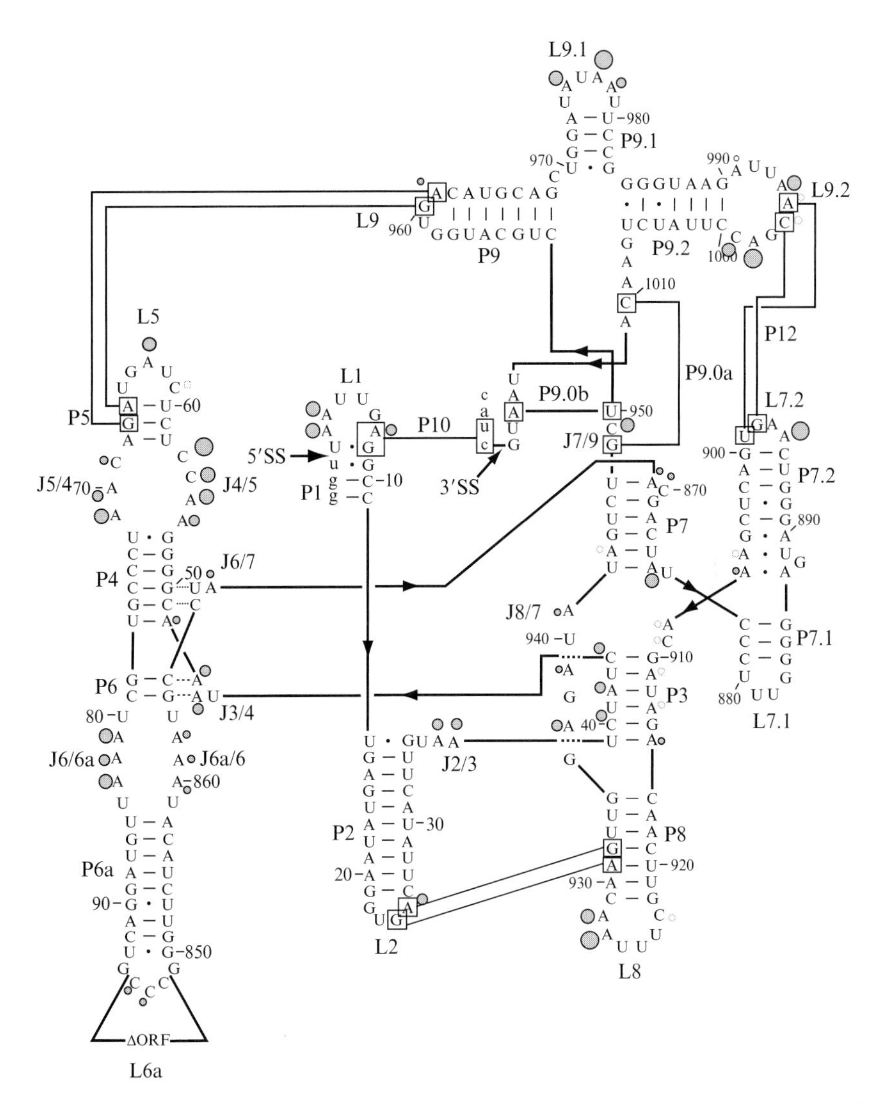

Figure 11.3 The intracellular structure of a catalytic RNA: the T4 phage-derived *td* group I intron. (A) Representative gels display intron residues accessible to DMS *in vivo*. Numbered residues highlighted by a gray box correspond to modified positions within the intron. The P3–P8 domain of the intron core (*left panel*), the center of the intron core covering the P7 stem as well as the P6–P6a element (*middle panel*), and the P4–P6 domain of the intron core together with the stem loops P1–P2 (*right panel*) are shown. A and C denote the sequencing lanes. (B) The DMS modification pattern was plotted onto the secondary structure map of the *td* intron. The size of the filled circles correlates with the relative modification intensity. This figure was adapted from Waldsich *et al.* (2002b).

7. Precipitate the cDNAs with 1 μl glycogen (10 mg/ml), 1 μl 0.5 *M* EDTA pH 8.0, and 45 μl ethanol/0.3 *M* NaOAc pH 5.0. Freeze the sample at −20 °C for 1 h (or overnight).
8. Centrifuge the sample at 4 °C for 30–60 min (18,000×*g*). Discard the supernatant, wash the pellet with 70% ethanol, and resuspend the dried pellet in 8 μl loading buffer [7 *M* urea, 25% sucrose, 0.25% bromphenol blue, 0.25% xylene cyanol in 1× TBE (0.089 *M* Tris base, 0.089 *M* boric acid, 2 m*M* EDTA)].

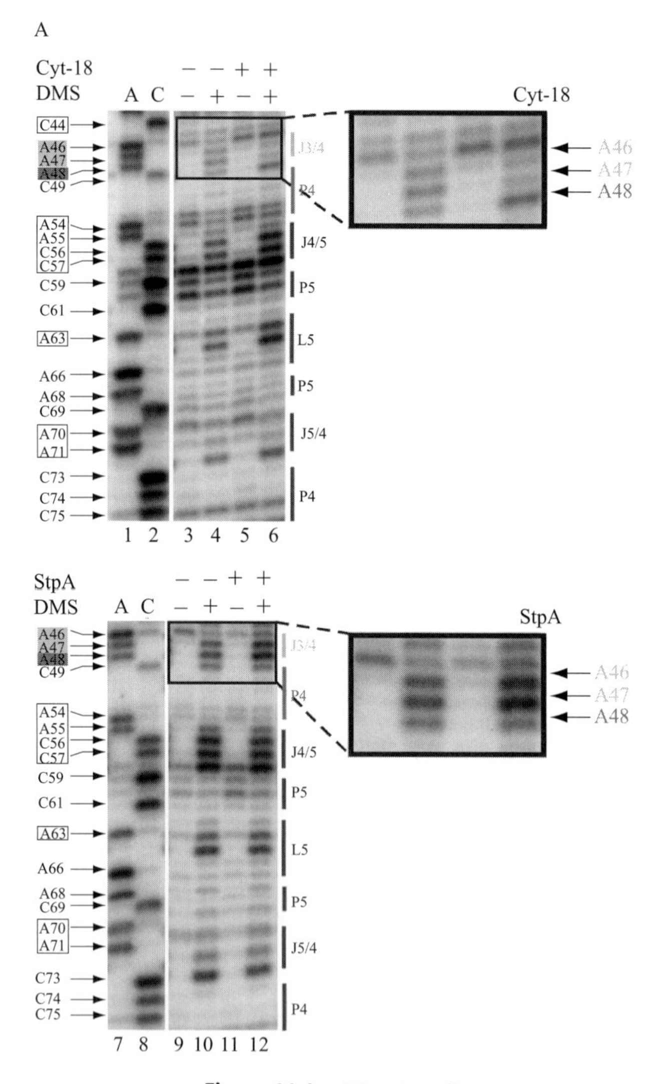

Figure 11.4 (Continued)

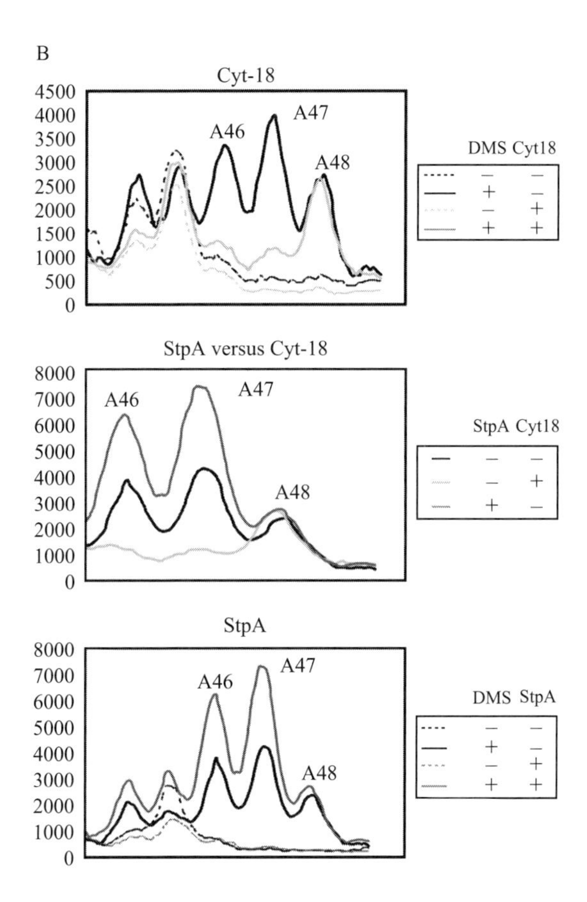

Figure 11.4 Protein-facilitated folding of the *td* intron. (A) The representative gels show the base-triple interaction between adenines in J3/4 and stem P6 within the *td* group I intron. Changes in the DMS modification pattern of the *td* intron *in vivo* due to the presence of the specific RNA-binding protein Cyt-18 are shown in the *upper panel* or due to the presence of the RNA chaperone StpA in the *lower panel*. Numbered nucleotides, which are highlighted by open boxes at the left of the gel, are modified by DMS, whereby filled boxes indicate a change in the DMS accessibility. The gel part boxed in black is outlined to point out the different effect of Cyt-18 compared to StpA on the residues A46 and A47 in J3/4. The sequencing lanes are labeled with A and C. In the presence of Cyt-18, the amount of *td* RNA is increased in the cells, as reflected by the increase of nonspecific stops in untreated samples (cf. lanes 3 and 5), as well as by the increased modification intensity of residues A55–C57 and A63 in lanes 4 and 6. (B) Quantification of the outlined gel segments in the presence of Cyt-18 (*upper panel*) or StpA (*lower panel*). The opposite effects of these proteins on the accessibility of the two adenines in J3/4 to DMS are summarized in the *middle panel*. This figure was adapted from Waldsich *et al.* (2002b).

9. Load half of the sample on a standard 8% denaturing polyacrylamide gel [7 *M* urea, 8% acrylamide/bisacrylamide (19:1) in 1× TBE]. Run the bromphenol blue dye to the bottom by applying 40 W (~1100 V, ~30 mA). Store the remainder of the sample at −20 °C.
10. Dry the gel and expose it in phosphorimager cassettes (Molecular Dynamics, GE Healthcare) over night. After scanning (on Storm 820 phosphorimager) the gel is analyzed using the ImageQuant software (Molecular Dynamics, GE Healthcare).

3. Application

While a few studies looked at the *in vivo* structure of RNAs (Adilakshmi *et al.*, 2006; Balzer and Wagner, 1998; Doktycz *et al.*, 1998; Mereau *et al.*, 1997; Wells and Ares, 1994; Zaug and Cech, 1995), intracellular RNA folding as well as associated intermediates and pathways have been studied to an even smaller extent (Donahue *et al.*, 2000; Mahen *et al.*, 2005; Waldsich *et al.*, 2002a,b). By employing *in vivo* DMS structural probing we first focused on determining the structure of the T4 phage-derived *td* group I intron in *E. coli* (Fig. 11.3; Waldsich *et al.*, 2002b) and of the *S. cerevisiae* coxI/ai5γ group II intron in yeast mitochondria (Fig. 11.2; A. Liebeg and C. Waldsich, unpublished data). The DMS pattern of both introns revealed that the majority of the intron population adopts the native state in accordance with the structural model derived from *in vitro* biochemical and phylogenetic data (Figs. 11.2 and 11.3; A. Liebeg and C. Waldsich, unpublished data; Waldsich *et al.*, 2002b). For example, the ε–ε' and λ–λ' tertiary interactions are critical active site constituents within group II introns (Boudvillain *et al.*, 2000). The DMS accessibility of the respective residues suggests that A115 indeed contacts D5 (G836) and C117 forms the H-bond with G3, implying that the majority of spliced intron molecules maintains the native conformation (A. Liebeg and C. Waldsich, unpublished data).

In case of the *td* group I intron (Fig. 11.3) the obtained results allowed the identification of a novel tertiary interaction and a refinement of the 3D model of this intron (Waldsich *et al.*, 2002b). Taking this approach a step further, we also analyzed the structure of *td* intron mutants, assuming that the mutations interfere with folding at different stages, thereby causing an arrest in the folding pathway. This strategy provided the first insight into the structure of folding intermediates and suggested an order of events in a hierarchical folding pathway *in vivo* (Waldsich *et al.*, 2002b).

Like other RNAs the *td* pre-mRNA is prone to misfolding, which results in a reduced splicing efficiency (Semrad and Schroeder, 1998). However, proteins with RNA chaperone activity or a group I intron splicing factor are capable of rescuing splicing of the *td* pre-mRNA (Clodi *et al.*, 1999; Semrad and Schroeder, 1998). To further our understanding of protein-facilitated RNA folding, we elucidated the mechanisms of action of an RNA chaperone (StpA) and of an RNA cofactor (Cyt-18), which specifically binds to RNA. Therefore, the impact of both StpA and Cyt-18 on the structure of the *td* intron was probed *in vivo*. As exemplified by the base-triple interactions of A46 and A47 in J3/4 with the stem P6, the obtained results provided direct evidence for protein-induced conformational changes within catalytic RNA *in vivo* (Fig. 11.4): While StpA resolves tertiary contacts enabling the RNA to refold, Cyt-18 guides folding of the *td* intron, thereby preventing misfolding and contributing to the overall compactness of the intron *in vivo* (Waldsich *et al.*, 2002a). Notably, protein-induced protections from DMS modification (like in case of Cyt-18) do not necessarily indicate a physical interaction of the RNA and protein, but could be the results of a structural change upon protein binding. In brief, the *in vivo* structural probing technique enabled us to gain profound insights into the mechanism of action of RNA chaperones.

4. Limitations

The most significant drawback of this *in vivo* structural probing technique is that the method detects only a few out of many functional groups of a ribonucleotide, which may participate in intra- and intermolecular interactions. This disadvantage can partially be overcome by complementing the DMS assay with other base-specific chemical probing techniques.

Another main obstacle, however, is that the obtained DMS modification pattern is averaged over the entire RNA population and over time. The lack of time resolution is especially problematic when it comes to study *in vivo* folding and conformational changes, which are time-related events, rendering these questions more difficult to be addressed within cells. This obstacle can be partly overcome by introducing mutations in tertiary contacts, which cause an arrest in the folding pathway. DMS structural probing of such mutant RNA molecules allows inferring information on intracellular folding pathways and associated intermediates.

Also, the target RNA does not necessarily represent a single main population, but might be partitioned among distinct species leading to mix RNA populations. It is, therefore, important to be able to differentiate, for example, between folded versus unfolded molecules, naked RNA versus RNA–protein particles, and spliced versus precursor RNA in order to assign the modification pattern and its concomitant interpretation to a specific population.

Despite the apparent limitations, which are simply due to the complexity of the cellular environment, the *in vivo* chemical probing technique has provided groundbreaking insights into the intracellular structure of natively folded RNA (Adilakshmi *et al.*, 2006; Ares and Igel, 1990; Balzer and Wagner, 1998; Climie and Friesen, 1988; Doktycz *et al.*, 1998; Harris *et al.*, 1995; Mereau *et al.*, 1997; Moazed *et al.*, 1988; Senecoff and Meagher, 1992; Wells and Ares, 1994; Zaug and Cech, 1995) and of folding intermediates (Waldsich *et al.*, 2002b) as well as in RNA–protein interactions and their associated conformational changes in living cells (Waldsich *et al.*, 2002a).

5. Conclusion

During the past years, it has become increasingly evident that RNA plays a central role in most cellular processes. To gain fundamental insights into the basic function of RNAs, it is utterly important to investigate RNA structure formation in living cells. However, the available methodologies are still limiting (Adilakshmi *et al.*, 2006; Lindell *et al.*, 2002; Moazed *et al.*, 1988; Szewczak, 2008; Waldsich *et al.*, 2002a). So far, chemical probing *in vivo* has proven to be the most powerful tool to explore RNA structure, folding intermediates, and ligand interactions. Nevertheless, there is a demand for developing novel techniques to derive rules governing RNA folding *in vivo* facilitated by proteins and other cellular factors. In particular, it is of great interest to reveal the role of protein cofactors in guiding proper folding and in providing stability for the RNA architecture—a very general and fundamental aspect for RNA to fold into its functional conformation.

ACKNOWLEDGMENTS

Phil Perlman is acknowledged for his generous gifts of yeast strains. We are grateful to Olga Fedorova, Georgeta Zemora, Oliver Mayer, and Michael Wildauer for helpful discussions and for critically reading this chapter. We thank Renée Schroeder for her support. This work was supported by Austrian Science Foundation FWF grant Y401 to C.W.

REFERENCES

Adilakshmi, T., Lease, R. A., and Woodson, S. A. (2006). Hydroxyl radical footprinting *in vivo*: Mapping macromolecular structures with synchrotron radiation. *Nucleic Acids Res.* **34,** e64.

Ares, M. Jr., and Igel, A. H. (1990). Lethal and temperature-sensitive mutations and their suppressors identify an essential structural element in U2 small nuclear RNA. *Genes Dev.* **4,** 2132–2145.

Balzer, M., and Wagner, R. (1998). Mutations in the leader region of ribosomal RNA operons cause structurally defective 30 S ribosomes as revealed by *in vivo* structural probing. *J. Mol. Biol.* **276,** 547–557.

Boudvillain, M., Delencastre, A., and Pyle, A. M. (2000). A new RNA tertiary interaction that links active-site domains of a group II intron and anchors them at the site of catalysis. *Nature* **406,** 315–318.

Brunel, C., and Romby, P. (2000). Probing RNA structure and RNA–ligand complexes with chemical probes. *Methods Enzymol.* **318,** 3–21.

Brunel, C., Romby, P., Westhof, E., Ehresmann, C., and Ehresmann, B. (1991). Three-dimensional model of *Escherichia coli* ribosomal 5S RNA as deduced from structure probing in solution and computer modeling. *J. Mol. Biol.* **221,** 293–308.

Climie, S. C., and Friesen, J. D. (1988). *In vivo* and *in vitro* structural analysis of the rplJ mRNA leader of *Escherichia coli*. Protection by bound L10–L7/L12. *J. Biol. Chem.* **263,** 15166–15175.

Clodi, E., Semrad, K., and Schroeder, R. (1999). Assaying RNA chaperone activity *in vivo* using a novel RNA folding trap. *EMBO J.* **18,** 3776–3782.

Collart, M. A., and Oliviero, S. (2001). Preparation of yeast RNA. *Curr. Protoc. Mol. Biol.* **13,** 13.12.1–13.12.5.

Doktycz, M. J., Larimer, F. W., Pastrnak, M., and Stevens, A. (1998). Comparative analyses of the secondary structures of synthetic and intracellular yeast MFA2 mRNAs. *Proc. Natl. Acad. Sci. USA* **95,** 14614–14621.

Donahue, C. P., Yadava, R. S., Nesbitt, S. M., and Fedor, M. J. (2000). The kinetic mechanism of the hairpin ribozyme *in vivo*: Influence of RNA helix stability on intracellular cleavage kinetics. *J. Mol. Biol.* **295,** 693–707.

Ehresmann, C., Baudin, F., Mougel, M., Romby, P., Ebel, J. P., and Ehresmann, B. (1987). Probing the structure of RNAs in solution. *Nucleic Acids Res.* **15,** 9109–9128.

Gallouzi, I. E., Brennan, C. M., Stenberg, M. G., Swanson, M. S., Eversole, A., Maizels, N., and Steitz, J. A. (2000). HuR binding to cytoplasmic mRNA is perturbed by heat shock. *Proc. Natl. Acad. Sci. USA* **97,** 3073–3078.

Harris, K. A. Jr., Crothers, D. M., and Ullu, E. (1995). *In vivo* structural analysis of spliced leader RNAs in *Trypanosoma brucei* and *Leptomonas collosoma*: A flexible structure that is independent of cap4 methylations. *RNA* **1,** 351–362.

Herschlag, D. (1995). RNA chaperones and the RNA folding problem. *J. Biol. Chem.* **270,** 20871–20874.

Leontis, N. B., and Westhof, E. (2001). Geometric nomenclature and classification of RNA base pairs. *RNA* **7,** 499–512.

Leontis, N. B., Lescoute, A., and Westhof, E. (2006). The building blocks and motifs of RNA architecture. *Curr. Opin. Struct. Biol.* **16,** 279–287.

Lindell, M., Romby, P., and Wagner, G. H. (2002). Lead (II) as a probe for investigating RNA structure *in vivo*. *RNA* **8,** 534–541.

Lindell, M., Brannvall, M., Wagner, E. G., and Kirsebom, L. A. (2005). Lead(II) cleavage analysis of RNase P RNA *in vivo*. *RNA* **11,** 1348–1354.

Mahen, E. M., Harger, J. W., Calderon, E. M., and Fedor, M. J. (2005). Kinetics and thermodynamics make different contributions to RNA folding *in vitro* and in yeast. *Mol. Cell* **19,** 27–37.

Mereau, A., Fournier, R., Gregoire, A., Mougin, A., Fabrizio, P., Luhrmann, R., and Branlant, C. (1997). An *in vivo* and *in vitro* structure–function analysis of the *Saccharomyces cerevisiae* U3A snoRNP: Protein–RNA contacts and base-pair interaction with the pre-ribosomal RNA. *J. Mol. Biol.* **273,** 552–571.

Moazed, D., and Noller, H. F. (1986). Transfer RNA shields specific nucleotides in 16S ribosomal RNA from attack by chemical probes. *Cell* **47,** 985–994.

Moazed, D., and Noller, H. F. (1989). Intermediate states in the movement of transfer RNA in the ribosome. *Nature* **342,** 142–148.

Moazed, D., Robertson, J. M., and Noller, H. F. (1988). Interaction of elongation factors EF-G and EF-Tu with a conserved loop in 23S RNA. *Nature* **334,** 362–364.

Pinol-Roma, S., Adam, S. A., Choi, Y. D., and Dreyfuss, G. (1989). Ultraviolet-induced cross-linking of RNA to proteins *in vivo*. *Methods Enzymol.* **180,** 410–418.

Pyle, A. M., Fedorova, O., and Waldsich, C. (2007). Folding of group II introns: A model system for large, multidomain RNAs? *Trends Biochem. Sci.* **32,** 138–145.

Schroeder, R., Grossberger, R., Pichler, A., and Waldsich, C. (2002). RNA folding *in vivo*. *Curr. Opin. Struct. Biol.* **12,** 296–300.

Semrad, K., and Schroeder, R. (1998). A ribosomal function is necessary for efficient splicing of the T4 phage *thymidylate synthase* intron *in vivo*. *Genes Dev.* **12,** 1327–1337.

Senecoff, J. F., and Meagher, R. B. (1992). *In vivo* analysis of plant RNA structure: Soybean 18S ribosomal and ribulose-1,5-bisphosphate carboxylase small subunit RNAs. *Plant Mol. Biol.* **18,** 219–234.

Szewczak, L. B. (2008). *In vivo* analysis of ribonucleoprotein complexes using nucleotide analog interference mapping. *Methods Mol. Biol.* **488,** 153–166.

Thirumalai, D. (1998). Native secondary structure formation in RNA may be a slave to tertiary folding. *Proc. Natl. Acad. Sci. USA* **95,** 11506–11508.

Thirumalai, D., Lee, N., Woodson, S. A., and Klimov, D. (2001). Early events in RNA folding. *Annu. Rev. Phys. Chem.* **52,** 751–762.

Treiber, D. K., and Williamson, J. R. (1999). Exposing the kinetic traps in RNA folding. *Curr. Opin. Struct. Biol.* **9,** 339–345.

Treiber, D. K., and Williamson, J. R. (2001). Beyond kinetic traps in RNA folding. *Curr. Opin. Struct. Biol.* **11,** 309–314.

Waldsich, C., Grossberger, R., and Schroeder, R. (2002a). RNA chaperone StpA loosens interactions of the tertiary structure in the *td* group I intron *in vivo*. *Genes Dev.* **16,** 2300–2312.

Waldsich, C., Masquida, B., Westhof, E., and Schroeder, R. (2002b). Monitoring intermediate folding states of the *td* group I intron *in vivo*. *EMBO J.* **21,** 5281–5291.

Wells, S. E., and Ares, M. Jr. (1994). Interactions between highly conserved U2 small nuclear RNA structures and Prp5p, Prp9p, Prp11p, and Prp21p proteins are required to ensure integrity of the U2 small nuclear ribonucleoprotein in *Saccharomyces cerevisiae*. *Mol. Cell. Biol.* **14,** 6337–6349.

Woodson, S. A. (2000a). Compact but disordered states of RNA. *Nat. Struct. Biol.* **7,** 349–352.

Woodson, S. A. (2000b). Recent insights on RNA folding mechanisms from catalytic RNA. *Cell. Mol. Life Sci.* **57,** 796–808.

Woodson, S. A. (2005). Metal ions and RNA folding: A highly charged topic with a dynamic future. *Curr. Opin. Chem. Biol.* **9,** 104–109.

Zaug, A. J., and Cech, T. R. (1995). Analysis of the structure of *Tetrahymena* nuclear RNAs *in vivo*: Telomerase RNA, the self-splicing rRNA intron, and U2 snRNA. *RNA* **1,** 363–374.

CHAPTER TWELVE

Structural Analysis of RNA in Living Cells by *In Vivo* Synchrotron X-Ray Footprinting

Tadepalli Adilakshmi,* Sarah F. C. Soper,† *and* Sarah A. Woodson‡

Contents

Abstract

Chemical footprinting methods are widely used to probe the solution structures of nucleic acids and their complexes. Among the many available modifying reagents, hydroxyl radical is exceptional in its ability to provide nucleotide-level information on the solvent accessibility of the nucleic acid backbone. Until

* Weis Center for Research, Geisinger Medical Center, Danville, Pennsylvania, USA

† Program in Cell, Molecular and Developmental Biology and Biophysics, Johns Hopkins University, Baltimore, Maryland, USA

‡ T.C. Jenkins Department of Biophysics, Johns Hopkins University, Baltimore, Maryland, USA

Methods in Enzymology, Volume 468

ISSN 0076-6879, DOI: 10.1016/S0076-6879(09)68012-5

recently, hydroxyl radical footprinting has been limited to *in vitro* experiments. We describe the use of synchrotron X-radiation to generate hydroxyl radicals within cells for effective footprinting of RNA–protein complexes *in vivo*. This technique gives results that are consistent with *in vitro* footprinting experiments, with differences reflecting apparent structural changes to the RNA *in vivo*.

1. Introduction

Hydroxyl radical footprinting has been utilized for over 30 years in the study of nucleic acids and their interactions (Burkhoff and Tullius, 1987; Tullius and Dombroski, 1986). Hydroxyl radicals cleave nucleic acids by abstracting a hydrogen atom from the ribose (Hertzberg and Dervan, 1984), providing a readout of the solvent accessibility of the nucleic acid backbone (Latham and Cech, 1989; Tullius and Greenbaum, 2005). While the earliest enzymatic footprinting experiments were designed to determine the sequence specificity of a DNA binding protein (Galas and Schmitz, 1978), footprinting techniques have evolved to provide detailed structural insight into nucleic acids and nucleic acid–protein complexes (Dhavan et al. 2003; Revzin, 1993). Recent advances in time-resolved hydroxyl radical footprinting and the development of new chemical footprinting reagents have enabled much larger molecular complexes to be analyzed (Shcherbakova *et al.*, 2008; Wilkinson *et al.*, 2006).

Despite the enormous value of *in vitro* footprinting experiments, questions often remain as to whether the structures observed in the test tube accurately reflect the complexes within living cells. Footprinting nucleic acid complexes *in vivo* would settle these uncertainties. Some footprinting probes have been used *in vivo*, but they can have serious limitations. For example, DNase I footprinting has been invaluable for defining DNA–protein contacts *in vivo*, but the cells require potentially disruptive pretreatment with ethanol to allow the entry of the nuclease (Cassler *et al.*, 1999). Other reagents, such as dimethylsulfate (DMS) (Climie and Friesen, 1988; Mayford and Weisblum, 1989; Waldsich *et al.*, 2002; Wells *et al.*, 2000) and kethoxal (Balzer and Wagner, 1998), require no permeabilization to probe the structures of nucleic acids *in vivo*, but have base specificities that limit their resolving power. Additionally, most reagents used *in vivo* require exposure times of several minutes, limiting their utility in detecting short-lived intermediates.

Synchrotron X-ray-dependent hydroxyl radical footprinting overcomes these limitations. Hydroxyl radicals cleave DNA and RNA equally well with little or no sequence specificity (Sclavi *et al.*, 1997). Their small size and lack of base specificity allow for acquisition of structure information at the single nucleotide level (Tullius and Dombroski, 1986; Tullius *et al.*, 1987). *In vitro*, hydroxyl radicals are most commonly generated through the Udenfriend variation of the Fenton reaction (Udenfriend *et al.*, 1954), where Fe.(II)-EDTA reacts with peroxide to create hydroxyl radicals. However, hydroxyl radicals can also be created and used for footprinting reactions through the ionization of intracellular water by X-rays (Hayes *et al.*, 1990; Ottinger and Tullius, 2000).

We describe a technique for rapidly footprinting RNA *in vivo* using a synchrotron X-ray beam. X-rays have been successfully used for the time-resolved footprinting of RNA folding *in vitro*, with exposure times of 1–10 ms (Sclavi *et al.*, 1998). We find that a 100 ms exposure to the high flux of a synchrotron X-ray beam is sufficient to probe the structure of intracellular RNAs, without the need to perturb cells with high concentrations of chemical reagents (Adilakshmi *et al.*, 2006). The combined penetrating power of X-rays and the utility of hydroxyl radicals as a structural probe together provide a significant advantage over nuclease or chemical footprinting techniques for probing nucleic acid dynamics inside cells.

2. Beamline Setup for *In Vivo* Footprinting

Hydroxyl radicals and free electrons are produced by the ionization of water (Klassen, 1987).

$$H_2O \xrightarrow{h\nu} H_2O^+ + e^-_{aq} \xrightarrow{H_2O} H_3O^+ + \bullet OH + e^-_{aq}$$

Hydroxyl radical generation by exposure to a ^{137}Cs gamma ray source for 15 min was used to footprint lambda repressor–DNA complexes *in vivo* (Ottinger and Tullius, 2000). However, a high flux "white light" synchrotron X-ray beam shortens the necessary exposure times to 100 ms or less. In our experiments, we use beamline X28C at the National Synchrotron Light Source (NSLS). X28C is a bending-magnet beamline administered and operated by the Case Western University Center for Synchrotron Biosciences that is dedicated to the footprinting of nucleic acids and proteins.

For X-ray footprinting experiments, the shape and flux density of the beam are important. The cross-section of the beam must be large enough to irradiate the entire surface area of the sample. However, an unnecessarily diffuse beam reduces the flux density, requiring longer exposure times. A beam with a configurable focusing mirror is ideal because these parameters can be varied for individual experiments. The X28C beamline utilizes a cylindrical toroidal mirror to focus the 6 mrad of horizontal radiation it receives from the NSLS X-ray storage ring. When configured to a point focus with a cross-section area of approximately 0.2 mm^2, the beam delivers a more than 350-fold increase in flux density as compared to the unfocused beam (Sullivan *et al.*, 2008). This increase in flux density translates to shorter exposure times.

For the purposes of *in vivo* synchrotron footprinting the narrowest (and most intense) beam that will irradiate the entire surface area of the sample should be used. Although sample heating is also increased at higher flux densities, this is generally offset by the shortened exposure time (Sullivan *et al.*, 2008). Equally important to the quality of the results is that samples are aligned with the most intense region of beam using a photodiode mounted behind the sample holder. The output from the diode is used to control the position of the motorized stage supporting the sample holder, via LabView software and a servo-motor controller (Gupta *et al.*, 2007).

For synchrotron footprinting applications, the exposure time must be controlled to within a few milliseconds. The X28C beamline utilizes a Uniblitz XRS6 shutter (Vincent Associates, NY) especially designed for X-ray applications, which allows beam extinction $>10^4$ up to 30 keV X-ray energy. The shutter can reliably open within 3.2 ms at a maximum rate of 50 exposures per second. The sample holder and the shutter are aligned to the peak of the beam intensity using a precision motorized table (Dhavan *et al.*, 2003).

To minimize thermal damage, cells are kept frozen at −34 to −38 °C during exposure to the beam through a cryostatted sample holder. This temperature is sufficient to hold the cytoplasm in a frozen state while still permitting the formation of hydroxyl radicals (Mazur, 1984). Although we cannot exclude the possibility of local heating during irradiation, no breakdown of the cells is observed upon irradiation. Recently, X28C has been upgraded with a cooled 23-sample holder mounted to a motorized stage that is controlled from outside the experimental hutch, allowing multiple samples to be exposed without reentering the hutch (Fig. 12.1) (S. Gupta, personal communication).

3. Preparation of Samples

Obtaining high quality, intact RNA is the most critical requirement for the success of synchrotron footprinting experiments. For this reason, the standard precautions for handling RNA must be followed at all times. RNase-free reagents and deionized (18 MΩ) water should be used in the preparation of all samples. Gloves must be worn, and plastic consumables should be used new from the package.

Frozen samples to be irradiated are prepared in advance and shipped to the beamline. Flash-freezing cells before irradiation makes it possible to take

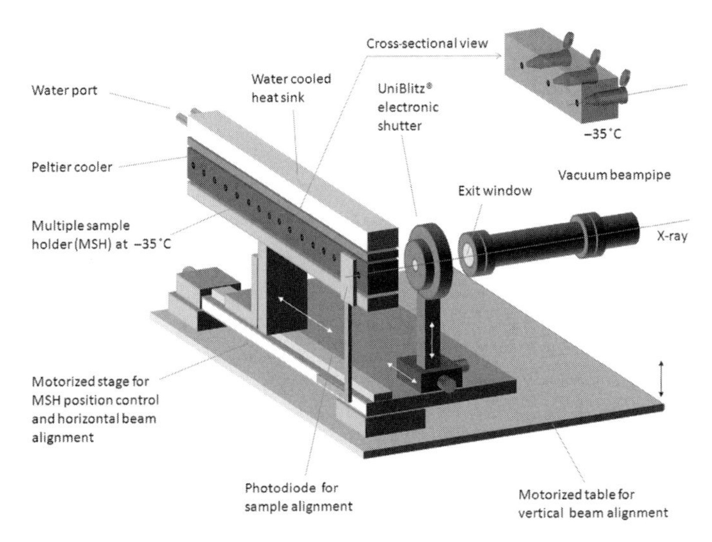

Figure 12.1 Multiple sample holder for irradiation of cell pellets. The block holds 23 samples and can be cooled to −38 °C. The sample holder was custom fabricated for use at beamline X28C. Samples are automatically aligned with the X-ray beam using a photodiode and motorized table. Exposure times are gated by an electronic shutter. Figure courtesy of S. Gupta.

"snapshots" of macromolecules in various stages of metabolism, while stabilizing cells against heat damage. Our experiments are performed with *Escherichia coli*, but the protocol could be adapted to other types of cells as long as suitable methods for RNA isolation and analysis are available.

To prepare samples, inoculate a single bacterial colony into 3–5 mL of appropriate medium and grow overnight at the appropriate temperature. The next day, dilute the culture 1:100 or 1:250 in 50 mL media and grow to OD600 of 0.6–1.0 (mid-log). This should take 2–3 h. If growth is too slow, reduce the dilution factor. When the culture has reached the desired density, rapidly cool the flask by swirling it vigorously in an ice/ethanol bath for 20–30 s. It is critical that the culture is cooled but not frozen at this stage, as freezing the culture will damage the cells.

Pellet the cells by centrifuging the culture in prechilled tubes at 6000 × *g* for 5 min at 4 °C. Carefully decant the supernatant, leaving the pellet as dry as possible. Wash the cell pellet with 300–500 μL TM buffer (10 m*M* Tris–HCl, pH 7.5 and 1 m*M* $MgCl_2$). This wash is essential for removing extracellular exonucleases. Resuspend the cell pellet in 300 μL of TM buffer (~6 μL/μg wet cells or 1.5×10^{11} cells/mL).

Dispense 5–10 μL aliquots of the cell suspension into sample tubes suitable for use at your beamline. The sample holder at X28C is designed for use with 0.2 mL BrandTech thin wall PCR tubes with attached caps

(Fisher catalog # 13-882-58). It is of utmost importance that the suspension be deposited at the very bottom of the tube, and that there are no droplets left clinging to the sides. The tubes cannot be centrifuged in a fixed angle rotor to collect the cell pellet, because the pellet will be displaced to one side and will not be aligned with the beam. A swinging bucket rotor may be used to collect the pellet at the bottom of the tube.

Snap-freeze the pellets in a dry ice/acetone bath (or liquid nitrogen) and store at -80 °C until further use. Inclusion of glycerol or DMSO for stabilizing cells under freezing conditions could be considered, but may increase the exposure times tremendously as these reagents are efficient free radical scavengers. We have also used purified 70S ribosomes (Nierhaus and Dohme, 1974), divided into 10 μL aliquots (1 μM) and snap-frozen in a dry ice/acetone bath as controls (Adilakshmi *et al.*, 2006).

Samples must be carefully packaged in a Styrofoam container with dry ice for overnight shipment to the beamline. Some samples should be kept aside for controls against the samples being damaged in shipping—these are "No shipping" controls. It is recommended to package the samples with at least 4 kg of dry ice to ensure they do not thaw *en route*. Samples that thaw at any time before RNA extraction should be discarded. Alternatively, a liquid nitrogen dry-shipper may be considered for shipping samples.

4. Exposure of Cells to X-Ray Beam

4.1. Fluorescent assay for calibration of beam intensity

As the condition of the beam can vary from day-to-day, and even within a day as the ring current decreases between injections, it is helpful to begin experiments by gauging the relative beam intensity with a fluorescent assay as described by Gupta *et al.* (2007). This is done by generating a dose-response curve using a fluorescent dye, which is photobleached upon exposure to the X-ray beam. The assay is performed with the samples at ambient temperature, before the sample holder has been cooled for use with the frozen cell samples. A 2 μM solution of fluorophore such as Alexa488 is prepared in sample buffer, such as TM (Gupta *et al.*, 2007). Six 5 μL aliquots are placed in 0.2 mL tubes and irradiated for 0, 10, 20, 30, 40, and 50 ms, respectively. The samples are diluted to 1 mL with distilled water and the fluorescence intensity measured with a standard fluorimeter or plate reader. The rate constant obtained by plotting fluorescence against exposure time can be compared against future experiments to ascertain the consistency of the beam over time and to allow the reevaluation of sample exposure times, if necessary.

4.2. Sample exposure

Following completion of the fluorophore dose-response curve the sample holder is cooled to −35 °C in preparation for exposing the frozen cells. Samples are brought to the beamline on dry ice and placed in the area outside the experimental hutch (or in a −80 °C freezer if convenient). Samples awaiting exposure should not be left in the hutch when the beam is on, as the high level of stray radiation within the hutch can lead to high background levels of cleavage. Some samples should be set aside and kept outside the hutch, as controls against stray radiation. We term these samples "No hutch" controls.

After the sample holder has been completely cooled, samples are uncapped and loaded for exposure. After loading samples the experimenter leaves the hutch, engaging the safety interlock to allow the beam to be activated. The position of the sample holder, the alignment, and the shutter, are operated from outside the hutch through a PC interfaced to the hardware components. This allows the experimenter to rapidly expose 23 samples at a time without reentering the hutch. To measure the level of background cleavage within the hutch, "mock exposed" samples are placed in the sample holder without opening the shutter. These are "in hutch" controls.

After all samples have been exposed, they should be repackaged for shipment to the home institution in the same way as they were shipped to the beamline. Alternatively, if laboratory facilities are available, the RNA can be extracted at the beamline site before shipment to the home institution for analysis. The purified RNA is less sensitive to degradation should samples thaw.

4.3. Determination of optimal exposure times

For reliable footprinting results, each molecule must be cleaved no more than once. This can be achieved by limiting the extent of cleavage of the target RNA or DNA fragment to 10–30% (Hsieh and Brenowitz, 1996). Many factors influence the absorption of X-rays by the sample, the production of hydroxyl radicals and the extent of their reaction with the target molecule. Therefore, the X-ray dose needed to achieve the optimal extent of cleavage must be determined empirically for each type of sample (Sclavi *et al.*, 1997).

As described above, free radicals are produced when water molecules within the cell absorb a photon. The radicals can diffuse within the cytoplasm up to ~10 Å before recombining with each other or reacting with another molecule (Dreyer and Dervan, 1985). In general, longer exposure times are needed to cleave nucleic acids in cells than in solution. First, the creation of hydroxyl radicals is limited by dehydration of the cell cytoplasm, while their diffusion is limited by molecular crowding (Mazur, 1984). Second, the cell interior is filled with free radical scavengers such as cysteine (Roberts *et al.*, 1995). Third, the slower diffusion of free radicals in frozen or supercooled water is expected to reduce the frequency of strand breaks in frozen cells (Huttermann *et al.*, 1992; Ohshima *et al.*, 1996). Despite this, in hydrated DNA at -196 °C most of the radiation damage is attributed to indirect action of hydroxyl radicals rather than the direct absorption of X-rays (Gregoli *et al.*, 1982; Huttermann *et al.*, 1992; Ohshima *et al.*, 1996).

To determine the correct exposure time, a dose-response curve is generated by exposing samples to the X-ray beam for varying times (Fig. 12.2). After extraction of total cellular RNA, the extent of cleavage is determined by extending a molar excess of $5'$-^{32}P-labeled primer complementary to the $3'$-end of the RNA with reverse transcriptase. By plotting the intensity of the cDNA corresponding to the full-length RNA versus the exposure time, and fitting the data to an exponential equation, the exposure time corresponding to 20% cleavage can be calculated.

A good dose-response curve will range from exposures that induce very little cleavage to exposures that cleave nearly all of the RNA within the cell. With the current X28C configuration, we find that an exposure range from 8 to 500 ms is satisfactory, with an optimal dose of 75 ms for

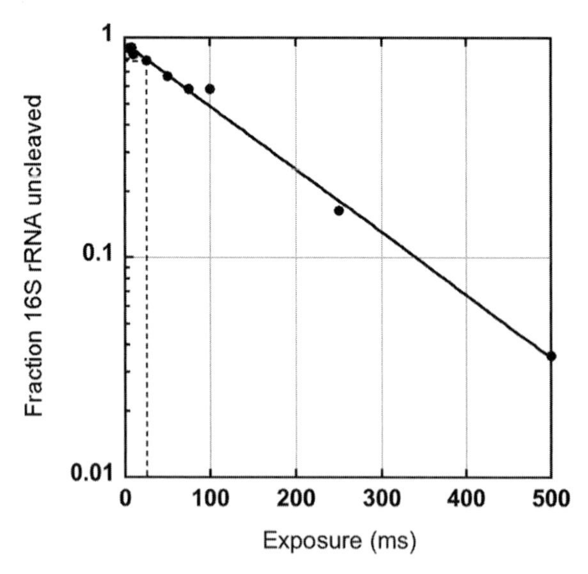

Figure 12.2 Determining the correct X-ray dose for footprinting. Dose-response curve showing relative fraction of intact 16S rRNA versus time of X-ray exposure of intact, frozen *E. coli* cells. The optimal dose in this experiment is obtained after 25–50 ms exposure. The relative amount of intact rRNA was determined by primer extension as described in the text.

approximately 20% cleavage (Fig. 12.2). Of course, it is not always possible to determine the appropriate exposure time in advance. Since the X28C multisample holder allows irradiation of hundreds of samples a day, however, it is possible to expose duplicate samples to the beam for different times, in order to ensure that at least one set of samples will provide RNA suitable for analysis.

5. Isolation of Total RNA from Irradiated Cells

Structural analysis by hydroxyl radical footprinting requires high-quality RNA templates, to reduce background signals from extraneous sources of RNA damage. The current generation of commercial kits specialized for prokaryotic total RNA extraction produce a good yield of high-quality, nick-free RNA from irradiated cell pellets. We have used the Qiagen RNeasy® Protect Bacteria Mini Kit and the Ambion RiboPure™-Bacteria Kit with satisfactory results. In our hands, the Qiagen RNeasy kit gives superior yields, with a typical recovery of 20 μg of RNA from a 10 μL cell pellet.

5.1. RNA isolation using the TRIzol method

RNA isolation using the TRIzol protocol also yields good-quality RNA suitable for primer extension analysis and can be less costly (Adilakshmi *et al.*, 2006). Add 100 μL TRIzol reagent directly to frozen cells and mix by pipeting until the cell pellet thaws. Repeat the freeze thaw cycle to ensure lysis of the cell wall. Alternatively, the cells can be resuspended in Max Bacterial Enhancement Reagent (Invitrogen) composed of chelating agents, detergent, and a buffer that inactivates endogenous RNases and promotes protein denaturation. This reagent is designed for use with TRIzol Reagent to improve the isolation of intact total RNA from Gram-positive and Gram-negative bacteria.

Following lysis, extract RNA with the addition of 20 μl of chloroform. Collect the aqueous phase by centrifugation at 12,000 $\times$ *g* for 15 min at 4 °C. Repeat the extraction of the sample with more TRIzol and chloroform to obtain high-quality RNA. Precipitate the RNA by incubating the sample with equal volume of isopropanol at room temperature for 10 min followed by centrifugation at 12,000 $\times$ *g* for 15 min at 4 °C. Wash the RNA pellet with 70% ethanol and resuspend in 20 μl RNase-free water. The quality of the RNA can be analyzed by loading 2 μl of the sample on 1.2% agarose gel prepared in 50 m*M* Tris-acetate and 1 m*M* EDTA. Estimate the RNA concentration by measuring its absorbance at 260 nm. Purified RNA can be stored at -80 °C for several weeks until further use.

5.2. Further considerations on RNA isolation from bacteria

The quality of the total RNA isolated from irradiated cells is critical for the success of *in vivo* footprinting experiments. If frozen cells are to be used for the experiments, it is essential that they remain completely frozen during the entire process, including shipping. Thawing and refreezing damages the cell membrane, resulting in breakdown of cellular RNA. RNA stabilization reagents such as RNA*later*™ (Ambion) do not interfere with X-ray cleavage. In our hands, such reagents do not significantly improve the quality of the RNA.

Cellular disruption is the most critical step, affecting the yield and quality of the isolated RNA. Slow disruption of the cell membrane, for example, by placing cells or tissue in lysis solution without any additional physical shearing, may result in RNA degradation by releasing endogenous RNases before proteins are fully denatured. On the other hand, incomplete disruption of cells decreases the yield of RNA.

One solution is to digest bacterial cell walls with lysozyme to form spheroplasts. Gram-positive bacteria usually require more rigorous digestion (increased incubation time, increased incubation temperature) than Gram-negative bacteria. The spheroplasts can then be easily lysed with vigorous vortexing or sonication in guanidine isothiocyanate buffer. Resuspend the pellet in 300 μL of 10 m*M* Tris–HCl, 1 m*M* EDTA containing lysozyme (freshly prepared). The final concentration of lysozyme should be 3 mg/mL for Gram-positive bacteria and 0.4 mg/mL for Gram-negative bacteria. Incubate at room temperature 5–10 min for Gram-positive bacteria and 3–5 min for Gram-negative bacteria.

6. Primer Extension

Synchrotron footprinting cleavage patterns are analyzed through gene-specific priming of cDNA synthesis by reverse transcriptase. The primer extension terminates at nicks in the RNA template upstream of the primer binding site. Traditionally, primers have been 5′-labeled with ^{32}P to produce labeled cDNA that may be separated by denaturing PAGE and analyzed through autoradiography. Recently, protocols have been established for using fluorophore labeled primers and separating products using capillary gel electrophoresis on the Beckman-Coulter CEQ-8000 (Mitra *et al.*, 2008). The protocol used by our laboratory for the analysis of rRNA with radiolabeled or dye-labeled primers and slab gels is given below for convenience. This protocol is adapted from similar protocols for RNA structure probing in the literature (Inoue and Cech, 1985; Merryman *et al.*, 1999; Moazed *et al.*, 1986; Pikielny and Rosbash, 1985).

6.1. Labeling primers

Primers annealing at different regions of the target RNA should be designed such that the entire RNA sequence can be analyzed. For use with traditional PAGE gels, primers should be spaced about every 200 nucleotides along the length of the RNA of interest in order to ensure full coverage.

Primers are either conjugated with a dye or 5′-[^{32}P]-labeled with T4 polynucleotide kinase according to standard protocols. To ensure adequate specific activity, label 10 pmol primer with 50 μCi gamma-[^{32}P]-ATP (3000 Ci/mmol) and 5–10 U enzyme for 30 min at 37 °C. Inactivate the kinase by incubating the reaction at 65 °C for 10 min. Next, dilute samples with 30 μL nuclease-free water and remove unincorporated nucleotides using a size-exclusion spin column (e.g., TE-10, BD Sciences) as per the manufacturer's instructions.

6.2. Analysis of cleaved RNA by primer extension

Total RNA (0.25–0.5 μg) is used for primer extension with 5′-labeled primers that anneal to various regions of the rRNA. We recommend the use of Northstar AMV reverse transcriptase (Cape Cod Associates) for the best extensions with minimal pausing. (Protocols for CEQ analysis use SuperScript(R) III, Invitrogen.) In addition to experimental samples and controls, dideoxy sequencing reactions and a pausing control reaction are performed for each RNA to be analyzed.

Prepare the following buffers:

10× RT-Mg: 500 m*M Tris-HCl* pH 8.3, 600 m*M NaCl*, 100 m*M* DTT

10× RT+Mg: 500 m*M Tris-HCl* pH 8.3, 600 m*M NaCl*, 100 m*M* DTT, 60 m*M* Mg-acetate

Also prepare:

5× ddNTPs (one for each nucleotide A, C, G, T): 20 μL 5 m*M* ddNTP, 10 μL 10× RT+Mg, 70 μL nuclease-free H_2O

5× dNTPs: 3 μL each 100 m*M* dATP, dCTP, dGTP, dUTP; 16 μL 10× RT+Mg; 132 μL nuclease-free water

1× RT+Mg: 10 μL 10× RT+Mg, 90 μL nuclease-free water

These reagents and buffers may be prepared in advance and stored at −20 °C for several months.

Next, prepare the following cocktails in the amounts needed to perform the desired number of primer extension reactions.

SL cocktail (for sequencing reactions). This is the master mix for the four sequencing lanes and one reverse transcriptase (RT) pausing control lane. Purified native RNA provides the best control for RT pausing, but an *in vitro* transcript may also be used.

Unmodified RNA (2 μM)	2.0 μL
Labeled primer	3.3 μL
10× RT-Mg	5.0 μL
Nuclease-free H_2O	33.7 μL

Primer cocktail. A buffer without Mg^{2+} is used for the annealing step, because heating RNA in the presence of Mg^{2+} will lead to degradation due to RNA hydrolysis. Make enough cocktail for the number of RNA samples you have, not including sequencing and pausing control lanes.

	Per sample (μL):
Labeled primer	0.66
10× RT-Mg	1.0
Nuclease-free H_2O	5.14

dNTP cocktail. Make enough cocktail for all your samples plus five for the sequencing and pause lanes.

	Per sample (μL):
Nuclease-free H_2O	1.2
1× RT+Mg	2.0
5× dNTP	1.0

Primer annealing. Apportion 2 μL (0.25–0.5 μg) of total RNA to each reaction tube or of a 96-well plate, keeping samples on ice. Add 6.8 μL of Primer cocktail to each RNA aliquot. Anneal the primers and RNA by heating the reaction tubes and the SL cocktail to 90 °C for 1 min, and then return the tubes to ice. For sequencing reactions, put 0.6 μL each 5× ddNTP solution into a separate tube (four total). Transfer 8.8 μL from the SL cocktail to each of the four sequencing tubes. The remaining SL cocktail will serve as the RT pausing control.

Reverse transcriptase. This sensitive enzyme should be diluted into the reaction buffer at the last moment. Dilute enough RT for each sample and the five sequencing tubes.

	Per sample (μL):
10× RT+Mg	0.1
AMV RT (20–30 U/μL)	0.1
Nuclease-free water	0.8

Add the diluted RT to the dNTP cocktail tube. Mix gently but thoroughly. Add 5.2 μL dNTP cocktail containing enzyme to all tubes (including sequencing reactions). Immediately incubate at 48 °C for 20–90 min, depending on the length of the expected product. Return to ice.

Add 1.5 μL 3 *M* sodium acetate, pH 5.2, and 45 μL 100% ethanol to each reaction, and chill 20 min at − 80 °C or overnight at − 20 °C. Pellet RNA by centrifugation at 12,000 × *g* for 20–30 min at 4 °C, wash pellet once with 45 μL 70% ethanol (crucial to remove excess salt) and dry briefly under vacuum. If cDNAs are not to be separated immediately, freeze the dry pellets at − 20 °C. When you are ready to run the gel, resuspend them in

5–10 μL formamide gel loading buffer (for 10 mL: 9.5 mL deionized formamide, 400 μL 10× TBE, 50 μL 2% xylene cyanol, 50 μL 2% bromophenol blue; store at −20 °C in 0.5 mL aliquots, avoiding excessive freeze–thaw cycles). Electrophorese heat-denatured samples on an 8% 1× TBE sequencing gel at 55 W according to standard protocols (Fig. 12.3). Gels may be run from 2 to 4 h to resolve different portions of the RNA sequence. Dry gels on Whatman 3MM paper. Dried gels are exposed to phosphor storage screens and scanned using an imager.

6.3. Alterations for capillary gel electrophoresis

A similar primer extension protocol (Mitra *et al.*, 2008) may be used with fluorophore-labeled primers designed for capillary gel electrophoresis on a Beckman CEQ-8000 Gene Analyzer. Use a 10 μ*M* working stock of the primer labeled with an appropriate fluorophore such as Cy5 or a Beckman WellRED dye. When handling primer stock solutions, avoid bright light that will bleach the fluorophore. A second 70% ethanol wash step following primer extension is recommended to remove excess unincorporated primer.

Unlike ^{32}P-labeled cDNA, fluorophore-labeled cDNA cannot be stored as a dry pellet. Samples should be resuspended in 40 μL sample loading solution (Beckman-Coulter) and separated immediately. If samples must be stored, cover each with a drop of mineral oil to minimize oxidation, then wrap in foil to protect from photobleaching. Protocols for separating products on the CEQ-8000 and analyzing resulting data may be found elsewhere (Mitra *et al.*, 2008).

7. Analysis of X-Ray Footprinting Experiments

We have used two methods for the quantitation of footprinting data in sequencing gels. Most simply, the intensities of electrophoretic bands present on the digital autoradiograms can be determined using the software provided with the scanner, such as ImageQuant (Molecular Dynamics). This method involves manually defining integration areas around individual bands in each lane, which can be time consuming and may introduce inconsistencies. An advantage of this method, however, is that the more compressed bands in the upper half of the gel can still be analyzed.

Semiautomated footprinting analysis software (SAFA) allows for faster and more consistent quantitation of band intensities (Das *et al.*, 2005), but works best on very good quality gels. In our experience, around 90 nucleotides may be integrated using SAFA from a high-quality gel. Using either method, band intensities are downloaded into a spreadsheet program such as Microsoft Excel for further analysis and for export to plotting programs.

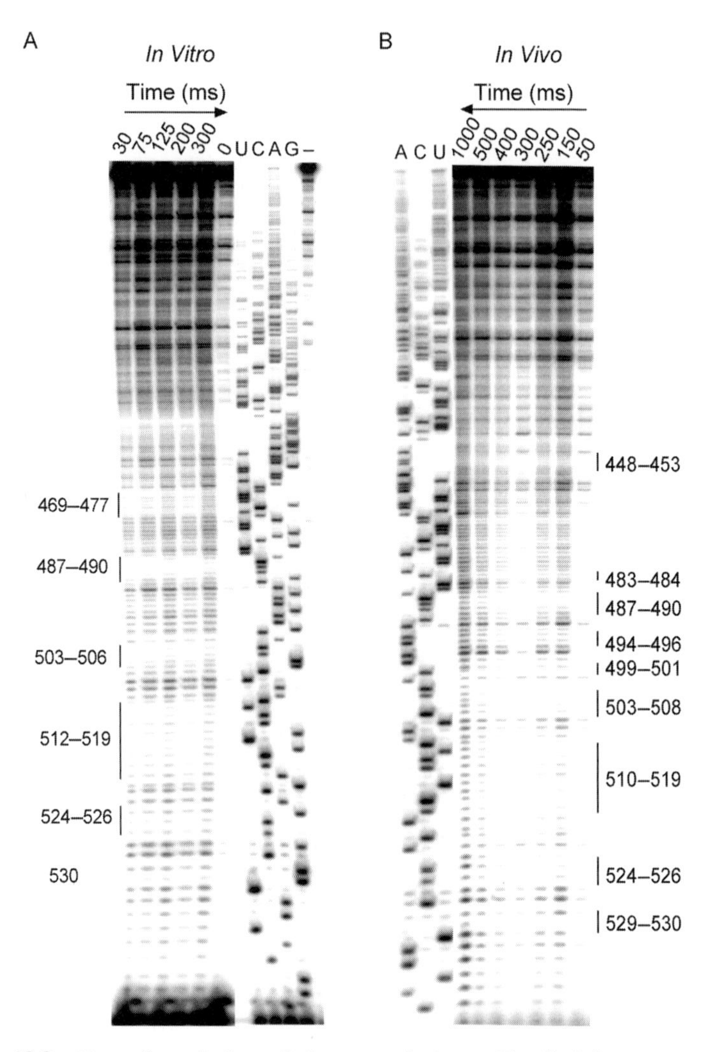

Figure 12.3 X-ray footprinting of ribosomes in intact *E. coli*. Primer extension of purified 70S ribosomes (A) or rRNA isolated from irradiated cells (B). Exposure times are given at the top of the gel. Radiolabeled primer anneals after nt 560 of the 16S rRNA. Product bands were assigned by comparison with dideoxy sequencing lanes. Redrawn from Adilakshmi *et al.* (2006).

The sequence of each band is identified by comparing the footprinting lanes to sequencing ladders run on the same gel.

It is important to compare exposed samples to No Shipping, No Hutch, and In Hutch controls to determine background rates of cleavage as well as intrinsic reverse transcriptase pause sites. If the background is too high, the source of the problem can be isolated by performing primer extension reactions on RNA transcribed *in vitro* or on linear plasmid DNA of the same sequence. High-quality primer extension reactions on these templates should produce mostly full-length cDNA with only a few pauses at regions of strong secondary structure or natural base modification. Suboptimal cDNA synthesis can result from a poorly designed primer, inhibition of the enzyme by salt or other impurities, inactive enzyme or an inadequate amount of template. If the problem is confined to cellular RNA templates, it can often be resolved by reisolating RNA from a duplicate set of irradiated cells.

8. Results on *E. coli* RNAs

We have used X-ray footprinting to probe the structure of ribosomal RNA in frozen *E. coli* cells. The conformation of the 16S rRNA was inferred from the pattern of cleavage intensities (Fig. 12.4). Strongly cleaved residues (relative to background) represent riboses that are exposed to the bulk solvent, while less cleaved or "protected" residues represent riboses that are inaccessible to solvent. At most positions, the hydroxyl radical cleavage rate also correlates well with the predicted solvent accessible surface area of C4′ and C5′ atoms, which were computed from crystallographic coordinates (Schuwirth *et al.*, 2005; Wimberly *et al.*, 2000) using the program Calc-surf (Gerstein, 1992). Deviations from the predicted cleavage pattern can provide insight into the mechanism of assembly *in vivo* or substrate interactions.

In our experience, the X-ray footprinting pattern of 16S rRNA in frozen *E. coli* cells agrees well with the footprints of purified 30S ribosomal subunits or 70S ribosomes obtained with Fe(II)-EDTA *in vitro* (Figs. 12.3 and 12.4), suggesting that the cleavage *in vivo* is primarily caused by attack of hydroxyl radical on RNA backbone (Adilakshmi *et al.*, 2006). Some base damage is expected due to the reactivity of nucleobases with hydroxyl radicals, which could contribute to the pattern of cDNA synthesis by reverse transcription. In practice, however, such base damage rarely results in cleavage of the phosphodiester backbone (Cadet *et al.*, 2004; Dizdaroglu, 1986; Duplaa and Teoule, 1985). The lack of sequence bias in the cleavage sites also suggests that the most common base damage events are by-passed by reverse transcriptase, as expected from studies of other DNA polymerases (Wallace, 1998).

While our work has primarily focused on rRNA, this technique is also useful for studying other complexes such as RNase P, which is one-tenth as abundant as ribosomal subunits during rapid growth (Dong *et al.*, 1996). *In vivo* protections of residues in P6, P17, P16, and L15 regions of the M1

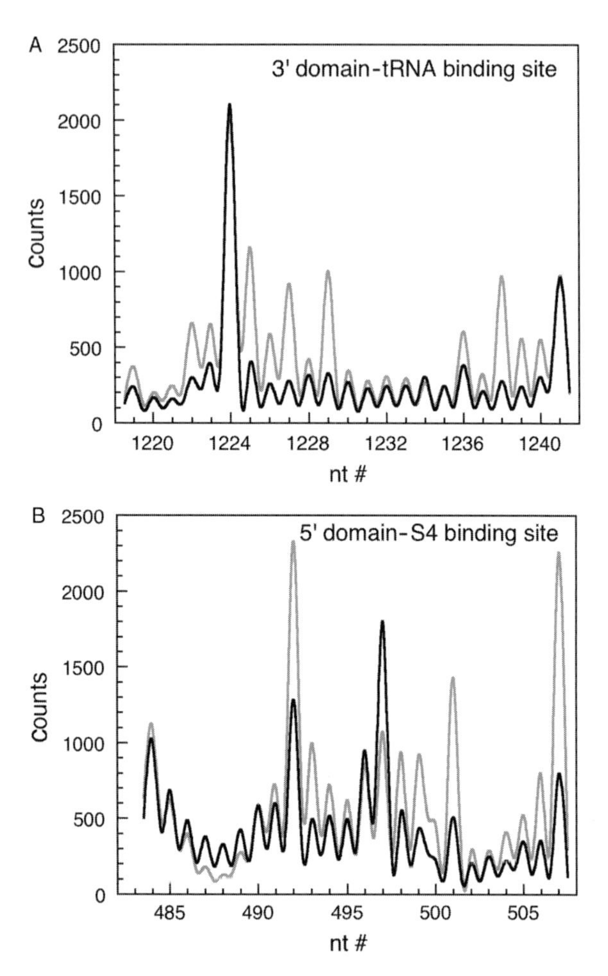

Figure 12.4 Solvent accessibility of rRNA by hydroxyl radical footprinting. Sequencing gels in Fig. 12.3 were quantified using a phosphorimager. In this region of the gel, peaks due to individual nucleotides are well resolved. Black, 70S ribosomes (*in vitro*); grey, *E. coli* cells (*in vivo*). (A) Region from 16S 3′-domain. (B) Region from 16S 5′-domain. Reprinted from Adilakshmi *et al.* (2006).

RNA subunit of *E. coli* RNase P (Adilakshmi *et al.*, 2006) correlated well with previous hydroxyl radical footprinting data in the presence of pre-tRNA (Torres-Larios *et al.*, 2005; Westhof *et al.*, 1996). We have also been able to footprint the pre-rRNA leader, despite the fact that the immature pre-rRNA is 10–100-fold less abundant than the mature rRNA (A.T., unpublished data).

9. Future of Footprinting

Because X-rays easily penetrate intact cells and even whole tissues, we expect this method to be applicable to a wide range of biological systems. The relatively short irradiation times and the ability to use frozen cells would enable this method to visualize changes in cell components after viral infection or during the course of development. The range of potential footprinting applications could be increased by the use of electrospray mass spectrometry to footprint proteins (Maleknia *et al.*, 1999) or ligation-mediated PCR to amplify footprinting patterns of single-copy genes or low-abundance mRNAs in eukaryotic cells (Grange *et al.*, 1997). Thus, *in vivo* X-ray-dependent hydroxyl radical footprinting offers a new approach for obtaining detailed information about the three-dimensional structure of nucleic acids *in situ*.

Acknowledgments

The authors thank Sayan Gupta, Mike Sullivan and Mark Chance for the development and operation of the footprinting beamline at X28, and Michael Brenowitz for communicating protocols for footprinting and data analysis. This work was supported by the NIH (GM60819). The Center for Synchtroton Biosciences and beamline X28C are supported by the NIH (P41-EB0001979) and DOE.

References

Adilakshmi, T., Lease, R. A., and Woodson, S. A. (2006). Hydroxyl radical footprinting *in vivo*: Mapping macromolecular structures with synchrotron radiation. *Nucleic Acids Res.* **34**(8), e64.

Balzer, M., and Wagner, R. (1998). A chemical modification method for the structural analysis of RNA and RNA–protein complexes within living cells. *Anal. Biochem.* **256**(2), 240–242.

Burkhoff, A. M., and Tullius, T. D. (1987). The unusual conformation adopted by the adenine tracts in kinetoplast DNA. *Cell* **48**(6), 935–943.

Cadet, J., Bellon, S., Douki, T., Frelon, S., Gasparutto, D., Muller, E., Pouget, J. P., Ravanat, J. L., Romieu, A., and Sauvaigo, S. (2004). Radiation-induced DNA damage: Formation, measurement, and biochemical features. *J. Environ. Pathol. Toxicol. Oncol.* **23**(1), 33–43.

Cassler, M. R., Grimwade, J. E., McGarry, K. C., Mott, R. T., and Leonard, A. C. (1999). Drunken-cell footprints: Nuclease treatment of ethanol-permeabilized bacteria reveals an initiation-like nucleoprotein complex in stationary phase replication origins. *Nucleic Acids Res.* **27**(23), 4570–4576.

Climie, S. C., and Friesen, J. D. (1988). *In vivo* and *in vitro* structural analysis of the rplJ mRNA leader of *Escherichia coli*. Protection by bound L10–L7/L12. *J. Biol. Chem.* **263**(29), 15166–15175.

Das, R., Laederach, A., Pearlman, S. M., Herschlag, D., and Altman, R. B. (2005). SAFA: Semi-automated footprinting analysis software for high-throughput quantification of nucleic acid footprinting experiments. *RNA* **11**(3), 344–354.

Dhavan, G. M., Chance, M. R., and Brenowitz, M. (2003). Kinetics analysis of DNA–protein interactions by time resolved synchrotron X-ray footprinting. *In* "Kinetic Analysis of Macromolecules: A Practical Approach," (K. A. Johnson, ed.), pp. 75–86. IRL Press, Oxford University, Oxford.

Dizdaroglu, M. (1986). Characterization of free radical-induced damage to DNA by the combined use of enzymatic hydrolysis and gas chromatography-mass spectrometry. *J. Chromatogr.* **367**(2), 357–366.

Dong, H., Kirsebom, L. A., and Nilsson, L. (1996). Growth rate regulation of 4.5 S RNA and M1 RNA the catalytic subunit of *Escherichia coli* RNase P. *J. Mol. Biol.* **261**(3), 303–308.

Dreyer, G. B., and Dervan, P. B. (1985). Sequence-specific cleavage of single-stranded DNA: Oligodeoxynucleotide-EDTA X Fe(II). *Proc. Natl. Acad. Sci. USA* **82**(4), 968–972.

Duplaa, A. M., and Teoule, R. (1985). Sites of gamma radiation-induced DNA strand breaks after alkali treatment. *Int. J. Radiat. Biol. Relat. Stud. Phys. Chem. Med.* **48**(1), 19–32.

Galas, D. J., and Schmitz, A. (1978). DNAse footprinting: A simple method for the detection of protein–DNA binding specificity. *Nucleic Acids Res.* **5**(9), 3157–3170.

Gerstein, M. (1992). A resolution-sensitive procedure for comparing protein surfaces and its application to the comparison of antigen-combining sites. *Acta Cryst.* **A48,** 271–276.

Grange, T., Bertrand, E., Espinas, M. L., Fromont-Racine, M., Rigaud, G., Roux, J., and Pictet, R. (1997). *In vivo* footprinting of the interaction of proteins with DNA and RNA. *Methods* **11**(2), 151–163.

Gregoli, S., Olast, M., and Bertinchamps, A. (1982). Radiolytic pathways in gamma-irradiated DNA: Influence of chemical and conformational factors. *Radiat. Res.* **89**(2), 238–254.

Gupta, S., Sullivan, M., Toomey, J., Kiselar, J., and Chance, M. R. (2007). The Beamline X28C of the Center for Synchrotron Biosciences: A national resource for biomolecular structure and dynamics experiments using synchrotron footprinting. *J. Synchrotron Radiat.* **14**(Pt. 3), 233–243.

Hayes, J. J., Kam, L., and Tullius, T. D. (1990). Footprinting protein–DNA complexes with gamma-rays. *Methods Enzymol.* **186,** 545–549.

Hertzberg, R. P., and Dervan, P. B. (1984). Cleavage of DNA with methidiumpropyl-EDTA-iron(II): Reaction conditions and product analyses. *Biochemistry* **23**(17), 3934–3945.

Hsieh, M., and Brenowitz, M. (1996). Quantitative kinetics footprinting of protein–DNA association reactions. *Methods Enzymol.* **274,** 478–492.

Huttermann, J., Rohrig, M., and Kohnlein, W. (1992). Free radicals from irradiated lyophilized DNA: Influence of water of hydration. *Int. J. Radiat. Biol.* **61**(3), 299–313.

Inoue, T., and Cech, T. R. (1985). Secondary structure of the circular form of the Tetrahymena rRNA intervening sequence: A technique for RNA structure analysis using chemical probes and reverse transcriptase. *Proc. Natl. Acad. Sci. USA* **82**(3), 648–652.

Klassen, N. V. (1987). Primary products in radiation chemistry. *In* "Radiation Chemistry: Principles and Applications," (I. Farhataziz and M. A. J. Rodgers, eds.), pp. 29–61. VCH Publishers, New York.

Latham, J. A., and Cech, T. R. (1989). Defining the inside and outside of a catalytic RNA molecule. *Science* **245**(4915), 276–282.

Maleknia, S. D., Brenowitz, M., and Chance, M. R. (1999). Millisecond radiolytic modification of peptides by synchrotron X-rays identified by mass spectrometry. *Anal. Chem.* **71**(18), 3965–3973.

Mayford, M., and Weisblum, B. (1989). Conformational alterations in the ermC transcript *in vivo* during induction. *EMBO J.* **8**(13), 4307–4314.

Mazur, P. (1984). Freezing of living cells: Mechanisms and implications. *Am. J. Physiol.* **247**(3 Pt. 1), C125–C142.

Merryman, C., Moazed, D., McWhirter, J., and Noller, H. F. (1999). Nucleotides in 16S rRNA protected by the association of 30S and 50S ribosomal subunits. *J. Mol. Biol.* **285**(1), 97–105.

Mitra, S., Shcherbakova, I. V., Altman, R. B., Brenowitz, M., and Laederach, A. (2008). High-throughput single-nucleotide structural mapping by capillary automated footprinting analysis. *Nucleic Acids Res.* **36**(11), e63.

Moazed, D., Stern, S., and Noller, H. F. (1986). Rapid chemical probing of conformation in 16 S ribosomal RNA and 30 S ribosomal subunits using primer extension. *J. Mol. Biol.* **187**(3), 399–416.

Nierhaus, K. H., and Dohme, F. (1974). Total reconstitution of functionally active 50S ribosomal subunits from *Escherichia coli*. *Proc. Natl. Acad. Sci. USA* **71**(12), 4713–4717.

Ohshima, H., Iida, Y., Matsuda, A., and Kuwabara, M. (1996). Damage induced by hydroxyl radicals generated in the hydration layer of gamma-irradiated frozen aqueous solution of DNA. *J. Radiat. Res. (Tokyo)* **37**(3), 199–207.

Ottinger, L. M., and Tullius, T. D. (2000). High-resolution *in vivo* footprinting of a protein–DNA complex using γ-radiation. *J. Am. Chem. Soc.* **122**(24), 5901–5902.

Pikielny, C. W., and Rosbash, M. (1985). mRNA splicing efficiency in yeast and the contribution of nonconserved sequences. *Cell* **41**(1), 119–126.

Revzin, A. (1993). Footprinting Techniques for Studying Nucleic Acid–Protein Complexes (A Volume of Separation, Detection, and Characterization of Biological Macromolecules). Academic Press, New York.

Roberts, J. C., Koch, K. E., Detrick, S. R., Warters, R. L., and Lubec, G. (1995). Thiazolidine prodrugs of cysteamine and cysteine as radioprotective agents. *Radiat. Res.* **143**(2), 203–213.

Schuwirth, B. S., Borovinskaya, M. A., Hau, C. W., Zhang, W., Vila-Sanjurjo, A., Holton, J. M., and Cate, J. H. (2005). Structures of the bacterial ribosome at 3.5 A resolution. *Science* **310**(5749), 827–834.

Sclavi, B., Woodson, S., Sullivan, M., Chance, M. R., and Brenowitz, M. (1997). Time-resolved synchrotron X-ray "footprinting", a new approach to the study of nucleic acid structure and function: Application to protein–DNA interactions and RNA folding. *J. Mol. Biol.* **266**(1), 144–159.

Sclavi, B., Sullivan, M., Chance, M. R., Brenowitz, M., and Woodson, S. A. (1998). RNA folding at millisecond intervals by synchrotron hydroxyl radical footprinting. *Science* **279**(5358), 1940–1943.

Shcherbakova, I., Mitra, S., Beer, R. H., and Brenowitz, M. (2008). Following molecular transitions with single residue spatial and millisecond time resolution. *Methods Cell Biol.* **84,** 589–615.

Sullivan, M. R., Rekhi, S., Bohon, J., Gupta, S., Abel, D., Toomey, J., and Chance, M. R. (2008). Installation and testing of a focusing mirror at beamline X28C for high flux X-ray radiolysis of biological macromolecules. *Rev. Sci. Instrum.* **79**(2 Pt. 1), 025101.

Torres-Larios, A., Swinger, K. K., Krasilnikov, A. S., Pan, T., and Mondragon, A. (2005). Crystal structure of the RNA component of bacterial ribonuclease P. *Nature* **437**(7058), 584–587.

Tullius, T. D., and Dombroski, B. A. (1986). Hydroxyl radical "footprinting": High-resolution information about DNA–protein contacts and application to lambda repressor and Cro protein. *Proc. Natl. Acad. Sci. USA* **83**(15), 5469–5473.

Tullius, T. D., and Greenbaum, J. A. (2005). Mapping nucleic acid structure by hydroxyl radical cleavage. *Curr. Opin. Chem. Biol.* **9**(2), 127–134.

Tullius, T. D., Dombroski, B. A., Churchill, M. E., and Kam, L. (1987). Hydroxyl radical footprinting: A high-resolution method for mapping protein–DNA contacts. *Methods Enzymol.* **155,** 537–558.

Udenfriend, S., Clark, C. T., Axelrod, J., and Brodie, B. B. (1954). Ascorbic acid in aromatic hydroxylation. I. A model system for aromatic hydroxylation. *J. Biol. Chem.* **208**(2), 731–739.

Waldsich, C., Grossberger, R., and Schroeder, R. (2002). RNA chaperone StpA loosens interactions of the tertiary structure in the td group I intron *in vivo*. *Genes Dev.* **16**(17), 2300–2312.

Wallace, S. S. (1998). Enzymatic processing of radiation-induced free radical damage in DNA. *Radiat. Res.* **150**(5 Suppl.), S60–S79.

Wells, S. E., Hughes, J. M., Igel, A. H., and Ares, M. Jr. (2000). Use of dimethyl sulfate to probe RNA structure *in vivo*. *Methods Enzymol.* **318,** 479–493.

Westhof, E., Wesolowski, D., and Altman, S. (1996). Mapping in three dimensions of regions in a catalytic RNA protected from attack by an Fe(II)-EDTA reagent. *J. Mol. Biol.* **258**(4), 600–613.

Wilkinson, K. A., Merino, E. J., and Weeks, K. M. (2006). Selective 2'-hydroxyl acylation analyzed by primer extension (SHAPE): Quantitative RNA structure analysis at single nucleotide resolution. *Nat. Protoc.* **1**(3), 1610–1616.

Wimberly, B. T., Brodersen, D. E., Clemons, W. M. Jr., Morgan-Warren, R. J., Carter, A. P., Vonrhein, C., Hartsch, T., and Ramakrishnan, V. (2000). Structure of the 30S ribosomal subunit. *Nature* **407**(6802), 327–339.

CHAPTER THIRTEEN

Determination of Intracellular RNA Folding Rates Using Self-Cleaving RNAs

Peter Y. Watson *and* Martha J. Fedor

Contents

Abstract

We have developed a system that relies on RNA self-cleavage to report quantitatively on assembly of RNA structures *in vivo*. Self-cleaving RNA sequences are inserted into mRNAs or snoRNAs and expressed in yeast under the control of a regulated promoter. Chimeric RNAs that contain self-cleaving ribozymes turn over faster than chimeric RNAs that contain a mutationally inactivated ribozyme by an amount that reflects the rate at which the ribozyme folds and self-cleaves. A key feature of this system is the choice of assay conditions that selectively monitor intracellular assembly and self-cleavage by suppressing further ribozyme activity during the analysis.

Departments of Chemical Physiology and Molecular Biology, The Skaggs Institute for Chemical Biology, The Scripps Research Institute, La Jolla, California, USA

Methods in Enzymology, Volume 468
ISSN 0076-6879, DOI: 10.1016/S0076-6879(09)68013-7

1. Introduction

Assembly and dissociation of RNA structures play essential roles in virtually all RNA-mediated processes *in vivo*. Full understanding of these processes requires detailed knowledge of the nature of the intermediates in RNA assembly and dissociation pathways and the relationship between RNA secondary and tertiary structures and the rates and equilibria of the conformational transitions that comprise these pathways. However, detailed information about specific conformational transitions can be challenging to obtain in a biological context; many components interact in complex pathways making individual steps difficult to resolve experimentally. Ribozyme activity reports on assembly of well-defined RNA structures, making RNA enzymes valuable models for studies of RNA folding. The application of quantitative enzymological methods to RNA enzymes has led to detailed insights into the conformational transitions that define RNA assembly and reaction pathways. We developed a system to exploit the simple cleavage and ligation reactions of hairpin ribozymes to investigate how structure–function relationships established from *in vitro* studies translate to the behavior of RNA in living cells (Donahue and Fedor, 1997).

The hairpin ribozyme belongs to the family of catalytic RNAs that catalyzes reversible phosphodiester cleavage through the same chemical mechanism used by degradative protein ribonucleases such as ribonuclease A (Fedor, 2009). Minimal hairpin ribozymes contain two essential helix-loop-helix elements, A and B, that dock to bring loops A and B together to create the active site (Fig. 13.1A). Cleavage within loop A gives 5′- and 3′-cleavage products that associate through intermolecular base pairs in the H1 helix. In plant satellite RNAs, where hairpin ribozymes participate in processing replication intermediates, two additional helices are located between the A and B domains, so that assembly occurs in the context of a four-way helical junction (Fig. 13.1B). Hairpin ribozymes also can assemble from separate ribozyme and substrate RNAs through intermolecular base pairs in helices H1 and H2 (Fig. 13.1C).

Conserved nucleotides in loops A and B have key roles in active site architecture and catalytic chemistry. The base-paired helices and the interdomain junction play no direct role in catalytic chemistry but these structures profoundly affect observed cleavage activity by influencing assembly and dissociation steps in the reaction pathway (Fig. 13.2). Cleavage products associate through intermolecular base pairs in the H1 helix and changes in the length or sequence of H1 have dramatic effects on product dissociation kinetics (Donahue *et al.*, 2000; Hegg and Fedor, 1995). Under standard reaction conditions *in vitro* that include high concentrations of Mg^{2+}, minimal ribozymes partition almost equally between extended and docked conformations (Walter *et al.*, 1999) and exhibit very little activity in

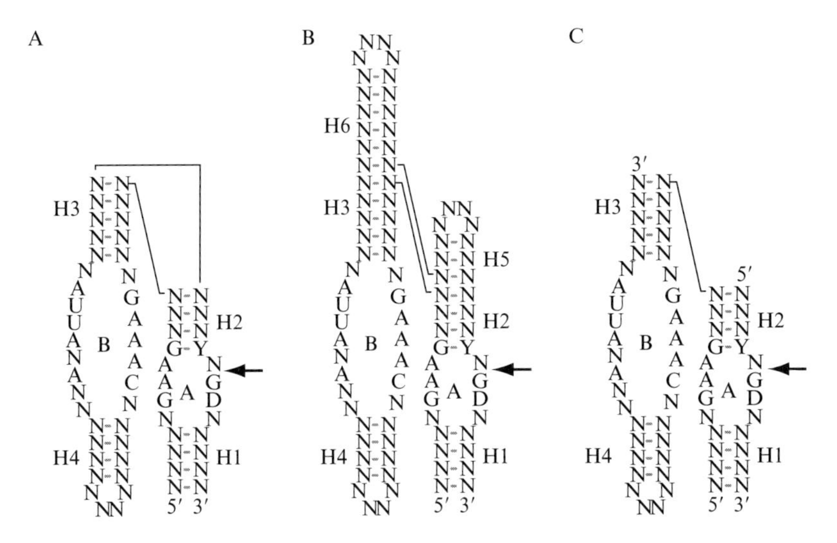

Figure 13.1 Hairpin ribozyme configurations used for RNA folding studies. (A) The minimal self-cleaving ribozyme contains two essential helix-loop-helix domains, with four helices, H1–H4, and two loops, A and B. (B) The natural form of the hairpin ribozyme contains two additional helices, H5 and H6, and assembles in the context of a four-way helical junction. (C) An intermolecular form of the hairpin ribozyme contains separate ribozyme and substrate RNAs that associate through complementary base pairs in H1 and H2. Reversible cleavage of the reactive phosphodiester in loop A indicated by the arrow generates 5′- and 3′-product RNAs in a complex stabilized by tertiary interactions within loop A, between loops A and B, and by intermolecular base pairs.

reactions with low Mg^{2+} concentrations that more closely approximate an intracellular ionic environment (Yadava *et al.*, 2001). Assembly in the context of a four-way helical junction stabilizes the docked tertiary structure (Murchie *et al.*, 1998; Walter *et al.*, 1998a,b,c, 1999) and supports rapid cleavage under less favorable ionic conditions (Yadava *et al.*, 2001). The enhanced tertiary structure stability conferred by a four-way helical junction also affects the internal equilibrium between cleavage and ligation of bound products (Fedor, 1999). Minimal ribozymes catalyze ligation of bound products about two times faster than they self-cleave under standard conditions (Nesbitt *et al.*, 1999). In the context of a four-way helical junction, the internal equilibrium shifts to favor ligation nearly 20-fold (Fedor, 1999).

Thus, observed cleavage kinetics are exquisitely sensitive to structural effects on assembly and dissociation steps in the hairpin ribozyme reaction pathway. Reaction conditions or structural changes that stabilize the docked tertiary structure reduce observed cleavage rates by promoting re-ligation of bound products. Likewise, stabilizing the base-paired H1 helix that forms between 5′-ribozyme and 3′-cleavage product RNAs reduces observed cleavage rates by promoting accumulation of a ribozyme–product complex that is able to undergo re-ligation. We have developed an approach to

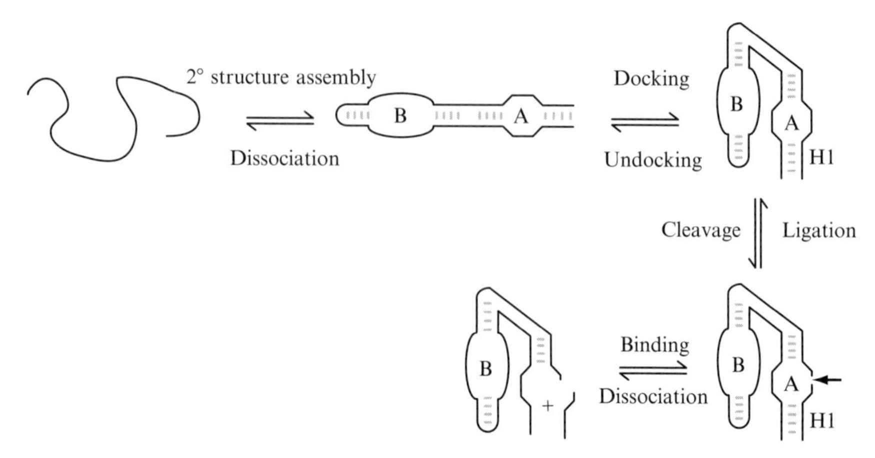

Figure 13.2 The hairpin ribozyme self-cleavage pathway includes RNA folding steps needed for secondary structure assembly, interdomain docking, and product release as well as the reversible catalytic step. Observed cleavage kinetics is exquisitely sensitive to the kinetics and equilibria of the folding steps in the reaction pathway. Reaction conditions, such as low temperatures and high divalent cations concentrations, and structural features that stabilize the docked tertiary structure shift the internal equilibrium between cleavage and ligation in favor of ligation, and reduce observed cleavage rates by promoting re-ligation of bound products. Reaction conditions and structural features that stabilize the intermolecular H1 helix that forms between cleavage product RNAs also reduce observed cleavage rates by promoting accumulation of the ribozyme–product complex that is poised to undergo re-ligation. Thus, observed cleavage rates provide a sensitive reporter of RNA tertiary and secondary structure stability.

investigate how these structure–function relationships that have been elucidated through *in vitro* studies translate to the behavior of RNAs in the complex environment of a cell. The hairpin ribozyme is an excellent model for RNA folding studies because it combines structural elements that are common to virtually all functional RNAs, including base-paired RNA helices, helix junctions, a ribose zipper hydrogen bonding motif, and a cross-strand stacking motif commonly found at RNA–RNA and RNA–protein interfaces (Fedor, 2009). Basic information about the kinetics and equilibria of conformational transitions in the reaction pathways of hairpin ribozymes are likely to have broad relevance to the behavior of all RNAs that include similar structural elements.

2. Using RNA Turnover Rates as a "Clock" for Measuring RNA Assembly Kinetics

All RNAs turn over in the cell with a characteristic half-life (Houseley and Tollervey, 2009). Our system makes use of the intrinsic RNA turnover rate through the endogenous RNA degradation machinery as a "clock" to gauge the kinetics of RNA assembly and cleavage *in vivo*. A chimeric RNA that contains a functional ribozyme disappears through endogenous degradation pathways, but also disappears though self-cleavage (Fig. 13.3). Consequently, chimeric self-cleaving RNAs disappear faster than chimeric RNAs that contain a mutationally inactivated ribozyme by an amount that corresponds to the intracellular self-cleavage rate. We chose to insert hairpin ribozymes into the yeast PGK1 gene for initial experiments due to the slow turnover and high abundance of its mRNA (Herrick *et al.*, 1990), which provides a sensitive gauge for ribozyme cleavage-mediated acceleration of mRNA decay (Donahue and Fedor, 1997). Ribozymes inserted into chimeric snoRNAs have also been used to monitor nucleolar RNA folding and complex formation pathways (Samarsky *et al.*, 1999; Yadava *et al.*, 2004).

To study intracellular ribozyme assembly and cleavage, RNA extraction and RNase protection assays have been optimized to ensure that the

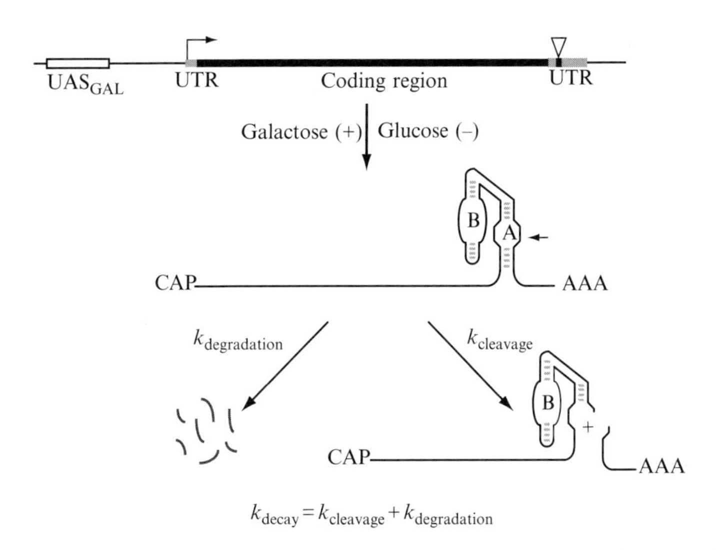

Figure 13.3 Intrinsic RNA decay rates provide a "clock" for quantification of intracellular ribozyme assembly and catalysis. Hairpin ribozyme variants are expressed in *S. cerevisiae* as chimeric PGK1 mRNAs under the control of the galactose-inducible, glucose-repressible upstream activation sequence, UAS$_{GAL}$. Self-cleaving and mutationally inactivated chimeric mRNAs both decay through the endogenous pathways responsible for RNA turnover at a rate, $k_{degradation}$, that is characteristic of PGK1 mRNA decay. Since self-cleaving RNAs disappear both through self-cleavage and through normal RNA turnover pathways, assembly and cleavage of a functional ribozyme accelerates the disappearance of chimeric RNA by an amount that corresponds to the intracellular cleavage rate, $k_{cleavage}$. Consequently, intracellular cleavage kinetics can be calculated from the difference between k_{decay} values measured for a chimeric RNA that contains a self-cleaving ribozyme and a chimeric RNA that contains a mutationally inactivated ribozyme.

ribozyme activity measured in yeast RNA truly arises from intracellular ribozyme activity and not from cleavage or ligation that occurs during isolation and analysis of the yeast RNA. Preliminary experiments in which radioactive self-cleaving RNAs were added to yeast to track RNA self-cleavage during an analysis showed that up to 20% of the added transcripts self-cleaved during conventional RNA extraction procedures. Self-cleavage extents approached 50% when RNA samples were subjected to repeated freezing and thawing, even in the presence of divalent cation chelators (Donahue and Fedor, 1997). Some assays commonly used to quantify RNA, such as PCR and primer extension assays, require pH, temperature, and ionic conditions that are nearly optimal for RNA self-cleavage. However, we found that virtually all *in vitro* ribozyme activity can be suppressed by extracting RNA from yeast at low pH, low salt concentrations, and low temperatures. In RNase protection assays, hybridization of a radioactive probe that is complementary to ribozyme sequences in chimeric RNAs immediately after extraction serves to prevent ribozymes from folding into functional structures and self-cleaving (Fig. 13.4).

Our method for quantitative analysis of intracellular RNA folding relies on intrinsic RNA turnover rates to serve as a "clock" for ribozyme assembly

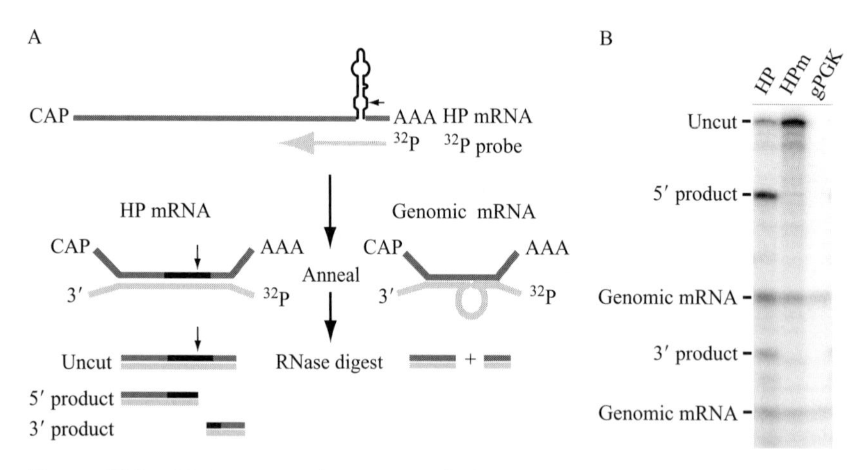

Figure 13.4 RNase protection assay of chimeric PGK1 mRNA self-cleavage. (A) Total RNA from yeast carrying a chimeric PGK1 gene with a ribozyme insertion in the 3′-untranslated region is annealed with an [α-^{32}P]-transcript of PGK1 sequences containing an insertion complementary to the ribozyme (^{32}P probe, light gray arrow). Unhybridized sequences are removed by digestion with a mixture of single-strand-specific RNases. (B) Protected fragments are fractionated on denaturing gels. Dashes mark fragments corresponding to uncleaved chimeric mRNA, cleavage products and genomic *PGK1* mRNA sequences detected in yeast carrying functional (HP) or inactive (HPm) hairpin ribozymes in the 3′-untranslated region of plasmid-born PGK1 genes, or carrying only genomic PGK1. Adapted from Donahue and Fedor (1997).

and self-cleavage (Fig. 13.3). Self-cleaving and mutationally inactivated chimeric RNAs both decay through the endogenous pathways responsible for RNA turnover, which occurs at rates, $k_{degradation}$, that are characteristic of individual RNAs. Self-cleaving RNAs disappear both through self-cleavage and through normal RNA turnover pathways, so assembly and self-cleavage of a functional ribozyme accelerates the disappearance of chimeric RNA by an amount that corresponds exactly to the intracellular cleavage rate, $k_{cleavage}$. Therefore, intracellular cleavage rates can be calculated from the difference between k_{decay} values measured for chimeric RNAs that contain a self-cleaving ribozyme and chimeric RNAs that contain a mutationally activated ribozyme.

Intracellular cleavage rates are quantified using two complementary assays that are expected to give identical results. In one method, decay rates for self-cleaving and mutationally inactivated chimeric RNAs are obtained from the rate of disappearance of chimeric RNAs over time after transcription inhibition, and intracellular cleavage rates are calculated from the difference between the two. Once an intrinsic degradation rate has been determined for a chimeric RNA containing a mutationally inactivated ribozyme using a time course experiment, the intracellular cleavage rate also can be calculated from the relative abundance of chimeric RNAs that contain a functional or an inactive ribozyme at steady state.

2.1. Expression of chimeric self-cleaving RNAs in yeast

To express self-cleaving RNAs in yeast, hairpin ribozyme sequences are inserted into the 3′-untranslated region of a PGK1 gene (open arrow, Fig. 13.3). Ribozymes also have been expressed in yeast as chimeric snoRNAs in which ribozyme sequences are substituted for the ribosomal RNA recognition sequence (Samarsky *et al.*, 1999; Yadava *et al.*, 2004). Genes encoding chimeric RNAs are fused to a galactose upstream activation sequence, GAL_{UAS}, which allows transcription to be induced by growth of yeast in galactose medium and inhibited rapidly when yeast are transferred into glucose medium (Johnston *et al.*, 1994). Intracellular decay rates can be calculated from the amount of RNA that remains at various times after transcription inhibition (Parker *et al.*, 1991). Chimeric mRNAs containing functional ribozymes (HP) are compared to the same chimeric mRNA in which the ribozyme sequence has a $G_{+1}A$ mutation that eliminates catalytic activity (HPm) to control for any effect the ribozyme structure might have on chimeric RNA stability in the absence of cleavage.

Genes encoding chimeric RNAs are propagated in yeast using single copy shuttle vectors (Sikorski and Hieter, 1989). We use derivatives of pRS316, a centromeric plasmid with a URA3 selectable marker, for most of our experiments and the detailed protocols described here specify conditions for URA3 selection. Plasmids typically are propagated in *Saccharomyces cerevisiae* strain HFY114 (*MATa ade2-1 his3-11,15 leu2-3, 112, trp1-1 ura3-1 can1-10*) (He *et al.*, 2008). Plasmid transformations and selections are carried out with yeast grown with glucose as a carbon source to maximize efficiency but it is important to allow yeast to adapt to growth in galactose over several days to ensure complete alleviation of glucose repression. Yeast are grown to mid-log phase (A_{600} ~0.6–0.8) in shaking culture flasks at 30 °C in minimal synthetic media (SG) supplemented with adenine, histidine, leucine, tryptophan, and 2% galactose for steady-state RNA measurements and transferred into media with 2% glucose for transcription inhibition (Amberg *et al.*, 2005).

2.1.1. SG medium supplemented with adenine, histidine, leucine, and tryptophan (1 L)

- 1.7 g yeast nitrogen base without amino acids or ammonium acetate
- 5 g ammonium sulfate
- 10 mL 2 mg/mL adenine
- 2 mL 10 mg/mL histidine
- 2 mL 10 mg/mL leucine
- 2 mL 10 mg/mL tryptophan
- Add 20 g agar for plates
- Bring to 900 mL with H_2O
- Autoclave 20 min at 121 °C (250 °F)
- Cool to 65 °C
- Add 100 mL filter sterilized 20% galactose

2.1.2. SD medium supplemented with adenine, histidine, leucine, and tryptophan (1 L)

- 1.7 g yeast nitrogen base without amino acids or ammonium acetate
- 5 g ammonium sulfate
- 10 mL 2 mg/mL adenine
- 2 mL 10 mg/mL histidine
- 2 mL 10 mg/mL leucine
- 2 mL 10 mg/mL tryptophan
- Add 20 g agar for plates
- Bring to 950 mL with H_2O
- Autoclave 20 min at 121 °C (250 °F)
- Cool to 65 °C
- Add 50 mL filter sterilized 40% glucose

2.2. Collection of yeast samples for analysis of RNA decay rates

In a typical experiment, three aliquots are collected from mid-log phase yeast cultures ($A_{600} = 0.6$–0.8) in SG medium that supports maximum transcription. Two of the three aliquots will be pooled and used to measure the abundance of chimeric RNA at steady state (t_0) and to confirm that RNase protection assays give an accurate measure of chimeric RNA abundance. The third aliquot will be used to monitor suppression of *in vitro* ribozyme activity using a coextracted radiolabeled ribozyme transcript. Nine more aliquots are collected at various intervals after yeast have been transferred from SG medium to SD medium to inhibit transcription to determine the RNA decay time course. A 50 mL culture at mid-log phase ($A_{600} = 0.6$–0.8) contains enough yeast to measure decay kinetics for RNAs as abundant as yeast PGK1 mRNA or U3 snoRNAs. Larger culture volumes, and proportionately larger RNA extraction volumes, can be used to quantify decay kinetics for chimeric RNAs that are less abundant than PGK1 mRNA. Experiments typically are carried out in duplicate to facilitate analyses of experimental uncertainty.

The time points for a decay time course experiment are selected based on estimated decay rates. Chimeric PGK1 mRNA turns over through the endogenous RNA degradation machinery at a rate of about 0.05 min^{-1} in yeast cultures grown in minimal synthetic media at 30 °C and decays with a $t_{1/2}$ of $0.693/k_{\text{degradation}}$, or approximately 15 min (Donahue *et al.*, 2000). Decay follows an exponential rate equation, and is nearly complete after 6 half-lives, or 90 min, for a chimeric RNA with an inactive ribozyme insertion (Table 13.1). To capture the full time course of RNA turnover through endogenous degradation pathways, yeast aliquots typically are collected starting at 1 min after transcription inhibition and the interval between time points is doubled between successive time points up to 120 min. A chimeric RNA that disappears both through self-cleavage and through the endogenous RNA turnover pathways can exhibit a decay rate as fast as

Table 13.1 Exponential decay time course

Number of half-lives	Fraction of RNA remaining	% of RNA remaining
0	1/1	100
1	1/2	50
2	1/4	25
3	1/8	12.5
4	1/16	6.25
5	1/32	3.125
6	1/64	1.56

0.5 min^{-1}, with a $t_{1/2}$ of 0.693/0.5 min^{-1}, or approximately 1.4 min, and decay can be complete in less than 10 min (Yadava *et al.*, 2001). To capture a full decay time course, yeast aliquots are collected after the shortest possible time following transcription inhibition, and the interval between successive time points is doubled through 10 or 20 min. When the expected decay rate for a chimeric RNA is uncertain, an initial time course is based on an estimate of $t_{1/2}$ and extra time points are added at the beginning and the end.

Yeast from a single colony from an SG plate are used to inoculate 3 mL SG medium with appropriate supplements (Section 2.1.1) and grown in a 15 mL round-bottom snap-cap polypropylene tube overnight at 30 °C in an orbital shaker at 200 rpm. The next day, aliquots of the overnight culture are used to inoculate three 50 mL cultures of SG medium in 250 mL culture flasks to an A_{600} of 0.025, 0.05, and 0.1 and grown overnight with shaking at 30 °C. Under these growth conditions, the HFY114 yeast strain doubles about every 2 h. Inoculation of three cultures with different amounts of yeast ensures that at least one culture will reach mid-log phase at a convenient time. Yeast strains with different doubling times will require different inoculation protocols.

1. Before the yeast cultures reach mid-log phase, set up a dry ice ethanol bath. The bath should bubble slowly and become viscous. Label 1.6 mL microfuge tubes for duplicate sets of steady state and time point samples in advance with an ethanol-resistant marker. Equilibrate a shaking water bath to 30 °C.
2. Transfer 6×1.5 mL aliquots from a 50 mL culture into 1.6 mL microfuge tubes. These samples will be used to measure chimeric RNA abundance at steady state, confirm the linearity of the RNase protection assay, and monitor the suppression of background cleavage with radiolabeled ribozyme transcript in a cleavage tracking control experiment.
3. Place the remaining 41 mL culture in a 50 mL polypropylene conical tube. Centrifuge at room temperature for 5 min at 2100×*g*. Decant as much supernatant as possible. Resuspend the pellet in the residual supernatant and transfer the suspension to a 1.6 mL microfuge tube.
4. Centrifuge the tubes containing yeast from the 1.5 mL samples at 15,000×*g* for 30 s in a microfuge and centrifuge the 41 mL aliquots at 15,000×*g* for 5 min in a tabletop centrifuge to pellet the yeast. Remove the supernatant from all of the tubes.
5. Freeze the yeast pellets from the t_0 samples in the dry ice/ethanol bath and store at −80 °C until the RNA extraction is performed. Resuspend the large pellet in 750 μL warm (30 °C) SD minimal media (Section 2.1.2), immediately start the timer, and record the time (t_0).
6. Transfer the yeast suspension into 50 mL 30 °C SD media in a 250 mL culture flask. Incubate in a 30 °C water bath, shaking at 200 rpm.

7. One minute before each time point, transfer 2×1.5 mL aliquots from the culture flask into 1.6 mL microfuge tubes, immediately returning the culture flask to the 30 °C water bath. Centrifuge for 30 s at 15,000×*g*. Quickly remove the supernatant with a disposable transfer pipette and discard. Place both tubes in the dry ice/ethanol bath at the chosen time (t_1).
8. Repeat step 6 for each time point.
9. Transfer the frozen pellets to a −80 °C freezer until the RNA extraction is carried out.

2.3. Extraction of yeast RNA and coprecipitation with hybridization probes

Total RNA is extracted from yeast using conditions that suppress ribozyme activity in order to ensure that the abundance of uncut chimeric RNA reflects intracellular ribozyme activity and is not complicated by additional cleavage or ligation that occurred during the analyses. The extraction is carried out in a cold room with the reagents, tubes chilled on ice, the microfuge precooled to −4 °C, and with the pH of the reagents adjusted to pH 4.0. The volume of extraction buffer is adjusted according to the initial A_{600} of the yeast culture to ensure that the buffer is sufficient to maintain low pH following yeast lysis and the final concentration of yeast RNA is the same for each sample. The amount of actin mRNA is quantified in each sample to use for normalization of chimeric RNA abundance to total yeast RNA.

Three yeast aliquots are collected from steady-state cultures in SG media, before yeast are transferred to SD media to inhibit transcription. The deproteinized yeast extracts from two of these aliquots are pooled and used to prepare three steady-state samples , t_0A, t_0B and t_0C, that will be used to confirm that the radioactive signal is proportional to the amount of chimeric RNA. The t_0A sample contains half as much yeast RNA as the t_0B sample, but contains the same amount of hybridization probe. Samples t_0B and t_0C contain the same amount of yeast RNA but the t_0C sample contains twice as much hybridization probe as the t_0B sample.

The third steady-state sample, t_0D, is used to monitor suppression of ribozyme activity during the assay. A cleavage tracking control experiment is included in every analysis in which a radiolabeled, uncleaved ribozyme transcript is added to one yeast pellet and subjected to the same extraction procedure in parallel with the yeast samples. Self-cleavage of the ribozyme transcript serves to signal any incomplete suppression of yeast RNA self-cleavage activity that might occur during the assay. A particularly robust variant of the ribozyme, LR43, is used for this *in vitro* cleavage tracking control (Yadava *et al.*, 2001). Less than 2% of this ribozyme is expected to self-cleave when low pH and low temperatures are effectively maintained during the extraction, and the results are discarded if more than 10% of the ribozyme is found to have self-cleaved during the extraction.

1. Chill aluminum dry blocks in ice in a cold room in advance. Label three sets of twelve 1.6-mL microfuge tubes. Each set includes three tubes for samples collected from SG media (steady-state RNA abundance, t_0) and one tube for each of the nine time points of the decay time course collected after the transfer into SD media.
2. For the amount of yeast corresponding to 1.5 mL of culture at $A_{600} = 0.66$ (1 A_{600}), add 500 μL PI (phenol:isoamyl alcohol, 24:1, vol:vol, preequilibrated with RNA extraction buffer, pH 4) to each tube.
3. To the second set of tubes, add 380 μL CI (chloroform:isoamyl alcohol, 24:1, vol:vol, equilibrated with RNA extraction buffer, pH 4). The volumes of PCI and CI should be adjusted proportionately for smaller or larger yeast pellets.
4. To the third set of tubes, add [α-^{32}P]-labeled hybridization probes complementary to the chimeric RNA and to actin mRNA along with 1 μL carrier RNA (Sigma, calf liver type IV, 7.5 mg/mL). 100,000 cpm will be a sufficient amount of probe to ensure that probe is in excess relative to the chimeric RNA in the samples when hybridization probes are prepared as described below and used within 2 weeks of the [α-^{32}P]ATP reference date. Add an additional 100,000 cpm of the actin mRNA and chimeric RNA probes to the t_0C tube. Bring the total volume in the third set of tubes to 200 μL with RNA extraction buffer.
5. Place all three sets of tubes (PCI, CI, and hybridization tubes) in chilled metal microfuge blocks and chill them in the cold room for at least 20 min.
6. Transfer tubes containing frozen yeast pellets from the -80 °C freezer to the cold room in chilled blocks in a -20 °C bench top cooler.
7. To each pellet containing yeast equivalent to 1 A_{600}, add 500 μL RNA extraction buffer, 560 μL PCI (phenol:chloroform:isoamyl alcohol, 25:24:1, vol:vol:vol, preequilibrated with RNA extraction buffer, pH 4), 63 μL 10% SDS, and 150 mg acid-washed glass beads.
8. To the t_0D sample, add approximately 10,000 cpm [α-^{32}P] labeled uncut ribozyme transcript for the cleavage tracking control.
9. Arrange tubes in a multitube vortex mixer rack and vortex at top speed for 1 min, taking care not to warm the samples with fingers. Centrifuge at -4 °C for 5 min to separate phases. We find that a refrigerated microfuge is essential to prevent warming the tubes to temperatures that permit cleavage.
10. Transfer the upper aqueous phase from each tube to the corresponding prechilled tube containing PCI. Take care to avoid transferring any material from the interface between the two phases.

11. Vortex and centrifuge as in step 4.
12. Transfer the upper aqueous phase from the PCI tube to the corresponding prechilled tube containing CI.
13. Vortex and centrifuge as in step 4.
14. Mix 200 μL of the upper aqueous phases from each of the t_0A and t_0B samples (that do not contain radiolabeled RNA). Add 75 μL of the mixed t_0 sample to tube 0A of the third set of tubes that contain hybridization probes. Add 75 μL extraction buffer to the t_0A sample to equalize the sample volumes. Add 150 μL of the mixed t_0 sample to tubes 0B and 0C.
15. Transfer 150 μL of the aqueous phase from the tube containing the [α-^{32}P]-labeled ribozyme transcript to a tube containing 200 μL prechilled RNA extraction buffer.
16. Transfer 150 μL of the aqueous phase from each of the remaining time point samples to the third set of prechilled tubes.
17. Add 700 μL ethanol at -20 °C, transport tubes from cold room to a -20 °C freezer in a prechilled -20 °C carrier and allow the nucleic acids to precipitate for 30 min.
18. Centrifuge at 20,000$\times g$ for 25 min at 4 °C. Return the tubes to chilled blocks.
19. Remove the supernatant in the cold room. Allow the remaining ethanol to evaporate for 5 min, and proceed immediately to the hybridization step.

2.3.1. RNA extraction buffer

- 100 m*M* Sodium acetate
- 10 m*M* EDTA
- Adjust pH to 4.0 with glacial acetic acid at 25 °C

2.4. RNase protection assay

Chimeric RNAs are quantified using RNase protection assays in which yeast RNA is annealed to complementary radioactive probes, the unhybridized probe is digested with single-strand specific ribonucleases, and protected probe fragments are fractionated by denaturing gel electrophoresis (Sambrook *et al.*, 1989) (Figs. 13.3 and 13.5). To prevent reassembly of functional ribozymes that might undergo further self-cleavage or ligation during the analysis, yeast RNA is annealed with complementary radioactive probes immediately after the RNA is extracted from yeast and deproteinized.

Hybridization probes are designed to anneal with chimeric RNA and give protected fragments corresponding to uncut chimeric RNA, cleavage

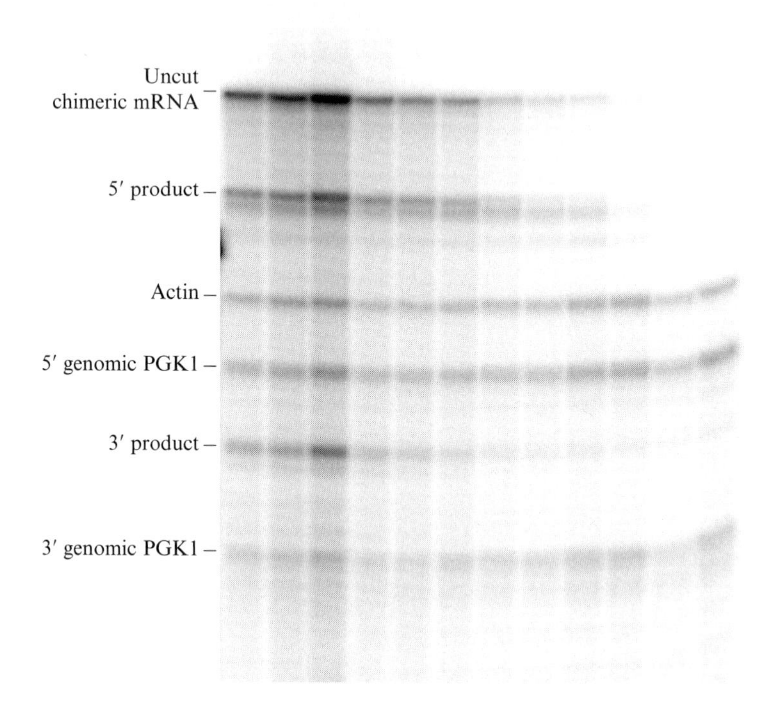

Figure 13.5 Chimeric PGK1 mRNA decay time course. Total RNA was extracted from yeast grown to mid-log phase in minimal medium with 2% galactose (t_0A, t_0B, and t_0C), and at various intervals after yeast were transferred to minimal medium with 2% glucose to inhibit transcription. Hybridization of yeast RNA with a radioactive probe that is complementary to chimeric PGK1 mRNA, followed by digestion with single-strand-specific ribonucleases, gives protected fragments corresponding to uncut chimeric RNA, 5′- and 3′-products of intracellular cleavage and genomic PGK1 sequences located upstream and downstream of the ribozyme insertion. The intensity of fragments corresponding to uncut chimeric mRNA is normalized to the abundance of actin mRNA and used to calculate chimeric PGK1 mRNA decay kinetics (Fig. 13.6). Sample in lanes 0A and 0B contained the same amount of hybridization probe but half as much yeast RNA was added to the sample in lane 0A. The sample in lane 0B contained the same amount of yeast RNA as the sample in lane 0C but twice as much radioactive probe was added to the sample in lane 0C.

products, genomic RNAs, and the actin mRNA used for normalization that can be distinguished from each other and from any residual undigested probe following denaturing gel electrophoresis (Fig. 13.3). Sequences encoding hybridization probes are inserted into the polylinker of pGEM plasmids (Promega) and radiolabeled probes are prepared by T7 RNA polymerase transcription in reactions with [α-^{32}P]ATP.

Comparisons of samples with different amounts of yeast and probe RNAs are carried out to confirm that radioactive probe concentrations are in excess relative to the concentrations of chimeric yeast RNAs. This comparison serves to confirm that the amount of probe that is protected from digestion by duplex formation with yeast RNA is proportional to the amount of chimeric yeast RNA in the sample.

2.4.1. Preparation of radiolabeled hybridization and cleavage tracking probes

[α-^{32}P]-Labeled hybridization probes that are complementary to chimeric RNAs and to actin mRNA are prepared by T7 RNA polymerase transcription of plasmids that have been linearized by restriction enzyme digestion and fractionated from undigested plasmid by agarose gel electrophoresis, as described (Milligan and Uhlenbeck, 1989). Reactions for synthesis of actin mRNA and chimeric RNA probes are prepared from a single transcription mix, which is divided before adding the linearized plasmid templates. This step ensures that probes used for quantification and normalization have the same specific activity.

Uncut ribozyme, LR43, for use as a cleavage tracking control is prepared by transcription of pTLR43 that has been linearized by digestion with Bgl II (Yadava *et al.*, 2001). Transcription of LR43 is carried out in the same reaction mixture used for transcription of hybridization probes but the ribozyme reaction is incubated at 30 °C for 30 min, followed by a 15-min incubation at 10 °C, and 10-fold dilution into gel loading buffer to inactivate the ribozyme. The low temperature incubation promotes ligation and maximizes recovery of uncut ribozyme.

2.4.2. Transcription buffer, 10×

- 400 m*M* Tris–Cl, pH 7.9
- 10 m*M* spermidine
- 50 m*M* dithiothreitol
- 0.1% Triton X-100
- 100 m*M* $MgCl_2$

2.4.3. Transcription reaction, 10 μL

- 1.0 μL 10× transcription buffer
- 1.5 μL H_2O
- 4.0 μL [α-^{32}P]ATP (3000 Ci/mmol)
- 2.0 μL 5× NTP mix (10 m*M* each UTP, GTP, CTP, 0.5 m*M* ATP)
- 1.0 μL 200–400 ng/μL linearized plasmid template
- 0.5 μL 0.2 mg/mL T7 RNA polymerase

1. Combine reaction components, adding template and T7 RNA polymerase last.
2. Incubate for 1 h at 37 °C.
3. Quench by adding 20 μL gel loading buffer.
4. Fractionate by denaturing gel electrophoresis (10% acrylamide, 19:1 acrylamide-bisacrylamide-8 *M* urea-TBE).
5. Locate probe transcripts by autoradiography and excise from the gel.
6. Elute transcripts from minced gel slices in approximately 5 volumes of 0.1 *M* sodium acetate–1 m*M* EDTA by shaking overnight in the cold room.
7. Filter through a disposable, fritted column (Biospin chromatography column, BioRad) to remove gel fragments.
8. Store at −20 °C.

2.4.4. Gel loading buffer

- 95% formamide
- 25 m*M* EDTA
- 0.025% bromophenol blue
- 0.025% xylene cyanol

2.4.5. Hybridization of radiolabeled probe to chimeric RNA

RNase protection assays are carried out generally as described (Sambrook *et al.*, 1989). Completely dissolving the pellet in hybridization buffer is a critical step that usually requires several minutes of vigorous mixing. Optimal hybridization conditions vary among different chimeric RNAs and must be determined empirically by identifying temperatures that maximize the intensity of specific products and minimize background. It is important to maintain the entire hybridization tube at the same temperature during the hybridization incubation to prevent water from evaporating from the hybridization solution and condensing on the lid. If water were allowed to evaporate, hybrids would be denatured in the concentrated formamide left behind. It is also important to maintain the appropriate hybridization temperature until samples are treated with ribonuclease. If tubes are transferred to the bench top or an ice bucket, lower temperatures can promote nonspecific annealing.

1. Add 15 μL hybridization solution (Section 2.4.7) to each tube, except for the tube that contains the ^{32}P-labeled ribozyme transcript for tracking background cleavage. Vortex vigorously and pipet up and down to dissolve the pellets completely.
2. Confirm that all of the radioactivity is in solution in the following way. Spin the tubes for 5 s at 15,000×*g* in a microfuge. Draw half of the solution into a pipet tip and estimate radioactivity using a Geiger counter. If the pellet is fully dissolved, the same amount of radioactivity

will be in the pipet tip as in the tube. If some RNA is undissolved, more radioactivity will remain in the tube.

3. Add 15 μL formamide and 15 μL gel loading buffer (Section 2.4.4) to dissolve the ribozyme transcript in the cleavage tracking tube. Store the tube at −20 °C until the gel is ready to load.
4. Place tubes in a thermal cycler and run a hybridization program. A typical program might include: 5 min each at 95, 90, 85, 80, 75, 70, 65, 60, and 55 °C, followed by an overnight incubation at 50 °C. Proceed with the RNase digestion step when the annealing program is complete.

2.4.6. 5× Hybridization buffer

- 200 m*M* Na PIPES, pH 6.5
- 2 *M* NaCl
- 5 m*M* EDTA

2.4.7. Hybridization solution, 500 μL

- 100 μL 5× hybridization buffer
- 400 μL deionized formamide

2.4.8. RNase digestion and fractionation of products

This protocol has been optimized for analyses of chimeric PGK1 mRNAs. Digestion conditions must be sufficient to degrade the portion of the hybridization probe that is not complementary to chimeric yeast RNA and remains single stranded, but not degrade the duplex formed by annealing radioactive probe and yeast RNA. Optimal conditions for quantification of other chimeric RNAs will vary and must be identified by surveying ribonuclease concentrations and digestion times and temperatures that give the maximum intensity of bands that represent uncut chimeric RNAs and cleavage products with minimal background.

1. Prepare enough RNase mix (Section 2.4.9) to have 150 μL per sample plus an additional 150 μL to compensate for pipetting inaccuracies. Keep the RNA hybridization tubes at 50 °C while preparing this mix.
2. Add 150 μL RNase mix to each hybridization tube. Mix gently by pipetting the solution up and down. Avoid trapping RNase solution in the lid where it might be inaccessible to the proteinase K digestion step that follows. Incubate at 25 °C for 30 min.
3. To each tube, add 10 μL freshly prepared proteinase K (10 mg/mL), 10 μL 10% SDS, and 3 μL 7.5 mg/ml carrier RNA. Spin briefly and mix again to ensure no ribonuclease solution on the sides of the tubes escapes proteinase K digestion. Incubate at 37 °C for 30 min.

4. Move tubes to a fume hood. Add 200 μL PCI (phenol:chloroform: isoamyl alcohol (25:24:1, vol:vol:vol), pH 8). Vortex until milky white, approximately 30 s. Centrifuge at $4000\times g$ for 5 min to separate phases.
5. Transfer 170 μL of the aqueous top phase to a 0.6 mL tube. Add 430 μL -20 °C ethanol and vortex briefly to mix. Allow nucleic acids to precipitate at -20 °C for at least 30 min.
6. Decant ethanol. Add 20 μL gel loading buffer (Section 2.4.4). Vortex vigorously and pipet up and down until RNA is completely dissolved.
7. Estimate the radioactivity in the t_0B sample using a Geiger counter. The sample should contain approximately 2000 cpm. Steps 8 and 9 describe preparation of electrophoresis markers with 1000 cpm. If the t_0B sample contains significantly more or less than 2000 cpm, the amount of marker should be adjusted accordingly to ensure that the marker and sample bands will have similar intensities.
8. Prepare a cleaved ribozyme marker for comparison with the cleavage tracking control sample. Add 1000 cpm (approximately 1 μL of a 1:50 dilution) of the [^{32}P]-labeled uncut ribozyme transcript that was used for the cleavage tracking control. Add 9 μL ribozyme reaction buffer (Section 2.4.10). Incubate at 37 °C for at least 15 min. Add 10 μL gel loading buffer (Section 2.4.4).
9. Add 15 μL gel loading buffer (Section 2.4.4) to four 0.6 mL tubes. Add approximately 1000 cpm actin probe, 1000 cpm chimeric RNA probe, and 1000 cpm uncut LR43 to separate tubes. Add [γ-^{32}P]-labeled electrophoresis size markers, such as a [γ-^{32}P]-labeled Msp I digest of pBR322 DNA, to contain approximately 1000 cpm/band.
10. Collect RNA in the bottom of the tubes with a brief microfuge spin. Heat to 95 °C for 2 min. Fractionate samples by denaturing gel electrophoresis using 6% (acrylamide:bisacrylamide, 19:1), 8 *M* urea, TBE, 40 cm × 22 cm × 0.1 cm for 2 h at 25 v/cm, after preelectrophoresis for at least 30 min.
11. Quantify full-length chimeric RNA and actin mRNA band intensities by radioanalytic imaging.

2.4.9. RNase mix

- 1.25 μg/mL RNase A
- 100 U/mL RNase T1
- 200 m*M* sodium acetate
- 10 m*M* Tris–Cl, pH 8
- 1 m*M* EDTA

2.4.10. Ribozyme reaction buffer

- 50 m*M* HEPES, pH 7.5
- 10 m*M* $MgCl_2$
- 0.1 m*M* EDTA

2.5. Quantification of intracellular RNA folding and catalysis rates

Three samples, t_0A, t_0B, and t_0C, are analyzed to determine the abundance of uncut chimeric RNA during steady-state growth in galactose. These steady-state RNA samples are prepared with different amounts of yeast RNA and hybridization probes. The intensities of bands in these samples are compared to ensure that intensity is proportional to the amount of yeast RNA and is not limited by an insufficient amount of hybridization probe. The t_0A sample contains half as much yeast RNA as the t_0B sample, but contains the same amount of hybridization probe. If the hybridization probe is in large excess relative to yeast RNA in both t_0A and t_0B samples, as intended, the t_0A sample will produce bands with half the intensities as the bands in the t_0B sample (Fig. 13.5, lanes 0A, 0B). Samples t_0B and t_0C contain the same amount of yeast RNA hybridization probe but the t_0C sample contains twice as much hybridization probe as the t_0B sample. Therefore, the bands in the t_0B and t_0C samples should have the same intensities, if yeast RNA is indeed limiting (Fig. 13.5, lanes 0B, and 0C).

The time required for ribozymes to assemble and cleave *in vivo* can be calculated from chimeric RNA decay kinetics determined from decay time course experiments and from the relative abundance of uncut chimeric RNAs that contain functional or mutationally inactivated ribozymes at steady state (Fig. 13.6). Each type of experiment is subject to different experimental artifacts. Failure to completely suppress ribozyme self-cleavage during the analysis, for example, could lead to underestimation of chimeric RNA abundance at steady state, but would have little effect on decay kinetics determined from time course experiments since uncut RNA in samples from each time point would be reduced by a similar amount. Conversely, errors in collecting samples at short intervals after transcription inhibition could affect measurement of decay kinetics in time course experiments without affecting measurements of chimeric RNA abundance at steady state. Agreement between time course and steady-state results provides important confirmation that neither experiment was invalidated by these potential problems.

2.5.1. Quantification of ribozyme activity from decay time courses

Band intensities are quantified by radioanalytic imaging, then normalized to the number of adenine residues in each radioactive probe fragment labeled during transcription with [α-^{32}P]ATP. Next, the abundance of uncut chimeric mRNA is normalized relative to actin mRNA to compensate for any variation in the total amount of yeast RNA in each sample. To measure the chimeric RNA decay rate, k_{decay}, the amount of uncut chimeric RNA is plotted as a function of time (Fig. 13.6A) and fit to an exponential decay equation:

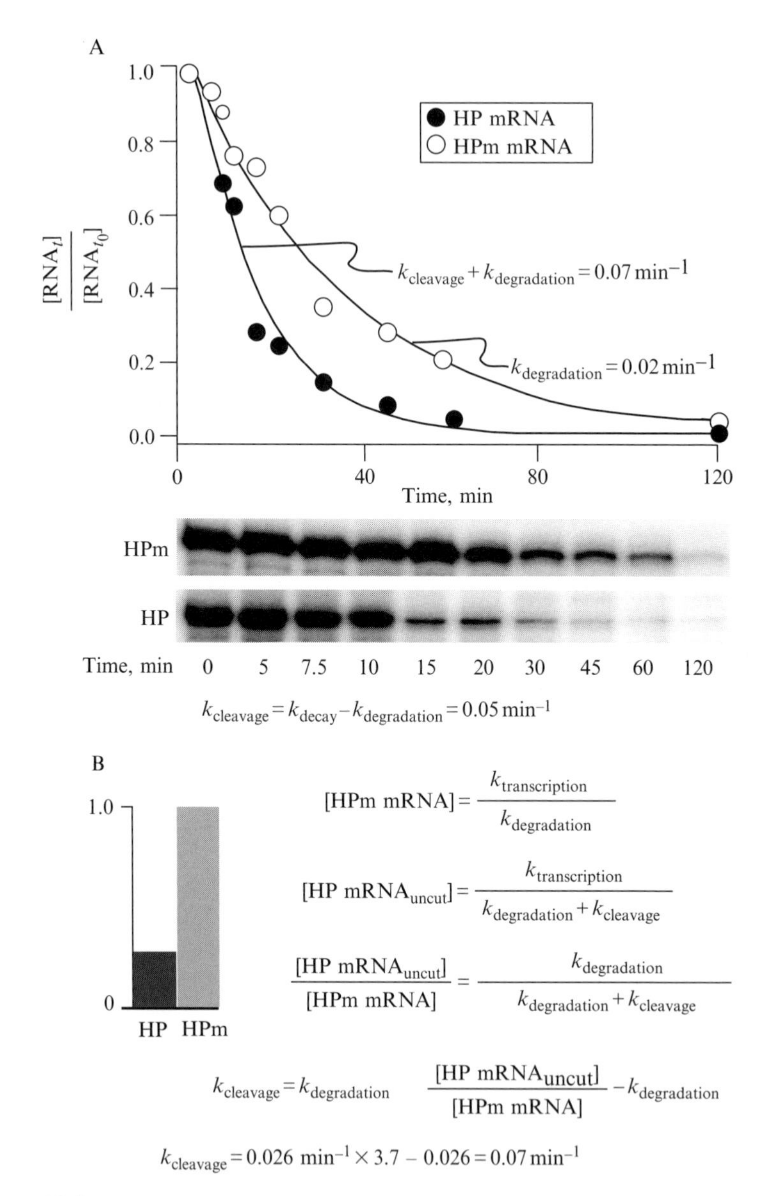

Figure 13.6 Quantification of intracellular ribozyme self-cleavage kinetics. (A) Calculating intracellular self-cleavage rates from rates of chimeric RNA decay after transcription inhibition. The abundance of uncut chimeric mRNA, normalized to the abundance actin mRNA, is plotted as a function of time. Decay rates for chimeric RNAs were calculated from the fit to an exponential rate equation, Eq. (13.1). The difference between the decay rate of 0.07 min^{-1} observed for a chimeric mRNA with a self-cleaving ribozyme (filled circles) and the rate of 0.02 min^{-1} observed for a chimeric mRNA with a mutationally inactivated ribozyme (open circles) gives an

$$\frac{[\mathrm{RNA}_t]}{[\mathrm{RNA}_{t0}]} = \frac{[\mathrm{RNA}_{t=\infty}]}{[\mathrm{RNA}_{t=0}]} \times (1 - \mathrm{e}^{k_{\mathrm{decay}}t}) \tag{13.1}$$

k_{decay} for a chimeric RNA that contains an active ribozyme represents the sum of self-cleavage and intrinsic RNA degradation rates. That is, $k_{\mathrm{decay,HP}} = k_{\mathrm{cleavage}} + k_{\mathrm{degradation}}$. For a chimeric RNA that has the same ribozyme insertion but with an inactivating mutation, k_{decay} represents the intrinsic degradation rate. That is, $k_{\mathrm{decay,HPm}} = k_{\mathrm{degradation}}$ (Figs. 13.4 and 13.6A). Therefore, the time needed for the ribozyme to assemble and cleave *in vivo*, k_{cleavage}, can be calculated from the difference between the decay rate measured for a self-cleaving chimeric RNA and the rate measured for decay of a chimeric RNA that contains an inactive ribozyme, according to Eq. (13.2):

$$k_{\mathrm{cleavage}} = k_{\mathrm{decay,HP}} - k_{\mathrm{decay,HPm}} \tag{13.2}$$

2.5.2. Quantification of ribozyme activity from RNA abundance at steady state

Once the intrinsic degradation rate has been determined for a chimeric RNA containing a mutationally inactivated ribozyme in a time course experiment, intracellular cleavage rates can also be calculated from the relative abundance of chimeric RNAs that contain functional and inactive ribozymes at steady state (Fig. 13.6B). RNA abundance at steady state reflects synthesis and decay rates. Self-cleaving and mutationally inactivated chimeric RNAs differ only by the presence of a G or an A at the cleavage site, so they are transcribed at identical rates but they decay at different rates due to the contribution that self-cleavage makes to decay of the self-cleaving RNA. The ratio of mutant and self-cleaving chimeric RNA abundance at steady state, combined with the rate of RNA turnover through endogenous RNA degradation pathways, can be used to calculate intracellular k_{cleavage} according to Eq. (13.3):

$$k_{\mathrm{cleavage}} = k_{\mathrm{degradation}} \times \frac{[\text{uncut HP RNA}]}{[\text{HPm RNA}]} - k_{\mathrm{degradation}} \tag{13.3}$$

intracellular self-cleavage rate of 0.05 min^{-1}. Adapted from Donahue and Fedor (1997). (B) Calculating intracellular self-cleavage rates from chimeric RNA abundance at steady state. Since chimeric self-cleaving RNAs are transcribed at the same rate as chimeric RNAs with an inactive ribozyme, the reduced abundance of self-cleaving RNAs at steady state can be attributed to the accelerated decay due to self-cleavage. Once the intrinsic degradation rate has been determined for an inactive chimeric RNA, the self-cleavage rate can be calculated from the relative abundance of self-cleaving and mutationally inactivated chimeric RNAs at steady state.

3. Applications

RNA catalysis provides a sensitive, quantitative signal that an RNA has assembled into a functional structure making RNA enzymes particularly useful for RNA structure–function studies. Quantitative enzymological studies of ribozymes have produced detailed insights into the structural transitions that comprise RNA assembly and reaction pathways *in vitro*. However, the complexity of biological reactions and lack of information about the intracellular environment limits the accuracy with which *in vitro* reactions can recapitulate biology. These methods were developed to learn how structure–function principles established from mechanistic studies of RNA enzymes *in vitro* translate to the complex environment of a living cell by exploiting the simple RNA cleavage and ligation reactions of hairpin ribozymes. This approach for quantitative analysis of intracellular ribozyme cleavage kinetics has been used to investigate several aspects of RNA folding *in vivo*. Although the complexities of an intracellular environment might be expected to affect RNA assembly in unpredictable ways, our studies have shown that kinetic and equilibrium parameters for intra- and intermolecular ribozyme reactions agree remarkably well with parameters measured *in vitro* provided that *in vitro* reaction conditions approximate an intracellular ionic environment. On the other hand, experiments designed to investigate mechanisms of RNA secondary structure assembly revealed striking differences between intracellular and *in vitro* reactions in partitioning of RNAs among alternative secondary structures. These applications of this method are summarized below.

3.1. Kinetic mechanism of intracellular hairpin ribozyme self-cleavage

Minimal and natural forms of the hairpin ribozyme catalyze the same chemical reaction but exhibit very different kinetic mechanisms (Fedor, 1999). The natural form of the hairpin ribozyme in viral satellite RNAs contains two helices, H5 and H6, located between the A and B domains that allow the ribozyme to assemble in the context of a four-way helical junction (Fig. 13.1B). Helices H5 and H6 are not essential for activity but support interdomain docking under less favorable ionic conditions (Murchie *et al.*, 1998; Walter *et al.*, 1998a,b,c, 1999) and support rapid cleavage and ligation under physiological salt conditions *in vitro* (Yadava *et al.*, 2001). Consistent with the enhanced activity of four-way junction variants under physiological ionic conditions *in vitro*, four-way junction ribozymes self-cleave much faster than minimal ribozymes in yeast (Yadava *et al.*, 2001).

To learn whether the enhanced tertiary structure stability conferred by the four-way junction also promotes intracellular ligation, we measured intracellular cleavage kinetics for four-way junction variants that contain H1 sequences with as few as one or as many as 20 base pairs. If the internal equilibrium favors ligation for ribozymes with a four-way helical junction *in vivo* as it does under "physiological" ionic conditions *in vitro*, we expected variation of H1 stability to have dramatic effects on intracellular self-cleavage kinetics. Variants with stable H1 sequences that dissociate slowly should display low intracellular cleavage activity because cleavage should be reversed by rapid re-ligation of bound products. Variants with H1 sequences that are stable enough to support assembly but weak enough to dissociate rapidly should exhibit maximum self-cleavage rates.

Our results agreed well with these predictions. A ribozyme with a four-way helical junction and three base pairs in H1, HP43, displayed the fastest intracellular cleavage rate and a variant with 20 base pairs in H1, HP420, exhibited the slowest self-cleavage rate. The reduced intracellular cleavage rate observed for the variant that forms the most stable ribozyme–product complex provides strong evidence that slow product dissociation allows re-ligation of bound products *in vivo* as it does *in vitro* (Fig. 13.2). A variant in which H1 contains four base pairs, HP44, self-cleaved at an intermediate rate, suggesting that this ribozyme–product complex partitions between dissociation and re-ligation of bound products *in vivo*. Comparison of HP43, HP44, and HP420 intracellular cleavage kinetics enabled us to estimate ligation and product dissociation rate constants by assuming that dissociation of an H1 helix with three base pairs is much faster than ligation and that slow self-cleavage of ribozymes with more stable H1 helices reflects re-ligation of bound products. Using values of 4 min^{-1} and 0.7 min^{-1} for intracellular $k_{ligation}$ and $k_{cleavage}$, respectively, and 0.28 min^{-1} for the intracellular HP44 mRNA self-cleavage rate, a value of 2.9 min^{-1} can be calculated for the rate of product dissociation *in vivo* (Fig. 13.7A). Thus, the ribozyme complex with four base pairs in H1 partitions nearly equally between dissociation and re-ligation *in vivo*.

These kinetic and equilibrium parameters measured for an intracellular ribozyme self-cleavage reaction agree remarkably well with the parameters measured for the same ribozyme variant *in vitro* provided that the *in vitro* reactions approximate intracellular ionic conditions (Yadava *et al.*, 2001). This quantitative analysis of the intracellular self-cleavage mechanism provides no evidence that any component of the intracellular environment significantly alters the intrinsic stability of the secondary and tertiary structure elements that comprise a functional ribozyme.

3.2. Kinetics of intracellular RNA complex formation

The kinetics and equilibria of intracellular RNA complex formation have been analyzed by expressing ribozymes and substrates as separate chimeric RNAs (Yadava *et al.*, 2004). In this system, cleavage requires ribozymes to

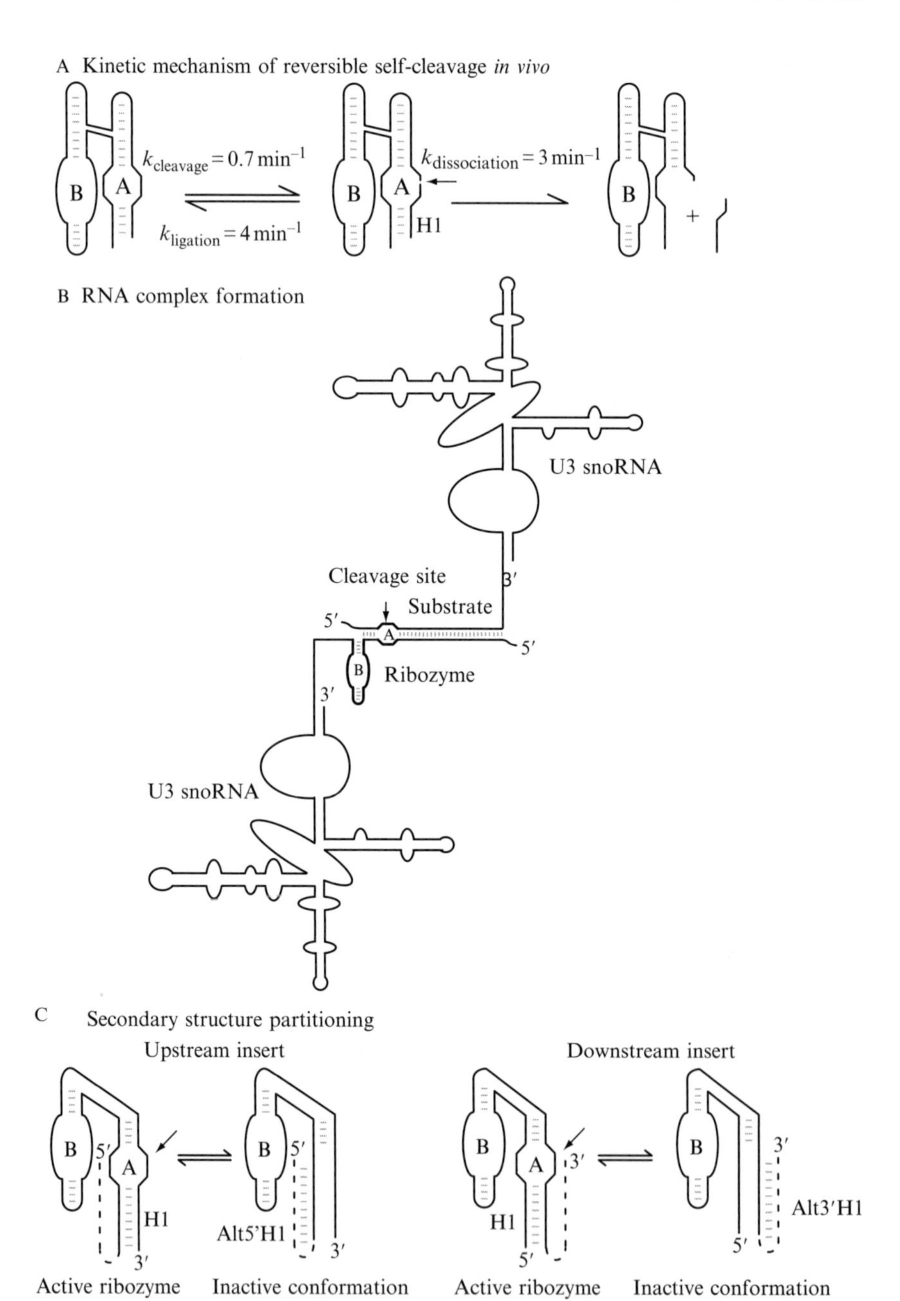

Figure 13.7 Applications. (A) Kinetic mechanism of hairpin ribozyme self-cleavage *in vivo*. Comparison of intracellular self-cleavage rates for a series of ribozyme variants that assemble in the context of a four-way helical junction and have different H1 sequences enabled the determination of kinetic parameters for intracellular cleavage,

recognize substrates and form functional complexes by intermolecular base pairing in helices H1 and H2 (Figs. 13.1C and 13.7B). Our first experiments were carried out with chimeric PGK1 mRNAs so that cleavage required assembly of chimeric mRNA complexes localized in the cytoplasm. However, we detected very little cleavage activity with chimeric PGK1 mRNAs, probably because cytoplasmic PGK1 mRNA concentrations were too low to drive ribozyme–substrate complex formation. We next expressed ribozymes and substrates as chimeric snoRNAs that localize in the nucleolus (Fig. 13.7B). Ribozyme and substrate sequences were substituted for the portion of U3 snoRNA that normally directs recognition of ribosomal RNA targets. Even though similar amounts of PGK1 mRNAs and snoRNAs were transcribed under galactose induction, nuclear snoRNA concentrations are about 1000-fold higher than cytoplasmic mRNA concentrations due to the small volume of the nucleolus.

Fusion of chimeric ribozyme snoRNA genes to UAS_{GAL} allowed us to modulate intracellular concentrations of ribozyme snoRNAs by varying the galactose concentration in growth media. Chimeric substrate snoRNAs were expressed from the normal U3 promoter and were not subject to galactose induction. At intracellular chimeric ribozyme snoRNA concentrations between 0.3 and 1.5 μM, observed cleavage rates varied from 0.009 to 0.024 min^{-1}. Fitting the data to the Michaelis–Menten equation gave an approximate k_{cleavage} value of 0.037 min^{-1} and a K_{M}^{S} value of 1.0 μM for intracellular substrate cleavage. This cleavage rate is not significantly different from the rate measured for intracellular chimeric snoRNP self-cleavage, evidence that a ribozyme–substrate complex that forms between two chimeric snoRNPs is nearly as functional as a self-cleaving ribozyme that assembles within a single snoRNA.

Although ribozymes and substrates with small regions of complementarity cleave at maximum rates in reactions with high divalent cation concentrations *in vitro*, large complementarity regions over 20 base pairs were required to maximize substrate binding kinetics *in vivo* and in reactions designed to mimic an intracellular ionic environment *in vitro*. Comparison of second-order rate constants for intermolecular cleavage with diffusion coefficients measured directly for U3 snoRNPs *in vivo* suggests that complex

ligation, and product dissociation for a hairpin ribozyme with four base pairs in the intermolecular H1 helix (Yadava *et al.*, 2001). (B) Intracellular RNA complex formation kinetics. Analyses of an intermolecular reaction in which ribozyme and substrate RNAs were expressed as separate chimeric snoRNAs enabled determination of intracellular RNA complex formation kinetics (Yadava *et al.*, 2004). (C) Partitioning among alternative mRNA secondary structures. Hairpin ribozyme variants with complementary sequences inserted adjacent to the ribozyme sequence were used to assess partitioning between functional ribozyme structures and nonfunctional stem-loops (Mahen *et al.*, 2005).

formation between chimeric ribozyme and substrate snoRNPs in yeast nuclei is diffusion limited. Our results suggest that the intracellular ionic environment is a major determinant of RNA complex formation efficiency, consistent with the conclusions drawn from our earlier investigations of intracellular self-cleavage reactions. These results also suggest that *in vitro* studies can have significant value in understanding intracellular RNA recognition and assembly mechanisms provided that *in vitro* reactions recapitulate intracellular ionic conditions.

3.3. Partitioning among alternative RNA secondary structures

To investigate how RNAs can adopt specific functional structures despite the capacity to form alternative nonfunctional structures with similar stabilities, we designed hairpin ribozymes with the potential to form defined stem-loop structures that are incompatible with assembly of a functional ribozyme (Mahen *et al.*, 2005) (Fig. 13.7C). In one variant, a complementary insert located upstream of the ribozyme can form 10 base pairs with the 5′-end of the ribozyme sequence to create a stem-loop, Alt5′H1, that is incompatible with assembly of the essential H1 helix of the ribozyme. In a second variant, an insert located downstream of the ribozyme can form 10 base pairs with 3′-terminal ribozyme nucleotides to form a stem-loop, Alt3′H1. Formation of either AltH1 stem-loop blocks assembly of a functional ribozyme, so self-cleavage activity can be used to monitor partitioning between functional ribozymes and nonfunctional stem-loops.

Ribozymes with upstream or downstream inserts displayed very different self-cleavage activity in cotranscriptional assembly reactions *in vitro*, consistent with a sequential folding mechanism. A ribozyme with an upstream insert self-cleaved very slowly, suggesting that a stem-loop that sequesters the upstream part of H1 blocks ribozyme assembly. In contrast, ribozymes retained significant activity when a complete ribozyme could assemble before a downstream insert was transcribed *in vitro*. The distinct effects of upstream and downstream inserts suggested that structures assemble sequentially *in vitro* allowing upstream structures that can assemble first to dominate the folding pathway.

Partitioning between functional ribozymes and nonfunctional stem-loops did not show evidence of sequential folding in yeast; a downstream insert was able to block ribozyme assembly completely even though the upstream ribozyme sequence was synthesized first. This evidence that a downstream stem-loop can compete effectively with ribozyme assembly suggests that some feature of the intracellular environment affects folding in some way that simple *in vitro* reactions do not recapitulate. Ongoing experiments focus on elucidating the mechanism through which this occurs.

ACKNOWLEDGMENTS

We thank Christine Donahue, Ramesh Yadava, Elisabeth Mahen, Jason Harger, and Elise Calderon whose efforts contributed to the development of these methods over the years, and Steffen Grimm for critical reading of the manuscript. Research in the authors' laboratory is supported by National Institutes of Health grants GM046422 and GM062277 to MJF.

REFERENCES

Amberg, D. C., *et al.* (2005). Methods in Yeast Genetics: A Cold Spring Harbor Laboratory Course Manual. Cold Spring Harbor Press, Cold Spring Harbor, NY.

Donahue, C. P., and Fedor, M. J. (1997). Kinetics of hairpin ribozyme cleavage in yeast. *RNA* **3,** 961–973.

Donahue, C. P., *et al.* (2000). The kinetic mechanism of the hairpin ribozyme *in vivo*: Influence of RNA helix stability on intracellular cleavage kinetics. *J. Mol. Biol.* **295,** 693–707.

Fedor, M. (2009). Comparative enzymology and structural biology of RNA self-cleavage. *Annu. Rev. Biophys.* **38,** 271–299.

Fedor, M. J. (1999). Tertiary structure stabilization promotes hairpin ribozyme ligation. *Biochemistry* **38,** 11040–11050.

He, F., *et al.* (2008). Qualitative and quantitative assessment of the activity of the yeast nonsense-mediated mRNA decay pathway. *In* "RNA Turnover in Eukaryotes: Analysis of Specialized and Quality Control RNA Decay Pathways," Vol. 449, pp. 127–147. Elsevier Academic Press, San Diego.

Hegg, L. A., and Fedor, M. J. (1995). Kinetics and thermodynamics of intermolecular catalysis by hairpin ribozymes. *Biochemistry* **34,** 15813–15828.

Herrick, D., *et al.* (1990). Identification and comparison of stable and unstable mRNAs in *Saccharomyces cerevisiae. Mol. Cell. Biol.* **10,** 2269–2284.

Houseley, J., and Tollervey, D. (2009). The many pathways of RNA degradation. *Cell* **136,** 763–776.

Johnston, M., *et al.* (1994). Multiple mechanisms provide rapid and stringent glucose repression of GAL gene expression in *Saccharomyces cerevisiae. Mol. Cell. Biol.* **14,** 3834–3841.

Mahen, E. M., *et al.* (2005). Kinetics and thermodynamics make different contributions to RNA folding *in vitro* and in yeast. *Mol. Cell* **19,** 27–37.

Milligan, J. F., and Uhlenbeck, O. C. (1989). Synthesis of small RNAs using T7 RNA polymerase. *Methods Enzymol.* **180,** 51–62.

Murchie, A. I., *et al.* (1998). Folding of the hairpin ribozyme in its natural conformation achieves close physical proximity of the loops. *Mol. Cell* **1,** 873–881.

Nesbitt, S. M., *et al.* (1999). The internal equilibrium of the hairpin ribozyme: Temperature, ion and pH effects. *J. Mol. Biol.* **286,** 1009–1024.

Parker, R., *et al.* (1991). Measurement of mRNA decay rates in *Saccharomyces cerevisiae. Methods Enzymol.* **194,** 415–423.

Samarsky, D. A., *et al.* (1999). A small nucleolar RNA:ribozyme hybrid cleaves a nucleolar RNA target *in vivo* with near-perfect efficiency. *Proc. Natl. Acad. Sci.* **96,** 6609–6614.

Sambrook, J., *et al.* (1989). Molecular Cloning: A Laboratory Manual. Cold Spring Harbor Laboratory Press, New York.

Sikorski, R. S., and Hieter, P. (1989). A system of shuttle vectors and yeast host strains designed for efficient manipulation of DNA in *Saccharomyces cerevisiae. Genetics* **122,** 19–27.

Walter, F., *et al.* (1998a). Global structure of four-way RNA junctions studied using fluorescence resonance energy transfer. *RNA* **4,** 719–728.
Walter, F., *et al.* (1998b). Folding of the four-way RNA junction of the hairpin ribozyme. *Biochemistry* **37,** 17629–17636.
Walter, F., *et al.* (1998c). Structure and activity of the hairpin ribozyme in its natural junction conformation: Effect of metal ions. *Biochemistry* **37,** 14195–14203.
Walter, N. G., *et al.* (1999). Stability of hairpin ribozyme tertiary structure is governed by the interdomain junction. *Nat. Struct. Biol.* **6,** 544–549.
Yadava, R. S., *et al.* (2001). Hairpin ribozymes with four-way helical junctions mediate intracellular RNA ligation. *J. Mol. Biol.* **309,** 893–902.
Yadava, R. S., *et al.* (2004). Kinetic analysis of ribozyme–substrate complex formation in yeast. *RNA* **10,** 863–879.

SECTION TWO

IDENTIFYING METAL ION INTERACTIONS IN RNA

CHAPTER FOURTEEN

Separation of RNA Phosphorothioate Oligonucleotides by HPLC

John K. Frederiksen*,†,1 and Joseph A. Piccirilli†,‡

Contents

Abstract

Phosphorothioate oligonucleotides are indispensable tools for probing nucleic acid structure and function and for the design of antisense therapeutics. Many applications involving phosphorothioates require site- and stereospecific substitution of individual *pro*-R_P or *pro*-S_P nonbridging oxygens. However, the traditional approach to phosphorothioate synthesis produces a mixture of R_P and S_P diastereomers that must be separated prior to use. High-performance

* The Pritzker School of Medicine, University of Chicago, Chicago, Illinois, USA
† Department of Biochemistry and Molecular Biology, University of Chicago, Chicago, Illinois, USA
‡ Department of Chemistry, University of Chicago, Chicago, Illinois, USA
1 Present address: Department of Pathology and Laboratory Medicine, University of Rochester Medical Center, Rochester, New York, USA

Methods in Enzymology, Volume 468
ISSN 0076-6879, DOI: 10.1016/S0076-6879(09)68014-9

liquid chromatography (HPLC) has proven to be a versatile method for effecting this separation, with both reversed phase (RP) and strong anion exchange (SAX) protocols yielding favorable results. In this chapter, we present several examples of successful separations of RNA phosphorothioate diastereomers by HPLC. We also report the use of complementary DNA oligonucleotides for the separation of poorly resolved phosphorothioate RNAs.

1. Introduction: Phosphorothioate Oligonucleotides and the Need for Separation

The unique properties of phosphorothioate oligonucleotides make them suitable for a wide range of biochemical and medicinal applications. In a phosphorothioate, sulfur replaces one of the two nonbridging oxygens of a phosphodiester linkage. This substitution confers chirality at the phosphorous center (Fig. 14.1) and shifts the electronic charge distribution of the linkage so that the formal negative charge localizes on sulfur (Frey and Sammons, 1985). The bulkier sulfur atom and the longer phosphorous–sulfur bond also alter the steric environment from that of the original phosphodiester linkage. As a result of these changes, biological macromolecules often interact preferentially with one of the two diastereomers,

Figure 14.1 RNA phosphorothioate diastereomers. Sulfur sizes and bond lengths not drawn to scale. (See Color Insert.)

making phosphorothioates useful as mechanistic and structural probes. Several papers have reported the use of phosphorothioates to determine the stereochemical course of protein- and RNA-catalyzed reactions (Burgers *et al.*, 1979; Eckstein *et al.*, 1972, 1981; McSwiggen and Cech, 1989; Moore and Sharp, 1993; Padgett *et al.*, 1994; Potter *et al.*, 1983a). Moreover, as sulfur preferentially coordinates softer metal ions than does oxygen, phosphorothioates have figured prominently in the investigations of metal ion catalysis in ribozymes via the metal ion rescue approach (Chen *et al.*, 1997; Christian *et al.*, 2000, 2006; Crary *et al.*, 2002; Forconi *et al.*, 2008; Gordon and Piccirilli, 2001; Hougland *et al.*, 2005; Osborne *et al.*, 2005; Shan *et al.*, 2001; Wang *et al.*, 1999; Warnecke *et al.*, 1996; Yoshida *et al.*, 1999; see also Chapter 15 in this volume). In the context of DNA, phosphorothioates confer resistance to nucleases and are incorporated into therapeutic antisense oligonucleotides to increase plasma half-life and improve pharmacokinetics (Akhtar *et al.*, 1991; Hoke *et al.*, 1991; Stein and Cheng, 1993). In addition to these biochemical applications, phosphorothioates also appear to play a physiologic role for some organisms. Recent work has demonstrated that certain bacteria incorporate phosphorothioates within their genomes, which may help to maintain the structural integrity of the bacterial chromosome (Wang *et al.*, 2007).

Many experiments involving phosphorothioates, especially those investigating catalytic reaction mechanisms, require site- and stereospecific phosphorothioate substitutions. An example encountered frequently in the field of ribozyme catalysis is the sulfur substitution of a *pro*-R_P or *pro*-S_P nonbridging oxygen suspected of coordinating a catalytic metal ion. The nonbridging oxygens are not equivalent stereochemically and must be substituted individually with sulfur to assess their potential catalytic roles. To this end, diastereomerically pure R_P and S_P phosphorothioate substrates or ribozymes must be constructed. Depending on the experiment, this can entail sequential splint-mediated enzymatic ligations of short, diastereospecific phosphorothioate oligonucleotides to *in vitro*-transcribed flanking RNAs (Moore and Sharp, 1993; Stark *et al.*, 2006; see also Chapter 2 of volume 469; Fig. 14.2).

One of the challenges associated with obtaining diastereomerically pure phosphorothioates is the method by which phosphorothioates are introduced during solid-phase oligonucleotide synthesis. In each synthetic cycle, the free 5′-OH of the immobilized oligonucleotide attacks the phosphorous atom of an incoming phosphoramidite (Fig. 14.3). To form the natural phosphodiester, the resulting trivalent phosphite is oxidized to the pentavalent phosphate by treatment with a mixture of iodine and water. In phosphorothioate synthesis, the phosphite is treated instead with one of several different sulfurization agents (Connolly *et al.*, 1984; Iyer *et al.*, 2002; Song *et al.*, 2003; Stec *et al.*, 1993; Vu and Hirschbein, 1991; Xu *et al.*, 1996). This oxidizes P(III) to P(V) as in standard synthesis, but with sulfur replacing one of the nonbridging oxygens, producing a mixture of R_P and S_P

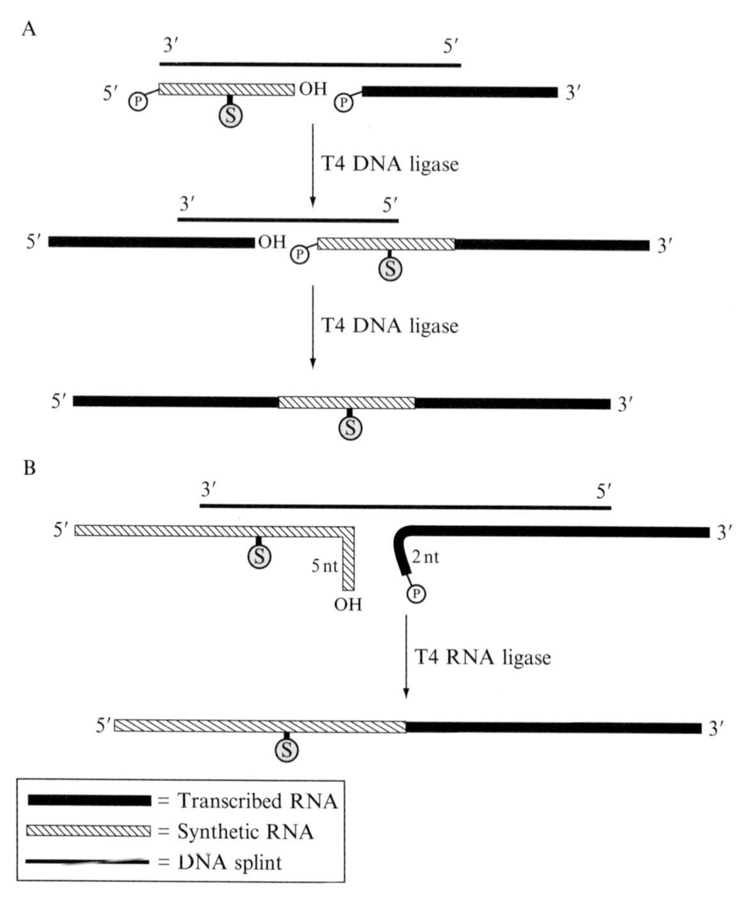

Figure 14.2 Semisynthetic construction of RNAs containing site- and diastereospecific phosphorothioate substitutions. (A) T4 DNA ligase-mediated ligation (Moore and Sharp, 1993). The RNAs to be joined together are held in apposition by a fully complementary bridging DNA splint. T4 DNA ligase recognizes the duplex as nicked, and catalyzes the attack of the free 3′-OH on the 5′-phosphate group. This reaction proceeds with relatively low turnover due to the slow rate of dissociation of the enzyme from the duplex. Consequently, preparation of large quantities of modified RNA often requires near-stoichiometric quantities of ligase. (B) T4 RNA ligase-mediated ligation (Stark *et al.*, 2006). For single-stranded DNA or RNA, this enzyme catalyzes the attack of any free 3′-OH on any 5′-phosphate group with multiple turnover. In ligation reactions, the RNAs to be joined together are held in place with a bridging DNA splint that places the ligation junction within a seven-nucleotide bulge. This bulge includes the last five nucleotides and free 3′-OH of the 5′-RNA, and the 5′-phosphate group and first two nucleotides of the 3′-RNA. T4 RNA ligase recognizes the free ends of the bulge as single stranded and ligates them together. To minimize cross reactions between the splint and 5′-phosphorylated RNA, these ligations may be performed with DNA splints containing a 3′-dideoxynucleotide.

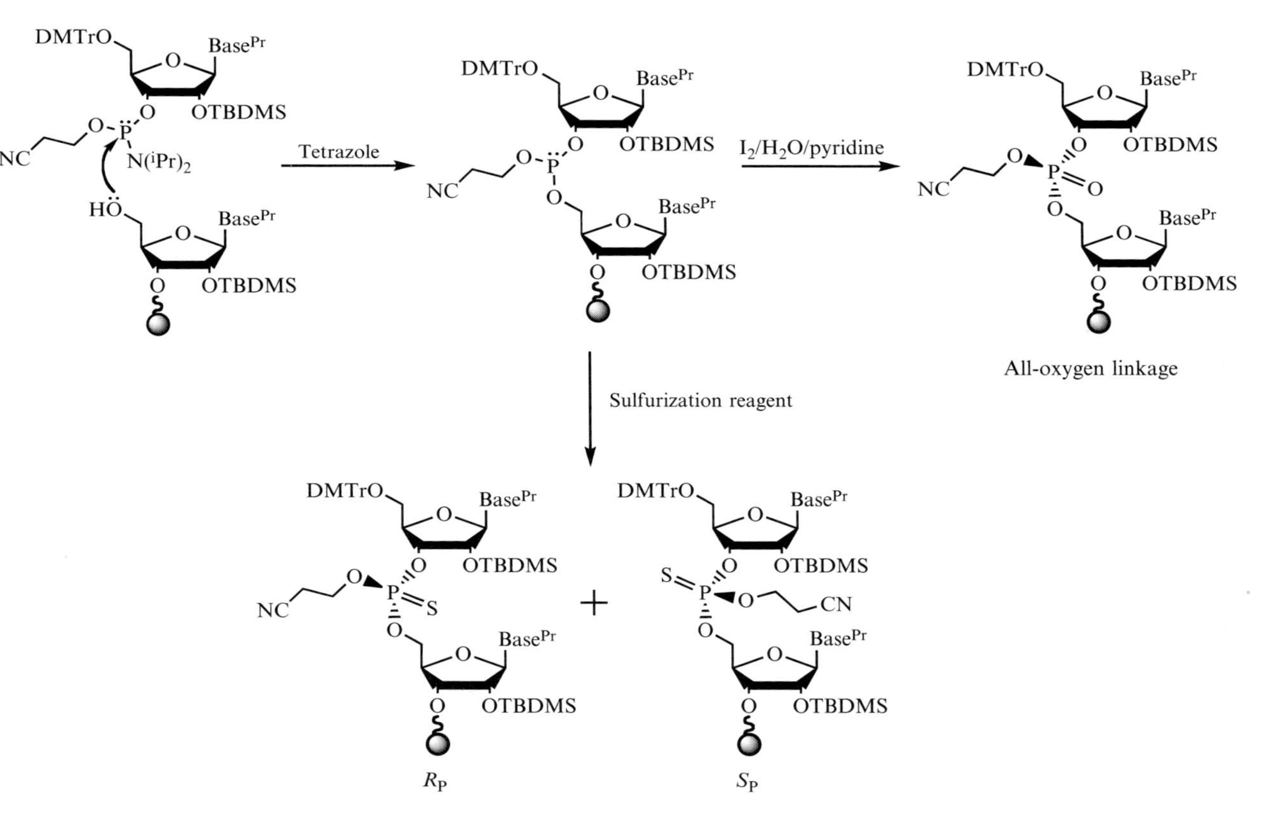

Tetrazole
I2/H2O/pyridine
All-oxygen linkage
Sulfurization reagent
Rp
Sp
Mixture of diastereomeric phosphorothioates

diastereomers that must be separated. Methods of introducing diastereospecific phosphorothioates do exist and have been adapted for use in automated synthesizers (Guga and Stec, 2003; Oka *et al.*, 2008, 2009; Fig. 14.4). However, these methods require the synthesis of new *P*-chiral monomer units and have not yet replaced the traditional approach outlined above. In time, they should prove instrumental for routine synthesis not only of single stereospecific phosphorothioates, but also of multiple such linkages within a single oligonucleotide. This is not easily achievable using standard synthetic protocols, since the introduction of each new phosphorothioate results in a more complex mixture of diastereomers whose separation becomes intractable.

2. HPLC Separation of Phosphorothioate Diastereomers

Once synthesis and deprotection of a phosphorothioate oligonucleotide is completed, HPLC is the method of choice for separation of the R_P and S_P diastereomers. Several examples exist in the literature for reversed phase (RP) HPLC separation of DNA phosphorothioates (Murakami *et al.*, 1994; Stec and Zon, 1984; Stec *et al.*, 1984, 1985). For RNA phosphorothioates, the work of Slim and Gait on synthetic hammerhead ribozyme substrates is often cited (Slim and Gait, 1991). This involved the synthesis and purification of a 13-nt RNA substrate containing a stereospecific phosphorothioate at the hammerhead cleavage site. The substrate was purified using both strong anion exchange (SAX) and RP HPLC protocols. With SAX using a potassium phosphate buffer system as an initial purification step, the diastereomers eluted as a single peak. In contrast, the R_P and S_P diastereomers eluted as two peaks separated by about 1 minute using RP HPLC with 0.1 *M* ammonium acetate and an increasing percentage of acetonitrile. Subsequent variations on this protocol, as well as Tris- and ammonium acetate-based SAX buffer systems, have proved successful for separating phosphorothioates of many different lengths and sequences (Table 14.1).

The need for multiple protocols and buffer systems underscores that while some general principles of HPLC may be applied to phosphorothioate

Figure 14.3 Solid-phase synthesis of phosphorothioate oligonucleotides. Following tetrazole-catalyzed addition of the incoming nucleoside phosphoramidite, treatment with a mixture of iodine and water in pyridine oxidizes the phosphite intermediate to form a phosphotriester linkage. Subsequent base-catalyzed removal of the β-cyanoethyl protecting group yields the natural phosphodiester (not shown). In phosphorothioate synthesis, a sulfurization reagent effects oxidation, producing a mixture of phosphorothioate diastereomers. Abbreviations: DMTr, dimethoxytrityl; $Base^{Pr}$, chemically protected nucleobase; TBDMS, *tert*-butyldimethylsilyl; $N(^iPr)_2$, *N,N*-diisopropylamino.

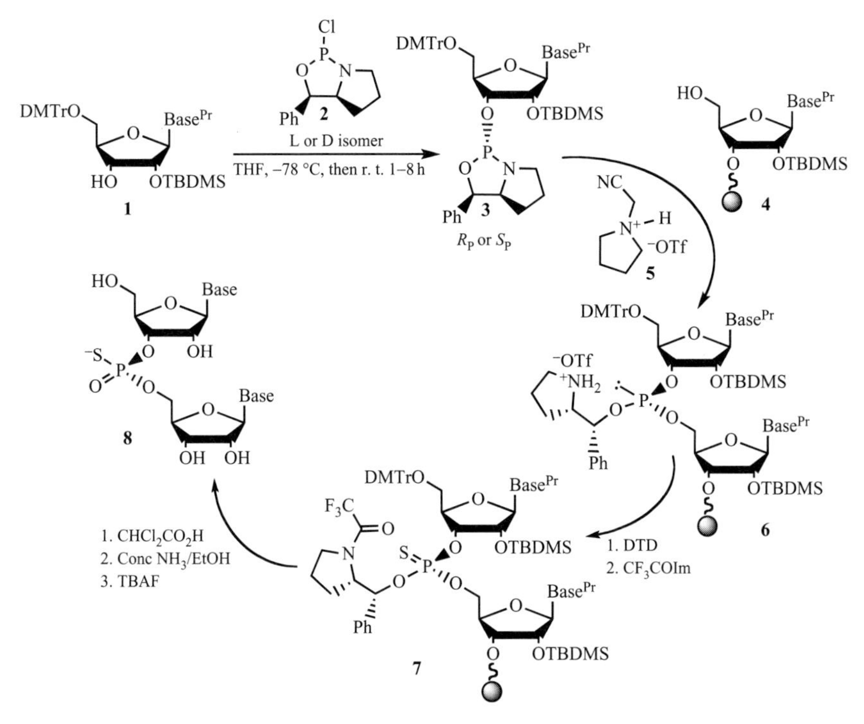

Figure 14.4 Synthesis of diastereospecific RNA phosphorothioates. *P*-chiral monomer units in the form of 3′-*O*-oxazaphospholidines (3) were synthesized by reacting 5′-DMTr- and 2′-TBDMS-protected nucleosides (1) with a stereospecific 2-chloro-1,3,2-oxazaphospholidine derivative 2. The monomers were coupled to nucleosides attached to a solid support (4) in the presence of the activator *N*-(cyanomethyl)pyrrolidinium triflate (CMPT) (5). Treatment of the resulting intermediate 6 with *N,N'*-dimethylthiuram disulfide (DTD) yielded the diastereospecific phosphorothioate. Acetylation of unreacted 5′-hydroxyl groups was accomplished with *N*-trifluoroacetylimidazole (CF_3COIm), which also acetylated the secondary amino group derived from CMPT to produce 7. Subsequent detritylation, cleavage from the solid support, and removal of TBDMS groups with tetra-*n*-butylammonium fluoride (TBAF) gave the free phosphorothioate oligonucleotide 8. Adapted from Oka *et al.*, 2008, 2009.

separations, establishing optimal conditions remains an empirical process. A successful separation depends on a multitude of variables—oligonucleotide length and sequence, phosphorothioate position, column type, buffer composition, slope of the elution gradient, and temperature, to name just a few. Consequently, whether and to what extent separation will occur cannot be determined with any degree of certainty, aside from the prediction that the phosphorothioates will have a longer retention time than the corresponding all-oxygen oligonucleotide. In our experience, separation of a seemingly inseparable phosphorothioate requires patience and a low threshold for switching to alternative HPLC conditions.

Table 14.1 Published HPLC conditions for separation of phosphorothioate oligoribonucleotides

Strong anion exchange		
Buffers	Length (nt)	Reference
A: 0.025 *M* tris-HCl pH 8.9 B: 0 to 900 m*M* NaCl	20	Hougland *et al.* (2005)
A: 0.010 *M* tris-HCl pH 9.3, 10% EtOH B: 20 to 1000 m*M* NaCl	17 19, 22	Wang *et al.* (1999) Gordon and Piccirilli (2001)
A: 0.1 *M* NH_4OAc, 2% CH_3CN, pH 8 B: 0 to 900 m*M* KCl	15, 16	Dertinger *et al.* (2000)
A: 0.025 *M* tris-HCl pH 7.4 B: 0 to 900 m*M* NaCl	8, 15	Forconi *et al.* (2008)
Reversed phase		
Buffers	Length (nt)	Reference
A: 0.1 *M* NH_4OAc B: 60% 0.1 *M* NH_4OAc/ 40% CH_3CN	7	Yoshida *et al.* (1999)
A: 0.1 *M* NH_4OAc B: 20% A/80% CH_3CN	11 13 17	Christian *et al.* (2000), Christian *et al.* (2006) Slim and Gait (1991) Peracchi *et al.* (1997), Wang *et al.* (1999)
A: 0.1 *M* NH_4OAc pH 7 B: CH_3CN	15, 16 19, 22	Dertinger *et al.* (2000) Gordon and Piccirilli (2001)
A: 0.1 *M* TEAAc pH 7 B: 20% 0.1 *M* TEAAc pH 7/80% CH_3CN	13	Warnecke *et al.* (1996)
A: 20 m*M* NH_4OAc pH 7, 2% CH_3CN B: 20 m*M* NH_4OAc pH 7, 20% CH_3CN	10	Crary *et al.* (2002)

2.1. Biochemical characterization

Basic biochemical characterization of a newly synthesized phosphorothioate prior to separation can save much time and vexation at the HPLC. On more than one occasion, we have tried to separate a particular phosphorothioate without success, only to discover subsequently that the oligonucleotide did not contain sulfur. Simple assays like alkaline hydrolysis, digestion with one or more RNases to assess sequence, and treatment with iodine to determine the position of the phosphorothioate linkage, can help avoid unnecessary optimization of HPLC conditions.

2.2. Phosphorothioate position

The position of the phosphorothioate within the oligonucleotide can influence the degree of separation achievable by HPLC. Generally, we have found that placing the phosphorothioate toward the center of the oligonucleotide tends to facilitate separation, although this is hardly a rigorous stipulation (see examples below). Some references report better resolution of DNA phosphorothioates when the phosphorothioate is located toward the 5′-end and the 5′-dimethoxytrityl (DMTr) protecting group is left in place (Stec and Zon, 1984; Zon, 1990). In RP HPLC, retaining the hydrophobic 5′-DMTr group increases the retention time of the full-length oligonucleotide, helping to differentiate it from the shorter side products associated with incomplete phosphoramidite coupling. While we have not tested this strategy for RNA phosphorothioates, the examples we present herein suggest that centrally and 3′-end-directed phosphorothioates can be readily separated.

2.3. Elution order

With regard to the order of elution from the column, the R_P diastereomer usually elutes before the S_P diastereomer on both SAX and RP HPLC. This observation has been made for diastereomers of both DNA (Bartlett and Eckstein, 1982; Potter *et al.*, 1983b; Romaniuk and Eckstein, 1982; Stec and Zon, 1984; Stec *et al.*, 1984; Uznanski *et al.*, 1982) and RNA (Burgers and Eckstein, 1978; Crary *et al.*, 2002; Peracchi *et al.*, 1997; Slim and Gait, 1991) phosphorothioates, and is often cited as empirical evidence of the correct assignment of configuration. This convention is not strictly rigorous, however, and in the case of DNA phosphorothioates, the elution order may be reversed if the 5′-DMTr group is retained to facilitate separation (Stec and Zon, 1984; Stec *et al.*, 1984). Ideally, the precise determination of configuration should be made on the basis of enzymatic digestion assays. In particular, R_P phosphorothioates are resistant to cleavage by nuclease P1

(Potter *et al.*, 1983b), while S_P phosphorothioates are resistant to cleavage by snake venom phosphodiesterase (SVP) (Burgers and Eckstein, 1978).

In our laboratory, we have separated many different phosphorothioate RNAs ranging in length from 6 to 25 nucleotides, using both SAX and RP HPLC. The remainder of this chapter describes several examples of these separations, along with their corresponding buffers and gradients. In addition, we present an example of how complementary DNA oligonucleotides can facilitate the separation of poorly resolved phosphorothioates.

3. Materials and Methods

3.1. Oligonucleotide sample preparation

Phosphorothioate oligonucleotides were purchased from Dharmacon, Inc. (Lafayette, CO) and deprotected according to the manufacturer's instructions. The lyophilized pellets were resuspended in 100 μL of water to a concentration of approximately 1–4 m*M*. For test runs, 500–1000 pmol of phosphorothioate oligonucleotide were mixed with water in a final volume of 100 μL and loaded onto the HPLC.

3.2. HPLC hardware

All separations were performed on either a Waters 600 or 2795 HPLC pump, connected to a Waters 996 or 2996 photodiode array, respectively. The pumps were controlled through a single computer running Waters Empower Pro software, which allowed both pumps to operate simultaneously. Oligonucleotide elution from the columns was followed by monitoring the absorbance of the eluent at 260 nm.

3.3. Columns

Separations by SAX employed a semipreparative Dionex DNAPac PA-100 9 × 250 mm anion exchange column running at a flow rate of 2 mL/min. For RP separations, a VyDAC small pore analytical C18 column (201SP54, 4.6 × 250 mm) was used at a flow rate of 1 mL/min. Although not described herein, some large-scale RP separations have been accomplished satisfactorily on a semipreparative VyDAC Dionex 201SP TM C18 column (201SP510, 10 × 250 mm) running at a flow rate of 2–3 mL/min.

Prior to each run, the columns were washed with 1–2 column volumes of buffer at the maximum eluent concentration associated with each system (SAX: 900 m*M* NaCl/25 m*M* Tris–HCl pH 8.9 or 900 m*M* KCl/100 m*M* NH_4OAc/2% CH_3CN pH 8.0; RP: 100% CH_3CN). The columns were then equilibrated with 1–2 column volumes of buffer at the starting conditions before the samples were loaded. Unless otherwise specified, all separations were conducted at room temperature (22–24 °C). In some instances, we have found that increasing the temperature to 45–65 °C can sharpen diastereomeric peaks and increase the separation time.

4. Examples of Phosphorothioate Oligonucleotide Separations

4.1. Solvent system 1: Strong anion exchange

- Solvent A: 0.25 *M* Tris–HCl pH 7.4–8.9
- Solvent B: H_2O
- Solvent C: 1 *M* NaCl

In this system, the percentage of solvent A is kept at 10% so that the background buffer concentration is 25 m*M* Tris–HCl. The gradient is created by gradually increasing the percentage of solvent C (the concentrated salt solution) in the mixture, with water making up the difference. Thus, the salt concentration can range from 0 to 0.9 *M* NaCl.

Figure 14.5 shows the successive optimization of HPLC conditions for the separation of a 25-nt RNA phosphorothioate. The initial gradient is relatively steep (+17.5 m*M* NaCl per minute, Fig. 14.3A) so as to determine the approximate salt concentration at which the diastereomers elute from the column. In most but not all instances, the observation of some separation even with this steep gradient indicates that further separation will be possible as conditions are refined. Following general HPLC principles, the separation time between the peaks should increase as the gradient becomes progressively shallower. In this case, starting at 450 m*M* NaCl and reducing the gradient to +2.5 m*M* NaCl per minute increased the separation time to 3.5 min (Fig. 14.5B). Further refinement of the gradient produced the trace in Fig. 14.5C, whose conditions were deemed suitable for preparative separation of the diastereomers.

Once the separation conditions have been optimized, larger quantities of phosphorothioate oligonucleotide may be loaded and separated. The peaks associated with these preparative runs are considerably broader and more asymmetrical compared to the peaks of the test runs (Fig. 14.6A). In addition to maximizing peak separation time during test runs, setting a baseline absorbance (usually 0.1–0.2 A_{260}) that defines the beginning and end of peak elution during preparative runs can help to avoid cross-contamination. Following preparative separation, excess salt must be removed from the pooled fractions of each phosphorothioate diastereomer. For oligonucleotides at least 10 bases in length, desalting may be accomplished by concentrating the oligonucleotides to dryness using a speedvac, resuspending in a

5′-P GGU AUG GUA *AUA AGC UGA CGG ACA U 3′

A

R_P

S_P

A_{260}

0.18 0.16 0.14 0.12 0.10 0.08 0.06 0.04 0.02 0.00

5.00 10.00 15.00 20.00 25.00 30.00 35.00 40.00 45.00 50.00

A: 0.25 *M* tris-HCl pH 8.9
B: H_2O
C: 1 *M* NaCl

Time (min)	%A	%B	%C
0	10	80	10
10	10	80	10
50	10	10	80

0.4 min separation

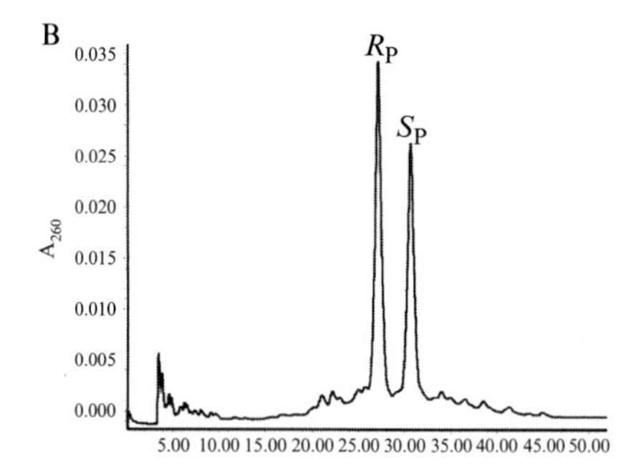

Time (min)	%A	%B	%C
0	10	45	45
10	10	45	45
50	10	35	55

3.5 min separation

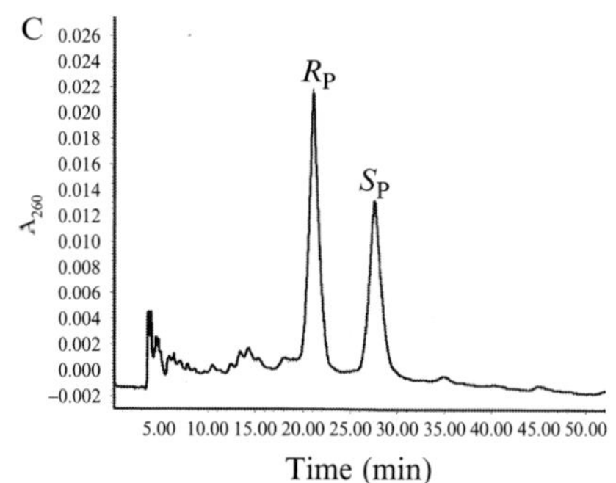

Time (min)	%A	%B	%C
0	10	43	47
10	10	43	47
50	10	40	50

6.5 min separation

Figure 14.5 Separation of phosphorothioate diastereomers of the 25-nt RNA 5′-P GGU AUG GUA ⋆AUA AGC UGA CGG ACA U-3′ (5′-P indicates the presence of a 5′-phosphate group, while ⋆ marks the position of the phosphorothioate linkage). A_{260} is the absorbance of the HPLC eluent measured at 260 nm. Conditions: Semi-preparative SAX column with a flow rate of 2 mL/min and a sodium chloride gradient of (A) 17.5 m*M*/min, (B) 2.5 m*M*/min, or (C) 0.75 m*M*/min. These gradients are approximations of the true rate of increase of the sodium chloride concentration, since the HPLC pumps utilized in these separations dispense solvents only in increments of 1%. Note that the axes do not show identical scales.

small volume of water ($\leq$1 mL), and passing over a gel filtration column such as a NAP-10 Sephadex G-25 column (GE Healthcare). Alternatively, the phosphorothioates may be desalted by passage through a Sep-Pak C18

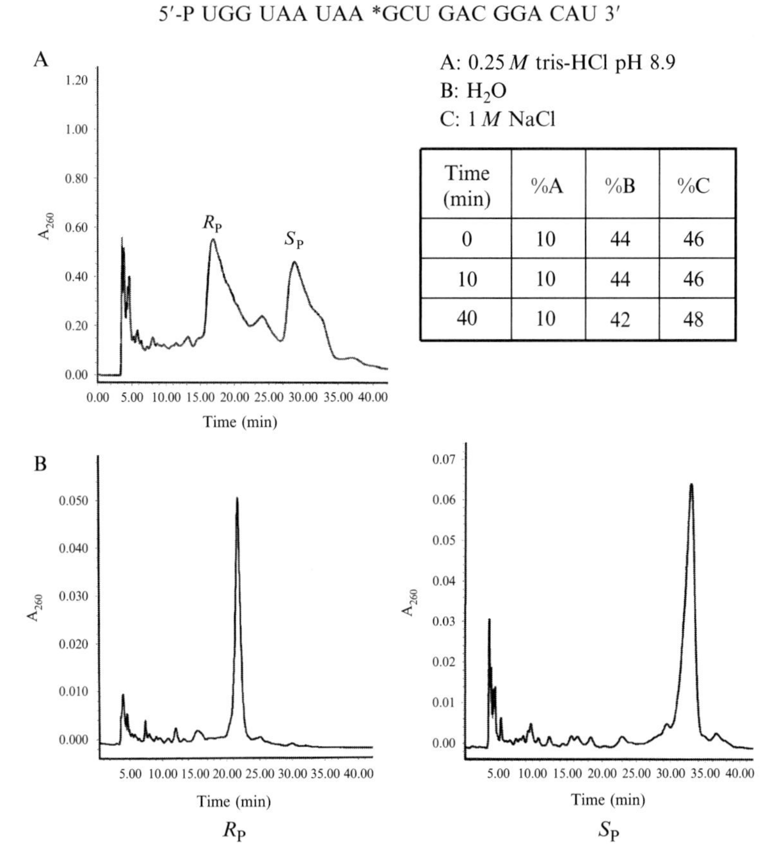

Time (min)	%A	%B	%C
0	10	44	46
10	10	44	46
40	10	42	48

Figure 14.6 Preparative separation of the 21-nt RNA 5′-P UGG UAA UAA ⋆GCU GAC GGA CAU-3′. Conditions: Semipreparative SAX column with a flow rate of 2 mL/min. (A) HPLC trace showing a representative preparative separation. (B) HPLC traces of individual R_P and S_P diastereomers following separation and desalting.

cartridge (Waters) or by dialyzing against 2 m*M* sodium citrate pH 6.5 at 4 °C (1:1000 volume differential with three buffer changes at 1–2 h intervals). These latter techniques are not as volume-dependent as gel filtration, and therefore may be more practical for desalting large volumes of eluted material. Oligonucleotides shorter than 10 bases should be desalted either via a Sep-Pak cartridge or by dialyzing, taking care to select dialysis tubing with an appropriate molecular weight cutoff value. After the desalted phosphorothioates have been concentrated, their purity should be assessed by HPLC (Fig. 14.6B).

As Fig. 14.7 shows, the gradients based on this solvent system may be adjusted to separate several different phosphorothioate oligonucleotides of

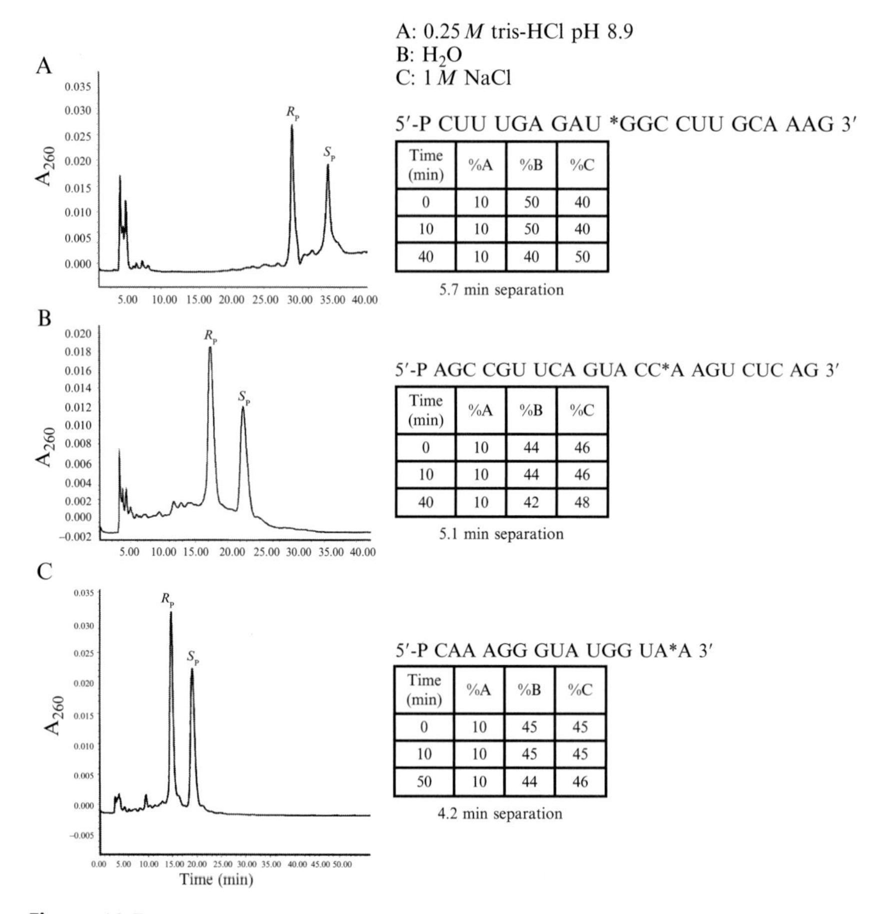

Time (min)	%A	%B	%C
0	10	50	40
10	10	50	40
40	10	40	50

Time (min)	%A	%B	%C
0	10	44	46
10	10	44	46
40	10	42	48

Time (min)	%A	%B	%C
0	10	45	45
10	10	45	45
50	10	44	46

Figure 14.7 Representative phosphorothioate oligonucleotides separated by using SAX HPLC. (A) 21 nt. (B) 23 nt. (C) 15 nt. Conditions: Semipreparative SAX column, flow rate 2 mL/min.

various sequences and lengths. We have used this system to separate phosphorothioates on the order of 15–25 nucleotides, although this range is relatively arbitrary.

4.2. Solvent system 2: Strong anion exchange

- Solvent A: 1 *M* NH_4OAc, 20% CH_3CN, pH 8.0
- Solvent B: H_2O
- Solvent C: 1 *M* KCl

This system may be appropriate for shorter phosphorothioates, up to about 15 nt (Dertinger *et al.*, 2000). Again, solvent A is kept at 10% throughout the run so that the background buffer concentration is 0.1 *M* NH_4OAc, 2% CH_3CN, pH 8.0. Figure 14.8 shows a representative separation for a 6-nt phosphorothioate. Concentration and desalting procedures are the same as for other SAX solvent systems, although NAP-10 columns should not be used for oligonucleotides shorter than 10 bases.

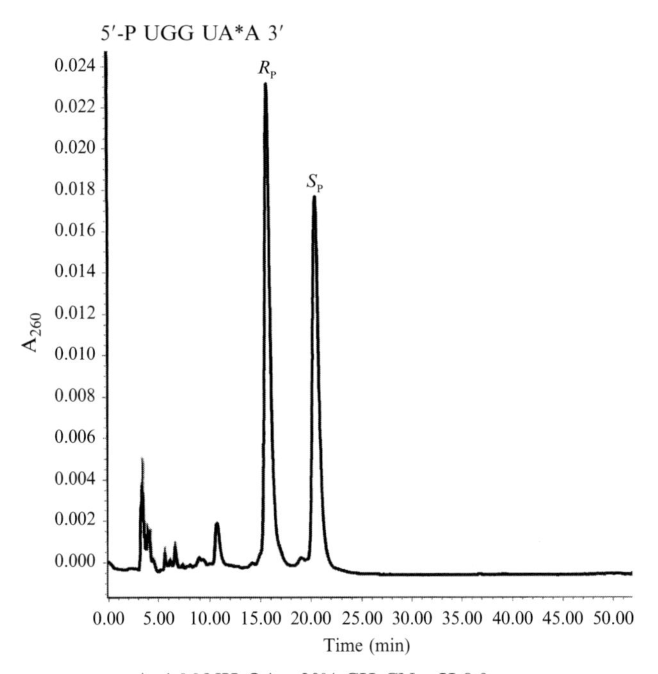

A: 1 *M* NH_4OAc, 20% CH_3CN, pH 8.0

B: H_2O

C: 1 *M* KCl

Time (min)	%A	%B	%C
0	10	70	20
10	10	70	20
50	10	63	27

4.8 min separation

Figure 14.8 Separation of phosphorothioate diastereomers of the 6-nt RNA 5′-PUGG UA*A-3′, using an alternative anion exchange solvent system. Conditions: Semipreparative SAX column, flow rate 2 mL/min.

4.3. Solvent system 3: Reversed phase

- Solvent A: 0.1 *M* NH_4OAc
- Solvent B: 20% A/80% CH_3CN

This is a widely used RP system that appears in the work of Slim and Gait (1991) and is suitable for separating phosphorothioates of many different lengths (Table 14.1). Figure 14.9 shows the RP preparative separation of a 15-nt phosphorothioate on an analytical C18 column. In this particular separation, elevating the temperature of the column to 45 °C improved the resolution of the diastereomers. Phosphorothioates separated by RP HPLC are ready to use following speedvac concentration and resuspension in water or buffer, and do not require a separate desalting procedure.

4.4. Using complementary DNA oligonucleotides to improve RNA phosphorothioate separation

In the event that suitable resolution of diastereomers cannot be achieved, we present an additional method that has yielded favorable results for certain intractable phosphorothioates. In this approach, first suggested by Daniel Herschlag and Alexander Kravchuk of Stanford University, the RNA phosphorothioate of interest is hybridized to a fully complementary DNA oligonucleotide prior to loading onto the HPLC (Kravchuk and Herschlag, unpublished results). The DNA and RNA oligonucleotides are combined such that the amount of complementary DNA strand equals or exceeds the amount of phosphorothioate RNA (for test runs, 500–1000 pmol of phosphorothioate oligonucleotide provide a satisfactory UV signal). One-tenth volume of TEN_{150} buffer (10 m*M* Tris–HCl pH 7.5, 1 m*M* EDTA, 150 m*M* NaCl) and water are added to a final buffer concentration of 1 m*M* Tris–HCl pH 7.5, 0.1 m*M* EDTA, and 15 m*M* NaCl (smaller volumes are preferable, on the order of 20–50 μL). The mixture is heated at 90 °C for 4 min, and then incubated at 37 °C for 1 h before loading onto the HPLC.

Figure 14.10 shows the SAX HPLC traces associated with a 15-nt RNA phosphorothioate both in the absence and in the presence of a complementary DNA oligonucleotide. Using the same gradient, the complementary DNA improves the separation of the phosphorothioate diastereomers dramatically. The peaks may then be collected, concentrated, and desalted in the same manner as described above for phosphorothioates separated via SAX HPLC. Following desalting, however, the complementary DNA is removed by treatment with RQ1 RNase-free DNAse (Promega) according to the manufacturer's instructions.

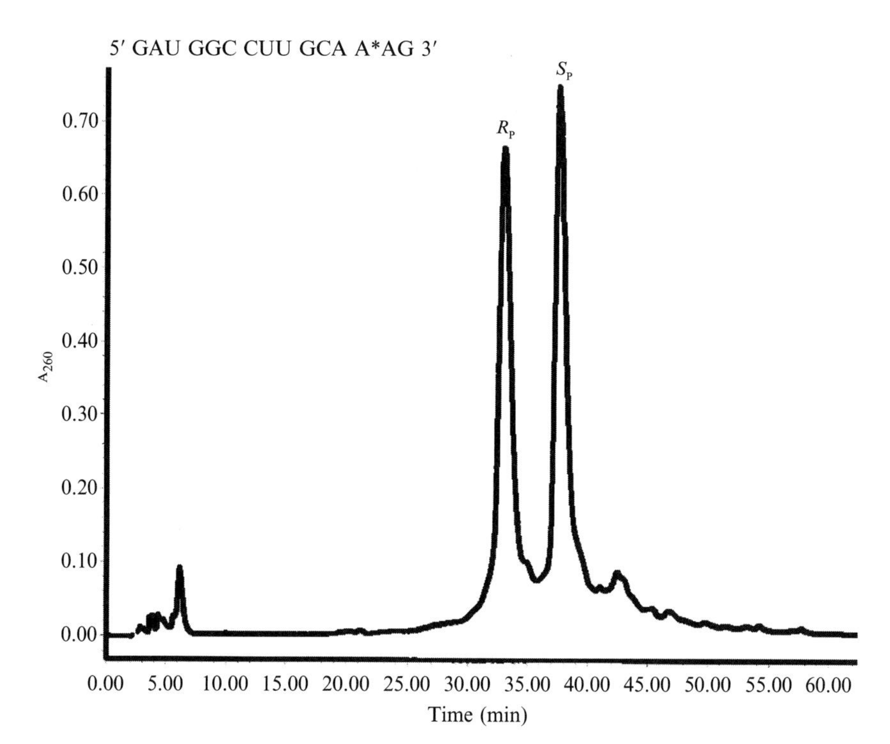

A: 0.1*M* NH_4OAc
B: 20% A/80% CH_3CN
45 °C

Time (min)	%A	%B
0	97	3
10	97	3
90	90	10

4.6 min separation

Figure 14.9 Preparative reversed phase separation of diastereomers of the 15-nt RNA 5′-GAU GGC CUU GCA A*AG-3′. Conditions: Analytical RP column, flow rate 1 mL/min, 45 °C.

4.5. Summary and Conclusion

Separation of oligonucleotide phosphorothioate diastereomers by HPLC is a trial-and-error process affected by a multitude of variables that no single set of conditions can control. We have therefore presented several different strategies that may be attempted should any one approach fail to resolve the

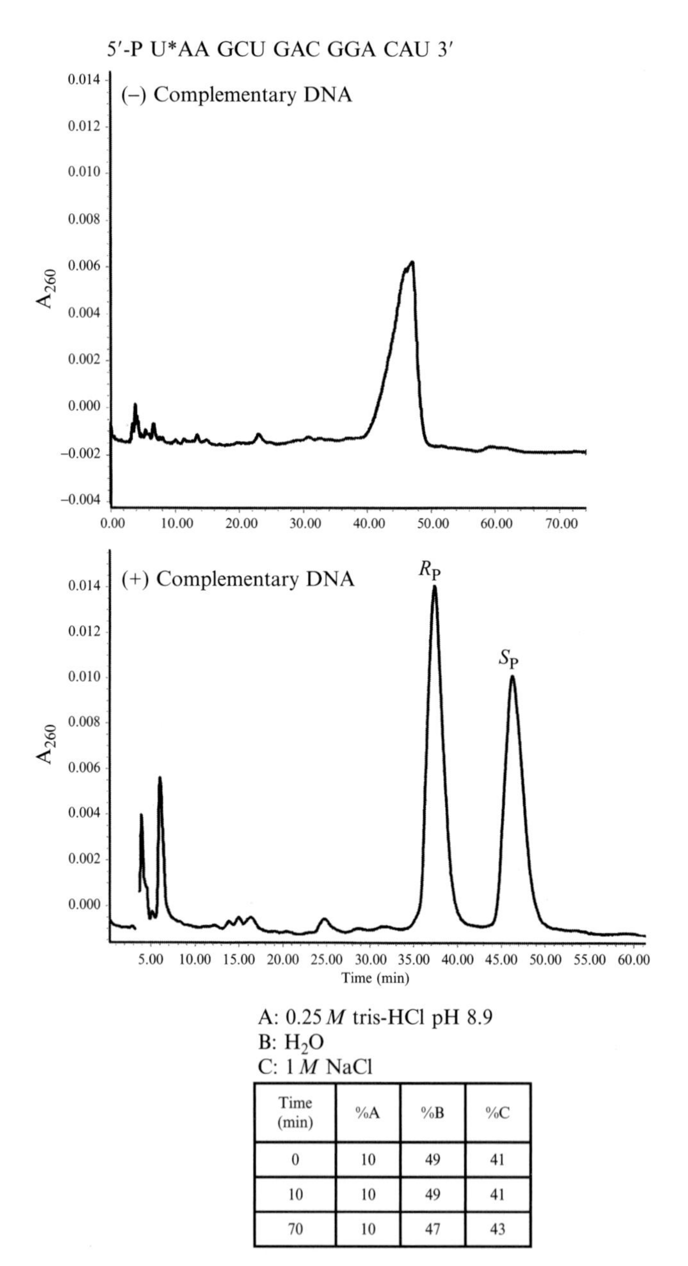

Time (min)	%A	%B	%C
0	10	49	41
10	10	49	41
70	10	47	43

Figure 14.10 Separation of phosphorothioate diastereomers of the 15-nt RNA 5′-P U*AA GCU GAC GGA CAU-3′. Both HPLC traces use the indicated gradient. Under these conditions, the diastereomers elute as a single asymmetric peak (top panel). However, the resolution improves considerably when the phosphorothioate is hybridized to a fully complementary DNA oligonucleotide prior to loading on the HPLC (bottom panel). While not apparent on this trace, the much shorter retention time of excess complementary DNA usually suffices to distinguish it from phosphorothioate RNA. Conditions: Semipreparative SAX column, flow rate 2 mL/min.

diastereomers. Our list is by no means comprehensive, but may be a convenient starting point from which optimal separation conditions may be found.

ACKNOWLEDGMENTS

The authors thank Dr. James Hougland, Dr. Edward Leung, Nikolai Suslov, and Nicole Tuttle for thoughtful discussions and comments on the manuscript. J.K.F. is supported by the Medical Scientist Training Program of the University of Chicago Pritzker School of Medicine.

REFERENCES

Akhtar, S., Kole, R., and Juliano, R. L. (1991). Stability of antisense DNA oligodeoxynucleotide analogs in cellular extracts and sera. *Life Sci.* **49,** 1793–1801.

Bartlett, P. A., and Eckstein, F. (1982). Stereochemical course of polymerization catalyzed by avian myeloblastosis virus reverse transcriptase. *J. Biol. Chem.* **257,** 8879–8884.

Burgers, P. M., and Eckstein, F. (1978). Absolute configuration of the diastereomers of adenosine 5′-O-(1thiotriphosphate): consequences for the stereochemistry of polymerization by DNA-dependent RNA polymerase from *Escherichia coli*. *Proc. Natl. Acad. Sci. USA* **75,** 4798–4800.

Burgers, P. M., Eckstein, F., and Hunneman, D. H. (1979). Stereochemistry of hydrolysis by snake venom phosphodiesterase. *J. Biol. Chem.* **254,** 7476–7478.

Chen, Y., Li, X., and Gegenheimer, P. (1997). Ribonuclease P catalysis requires Mg^{2+} coordinated to the *pro-Rp* oxygen of the scissile bond. *Biochemistry* **36,** 2425–2438.

Christian, E. L., Kaye, N. M., and Harris, M. E. (2000). Helix P4 is a divalent metal ion binding site in the conserved core of the ribonuclease P ribozyme. *RNA* **6,** 511–519.

Christian, E. L., Smith, K. M., Perera, N., and Harris, M. E. (2006). The P4 metal binding site in RNase P RNA affects active site metal affinity through substrate positioning. *RNA* **12,** 1463–1467.

Connolly, B. A., Potter, B. V., Eckstein, F., Pingoud, A., and Grotjahn, L. (1984). Synthesis and characterization of an octanucleotide containing the EcoRI recognition sequence with a phosphorothioate group at the cleavage site. *Biochemistry* **23,** 3443–3453.

Crary, S. M., Kurz, J. C., and Fierke, C. A. (2002). Specific phosphorothioate substitutions probe the active site of *Bacillus subtilis* ribonuclease P. *RNA* **8,** 933–947.

Dertinger, D., Behlen, L. S., and Uhlenbeck, O. C. (2000). Using phosphorothioate-substituted RNA to investigate the thermodynamic role of phosphates in a sequence specific RNA-protein complex. *Biochemistry* **39,** 55–63.

Eckstein, F., Schulz, H. H., Ruterjans, H., Haar, W., and Maurer, W. (1972). Stereochemistry of the transesterification step of ribonuclease T 1. *Biochemistry* **11,** 3507–3512.

Eckstein, F., Romaniuk, P. J., Heideman, W., and Storm, D. R. (1981). Stereochemistry of the mammalian adenylate cyclase reaction. *J. Biol. Chem.* **256,** 9118–9120.

Forconi, M., Lee, J., Lee, J. K., Piccirilli, J. A., and Herschlag, D. (2008). Functional identification of ligands for a catalytic metal ion in group I introns. *Biochemistry* **47,** 6883–6894.

Frey, P. A., and Sammons, R. D. (1985). Bond order and charge localization in nucleoside phosphorothioates. *Science* **228,** 541–545.

Gordon, P. M., and Piccirilli, J. A. (2001). Metal ion coordination by the AGC triad in domain 5 contributes to group II intron catalysis. *Nat. Struct. Biol.* **8,** 893–898.

Guga, P., and Stec, W. J. (2003). Synthesis of phosphorothioate oligonucleotides with stereodefined phosphorothioate linkages. *In* "Current Protocols in Nucleic Acid Chemistry," (S. L. Beaucage, ed.), Chp. 4, Unit 4.17. John Wiley & Sons, New York.

Hoke, G. D., Draper, K., Freier, S. M., Gonzalez, C., Driver, V. B., Zounes, M. C., and Ecker, D. J. (1991). Effects of phosphorothioate capping on antisense oligonucleotide stability, hybridization and antiviral efficacy versus herpes simplex virus infection. *Nucleic Acids Res.* **19,** 5743–5748.

Hougland, J. L., Kravchuk, A. V., Herschlag, D., and Piccirilli, J. A. (2005). Functional identification of catalytic metal ion binding sites within RNA. *PLoS Biol.* **3,** e277.

Iyer, R. P., Phillips, L. R., Egan, W., Regan, J. B., and Beaucage, S. L. (2002). The automated synthesis of sulfur-containing oligodeoxyribonucleotides using 3H-1,2-benzodithiol-3-one 1,1-dioxide as a sulfur-transfer reagent. *J. Org. Chem.* **55,** 4693–4699.

Kravchuk, A. V., and Herschlag, D., unpublished results.

McSwiggen, J. A., and Cech, T. R. (1989). Stereochemistry of RNA cleavage by the Tetrahymena ribozyme and evidence that the chemical step is not rate-limiting. *Science* **244,** 679–683.

Moore, M. J., and Sharp, P. A. (1993). Evidence for two active sites in the spliceosome provided by stereochemistry of pre-mRNA splicing. *Nature* **365,** 364–368.

Murakami, A., Tamura, Y., Wada, H., and Makino, K. (1994). Separation and characterization of diastereoisomers of antisense oligodeoxyribonucleoside phosphorothioates. *Anal. Biochem.* **223,** 285–290.

Oka, N., Yamamoto, M., Sato, T., and Wada, T. (2008). Solid-phase synthesis of stereoregular oligodeoxyribonucleoside phosphorothioates using bicyclic oxazaphospholidine derivatives as monomer units. *J. Am. Chem. Soc.* **130,** 16031–16037.

Oka, N., Kondo, T., Fujiwara, S., Maizuru, Y., and Wada, T. (2009). Stereocontrolled synthesis of oligoribonucleoside phosphorothioates by an oxazaphospholidine approach. *Org. Lett.* **11,** 967–970.

Osborne, E. M., Schaak, J. E., and Derose, V. J. (2005). Characterization of a native hammerhead ribozyme derived from schistosomes. *RNA* **11,** 187–196.

Padgett, R. A., Podar, M., Boulanger, S. C., and Perlman, P. S. (1994). The stereochemical course of group II intron self-splicing. *Science* **266,** 1685–1688.

Peracchi, A., Beigelman, L., Scott, E. C., Uhlenbeck, O. C., and Herschlag, D. (1997). Involvement of a specific metal ion in the transition of the hammerhead ribozyme to its catalytic conformation. *J. Biol. Chem.* **272,** 26822–26826.

Potter, B. V., Romaniuk, P. J., and Eckstein, F. (1983a). Stereochemical course of DNA hydrolysis by nuclease S1. *J. Biol. Chem.* **258,** 1758–1760.

Potter, B. V., Connolly, B. A., and Eckstein, F. (1983b). Synthesis and configurational analysis of a dinucleoside phosphate isotopically chiral at phosphorus. Stereochemical course of Penicillium citrum nuclease P1 reaction. *Biochemistry* **22,** 1369–1377.

Romaniuk, P. J., and Eckstein, F. (1982). A study of the mechanism of T4 DNA polymerase with diastereomeric phosphorothioate analogues of deoxyadenosine triphosphate. *J. Biol. Chem.* **257,** 7684–7688.

Shan, S., Kravchuk, A. V., Piccirilli, J. A., and Herschlag, D. (2001). Defining the catalytic metal ion interactions in the Tetrahymena ribozyme reaction. *Biochemistry* **40,** 5161–5171.

Slim, G., and Gait, M. J. (1991). Configurationally defined phosphorothioate-containing oligoribonucleotides in the study of the mechanism of cleavage of hammerhead ribozymes. *Nucleic Acids Res.* **19,** 1183–1188.

Song, Q., Wang, Z., and Sanghvi, Y. S. (2003). A short, novel, and cheaper procedure for oligonucleotide synthesis using automated solid phase synthesizer. *Nucleosides Nucleotides Nucleic Acids* **22,** 629–633.

Stark, M. R., Pleiss, J. A., Deras, M., Scaringe, S. A., and Rader, S. D. (2006). An RNA ligase-mediated method for the efficient creation of large, synthetic RNAs. *RNA* **12,** 2014–2019.

Stec, W. J., and Zon, G. (1984). Synthesis, separation, and stereochemistry of diastereomeric oligodeoxyribonucleotides having a 5′-terminal internucleotide phosphorothioate linkage. *Tetrahedron Lett.* **25,** 5275–5278.

Stec, W. J., Zon, G., and Egan, W. (1984). Automated solid-phase synthesis, separation, and stereochemistry of phosphorothioate analogs of oligodeoxyribonucleotides. *J. Am. Chem. Soc.* **106,** 6077–6079.

Stec, W. J., Zon, G., and Uznanski, B. (1985). Reversed-phase high-performance liquid chromatographic separation of diastereomeric phosphorothioate analogues of oligodeoxyribonucleotides and other backbone-modified congeners of DNA. *J. Chromatogr.* **326,** 263–280.

Stec, W. J., Uznanski, B., Wilk, A., Hirschbein, B. L., Fearon, K. L., and Bergot, B. J. (1993). Bis(O,Odiisopropoxy phosphinothioyl) disulfide -a highly efficient sulfurizing reagent for cost-effective synthesis of oligo(nucleoside phosphorothioate)s. *Tetrahedron Lett.* **34,** 5317–5320.

Stein, C. A., and Cheng, Y. C. (1993). Antisense oligonucleotides as therapeutic agents–is the bullet really magical? *Science* **261,** 1004–1012.

Uznanski, B., Niewiarowski, W., and Stec, W. J. (1982). The chemical synthesis of the RP and SP diastereomers of thymidyl(3′-5′)thymidyl 0,0-phosphorothioate. *Tetrahedron Lett.* **23,** 4289–4292.

Vu, H., and Hirschbein, B. L. (1991). Internucleotide phosphite sulfurization with tetraethylthiuram disulfide. Phosphorothioate oligonucleotide synthesis via phosphoramidite chemistry. *Tetrahedron Lett.* **32,** 3005–3008.

Wang, S., Karbstein, K., Peracchi, A., Beigelman, L., and Herschlag, D. (1999). Identification of the hammerhead ribozyme metal ion binding site responsible for rescue of the deleterious effect of a cleavage site phosphorothioate. *Biochemistry* **38,** 14363–14378.

Wang, L., Chen, S., Xu, T., Taghizadeh, K., Wishnok, J. S., Zhou, X., You, D., Deng, Z., and Dedon, P.C (2007). Phosphorothioation of DNA in bacteria by dnd genes. *Nat. Chem. Biol.* **3,** 709–710.

Warnecke, J. M., Furste, J. P., Hardt, W. D., Erdmann, V. A., and Hartmann, R. K. (1996). Ribonuclease P (RNase P) RNA is converted to a Cd2+-ribozyme by a single RP-phosphorothioate modification in the precursor tRNA at the RNase P cleavage site. *Proc. Natl. Acad. Sci. USA* **93,** 8924–8928.

Xu, Q., Barany, G., Hammer, R. P., and Musier-Forsyth, K. (1996). Efficient introduction of phosphorothioates into RNA oligonucleotides by 3-ethoxy-1,2,4-dithiazoline-5-one (EDITH). *Nucleic Acids Res.* **24,** 3643–3644.

Yoshida, A., Sun, S., and Piccirilli, J. A. (1999). A new metal ion interaction in the Tetrahymena ribozyme reaction revealed by double sulfur substitution. *Nat. Struct. Biol.* **6,** 318–321.

Zon, G. (1990). Purification of synthetic oligodeoxyribonucleotides. *In* "High Performance Liquid Chromatography in Biotechnology," (W. S. Hancock, ed.), pp. 301–397. J. Wiley and Sons, New York.

CHAPTER FIFTEEN

Use of Phosphorothioates to Identify Sites of Metal-Ion Binding in RNA

Marcello Forconi* *and* Daniel Herschlag†

Contents

Abstract

Single atom substitutions provide an exceptional opportunity to investigate RNA structure and function. Replacing a phosphoryl oxygen with a sulfur represents one of the most common and powerful single atom substitutions and can be used to determine the sites of metal-ion binding. Using functional assays of ribozyme catalysis, based on pre-steady-state kinetics, it is possible to extend this analysis to the transition state, capturing ligands for catalytic metal ions in this fleeting state. In conjunction with data determined from X-ray crystallography, this technique can provide a picture of the environment surrounding catalytic metal ions in both the ground state and the transition state at atomic resolution. Here, we describe the principles of such analysis, explain limitations of the method, and provide a practical example based on our results with the *Tetrahymena* group I ribozyme.

* Department of Biochemistry, Stanford University, Stanford, California, USA
† Departments of Biochemistry and Chemistry, Stanford University, Stanford, California, USA

Methods in Enzymology, Volume 468
ISSN 0076-6879, DOI: 10.1016/S0076-6879(09)68015-0

1. Introduction

The ability of RNA to express its biological function depends on the presence of divalent metal ions. Most of the divalent metal ions associated with RNA are loosely bound, in what is commonly referred to as the "ion atmosphere." These ions provide most of the positive charge needed to neutralize the enormous negative charge present on RNA. A small number of metal ions localize to particular regions of the RNA molecule and typically stabilize particular motifs and long-range contacts. For RNA enzymes, or ribozymes, metal ions may participate to the chemical transformation either directly or indirectly (see Chapter 5 of this volume).

Understanding how metal ions contribute to ribozyme function requires the identification of specific interactions made between metal ions and the RNA molecule and the assessment of the importance and role of these contacts for RNA function. X-ray crystallography provides a powerful technique to identify interactions, allowing the detection of multiple metal ions and their interactions in a single picture. Nevertheless, this picture often changes when conditions are changed, and it is not possible to establish the functionally relevant structure or the functional importance of a particular contact on the basis of structural data alone.

To address these key points functional data are needed. The use of single phosphorothioate-containing ribozymes, in conjunction with kinetic assays that assess the reactivity of these ribozymes in the presence of metal ions with high affinity for sulfur, provides a powerful means to address these key points, complementing and expanding the information obtained from structural methods. These approaches, widely applied to RNA enzymes (e.g., see Christian, 2005), were originated for studies of protein enzymes by Cohn *et al.* (1982) and Eckstein (1983).

2. Use of Phosphorothioate-Containing Ribozymes to Identify Sites of Metal-Ion Binding

2.1. Assays based on phosphorothioate substitutions on the ribozyme

Mg^{2+} is the most common cation found in RNA structures and is often found to contact an oxygen atom, typically the 2′-OH group or the negatively charged nonbridging oxygen atoms of RNA phosphoryl groups. This is not surprising, as studies with model compounds have shown that Mg^{2+} strongly prefers oxygen ligands over other type of ligands. It has been estimated that Mg^{2+} prefers oxygen over sulfur by a factor of ~31,000-fold (Pecoraro *et al.*, 1984). This behavior is recapitulated by the hard and soft acids and bases (HSAB) theory developed by Pearson (Parr and Pearson, 1983; Pearson, 1963), stating that small, nonpolarizable ("hard") Lewis acids strongly interact with small, nonpolarizable Lewis bases, while large, highly polarizable ("soft") Lewis acids interact more strongly with large, highly polarizable Lewis bases. Mg^{2+} is an example of a hard Lewis acid, and oxygen ligands are hard Lewis bases; conversely, nitrogen- and sulfur-containing ligands are softer Lewis bases than those containing oxygen and interact preferentially with soft metal ions.

Because of this marked preference, a logical strategy to test putative contacts between a specific phosphoryl oxygen atom and a Mg^{2+} ion is to substitute the oxygen ligand with sulfur, introducing a thiophosphoryl group on the RNA backbone. This substitution may prevent Mg^{2+} binding, and in this case the simplest expectation is that the phosphorothioate-containing ribozyme would be significantly less reactive than the unmodified ribozyme. This situation is shown in Fig. 15.1A, where the empty circle represents the lack of a metal ion in the ground state and in the transition state of a hypothetical reaction. However, a decrease in reactivity of the modified ribozyme can arise from factors other than perturbed metal-ion binding. For example, the modified ribozyme may be defective in reactivity because of steric clashes introduced by the sulfur atom, as sulfur is larger than the native oxygen (Shannon, 1976), or because of the different hydrogen-bond properties of oxygen and sulfur (Pecoraro *et al.*, 1984; Platts *et al.*, 1996; Zhou *et al.*, 2009). In other words, there are many ways to compromise activity.

To test for effects not specific to the metal ion–sulfur contact, the reactivity of the phosphorothioate-containing ribozyme is followed as a function of a metal ion for which the preference for oxygen is not as marked as it is for Mg^{2+} or a metal ion that prefer sulfur ligands over oxygen ligands. Mn^{2+} and Cd^{2+} are two metal ions representative of these classes, respectively (Pecoraro *et al.*, 1984). Usually, these metal ions are referred to as "more thiophilic" (or "softer") than Mg^{2+}. Other metal ions used in metal ion rescue experiments include Co^{2+} and Zn^{2+}. If the functional effect is due to the lack of metal ion, because of the poor affinity of Mg^{2+} for sulfur, more thiophilic metal ions can rescue the reactivity; this rescue depends on the metal ion being able to bind in place of Mg^{2+} to the modified site. This expectation is often fulfilled in ribozymes. In the following paragraphs, we will use Cd^{2+} to describe the metal ion rescue experiments for simplicity and because this cation has been widely used (along with Mn^{2+}) in such experiments.

The simple scenario of Fig. 15.1A does not take into account the complexity of an RNA molecule. RNA-bound metal ions often have more than one inner-sphere ligands, and substituting one phosphoryl oxygen with sulfur may not be enough to remove Mg^{2+} from its binding site.

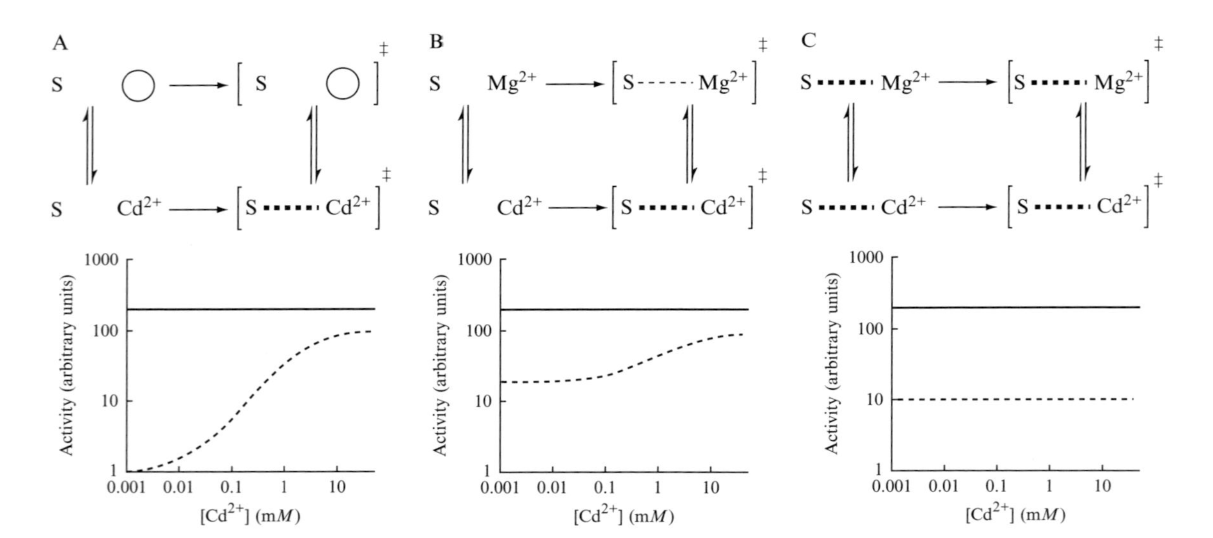

Figure 15.1 Possible situations upon the introduction of a thiophosphoryl group in the Mg^{2+}-binding site of a catalytic RNA. Upper panels: schematic representation of the interaction between the metal ion (if present) and the sulfur atom of the thiophosphoryl group in the ground state and the transition state of the reaction. Lower panels: predicted activity (arbitrary units) as a function of Cd^{2+} for the ribozyme containing the thiophosphoryl group (dashed line); the activity of the unmodified ribozyme is shown for comparison as a solid line. (A) The ribozyme containing the thiophosphoryl group is not able to bind Mg^{2+} at its original site. (B) The ribozyme containing the thiophosphoryl group can bind either Mg^{2+} or Cd^{2+}, and the contact between the metal ion and the sulfur atom is formed in the transition state only; the thicker dashes represent the stronger interaction of sulfur with Cd^{2+}. (C) The ribozyme containing the thiophosphoryl group can bind either Mg^{2+} or Cd^{2+}, and the same contact between the sulfur atom and the metal ion is formed in both the ground state and the transition state.

In this case, two more scenarios are possible, as shown in Fig. 15.1B and C. In Fig. 15.1B, Mg^{2+} is not removed upon the phosphorothioate substitution, but the contact between the sulfur atom of the ligand and this metal ion becomes stronger in the transition state (compared to the ground state). This could be due to a movement of the ligand as the reaction proceeds or, in the extreme case, to a contact that is formed only in the transition state. When Cd^{2+} occupies the metal-ion binding site, there is advantage over Mg^{2+} in the transition state, because of the stronger interaction that can be formed between Cd^{2+} and sulfur, compared to Mg^{2+} and sulfur. In this case, the activity of the modified ribozyme in Cd^{2+} is predicted to increase compared to Mg^{2+}.

In the scenario of Fig. 15.1C, the contact between the metal ion and the sulfur atom is formed in both the ground state and the transition state, regardless of the nature of the metal ion present in the binding site. In this case, there may be no advantage in having Cd^{2+} over Mg^{2+} in the metal-ion binding site, and the activity of the modified ribozyme is predicted, in the simplest case, to remain the same even if Cd^{2+} binds to the Mg^{2+}-binding site. This lack of a Cd^{2+} effect on the activity is analogous to what expected if the introduced sulfur atom does not contact a metal ion or if the metal ion cannot bind anymore at its site because of the introduction of a sulfur atom. Therefore, an alternative approach has to be undertaken to distinguish between these situations and to functionally identify the site in question as of metal-ion binding. One such approach is described in Section 2.2.

Thiophilic metal ions can affect steps other than the chemical step. For example, in the *Tetrahymena* group I ribozyme, Mn^{2+} stimulates a conformational change, speeding reactions that involve this conformational change (Shan and Herschlag, 2000). Further, in some cases a nonchemical rate-limiting step, which is rescued by Cd^{2+}, has been observed for the modified but not the unmodified ribozymes (Hougland *et al.*, 2005; M.F. and D.H. unpublished results). In these cases, the observation of an increase in the activity of the modified ribozyme when Mn^{2+} or Cd^{2+} is added may not be indicative of involvement of a metal ion important for the chemical transformation, as these ribozymes may react from a ground state that is different from that of the unmodified ribozyme or have an additional nonchemical rate-limiting step. Thus, it is crucial to perform appropriate controls to test that the unmodified and modified ribozymes react from the same ground state, and detailed pre-steady-state kinetic analysis is typically required.

Population experiments, such as nucleotide analog interference mapping (NAIM), whereby pools of ribozymes containing a single phosphorothioate substitution are prepared by *in vitro* transcription and their activity assayed in "standard conditions", have been extremely powerful for initial screening of candidates for metal-ion binding (Christian and Yarus, 1993; Harris and Pace, 1995; Jansen *et al.*, 2006; Kazantsev and Pace, 1998; Luptak and

Doudna, 2004; Sood *et al.*, 1998; Strauss-Soukup and Strobel, 2000). Indeed, such information can be obtained even when crystal structures are lacking. Nevertheless, to strongly implicate a particular phosphoryl oxygen in metal-ion binding, the phosphorothioate-containing ribozyme must be investigated in detailed kinetics experiments that take into account the concerns about the individual reaction steps being monitored. In principle, such experiments can be carried out on populations of molecules (Hertel *et al.*, 1996), but analysis of such data would be difficult and has not been fully developed.

2.2. Coupling phosphorothioate substitution on the ribozyme with atomic substitutions on the substrate

As described above, a situation in which the phosphorothioate substitution does not disrupt binding of Mg^{2+}, shown in Fig. 15.1C, is predicted to result in no change of activity of the modified ribozyme over increasing Cd^{2+} concentration. In terms of reactivity, this situation would be equivalent to a situation in which the sulfur atom introduces steric clashes, but does not contact a metal ion and thus the reactivity of the modified ribozyme cannot be rescued by Cd^{2+} addition, or a situation in which Cd^{2+} cannot substitute for Mg^{2+} at the metal-ion binding site. To distinguish among these possibilities, one needs a way to measure the affinity of the rescuing Cd^{2+} and compare with the unmodified ribozyme. If the sulfur atom of the introduced thiophosphoryl group contacts the rescuing Cd^{2+}, this affinity is expected to be higher for the phosphorothioate-containing ribozyme (compared to the unmodified ribozyme) because of the favorable Cd^{2+}–sulfur contact. Conversely, if the sulfur atom introduced on the ribozyme is deleterious for reasons other than metal-ion coordination, the affinity of the rescuing Cd^{2+} is predicted to be the same of that of the unmodified ribozyme, because no change in ligand coordination is expected between the modified and unmodified ribozymes. Nevertheless, more complex scenarios are also possible that include, for example, alterations in binding site geometry.

One way to measure the affinity for such "hidden" metal ions, and to obtain additional atomic-level information, involves coupling phosphorothioate substitution on the ribozyme with an analogous substitution on the substrate. An example is given below with the *Tetrahymena* group I ribozyme.

2.2.1. Thermodynamic Fingerprint Analysis (TFA) using the *Tetrahymena* group I ribozyme

The *Tetrahymena* group I ribozyme catalyzes a nucleotidyl transfer reaction from an oligonucleotide substrate (S), which mimics the natural 5′-splice site, to an exogenous guanosine (G), which serves as the nucleophile; the reaction is analogous to the first step of group I intron self-splicing (Eq. (15.1); Herschlag and Cech, 1990; Zaug *et al.*, 1988).

$$\underset{(S)}{\text{CCCUCUA}} + \text{G}_{\text{OH}} \rightarrow \underset{(P)}{\text{CCCUCU}_{\text{OH}}} + \text{GA} \tag{15.1}$$

Early metal-ion rescue experiments identified four transition state contacts between metal ions and atoms on the substrates through use of thio- and amino-substituted substrates and provided evidence against metal ion interactions at other positions (Liao *et al.*, 2001; Piccirilli *et al.*, 1993; Rajagopal *et al.*, 1989; Shan and Herschlag, 1999; Sjogren *et al.*, 1997; Weinstein *et al.*, 1997; Yoshida *et al.*, 1999, 2000). The substrate atoms involved in metal ion coordination are shown in Fig. 15.2.

The number of metal ions making contacts with these atoms can be determined by a quantitative approach called "Thermodynamic Fingerprint Analysis" (Shan and Herschlag, 1999; Shan *et al.*, 2001; Wang *et al.*, 1999), or TFA, which is described below.

Consider two substrates, S_o and S_x, corresponding to an unmodified and sulfur- or nitrogen-modified substrate. For example, S_x can represent an oligonucleotide substrate with a 3′-bridging sulfur on the leaving group (Piccirilli *et al.*, 1993). Schemes 15.1A and B show generic reaction frameworks for the unmodified (or cognate) and modified substrates, respectively. In Schemes 15.1A and B, the reaction rate reflects the fraction of ribozyme

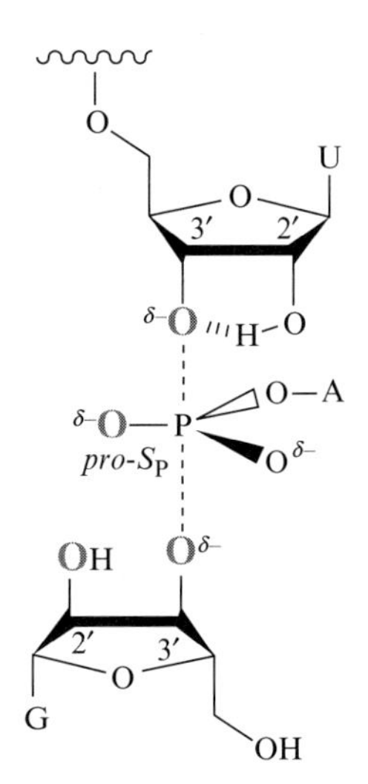

Figure 15.2 Schematic representation of the transition state for the reaction of the *Tetrahymena* group I ribozyme. Oxygen atoms indicated in bold have been linked to metal ion coordination by sulfur substitution or, for the 2′-OH of guanosine, by amino group substitution.

A

$$
\begin{array}{ccc}
{}^{Mg}E + S_o & \xrightleftharpoons{{}^{Mg}k_{S_o}} & [{}^{Mg}E \cdot S_o]^{\ddagger} \longrightarrow \text{Products} \\
K_{d,app}^{Cd} \updownarrow & & \updownarrow K_{S_o}^{Cd\ddagger} \\
{}^{Cd}E + S_o & \xrightleftharpoons{\alpha * {}^{Mg}k_{S_o}} & [{}^{Cd}E \cdot S_o]^{\ddagger} \longrightarrow \text{Products}
\end{array}
$$

B

$$
\begin{array}{ccc}
{}^{Mg}E + S_x & \xrightleftharpoons{{}^{Mg}k_{S_x}} & [{}^{Mg}E \cdot S_x]^{\ddagger} \longrightarrow \text{Products} \\
K_{d,app}^{Cd} \updownarrow & & \updownarrow K_{S_x}^{Cd\ddagger} \\
{}^{Cd}E + S_x & \xrightleftharpoons{\beta * {}^{Mg}k_{S_x}} & [{}^{Cd}E \cdot S_x]^{\ddagger} \longrightarrow \text{Products}
\end{array}
$$

Scheme 15.1 General reaction frameworks for the unmodified (S_O, panel A) and modified (S_X, panel B) substrates with and without bound Cd^{2+}, in a background of Mg^{2+} . Charges on metal ions are omitted for clarity.

that has Cd^{2+} bound, with a rate enhancement with saturating Cd^{2+} given by the parameters α and β for S_o and S_x, respectively. The Cd^{2+} concentration dependence of the observed rate constants gives the dissociation constant for Cd^{2+}, $K_{d,app}^{Cd}$, as shown in Eqs. (15.2a) and (15.2b). Because experiments are carried out in a background of Mg^{2+} to ensure proper folding of the ribozyme, this value is an "apparent affinity" that depends on the Mg^{2+} concentration present in solution.

$$k_{obs}^{S_o} = {}^{Mg}k_{S_o} \left[\frac{K_{d,app}^{Cd} + \alpha[Cd^{2+}]}{K_{d,app}^{Cd} + [Cd^{2+}]} \right] \tag{15.2a}$$

$$k_{obs}^{S_x} = {}^{Mg}k_{S_x} \left[\frac{K_{d,app}^{Cd} + \beta[Cd^{2+}]}{K_{d,app}^{Cd} + [Cd^{2+}]} \right] \tag{15.2b}$$

There are two critical features of this analysis. First, $K_{d,app}^{Cd}$ represents binding of Cd^{2+} to the reaction's ground state, even though it is determined by following a reaction rate that represents transient attainment of the reaction's transition state; nevertheless, $K_{d,app}^{Cd}$ is linked to the transition state affinities, $K_{S_O}^{Cd\ddagger}$ and $K_{S_x}^{Cd\ddagger}$, via the thermodynamic cycle shown in Scheme 15.1, such that $K_{S_O}^{Cd\ddagger} = K_{d,app}^{Cd}/\alpha$ and $K_{S_X}^{Cd\ddagger} = K_{d,app}^{Cd}/\beta$. Second, the reaction's ground state is the free ribozyme and free substrates so that the substrate modification has no affect on the affinity of Cd^{2+} for the ribozyme. In practice, experiments are sometimes carried out with certain substrates bound, provided that control experiments have determined that the association of those substrates is not coupled to the binding of the rescuing metal ion, that is, that the binding of such substrates does not affect the dissociation constant for the rescuing metal ion (Forconi *et al.*, 2008; Hougland *et al.*, 2005, 2006; Shan *et al.*, 1999).

A complication in metal-ion rescue experiments with RNA arises from the large number of associated metal ions. Because of this, metal ions can bind at sites other than the one involved in rescue and affect activity. In many cases, high concentrations of thiophilic metal ions cause inhibition of the normal reaction, apparently by binding at one or more alternate sites.

There can also be stimulatory effects from occupancy of other sites. If these effects are the same with the normal and the modified substrate, they are readily eliminated by using a *relative* rate constant, k_{rel}, which is simply the ratio of rate constants for the modified and unmodified substrates at each Cd^{2+} concentration, as shown in Eq. (15.3). This situation often holds, and this approach has been used on multiple occasions (Forconi *et al.,* 2008; Gordon and Piccirilli, 2001; Shan and Herschlag, 1999; Shan *et al.,* 2001; Sun and Harris, 2007; Wang *et al.,* 1999). Further, experiments are usually carried out in a background of high Mg^{2+} to minimize the effects from binding of the soft metal ions at other sites. Nevertheless, the possibility of binding of thiophilic metal ions at alternate sites that give differential effects on reactions of the cognate and modified substrates often cannot be eliminated and cannot be accounted for in a straightforward manner. In such cases, rescue experiments must look for consistency among multiple probes and experiments.

$$k_{rel} = {}^{Mg}k_{rel}\left[\frac{K_{d,app}^{Cd} + \beta[Cd^{2+}]}{K_{d,app}^{Cd} + \alpha[Cd^{2+}]}\right]; \quad {}^{Mg}k_{rel} = \frac{{}^{Mg}k_{S_x}}{{}^{Mg}k_{S_o}} \tag{15.3}$$

We return to the straightforward case of an unmodified ribozyme in which Cd^{2+} does not affect the reactivity of the unmodified substrate, S_o, except for nonspecific inhibition at high Cd^{2+} concentration that can be taken into account by the use of k_{rel} (Shan *et al.,* 2001); therefore, the parameter α in Scheme 15.1 and Eq. (15.3) is equal to 1. This simplifies Eq. (15.3) to Eq. (15.4). In Eq. (15.4), ${}^{Mg}k_{rel}$ and β can be easily determined, because they represent the value of k_{rel} in the absence of Cd^{2+} and the increase of k_{rel} at saturating Cd^{2+} concentration, respectively. Further, the value of $K_{d,app}^{Cd}$ can be precisely determined by plotting k_{rel} versus Cd^{2+} concentration and fitting to Eq. (15.4)).

$$k_{rel} = {}^{Mg}k_{rel}\left[\frac{K_{d,app}^{Cd} + \beta[Cd^{2+}]}{K_{d,app}^{Cd} + [Cd^{2+}]}\right]; \quad {}^{Mg}k_{rel} = \frac{{}^{Mg}k_{S_x}}{{}^{Mg}k_{S_o}} \tag{15.4}$$

For each substrate modification, the value of $K_{d,app}^{Cd}$ provides a "fingerprint" for the binding to the free ribozyme of the metal ion that rescues that particular substrate modification. If the rescue profiles obtained for two different substrate modifications have different $K_{d,app}^{Cd}$, this is a strong indication that distinct metal ions rescue each modification. If the values are the same, it is possible that the same site is involved or that different sites coincidentally have similar Cd^{2+} affinities. In this case, additional experiments are carried out to determine if one or two Cd^{2+} ions are required to rescue the two substrate modifications (Shan *et al.*, 2001; Wang *et al.*, 1999). Additional evidence can be obtained by perturbing individual metal-ion binding sites, as described below.

2.2.2. Introducing phosphorothioate substitution on the ribozyme

To determine ribozyme ligands for a particular metal ion, one most simply determines whether a phosphorothioate substitution on the ribozyme results in a higher affinity of the rescuing Cd^{2+}. In graphical terms, a lower $K_{d,app}^{Cd}$ (higher affinity) will manifest itself in a shift of the rescuing profile to lower Cd^{2+} values (Fig. 15.3; compare the gray lines to the black line). This effect holds regardless of the effect of Cd^{2+} on the reaction of the unmodified substrate (S_o) with the modified ribozyme, which can be neutral (Fig. 15.1C, corresponding to $\alpha = 1$ in Scheme 15.1A and resulting in the rescue profile shown in Fig. 15.3A), inhibitory (e.g., because the contact between the Cd^{2+} ion and the sulfur atom on the ribozyme needs to be broken to attain the transition state, corresponding to $\alpha < 1$ in Scheme 15.1A and resulting in the rescue profile shown in Fig. 15.3B), or stimulatory (Fig. 15.1A and B, corresponding to $\alpha > 1$ in Scheme 15.1A and resulting in the rescue profile shown in Fig. 15.3C).

Although the shape of the curves is the same in Fig. 15.3A–C, fitting the experimental k_{rel} values to Eq. (15.4) will give a misleading value of $K_{d,app}^{Cd}$ in the situations described by Fig. 15.3B and C (i.e., when Cd^{2+} affects the reactivity of the unmodified substrate in the phosphorothioate-containing ribozyme). In the examples shown in Fig. 15.3, fitting the gray curves in Fig. 15.3A–C to Eq. (15.4) gives values of $K_{d,app}^{Cd} = 0.2$, 2, and 0.002 mM, respectively, although the true value of $K_{d,app}^{Cd}$ is 0.2 mM in all of the three cases. Put another way, the value of $K_{d,app}^{Cd}$ determined from fitting to Eq. (15.4) corresponds to the true $K_{d,app}^{Cd}$ divided by α. Thus, if Cd^{2+} affects the reactivity of the unmodified substrate, S_o, in the phosphorothioate-containing ribozyme, Eq. (15.3) must be used to determine the correct value of $K_{d,app}^{Cd}$. However, in the absence of other information that allow to determine which Cd^{2+} ion is responsible for the stimulatory (or inhibitory) effect on the reaction of the unmodified substrate, it is often not possible to determine the value of α and therefore the value of $K_{d,app}^{Cd}$. Nevertheless, the shift in the rescuing profile remains a signature for an interaction between the sulfur atom on the ribozyme and the rescuing Cd^{2+} ion (Fig. 15.3A–C).

In ideal situations, the affinity of the rescuing metal ion is affected only by direct effects. In practice, indirect effects from altered charge distributions, steric clashes, and structural rearrangements upon substitution of sulfur for oxygen can result in changes in the affinity for a rescuing metal ion. Nevertheless, direct contacts will generally induce larger changes in the metal ion affinity compared to indirect effects. Thus, it is also important to determine the magnitude of effects arising from sulfur substitutions at a variety of positions.

TFA coupled with the introduction of phosphorothioate substitutions on the ribozyme has been successfully used in the *Tetrahymena* group I ribozyme to establish ligands for the catalytic metal ions. Among the many

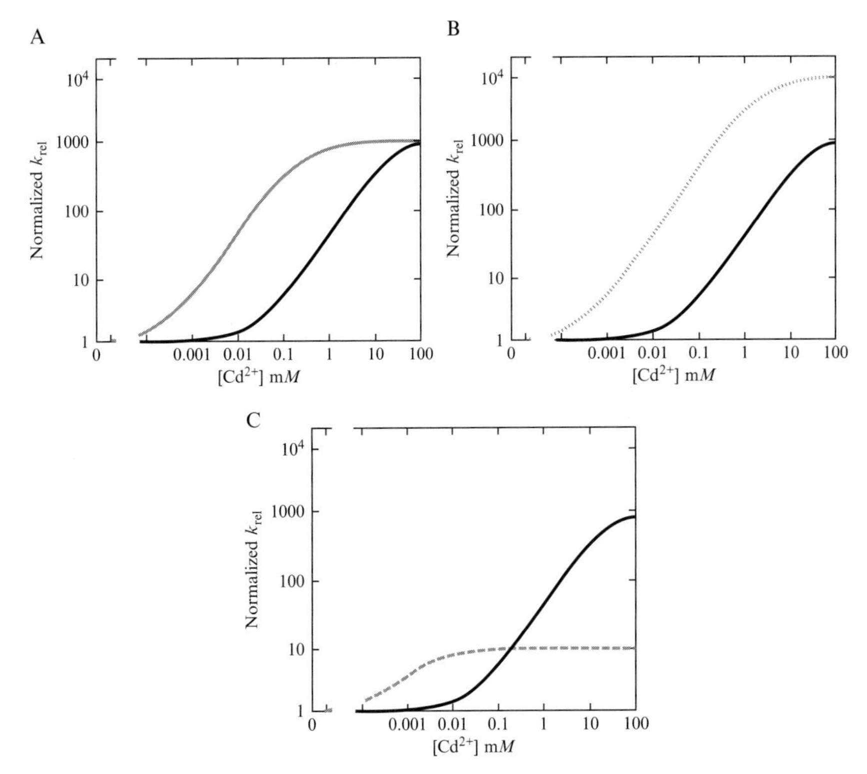

Figure 15.3 Expected trend of k_{rel} profiles (see text for details) for the unmodified ribozyme (black line) and for a ribozyme in which the introduced thiophosphoryl group makes a contact with a metal ion in both the ground state and the transition state (gray lines). k_{rel} is normalized for each ribozyme such that its value is equal to 1 in the absence of Cd^{2+}. In the simulations, β, which represents the rate enhancement of S_x with the metal-ion binding site saturated with Cd^{2+}, is set to 1000 for all ribozymes; $K_{d,app}^{Cd} = 20$ m*M* and 0.2 m*M* for the modified and unmodified ribozymes, respectively. (A) The Cd^{2+} that rescues reactivity of the modified substrate, S_x, has no effect on the reactivity of the unmodified substrate, S_o, in the modified ribozyme (solid gray line, $\alpha = 1$, Eq. (15.3)). (B) The Cd^{2+} that rescues reactivity of the modified substrate, S_x, has an inhibitory effect on the reactivity of the unmodified substrate, S_o, in the modified ribozyme (dotted gray line, $\alpha = 0.1$, Eq. (15.3)). (C) The Cd^{2+} that rescues reactivity of the modified substrate, S_x, has an inhibitory effect on the reactivity of the unmodified substrate, S_o, in the modified ribozyme (dashed gray line, $\alpha = 100$, Eq. (15.3)). The black line is the same in all the three panels, and is repeated for ease of comparison.

phosphorothioate substitutions introduced in the RNA backbone of the *Tetrahymena* ribozyme, only the ones predicted from crystal structures (Golden *et al.*, 2005; Guo *et al.*, 2005; Lipchock and Strobel, 2008; Stahley and Strobel, 2005) to contact a metal ion showed increasing Cd^{2+} ground state and transition state affinity compared to the unmodified ribozyme (Forconi *et al.*, 2007, 2008; Hougland *et al.*, 2005; M.F and

D.H., unpublished results). Analogous studies have been performed with the hammerhead (Wang *et al.*, 1999), the RNase P (Christian *et al.*, 2002), and the group II (Gordon and Piccirilli, 2001) ribozymes. In the following protocol, we describe how we have successfully introduced the phosphorothioate modification in the *Tetrahymena* ribozyme, performed kinetic tests and determined the ground state and transition state affinities for Cd^{2+}. This approach can be extended to other ribozymes, providing that:

1. The chemical step is followed in all the reactions monitored;
2. Proper controls are taken to ensure that the ground states of the unmodified and modified ribozymes are the same.

3. Protocols

3.1. Preparation of phosphorothioate-containing ribozymes

Working with RNA presents serious challenges, because of the prevalence of nucleases. Therefore, it is essential to work in "RNase-free" conditions. This involves the use of certified RNase-free plastic tubes and pipette tips, filtering water and reagents through a 0.2 μm filter, and changing gloves often.

3.1.1. General considerations on construction of phosphorothioate-containing RNA

Short (<80 nucleotides) RNAs containing a single phosphorothioate modification can be obtained by chemical synthesis.

Larger RNAs with a single phosphorothioate substitution are obtained by ligating a chemically synthesized RNA oligonucleotide (that contains the phosphorothioate modification) to the rest of the RNA molecule (Forconi *et al.*, 2007, 2008; Moore and Sharp, 1992). Depending on the position of the site of the modification and the length of the target RNA, this ligation can involve two or three pieces. Ligations with more pieces are also possible, if two or more modifications need to be introduced simultaneously. However, yields of fully ligated RNA decrease significantly with increasing number of pieces. Here, we describe a protocol for a three-piece ligation (Fig. 15.4), using transcribed RNA on the flanking regions and a chemically synthesized oligonucleotide, containing the phosphorothioate modification, in the middle piece. This protocol can be easily adapted to other situations.

Regardless of the number of pieces, RNA pieces can usually be ligated in a one-pot reaction using DNA ligase. DNA ligase joins RNA containing a 3'-hydroxyl group to RNA containing a 5′-phosphoryl group in a splint-assisted manner. Therefore, the pieces used in the ligation have to be prepared taking into account these requirements. Recent results (Ho and

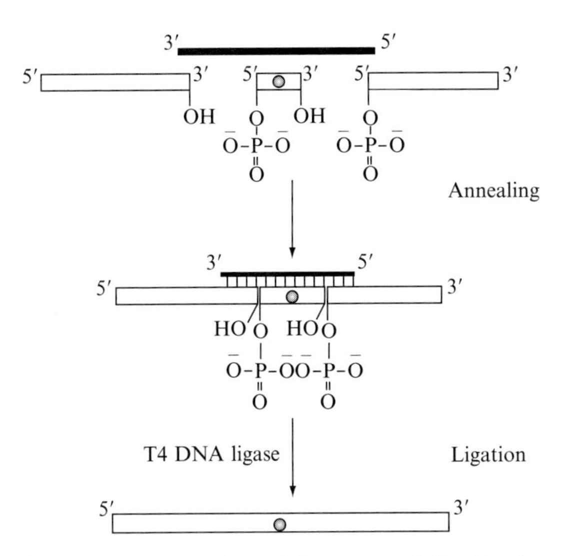

Figure 15.4 Schematic representation of the protocol for ligation of a fragment containing the phosphorothioate modification (represented by a gray sphere).

Shuman, 2002) suggest that RNA ligase 2, which also ligates RNA in a splint-assisted manner, can be used instead of DNA ligase, but we have not evaluated the efficiency and fidelity of this enzyme.

Ideally, the piece containing the phosphorothioate should be long enough to be specifically annealed to the splint, but shorter than 20 nucleotides so that the two phosphorothioate diastereoisomers can be easily separated by HPLC (see Section 3.1.6 and Chapter 14 of this volume). Diastereoisomers of oligonucleotides containing a phosphorothioate near the 5′-end are generally easier to separate by HPLC than diastereoisomers of oligonucleotides with a phosphorothioate near the 3′-end.

3.1.2. Transcription of the 5′-piece

Always run a test transcription to optimize reaction conditions for each construct, varying concentration of DNA template, quantity of RNA polymerase, and time of reaction. Once conditions are optimized, run a large-scale transcription. The values below are used for transcription of the region corresponding to nucleotides 22–296 of the *Tetrahymena* ribozyme and represent guidelines for other constructs. To ensure homogeneity of the 3′-ends, the transcript should have a 3′-hammerhead or HDV ribozyme (contained in the DNA template), that will self-cleave in transcription conditions; other cleavage strategies are also possible (Morl *et al.*, 2005). These cleavage elements leave an RNA with a terminal 2′,3′-cyclic phosphate; removal of this group is described in Section 3.1.5.

	Final concentration	Test transcription	Large-scale transcription
10× transcription buffer (400 m*M* Tris–HCl, pH 7.9; 60 m*M* $MgCl_2$; 100 m*M* NaCl; 20 m*M* spermidine)	1×	10 μl	1.0 ml
1 *M* dithiothreitol	40 m*M*	4 μl	0.4 ml
Mixture of NTPs, 10 m*M* each	2 m*M*	20 μl	2.0 ml
DNA template, ~20 μg/ml	1 μg/ml	5 μl	0.5 ml
T7 RNA polymerase, 20,000 units/ml	200 units/ml	1 μl	0.1 ml
H_2O		60 μl	6.0 ml
Total		100 μl	10.0 ml

(a) Combine reaction components, adding the DNA at last.
(b) Incubate at 37 °C for at least 3 h. For a quick test to determine whether RNA is being transcribed or not, run an aliquot of the reaction on a 2% denaturing agarose gel.
(c) To quench the reaction, add 250 m*M* EDTA, pH 8.2, to a final concentration of 50 m*M*.
(d) Precipitate the RNA by adding 0.5 volumes of 7.5 *M* ammonium acetate and 3 volumes of cold ethanol. Incubate for at least 30 min at −20 °C. Longer incubation times (overnight) can also be used.
(e) Spin the solution at 14,000 rpm for 15–30 min.
(f) Remove the supernatant and store it. Dissolve the pellet in an appropriate volume of loading buffer (10 m*M* EDTA, 90% formamide, 0.01% bromophenol blue, 0.01% xylene cyanol) and purify the RNA by gel electrophoresis, as described in Section 3.1.4.

3.1.3. Transcription of the 3′-piece

For splint ligations, the 3′-piece needs to be 5′-phosphorylated. This form can be achieved by adding at least a fivefold excess of 5′-GMP over GMP to the reaction in Eq. (15.1). The reaction protocol is otherwise the same.

In this particular case, homogeneity of the 3′-end of this piece is usually not critical; otherwise, methods analogous to those described for the homogeneity of the 5′-piece (see Section 3.1.2) can be used.

3.1.4. Gel purification of transcripts

Prepare the polyacrylamide gel in 7.5 *M* urea/1× TBE (90 m*M* Tris–borate, pH 8.5, 2.5 m*M* EDTA). Adjust the percentage of acrylamide depending on the size of the RNA to be purified. The thickness of the gel should be adjusted according to the amount of RNA to be loaded. For the particular 10 ml transcription mentioned in Section 3.1.2, we use 8% acrylamide, 3 mm thick gels, and load the RNA on one or two gels containing two 35-mm-long wells. Yields from a 10 ml transcription are typically 20–50 nmol after gel purification.

(a) Pre-run the gel using 1× TBE buffer for at least 30 min (65 W for a 3 mm thick gel). Use a metal plate to distribute heat evenly on the glass plate and prevent it from cracking.
(b) Load the sample (with the gel off) and run the gel at 65 W for the appropriate amount of time. Use charts (Maniatis *et al.,* 1975) to determine where the RNA runs compared to the dyes present in the loading buffer.
(c) Separate the plates and place the gel between saran wrap (both sides).
(d) Visualize the RNA bands using UV shadowing. Use a pen to indicate the location of the bands. Avoid damage of RNA by the UV light by minimizing its exposure.
(e) Cut out the bands using a sterilized razor blade or scalpel. It is possible that some of the hammerhead-containing construct is still present on the gel. This species migrates slower than the cleaved RNA. In this case, cut both bands and analyze them afterwards on denaturing PAGE. Place the gel slices in a conical tube. Use ethanol-washed, RNase-free glass rods to crush the RNA. Use a different glass rod for each different RNA.
(f) Add 1–2 volumes of elution buffer (50 m*M* Tris–HCl, pH 7.5; 10 m*M* EDTA; 0.3 *M* NaCl) to the excised bands. Elute the RNA overnight at 4 °C using a tube rotator.
(g) Spin down the samples using a swinging bucket centrifuge (5000×*g*). Immediately separate the supernatant from the gel pieces by using a syringe and a 0.2 or 0.45 μm filter. To increase the yields, repeat the extraction by adding more elution buffer to the gel pieces and eluting overnight as in step (f).
(h) Ethanol precipitate the RNA by adding 0.5 volumes of 3 *M* NaCl and 2.5 volumes of ice-cold ethanol. Incubate at −20 °C for at least 30 min.
(i) Spin the solution at 14,000 rpm for 15–30 min.
(j) Wash the pellet with 70% ethanol and dissolve it in water or buffer.

(k) Determine the concentration of the transcripts by UV spectrophotometry. Check the length and homogeneity of the construct by denaturing PAGE, alongside standards of known length.

3.1.5. Removal of 2′,3′-cyclic phosphate from the 5′-piece

Several methods are available for this operation (Morl *et al.*, 2005). Here, we describe a method that has worked with our constructs. The reaction can be scaled down depending on the quantity of RNA. To scale up, we typically run multiple reactions.

Set up the following reaction:

	Final concentration	Volume (μl)
1 *M* imidazole–HCl, pH 6	100 m*M*	63
1 *M* $MgCl_2$	10 m*M*	6.3
RNA, ~50 μ*M*	16 μ*M*	200
10 m*M* 2-mercaptoethanol	70 μ*M*	4.4
100 μg/ml BSA (RNase free)	19 μg/ml	120
T4 polynucleotide kinase, (10,000 units/ml)	200 units/ml	13.0
H_2O		223
Total		630

(a) Incubate at 37 °C for 6 h.
(b) Check the pH of the solution. If it is too low (<6), raise it to 7–8 by using an appropriate buffer; otherwise, the RNA will not be extracted correctly in the following step.
(c) Split the solution in an appropriate number of tubes and add 1 volume of phenol/chloroform/isoamyl alcohol solution (25:24:1, pH 8.0).
(d) Vortex and spin the samples. Separate the aqueous and organic layers.
(e) Mix each aqueous layer with 1 volume of chloroform.
(f) Vortex and spin the samples. Separate the aqueous layers.
(g) Pool together the aqueous layers of the same RNA.
(h) Back extract the phenol/chloroform/isoamyl alcohol solution with water or 10 m*M* Tris, pH 7.5. Vortex and spin the samples. Use this aqueous layer to back extract the RNA from the chloroform layer. Pool together the aqueous layers of the same RNA from the back extractions.
(i) Add 0.5 volumes of NaCl and 2.5 volumes of cold ethanol to each aqueous layer. Precipitate the RNA at −20 °C overnight.
(j) Spin the samples for 15 min at 15,000 rpm.
(k) Redissolve the RNA pellet in water and determine the concentration of RNA by UV spectrophotometry.

3.1.6. HPLC purification of phosphorothioate-containing oligonucleotides

Oligonucleotides should contain a 5′-phosphoryl group. The phosphorothioate diastereoisomers are purified using ion-exchange HPLC. Typical buffers used are:

Buffer A: 20 m*M* Tris, pH 7.6

Buffer B: 20 m*M* Tris, pH 7.6; 1 *M* NaCl

Elution times vary with the RNA construct. For a 15-mer with a phosphorothioate near the 5′-end, a gradient of 37–50% of B over 30 min affords good separation of the diastereoisomers.

Diastereoisomers of oligonucleotides containing a phosphorothioate near the 3′-end may not be separable using this technique. In this case, we have successfully separated the isomers by annealing to a oligodeoxyribonucleotide of complementary sequence (Forconi *et al.*, 2007). Separation of the DNA and RNA oligonucleotides was achieved by reverse-phase HPLC under denaturing conditions (Forconi *et al.*, 2007).

3.1.7. Ligation reaction

Ligation reactions are carried out in two parts. First, the RNA pieces are annealed on the DNA splint. This is achieved by heating the RNA to 90 °C, disrupting base pairs within the RNA, and slowly cooling down to room temperature. Second, the annealed solutions are mixed with the enzyme and left to react.

Always prepare a variant without the phosphorothioate modification (unmodified ribozyme) and compare its kinetic properties to those of the transcribed ribozyme. These properties should be identical.

3.1.7.1. Annealing step

	Final concentration	**Test ligation (μl)**	**Large-scale ligation (μl)**
3′-piece, 100 μ*M*	10 μ*M*	2.5	60
Middle piece (with phosphorothioate), 100 μ*M*	10 μ*M*	2.5	60
5′-piece, 100 μ*M*	10 μ*M*	2.5	60
DNA splint, 100 μ*M*	10 μ*M*	2.5	60
0.5 *M* NaCl	100 m*M*	5.0	–
3 *M* NaCl	100 m*M*	–	20
100 m*M* Tris, pH 7.5; 10 m*M* EDTA	10 m*M* Tris, 1 m*M* EDTA	2.5	60
H_2O		7.5	280
Total		25	600

(a) Prepare the above solution in a single Eppendorf tube. Gently vortex and spin.
(b) Heat the solution at 90 °C for 1 min.
(c) Allow the solution to cool slowly to room temperature.
(d) Steps (b) and (c) can be performed using a PCR machine. In this case, split large ligations in several tubes (maximum 100 μl of solution in each tube) to allow proper annealing.

3.1.7.2. Ligation

	Test ligation (μl)	Large ligation (μl)
10× ligase buffer (660 mM Tris–HCl, pH 7.6; 80 mM $MgCl_2$; 100 mM DTT; 1 mM ATP; 0.04% Triton X-100)	2.0	600
Annealed solution from previous step	2.0	600
RNasin 0.2 units/μl	2.0	–
RNasin 20 units/μl	–	6.0
T4 DNA ligase, 10 μM	4.0	–
T4 DNA ligase, 100 μM	–	120
H_2O	10	4670
Total	20	6000

(a) Prepare the reaction solution. Mix by pipetting up and down.
(b) Incubate at room temperature overnight (12–18 h). Longer or shorter times can be used; for longer times, fresh RNasin added to the solution after overnight incubation is sometimes helpful. However, RNA degradation may become an issue after 20 h.
(c) Precipitate the RNA and purify it on polyacrylamide denaturating gel as described for transcription reactions in Section 3.1.2.

Always consider the following:

- The concentration of T4 DNA ligase should be equal to the RNA concentration multiplied by the number of junctions. However, different preparations of T4 DNA ligase can have significantly different activity, and it is essential to adjust the quantity of ligase accordingly, based on the yields of test reactions.

- Different junctions ligate differently. Changing ligase concentration, oligonucleotides concentration, ATP concentration, or pH may help. We have found that for some junctions addition of 30% glycerol dramatically improves the yield.
- T4 DNA ligase often purifies in its adenylated form. Adjust the concentration of ATP in the ligase buffer to achieve the best ligation efficiency.

3.2. Kinetic assay

The phosphorothioate-containing ribozymes can be used to test for a contact with a metal ion, as described in Section 2.2. To extract the affinity of Cd^{2+} for the modified ribozyme and compare it to that for the wt ribozyme (see Scheme 15.1), we measure the observed rate constant for the reaction in Eq. (15.1) as a function of Cd^{2+} concentration. Typically 6–12 Cd^{2+}concentrations are used. For each ribozyme, reactions of the unmodified and sulfur- or amino-containing substrates are followed side-by-side at each Cd^{2+} concentration. All experiments are performed under single-turnover conditions, with ribozyme in excess of the oligonucleotide substrate S, which is added in trace as a radioactive species.

The following protocol is an example for reactions of the *Tetrahymena* group I ribozyme. Change accordingly to your system.

(a) Before starting the protocol, it is essential to carefully plan the timing of the kinetic assays. Prepare enough tubes containing stop solution (50 m*M* EDTA, 90% formamide, 0.01% bromophenol blue, 0.01% xylene cyanol) before starting the experiment. Plan to take at least 5–6 time points for each reaction.
(b) Prefold the wild type and modified ribozymes in 10 m*M* Mg^{2+}, 50 m*M* NaMOPS, pH 7.0, at 50 °C for 30 min. Handling is easiest if the solution is at least 20 μl.
(c) Allow the ribozyme solutions to cool to room temperature for 5 min.
(d) Add the appropriate quantity of the ribozyme solution to a mixture containing the rest of the reagents (*except the radiolabeled oligonucleotide substrate S*), including different amounts of Cd^{2+}, for a final volume of 8–18 μl.
(e) Allow the solutions to equilibrate at 30 °C, or at the temperature the reactions are carried out.
(f) Start the reactions by adding 2 μl of radiolabeled oligonucleotide substrate (S) to the solutions.
(g) Monitor reactions by withdrawing 1.5–2.0 μl of each reaction at different times, and quench them by addition to 4 μl of stop solution (50 m*M* EDTA, 90% formamide, 0.01% bromophenol blu, 0.01% xylene cyanol). Prepare enough tube containing stop solution. The quenched solutions are stable at room temperature, and can be stored until the

last time point is taken. Reactions should be followed for ~3-half lives whenever possible.

(h) Load each quenched reactions, each representing a time point, on denaturing gels containing 20% acrylamide/1× TBE to separate substrates and products by gel electrophoresis. Run reactions of modified and unmodified ribozymes one next to the other to ensure that the products are the same for the two ribozymes, at least as determined from gel mobility. Use bromophenol blue and xylene cyanol as references to determine for how long the gel should run. Important! In case of free substrates with free thiols, all the unreacted acrylamide must be destroyed prior to loading the sample. This is achieved by pouring the gels at least 10 h in advance, and including 5 m*M* DTT in the gel mix. In this case, remember to add the same concentration of DTT to the running buffer.

(i) Dry gels under vacuum, 80 °C, for 1 h.

(j) Place dried gels in a phosphorimager cassette for an appropriate time (do not saturate the signal).

(k) Scan the screen with a phosphorimager to detect the substrate (S) and product (P) bands. Ideally, there should be only two spots per reaction, corresponding to these two species.

(l) Quantify the intensities of the bands corresponding to S and P by using dedicated software.

(m) For each reaction, the pseudo-first-order rate constant for substrate disappearance (k_{obs}) is obtained by plotting the fraction of substrate present at each time point as a function of time, and fitting the data to the following equation:

$$\frac{[S]_t}{[S]_t + [P]_t} = a + (1 - a)e^{-k_{obs}t} \tag{15.5}$$

in which $[S]_t$ and $[P]_t$ are the concentrations of S and P present at time t, and a is the fraction of substrate present at infinite time. For slow reactions, estimate a on the basis of the reaction going to completion. This value should be close to zero.

(n) Values of k_{rel} are determined for each ribozyme at each Cd^{2+} concentration by dividing the observed rate constant for the sulfur-containing substrate by the rate constant of the oxygen-containing substrate.

(o) k_{rel} for each ribozyme are then plotted as a function of Cd^{2+}, and the results fit to Eq. (15.4) to obtain values of $K_{d,app}^{Cd}$.

ACKNOWLEDGMENTS

We thank the numerous researchers who have contributed to and used these approaches, especially Mildred Cohn and Fritz Eckstein for their pioneering efforts. This research was supported by a grant from the NIH (GM 49243) to D. H.

REFERENCES

Christian, E. L. (2005). Identification and characterization of metal ion binding by thiophilic metal ion rescue. *In* "Handbook of RNA Biochemistry, Vol. 1," (R. K. Hartmann, ed.), Wiley-VCH Verlag GmbH & Co., Weinheim.

Christian, E. L., and Yarus, M. (1993). Metal coordination sites that contribute to structure and catalysis in the group I intron from *Tetrahymena*. *Biochemistry* **32,** 4475–4480.

Christian, E. L., Kaye, N. M., and Harris, M. E. (2002). Evidence for a polynuclear metal ion binding site in the catalytic domain of ribonuclease P RNA. *EMBO J.* **21,** 2253–2262.

Cohn, M., Shih, N., and Nick, J. (1982). Reactivity and metal-dependent stereospecificity of the phosphorothioate analogs of ATP in the arginine kinase reaction. Structure of the metal-nucleoside triphosphate substrate. *J. Biol. Chem.* **257,** 7646–7649.

Eckstein, F. (1983). Phosphorothioate analogs of nucleotides—Tools for the investigation of biochemical processes. *Angew. Chem. Int. Ed.* **22,** 423–439.

Forconi, M., Piccirilli, J. A., and Herschlag, D. (2007). Modulation of individual steps in group I intron catalysis by a peripheral metal ion RNA. **13,** 1656–1667.

Forconi, M., Lee, J., Lee, J. K., Piccirilli, J. A., and Herschlag, D. (2008). Functional identification of ligands for a catalytic metal ion in group I introns. *Biochemistry* **47,** 6883–6894.

Golden, B. L., Kim, H., and Chase, E. (2005). Crystal structure of a phage Twort group I ribozyme–product complex. *Nat. Struct. Biol.* **12,** 82–89.

Gordon, P. M., and Piccirilli, J. A. (2001). Metal ion coordination by the AGC triad in domain 5 contributes to group II intron catalysis. *Nat. Struct. Biol.* **8,** 893–898.

Guo, F., Gooding, A. R., and Cech, T. R. (2005). Structure of the *Tetrahymena* ribozyme: Base triple sandwich and metal ion at the active site. *Mol. Cell* **16,** 351–362.

Harris, M. E., and Pace, N. R. (1995). Identification of phosphates involved in catalysis by the ribozyme RNase P RNA. *RNA* **1,** 210–218.

Herschlag, D., and Cech, T. R. (1990). Catalysis of RNA cleavage by the *Tetrahymena Thermophila* ribozyme. 1. Kinetic description of the reaction of an RNA substrate complementary to the active site. *Biochemistry* **29,** 10159–10171.

Hertel, K. J., Herschlag, D., and Uhlenbeck, O. C. (1996). Specificity of hammerhead ribozyme cleavage. *EMBO J.* **15,** 3751–3757.

Ho, C. K., and Shuman, S. (2002). Bacteriophage T4 RNA ligase 2 (gp24.1) exemplifies a family of RNA ligases found in all phylogenetic domains. *Proc. Natl. Acad. Sci. USA* **99,** 12709–12714.

Hougland, J. L., Kravchuk, A. V., Herschlag, D., and Piccirilli, J. A. (2005). Functional identification of catalytic metal ion binding sites within RNA. *PLoS Biol.* **3,** 1536–1548.

Hougland, J. L., Piccirilli, J. A., Forconi, M., Lee, J., and Herschlag, D. (2006). How the group I intron works: A case study of RNA structure and function. *In* "The RNA World," (R. F. Gesteland, T. R. Cech, and J. F. Atkins, eds.), pp. 133–206. Cold Spring Harbor Laboratory Press, Cold Spring Harbor, New York.

Jansen, J. A., McCarthy, T. J., Soukup, G. A., and Soukup, J. K. (2006). Backbone and nucleobase contacts to glucosamine-6-phosphate in the glmS ribozyme. *Nat. Struct. Biol.* **13,** 517–523.

Kazantsev, A. V., and Pace, N. R. (1998). Identification by modification-interference of purine N-7 and ribose 2′-OH groups critical for catalysis by bacterial ribonuclease P. *RNA* **4,** 937–947.

Liao, X., Anjaneyulu, P. S. R., Curley, J. F., Hsu, M., Boehringer, M., Caruthers, M. H., and Piccirilli, J. A. (2001). The *Tetrahymena* ribozyme cleaves a 5′-methylene phosphonate monoester approximately 10^2-fold faster than a normal phosphate diester: Implications for enzyme catalysis of phosphoryl transfer reactions. *Biochemistry* **40,** 10911–10926.

Lipchock, S. V., and Strobel, S. A. (2008). A relaxed active site after exon ligation by the group I intron. *Proc. Natl. Acad. Sci. USA* **105,** 5699–5704.

Luptak, A., and Doudna, J. A. (2004). Distinct sites of phosphorothioate substitution interfere with folding and splicing of the *Anabaena* group I intron. *Nucleic Acids Res.* **32,** 2272–2280.

Maniatis, T., Jeffrey, A., and van deSande, H. (1975). Chain length determination of small double- and single-stranded DNA molecules by polyacrylamide gel electrophoresis. *Biochemistry* **14,** 3787–3794.

Moore, M. J., and Sharp, P. A. (1992). Site-specific modification of pre-mRNA—The 2′-hydroxyl groups at the splice sites. *Science* **256,** 992–997.

Morl, M., Lizano, E., Willkomm, D. K., and Hartmann, R. K. (2005). Production of RNAs with homogeneous 5′ and 3′ ends. *In* "Handbook of RNA Biochemistry, Vol 1," (R. K. Hartmann, ed.), pp. 22–35. Wiley-VCH Verlag GmbH & Co., Weinheim.

Parr, R. G., and Pearson, R. G. (1983). Absolute hardness—Companion parameter to absolute electronegativity. *J. Am. Chem. Soc.* **105,** 7512–7516.

Pearson, R. G. (1963). Hard and soft acids and bases. *J. Am. Chem. Soc.* **85,** 3533–3539.

Pecoraro, V. L., Hermes, J. D., and Cleland, W. W. (1984). Stability constants of magnesium and cadmium complexes of adenine nucleotides and thionucleotides and rate constants for formation and dissociation of magnesium-ATP and magnesium-ADP. *Biochemistry* **23,** 5262–5271.

Piccirilli, J. A., Vyle, J. S., Cartuhers, M. H., and Cech, T. R. (1993). Metal-ion catalysis in the *Tetrahymena* ribozyme reaction. *Nature* **361,** 85–88.

Platts, J. A., Howard, S. T., and Bracke, B. R. F. (1996). Directionality of hydrogen bonds to sulfur and oxygen. *J. Am. Chem. Soc.* **118,** 2726–2733.

Rajagopal, J., Doudna, J. A., and Szostak, J. W. (1989). Stereochemical course of catalysis by the *Tetrahymena* ribozyme. *Science* **244,** 692–694.

Shan, S., and Herschlag, D. (1999). Probing the role of metal ions in RNA catalysis: Kinetic and thermodynamic characterization of a metal ion interaction with the 2′-moiety of the guanosine nucleophile in the *Tetrahymena* group I ribozyme. *Biochemistry* **38,** 10958–10975.

Shan, S., and Herschlag, D. (2000). An unconventional origin of metal-ion rescue and inhibition in the *Tetrahymena* group I ribozyme reaction. *RNA* **6,** 795–813.

Shan, S., Yoshida, A., Piccirilli, J. A., and Herschlag, D. (1999). Three metal ions at the active site of the *Tetrahymena* group I ribozyme. *Proc. Natl. Acad. Sci. USA* **96,** 12299–12304.

Shan, S., Kravchuk, A. V., Piccirilli, J. A., and Herschlag, D. (2001). Defining the catalytic metal ions interactions in the *Tetrahymena* ribozyme reaction. *Biochemistry* **40,** 5161–5171.

Shannon, R. D. (1976). Revised effective ionic radii and systematic studies of interatomic distances in halides and chalcogenides. *Acta Crystallogr.* **A32,** 751–767.

Sjogren, A. J., Petterson, E., Sjoberg, B. M., and Stromberg, R. (1997). Metal Ion interaction with cosubstrate in self-splicing group I introns. *Nucleic Acids Res.* **25,** 648–653.

Sood, V. D., Beattie, T. L., and Collins, R. A. (1998). Identification of phosphate groups involved in metal binding and tertiary interactions in the core of the *Neurospora* vs ribozyme. *J. Mol. Biol.* **282,** 741–750.

Stahley, M. R., and Strobel, S. A. (2005). Structural evidence for a two-metal-ion mechanism of group I intron splicing. *Science* **309,** 1587–1590.

Strauss-Soukup, J. K., and Strobel, S. A. (2000). A chemical phylogeny of group I introns based upon interference mapping of a bacterial ribozyme. *J. Mol. Biol.* **302,** 339–358.

Sun, L., and Harris, M. E. (2007). Evidence that binding of C5 protein to P RNA enhances ribozyme catalysis by influencing active site metal ion affinity. *RNA* **13,** 1505–1515.

Wang, S., Karbstein, K., Peracchi, A., Beigelman, L., and Herschlag, D. (1999). Identification of the hammerhead ribozyme metal ion binding site responsible for rescue of the deleterious effect of a cleavage site phosphorothioate. *Biochemistry* **38,** 14363–14378.

Weinstein, L. B., Jones, B., Cosstick, R., and Cech, T. R. (1997). A second catalytic metal ion in a group I ribozyme. *Nature* **388,** 805–808.

Yoshida, A., Sun, S. G., and Piccirilli, J. A. (1999). A new metal ion interaction in the *Tetrahymena* ribozyme reaction revealed by double sulfur substitution. *Nat. Struct. Biol.* **6,** 318–321.

Yoshida, A., Shan, S., Herschlag, D., and Piccirilli, J. A. (2000). The role of the cleavage site 2′-hydroxyl in the *Tetrahymena* group I ribozyme reaction. *Chem. Biol.* **7,** 85–96.

Zaug, A. J., Grosshans, C. A., and Cech, T. R. (1988). Sequence-specific endoribonuclease activity of the *Tetrahymena* ribozyme—Enhanced cleavage of certain oligonucleotide substrates that form mismatched ribozyme substrate complexes. *Biochemistry* **27,** 8924–8931.

Zhou, P., Tian, F. F., Lv, F. L., and Shang, Z. C. (2009). Geometric characteristics of hydrogen bonds involving sulfur atoms in proteins. *Proteins: Struct. Funct. Bioinformatics* **76,** 151–163.

CHAPTER SIXTEEN

EPR Methods to Study Specific Metal-Ion Binding Sites in RNA

Laura Hunsicker-Wang,* Matthew Vogt,† *and* Victoria J. DeRose‡

Contents

Abstract

The properties of metal-ion interactions with RNA can be explored by spectroscopic methods. In this chapter, we describe the use of paramagnetic Mn^{2+} ions and electron paramagnetic resonance (EPR)-based techniques to monitor the association of Mn^{2+} with RNA and related nucleotides. Solution EPR methods are used to determine the numbers of Mn^{2+} ions associating with RNA. For RNA poised with a single-bound Mn^{2+}, low-temperature EPR characteristics provide information about the asymmetry of the Mn^{2+} coordination site. To identify the RNA groups coordinating to the Mn^{2+} ion, ENDOR (electron nuclear double

* Department of Chemistry, Trinity University, San Antonio, Texas, USA
† Laboratory of Pathology, National Cancer Institute, Bethesda, Maryland, USA
‡ Department of Chemistry, University of Oregon, Eugene, Oregon, USA

Methods in Enzymology, Volume 468
ISSN 0076-6879, DOI: 10.1016/S0076-6879(09)68016-2

resonance) and ESEEM (electron spin echo envelope modulation) methods are applied. Both continuous-wave (CW) and electron spin echo (ESE)-detected ENDOR methods are described. This chapter includes practical details for RNA sample preparation, including isotope substitution and cryoprotection, and an overview of data acquisition and analysis methods used in these techniques, as well as examples from the current literature.

1. Introduction

A growing body of evidence indicates that site-bound metal ions play critical roles in RNA function. Site-bound metal ions can be analyzed via effects on function, and predicted based on high occupancies in X-ray crystallographic studies. In order to further determine the properties of metal ions that are important to RNA function, including their affinities, RNA ligands, level of hydration, and possible changes during activity, it is ideal to have a spectroscopic probe for the ion. For RNA function, it is generally assumed that Mg^{2+} is the biologically active divalent cation. Unfortunately, with exception of challenging ^{25}Mg NMR methods (Grant *et al.*, 2000), Mg^{2+} is spectroscopically silent. A potential substitute for this ion is Mn^{2+}, which has similar though not identical ion properties and often supports RNA activity. Mn^{2+} is paramagnetic and can be observed using electron paramagnetic resonance (EPR) spectroscopy and related techniques.

This chapter will provide practical information regarding the types of information that can be obtained from EPR spectroscopy of Mn^{2+} ions in RNA, as well as practical methods for preparing samples and acquiring data. The chapter will cover standard EPR spectroscopy, and two more advanced techniques that can be used to identify the ligands to the RNA-bound metal ion: ESEEM (electron spin echo envelope modulation) and ENDOR (electron-nuclear double resonance spectroscopy). We will focus solely on Mn^{2+}, but the methods described here would be applicable to other EPR-active metal ions such as Cu(II) (Santangelo *et al.*, 2007) and vanadyl $(VO)^{2+}$ ions (Mustafi *et al.*, 2003; Smith *et al.*, 2002). EPR spectroscopy has also been used to measure the properties of nitroxide organic radicals introduced in site-directed spin labeling (SDSL) of RNA, and those methods are covered in other chapters of this volume. This chapter is aimed at an audience interested in biophysical studies of RNA, most of whom will likely form collaborative interactions for the specialty ESEEM and ENDOR methods. More detailed reviews of those methods are available (Britt, 2003; DeRose and Hoffman, 1995; Prisner *et al.*, 2001).

1.1. Mn^{2+} as a functional substitute and spectroscopic probe for metal ions in RNA

The Mn(II) ion has a similar ionic radius to that of Mg^{2+} (~0.7 Å for Mg^{2+} vs. 0.8 Å for Mn^{2+}), meaning that its "charge-to-radius" ratio should provide similar electrostatic properties. The two ions are not identical, however, and can differ significantly in affinities to RNA-relevant ligands (reviewed in DeRose, 2008; DeRose *et al.*, 2003; Freisinger and Sigel, 2007; Martell and Smith, 1971–1974). [*Note*: Much of the RNA literature describes "ligands" as small molecules that bind to RNA. Here, "ligand" is used as the classic inorganic descriptor of an atom (possibly from the RNA) that is coordinated to a metal ion.] Mg^{2+} has a [Ne] electronic configuration, with filled s- and p-orbitals and no d-orbital contributions to bonding. Mg^{2+} is classified as a "hard" ion on the Pearson scale and forms mainly ionic interactions, with a preference for "hard" ligands such as oxygen. Mn^{2+} (Mn(II)) has an electron configuration of $[Ar]3d^5$, and the half-filled d-orbital shell contributes to its ligand preferences. Mn^{2+} is nearly always high-spin and takes on a six-coordinate, octahedral arrangement of ligands. Unlike other transition ions such as Cu(II), Co(II), or Zn(II), in Mn(II) complexes there is little energetic preference for an axially distorted geometry, or for a four-coordinate rather than six-coordinate environment. Mn^{2+} is classified as a "borderline" metal ion and interacts strongly with both oxygen and nitrogen ligands. The main differences between Mn^{2+} and Mg^{2+} ions that are relevant to RNA coordination is that Mn^{2+} binds more strongly to nitrogen ligands from nucleobases, and has a higher affinity for phosphodiester oxygen ligands than does Mg^{2+}, while having a slightly lower affinity for aqua ligands. Thus, Mn^{2+} may replace Mg^{2+} in functional sites, but may bind directly to a nitrogen ligand of a nucleobase instead of indirectly through a hydrogen-bonded water ligand. The apparent affinity of Mn^{2+} for an RNA site might be higher than that of Mg^{2+} (Hunsicker and DeRose, 2000a; Travers *et al.*, 2007), and in general Mn^{2+} stabilizes RNA structures at a lower concentration than is required for Mg^{2+}. X-ray crystallography studies have shown similar if not identical metal coordination sites for the two ions in high-occupancy sites (Ennifar *et al.*, 2003; Juneau *et al.*, 2001; Shi and Moore, 2000).

1.2. EPR spectroscopy of Mn(II)

EPR spectroscopy monitors transitions between energy levels of unpaired electrons in a magnetic field. A single isolated unpaired electron, with $S = 1/2$, gives rise to a single line in an EPR spectrum with a peak position that is characterized by the proportionality value "g," similar to an NMR chemical shift value. The most common method of recording EPR spectra is via "continuous-wave," or CW spectroscopy, and the most common

CW EPR spectrometers operate with an electromagnet that provides magnetic fields of 0–5000 Gauss, and an "X-band" or 9 GHz microwave source for excitation of the electrons. CW EPR spectra are generally recorded as derivatives of the absorption lineshape, meaning that a "single line" appears as a derivative that crosses zero at the peak position of the absorption line. An alternative method of recording EPR signals is via creation and detection of an electron-spin echo (ESE), a technique more similar to standard pulsed NMR methods. Pulsed or ESE-detected EPR spectroscopy is technically challenging due to the requirements for high-power, short microwave pulses, but it offers opportunities to manipulate the system that are not available in standard CW spectroscopy. For Mn^{2+}, pulsed EPR methods require low temperatures and frozen solutions. Detection via pulsed EPR methods forms the basis for the ESEEM and pulsed ENDOR techniques described in later sections.

Energy levels of electron spin states are influenced by their environment, which includes other electrons as well as nuclear spins with $I > 0$. The spin–spin interaction between electron and nuclear spins is the hyperfine interaction, and this can be quite informative for site identification. The magnitude of the electron–nuclear hyperfine interaction is sensitive to both through-space (dipolar) and through-bond contributions. If the unpaired electron is in an orbital with significant overlap with a nucleus that has a nuclear spin $I > 0$, the EPR signal might exhibit splitting from the electron–nuclear hyperfine interaction. An observable hyperfine interaction with a single $I = 1/2$ nucleus splits the EPR line into two, or $(2n\mathrm{I}+1)$ lines where n equals the number of equivalent nuclei. For paramagnetic metal ions, the metal nucleus itself is the most common source of EPR line splittings. In the case of high-spin Mn(II), the situation is nicely complex with a combination of electron spin $S = 5/2$ and nuclear spin $I(^{55}\mathrm{Mn}) = 5/2$ (Markham *et al.*, 1979; Reed and Poyner, 2000; Stich *et al.*, 2007) (Fig. 16.1). The energy levels from the unpaired electrons are all split by the electron–nuclear hyperfine interaction with the Mn nucleus, giving in theory 36 possible transitions (or more, if one considers second-order effects). For the majority of Mn(II) sites, however, there is significant degeneracy between energy levels and the EPR signal collapses into just six major lines that are separated by the electron–nuclear hyperfine interaction of around 90 Gauss. When viewed at low temperature, there is "substructure" between the major lines that is due to forbidden environments and is sensitive to ligand environment, as described below.

The Mn(II) EPR signal has three main properties that are useful in examining metal sites in RNA. When the Mn^{2+} ion coordinates to a biomolecule, changes in the ligand environment can result in altered EPR properties that (a) allow quantitation of "bound" versus "free" Mn^{2+}; (b) can be analyzed for information about the biomolecular site; and (c) form the basis of advanced methods of ENDOR and ESEEM that can

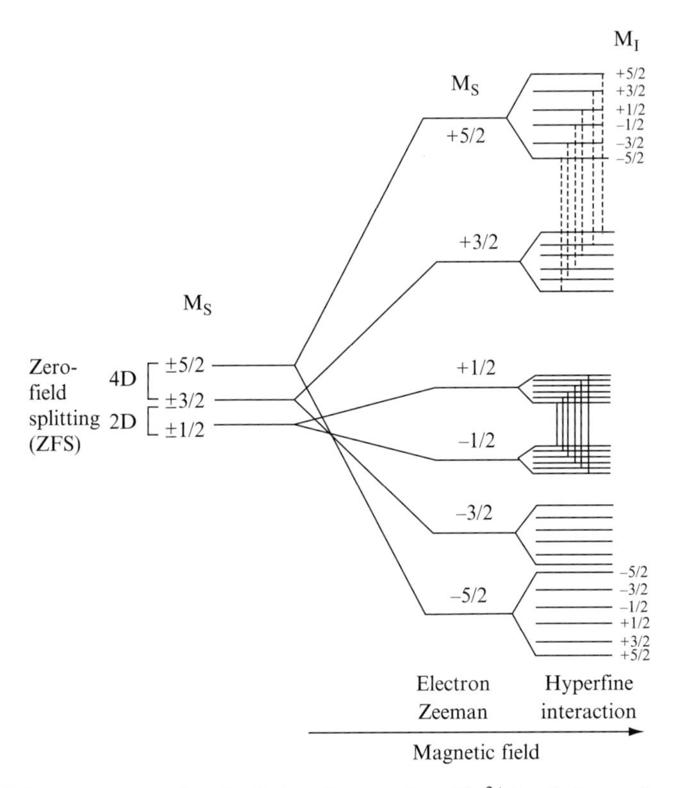

Figure 16.1 EPR energy-level splitting diagram for a Mn^{2+} ion in increasing external magnetic field. The influence of axial zero-field splitting (ZFS) value D is shown. Sets of six allowed EPR transitions take place between energy levels with $\Delta M_S = 1$, $\Delta M_I = 0$ and are indicated as vertical solid lines for the central "inner" $M_2 = \pm 1/2$ levels that are unaffected by ZFS, and as dashed lines for one of the other four possible ΔM_S sets that are known as "outer" transitions. For electronically symmetric species, all five sets of allowed EPR transitions occur at a similar energy, giving rise to a six-line EPR spectrum. Increasing asymmetry broadens the "outer" transitions, leading to broad underlying features in the low-temperature EPR signal as well as changes in "forbidden" transitions that are not shown here.

identify the metal-ion ligands. The bases for these three options are described briefly below, and practical details about the techniques and examples are provided in ensuing sections.

1.3. Preparation of RNA samples for EPR spectroscopy

1.3.1. 5′-end identity

As described below, EPR spectroscopy requires nanomole quantities of RNA. Methods for high-yield RNA synthesis are described in an earlier chapter of this volume. RNA synthesized by *in vitro* transcription carries a

5′-triphosphate, whereas RNA synthesized by solid-phase methods has a 5′-OH unless otherwise specified. Mn^{2+} and other multivalent cations bind avidly to triphosphates and particularly strongly to guanosine triphosphate, the usual 5′-end nucleobase used for *in vitro* transcription. For quantitative measurements, and in particular experiments designed at isolating individual metal sites, a 5′-OH terminus is optimal in order to avoid competitive binding with the chelating terminal triphosphate. Thus, it is advisable to dephosphorylate *in vitro* transcribed RNA for quantitative measurements and/or measurements in which population of single Mn^{2+} sites is desired. An alternative is to perform *in vitro* transcription in an excess of GMP, which can initiate transcription but not be elongated (Milligan and Uhlenbeck, 1989; Morrissey *et al.*, 1999) and provides a weaker chelate than the triphosphate.

1.3.2. Choice of buffer

A nonchelating buffer is required for these studies. We have used 5 m*M* triethanolamine (TEA), pH 7.5–7.8. HEPES, MES, and cacodylate are all standard nonchelating buffers. Tris has weak association with transition metal ions and we have avoided high concentrations of this buffer.

1.3.3. Importance of EDTA removal

EDTA is ubiquitous in RNA purification methods, being present in high concentrations in standard TBE (Tris–Borate–EDTA) gel running buffers. EDTA is present in lower micromolar concentrations in many storage buffers in order to chelate residual contaminating metal ions and slow nonspecific RNA hydrolysis. EDTA tightly chelates Mn^{2+}, and the resulting complex is EPR-silent at room temperature and gives a very broad EPR signal at low temperature. Thus, Mn–EDTA chelates will interfere with all experiments described below. For this reason, for quantitative EPR measurements it is important to dialyze RNA samples against several changes of an EDTA-free buffer. Our standard RNA preparation protocol involves gel purification and electroelution followed by dialysis against an EDTA-free buffer for 48–72 h, 4 °C, with 5–7 reservoir changes. RNA is then concentrated (Centricon, YM-3000), ethanol-precipitated, and the washed pellet resuspended in either autoclaved water or buffer to form a stock solution.

1.3.4. Mn^{2+} solutions

"Ultrapure"-grade Mn^{2+} sold as 1 *M* solutions in water (available from USB and Sigma) is the most convenient form of pure Mn^{2+}. Concentrated stock solutions can be stored as aliquots, frozen, until needed. In aqueous solution, Mn^{2+} slowly oxidizes to Mn^{3+}, which then disproportionates in solution to form Mn(IV) and Mn(II). Mn(III) is pale pink in color, and the Mn(IV) oxide is a brown precipitate, and both may become visible in solutions stored for ~1 week at benchtop. To avoid this instability, Mn^{2+}-buffer solutions are made fresh when needed from the frozen stock solutions, and not stored for more than a few days.

1.3.5. Cryoprotectants

Low-temperature EPR, ENDOR, and ESEEM experiments are performed on frozen solutions of RNA. It is standard to use a cryoprotectant in order to minimize ice crystal formation and macromolecule aggregates. In addition to raising doubts about the integrity of molecular structure, aggregates are a particular problem for the "advanced" EPR methods because aggregated samples have shortened relaxation times that interfere with the measurements. Some standard biomolecule cryoprotectants are 0.4 *M* sucrose, 20% ethylene glycol (by volume), or 50% glycerol (by volume). In Hammerhead Ribozyme (HHRz) activity studies, we found that neither 20% ethylene glycol nor 0.4 *M* sucrose interfered with either activity or Mn^{2+} association as monitored by EPR spectroscopy, but that glycerol was inhibitory. Ethylene glycol is a potential metal coordinating compound, however, and Hoogstraten and Britt reported a slight reduction in Mn-coordinated aqua ligands in the presence of 20% ethylene glycol (Hoogstraten and Britt, 2002). A more significant loss was observed in glycerol, however. Taken together, glycerol should be avoided as a cryoprotectant for these studies (Hoogstraten and Britt, 2002), and ethylene glycol used with some caution.

2. Room Temperature EPR Spectroscopy to Quantify Mn^{2+} Bound to RNA

EPR spectra are quantitative, meaning that Mn^{2+} ions in solution can be quantified based on the integrated area or estimated based on the height of the EPR signal. The "slow-exchange" limit for EPR spectroscopy of liquid samples, or the timeframe in which independent spectra are observed from each type of species in equilibrium, occurs for species exchanging on the timescale of microseconds. In experiments performed at room temperature, metal site populations that may give rise to "fast-exchange," averaged NMR spectra can result in superimposed EPR spectra from each individual state that is present in solution.

When the Mn(II) is bound to a large biomolecule, an interesting phenomenon takes place that results in significant broadening of the Mn(II) EPR line when it is recorded at ambient (room) temperatures. This effect is related to small changes in the symmetry of ligands around the metal ion that occur when the metal moves from a fully hydrated environment to one with a single or multiple new ligands. The effect can

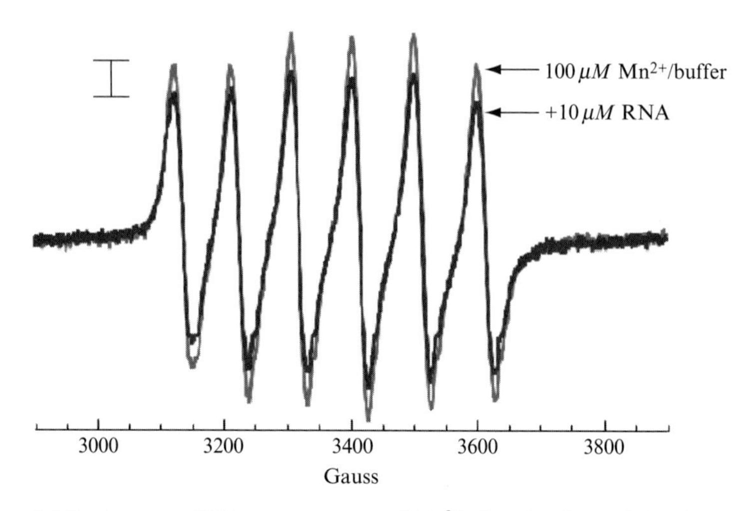

Figure 16.2 Aqueous EPR spectroscopy of Mn^{2+} showing loss of signal amplitude when Mn^{2+} binds to RNA. Addition of the RNA causes a decrease in the signal intensity that is proportional to the amount of Mn^{2+} ions interacting with the RNA molecule. The loss in signal intensity is quantified as "bound" Mn^{2+}. Samples: 5 m*M* TEA pH 7.8, 0.1 *M* NaCl, 0.1 m*M* $MnCl_2$ (top spectrum) and same with addition of 10 μ*M* mHHRz RNA (47 nucleotides, see Fig. 16.4), 50 μL samples. EPR spectra parameters: room temperature, 0.1 mW microwave power, 100 kHz modulation with 15 G MA, sum of 5 scans each.

occur for small molecules such as EDTA, which has an asymmetric $[N_2O_4]$ set of ligands. The broadening effect is enhanced when the Mn ion samples a site in a larger biomolecule because of the longer rotational correlation time of the ion when sampling the biomolecular site. The resulting diminution in solution EPR signal can be quantified as "bound Mn^{2+}" (Fig. 16.2). This method of quantifying Mn(II) binding to macromolecules was originally applied to proteins (Cohn and Townsend, 1954; Woody *et al.*, 1996) and has been used with other large molecules (Ottaviani *et al.*, 1996). The association of Mn^{2+} with tRNA (Danchin and Gueron, 1970), G-quartets (Marathias *et al.*, 1996), the P5abc subdomain of the Group I intron (S. Burns, Fig. 16.3) and with the hammerhead ribozyme (Edwards and Sigurdsson, 2005; Horton and DeRose, 2000; Horton *et al.*, 1998; Hunsicker and DeRose, 2000b; Kisseleva *et al.*, 2005) has been quantified using this property.

2.1. Protocol for Mn-binding titrations by EPR spectroscopy

2.1.1. Sample preparation

In a standard experiment, 10–50 μ*M* oligonucleotide in 5 m*M* TEA buffer at pH 7.8 and desired NaCl concentration were heated to 90 °C for 90 s and cooled on ice for 30 min. After cooling, divalent metal ($MnCl_2$) was added to the desired final concentration, using metal stock solutions made fresh for each day's experiment. The required volume depends on the sample tubes, described below. The RNA concentrations are chosen based on the balance of the apparent metal-ion affinities and the signal:noise requirements of the experiment, which relies on sensitive detection of the loss of Mn^{2+} EPR signal when the Mn^{2+} is bound to the RNA. For example, if the RNA sample has a high-affinity site with apparent $K_d \sim 10\ \mu M$, then at the low end of the titration $[Mn^{2+}_{free}]$ values of $<10\ \mu M$ must be detected. At the high end of the titration, a reduction of 10–50 μM from 1 mM $[Mn^{2+}_{free}]$, or a 1–5% difference, becomes difficult to accurately measure.

Samples are loaded into appropriate tubes for solution EPR spectroscopy. For aqueous solutions, the inner diameter of the EPR tube must be kept small in one dimension in order to minimize the interaction of the microwave electric field with the aqueous solution. Standard X-band EPR tubes have a 3 mm inner diameter (ID) and are too wide for this purpose. Capillary tubes with an inner diameter of 1.5 mm or less are appropriate. Alternatively, special "flat" cells with a rectangular shape that minimizes the width of the aqueous sample in one dimension, but allows an overall higher volume of sample to be measured, are available. The flat cell is placed with the shortest width between the poles of the electromagnet. An important aspect of quantitative EPR spectroscopy is that the height of the sample should be above the "active height," or height sampled by the resonator. For most EPR cavities, a height of 2 cm is sufficient. In our experiments, capillary tubes from Vitrocom with an ID of 1.5 mm were used, requiring 50 μL volume for an appropriate sample height. Bruker currently delivers a high-sensitivity cavity/sample tube system that is optimized for aqueous samples, and requires only 5 μL sample or less. EPR spectroscopy is very sensitive to radicals and other paramagnetic impurities. If sample tubes are to be reused, they should be cleaned carefully. A procedure that has worked is to soak them in dilute nitric or hydrochloric acid, rinse in deionized water, rinse in ethanol, and dry at benchtop or a glass oven. It is a helpful practice to check for residual EPR signals in empty tubes before loading precious RNA samples.

2.1.1.1. EPR data collection Tuning the instrument and data acquisition is specific to each model of EPR instrument, and will not be detailed here. However, certain aspects that are important to quantitative EPR spectroscopy will be mentioned. In addition to amount of detectable Mn^{2+}, an EPR signal amplitude depends sensitively on two additional parameters: the microwave power and the modulation amplitude (MA). These must be held constant during the Mn^{2+} titration. The microwave power is important, because if it is too high, the signal is partially saturated and the amplitude is no longer linear with concentration. Increasing the MA increases the signal height until the MA is close to the signal width, and after that point increasing the MA causes the signal to broaden. Optimal parameters that maximize signal:noise without distorting the EPR signal, then, are to increase both the microwave power and MA as high as possible without either saturating or broadening the EPR signal, respectively. Standard conditions used in our laboratory for X-band EPR detection of aqueous Mn^{2+} are 0.2 mW microwave power and 15–25 G MA, averaging 1–5 scans. Since EPR spectrometers can differ, however, it is suggested that the operator perform their own microwave power saturation and MA-saturation experiments on trial Mn^{2+} samples to find conditions of optimal sensitivity with linear response to concentration. Finally, it is noted that a uniform tuning procedure is important, since differently tuned samples can have differing amplitudes.

Temperature control can be important because continual operation of microwave cavities creates a heating effect. For this reason, a temperature readout and a flow of N_2 gas over the sample are also helpful.

2.1.1.2. Data analysis In these binding titrations, the "unbound" $[Mn^{2+}_{free}]$ is quantified from the amplitude of the EPR signal in comparison with a set of standards. A standard curve is constructed by obtaining EPR signal amplitudes for samples with known Mn^{2+} concentrations, using a range appropriate to the study (generally 10 μM–1 mM Mn^{2+}). EPR signal amplitudes can be estimated as the height change between the trough and the peak of a given EPR line. For Mn^{2+}, the most robust method calculates the amplitudes of all six lines, but using the middle four is sufficient (always use the same lines, and note whether the signal amplitude has been normalized by the gain or amplification of the instrument). The "bound" Mn^{2+} is then calculated as $[\text{added } Mn^{2+}]-[Mn^{2+}_{free}]$. A standard binding isotherm is created as a plot of $[Mn^{2+}_{bound}]/[RNA]$ versus $[Mn^{2+}_{free}]$. These data can be fit to binding models.

There is a significant, important discussion concerning the validity of various models for describing "binding" of a cation to a complex polyelectrolyte such as RNA (Das *et al.*, 2005; Misra *et al.*, 2003). When monitoring an effect of metal ions on the RNA itself, such as a metal-induced structural change, the experiment cannot be used with full confidence to describe the numbers or affinities of the cations because the cations are also influencing the global electrostatic properties of the system. In this EPR experiment, however, as in a dialysis experiment, the partitioning of the Mn^{2+} ions between a bound and unbound state is being directly quantified. The simplest binding model is a sum of noninteracting sites that each bind n_i ions with apparent $K_{d,i}$ values:

$$[Mn^{2+}_{bound}]/[RNA] = \sum_{i=1}^{j} n_i[Mn^{2+}_{free}]/(K_{d,i} + [Mn^{2+}_{free}])$$

Cooperativity is better visualized on a plot with $\log[\mathrm{Mn}^{2+}_{\mathrm{free}}]$ as the x-axis. In the case of cooperative Mn^{2+} ion association, the data will fit to a Hill equation:

$$\frac{[\mathrm{Mn}^{2+}_{\mathrm{bound}}]}{[\mathrm{RNA}]} = \sum_{i=1}^{j} \frac{[\mathrm{Mn}^{2+}_{\mathrm{free}}]^{n_i}}{(K_{\mathrm{d},i} + [\mathrm{Mn}^{2+}_{\mathrm{free}}]^{n_i})}$$

2.1.1.3. Influence of monovalent cations One method to isolate "specific" divalent cation sites involves adding high concentrations of monovalent cations such as Na^+ or K^+ to compete with nonspecific cation sites, and support global folding of the RNA. A current picture of cation-induced RNA structure indicates that complex RNAs may fold to near-native conformations in high concentrations of monovalent cations, but that specific divalent ion contacts are required to stabilize tertiary interactions that are also important to function. To isolate important divalent cation sites, high concentrations (2 *M* or higher) of monovalent cations may be desirable. This technique has been used in some studies that isolated the A9/G10.1 site in the HHRz (Horton *et al.*, 1998; Morrissey *et al.*, 1999, 2000; Vogt *et al.*, 2006), and high-affinity metal sites in P4–P6 domain of the Group I intron (Das *et al.*, 2005; Travers *et al.*, 2007). As shown below, increasing monovalent cation concentrations reduces the level of Mn^{2+} association with a 63-nucleotide P5abc RNA, reduces the apparent cooperativity values for bound Mn^{2+} ions (as Na^+ stabilizes global folding), and increases the apparent Mn^{2+} K_{d} values (due to competition with increasing Na^+) (S. Burns, Fig. 16.3).

2.1.1.4. Influence of anions We would like to provide a cautionary note concerning the counterion used for very high (>1 *M*) concentrations of monovalent cations. Cl^- and other halides can coordinate Mn^{2+} (as well as other cations). In our hands, clear evidence for Cl^- coordination to Mn^{2+}, with methanol as solvent, occurred at 3 *M* LiCl (Vogt, 2004). As a competitive ligand, methanol is significantly weaker than water, but this observation suggests that a more weakly coordinating counterion such as $\mathrm{NO_3}^-$ might be considered for Mn^{2+} binding studies in very high monovalent cation concentrations.

3. Low-Temperature EPR Spectroscopy of Mn^{2+} Ions Bound to RNA

As mentioned above, when a Mn^{2+} ion moves from an aquated, hexahydrated environment to an RNA environment in which one or more of the aqua ligands has exchanged for an RNA ligand, the symmetry

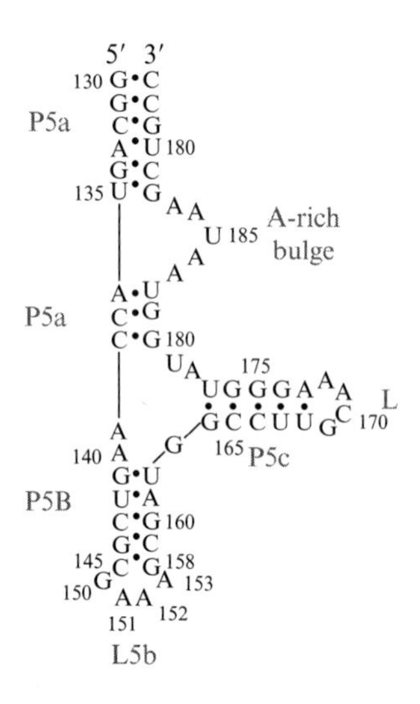

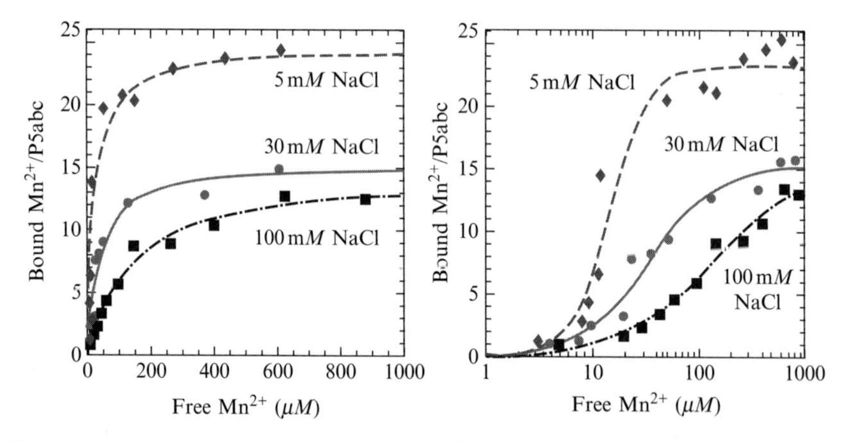

Figure 16.3 Effect of added [NaCl] on Mn^{2+} association with P5abc RNA (left). Bound Mn^{2+} are quantified using room temperature EPR spectroscopy (see Fig. 16.2 and text). Increasing [Na^+] decreases the number of bound Mn^{2+} as well as apparent affinities and cooperativity. Data obtained with 10 *μM* RNA in 5 m*M* TEA pH 7.8 and [NaCl] as indicated on plot, with EPR spectrometer parameters as given in Fig. 16.2.

of the Mn^{2+} ion has been altered. If detectable, the features of the EPR signal from the bound Mn^{2+} ion may reflect the change in symmetry, which manifest as a change in the ZFS parameters D and/or E and also a new pathway for electron spin relaxation. In aqueous solutions, this increase in T_2 along with the increased rotational correlation time caused by Mn^{2+} association broaden the signal enough such that it is no longer readily evident at room temperature. Lowering the temperature, however, allows the EPR signal of the bound Mn^{2+} ion to be detected along with that of any "unbound" or hexahydrate Mn^{2+}. This generally requires freezing the sample, with a judicious choice of cryoprotectant. If the sample can be poised under conditions of stoichiometric binding, such that all added Mn^{2+} are bound to the RNA, then the low-temperature EPR signal of Mn^{2+} ions bound to the RNA can be detected and analyzed.

The changes that occur in the Mn^{2+} EPR signal upon alterations in the Mn^{2+} ligand environment can be fairly subtle. In most situations, the difference is observed in the relative intensities of the "substructure" lines that occur between the major six-line features (Markham *et al.*, 1979; Reed and Poyner, 2000; Stich *et al.*, 2007). This substructure is ascribed to the intensities of "semiforbidden" transitions that become more allowed as the electronic symmetry around the ion is lowered. In the Hamiltonian that describes energy levels of the Mn^{2+} unpaired electrons, ligand field symmetry is described by the zero-field splitting terms D and E. These terms are generally very small for the mainly ionic binding around Mn^{2+}. For example, in the A9/G10.1 site of the HHrz, Mn^{2+} is coordinated to two RNA ligands (a nonbridging phosphodiester oxygen and a guanine N7 imino nitrogen ligand), and four aqua ligands. The resulting EPR signal is altered subtly in the substructure between the fifth and sixth ^{55}Mn hyperfine lines in comparison with hexaaqua Mn^{2+} (Morrissey *et al.*, 2000). Simulations indicate that this difference can be ascribed to a $\sim$50 cm^{-1} difference in E, the rhombic zero-field splitting term (Vogt, 2004) (Fig. 16.4).

Larger changes in Mn^{2+} EPR signals can occur in more electronically asymmetric environments. For example, Mn^{2+} bound to EDTA has a significantly more altered EPR signal. The Mn–EDTA complex is sufficiently distorted that it is not observed at room temperature, even though the complex is low in molecular weight. For Mn–EDTA, the low-temperature EPR signal shows very broad signals due to large ZFS parameters (Stich *et al.*, 2007). A similar situation occurs for Mn^{2+} coordinated to the protein fosfomycin (Smoukov *et al.*, 2002; Walsby *et al.*, 2005). The calculated ZFS parameters for these Mn^{2+} centers ($D \sim$ 1000–2000 cm^{-1}) are $\sim$10 times those of Mn^{2+} in the HHRz ($D \sim$ 100–200 cm^{-1}) (Stich *et al.*, 2007; Vogt, 2004).

3.1. Protocols for low-temperature EPR spectroscopy of Mn^{2+} bound to RNA

3.1.1. Sample preparation

The goal of most low-temperature EPR experiments is to obtain an accurate lineshape for a single populated Mn^{2+} site. In ensuing sections, this EPR signal will form the basis for ENDOR and ESEEM measurements. Standard low-temperature EPR experiments were performed with a final Mn^{2+} concentration of 200–500 μM Mn^{2+} and slightly higher concentrations

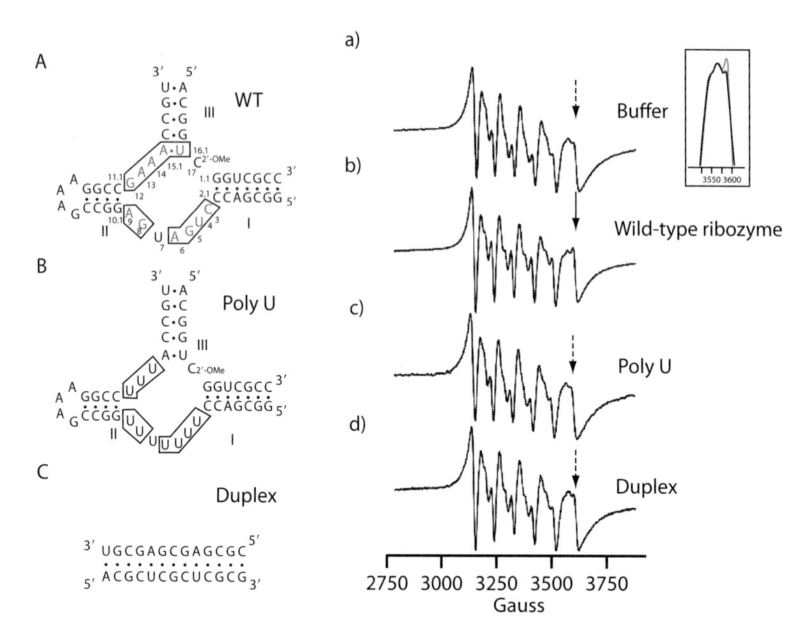

Figure 16.4 Low-temperature EPR spectra of Mn^{2+} bound to the mHHRz, a "poly-U" variant lacking the conserved core, and a 13-nucleotide RNA duplex. Mn^{2+} coordination to a site within the wild-type mHHRz results in a small change in spectral lineshape, observed in the expanded spectrum at upper right and simulated in Fig. 16.5. EPR samples are 250 μM RNA and 250 μM Mn^{2+}, in 0.1 M NaCl, 5 mM TEA pH 7.8, 20% (v/v) ethylene glycol. EPR data collected at a temperature of 10 K, 0.2 mW microwave power, 15 G MA, 1–5 scans each.

of RNA. In a saturation study with 250 μM mHHRz, we found that 0.75:1 Mn:RNA resulted in the clearest difference in lineshape, suggesting optimal population of the specific Mn^{2+} site (Vogt, 2004). (While it may be tempting to analyze this behavior in terms of metal-ion affinities, this is not recommended given the presence of cryoprotectant and low temperatures). Final volumes of the samples are 70–120 μL, depending on the sample tube. In a standard preparation, 50–100 μL of 250–500 μM oligonucleotide in 5 mM TEA buffer at pH 7.8 and desired NaCl concentration is heated to 90 °C for 90 s and cooled on ice for 30 min. After cooling, divalent metal ($MnCl_2$) was added to the desired final concentration, using metal stock solutions made fresh for each day's experiment. Cryoprotectant (in appropriate buffer) is then added to a final value of either 0.4 M sucrose or 20% (v/v) ethylene glycol (see above discussion concerning choice of cryoprotectant). Gentle mixing with the pipette occurs after addition of metal and of cryoprotectant.

Standard EPR tubes for spectroscopy of frozen aqueous solutions, available from Wilmad, are 3 mm ID and are ~8 in. high, and are sufficiently filled to ~2 cm height with ~120 μL of sample. Filling these tubes with the viscous cryoprotected RNA sample (and leaving no bubbles) is best done with a long syringe needle, or a long piece of tubing attached to a standard syringe. For ease of sample manipulation and for Q-band ENDOR measurements, we sometimes use shorter 3 in. high, thin-walled Q-band tubes with 2 mm ID. These tubes can be slid into regular long X-band tubes for standard spectroscopy and can be used directly in smaller Q-band cavities.

Samples are frozen in liquid nitrogen before insertion into the cold EPR cavity, and this can conveniently be done in a small dewar or Styrofoam cup. In order to avoid condensing atmospheric O_2 and CO_2, the sample should be capped when being frozen. O_2 is paramagnetic and can give rise to broad background signals. Particularly important for the larger X-band tubes, the caps should be removed before thawing the sample. Condensed gases will expand upon thawing, sometimes explosively, making it a good practice to wear safety glasses for this process and around cryogens in general.

3.1.2. EPR data collection

For Mn^{2+} spectra of frozen solutions, standard conditions used in our laboratory include 10–20 K temperature (liquid He), <1 mW microwave power and 15–25 G MA, averaging 1–5 scans.

3.1.3. EPR lineshape simulations

There are many programs available for simulating EPR spectra. Simulation programs can differ substantially in rigor. An excellent MatLab-based suite is "EasySpin," created and maintained by Dr. Stefan Stoll (Stoll and Schweiger, 2006) and available free of charge at: http://www.easyspin.org/. This full-Hamiltonian program is particularly suited for spectral simulations of Mn(II), for which important factors include the ability to calculate semiforbidden transitions based on zero-field splitting parameters. EasySpin also allows inclusion of different broadening parameters, also known as "strain" parameters. "Strain" or slight heterogeneity in g-, A-, and D,E can all arise from tiny site-to-site variations in individual Mn^{2+} environments.

The Mn(II) spectra shown below were simulated in EasySpin. It should be noted that the parameters shown below are not necessarily unique, because the spectral linewidths and amplitudes are sensitive to both "strain" and "semiforbidden" transitions, and we have not rigorously evaluated all possible parameters. However, the simulations show that the more pronounced forbidden transitions in the mHHRz Mn^{2+} EPR spectrum can be simulated by an increase in E, the rhombic ZFS parameter of 173 MHz in comparison to $E = 90$ MHz for Mn–GMP (both at $D = 520$ MHz). Data obtained at much higher frequency, D-band (140 GHz) have been simulated with similar parameters of D and $E = 550$–600, 130

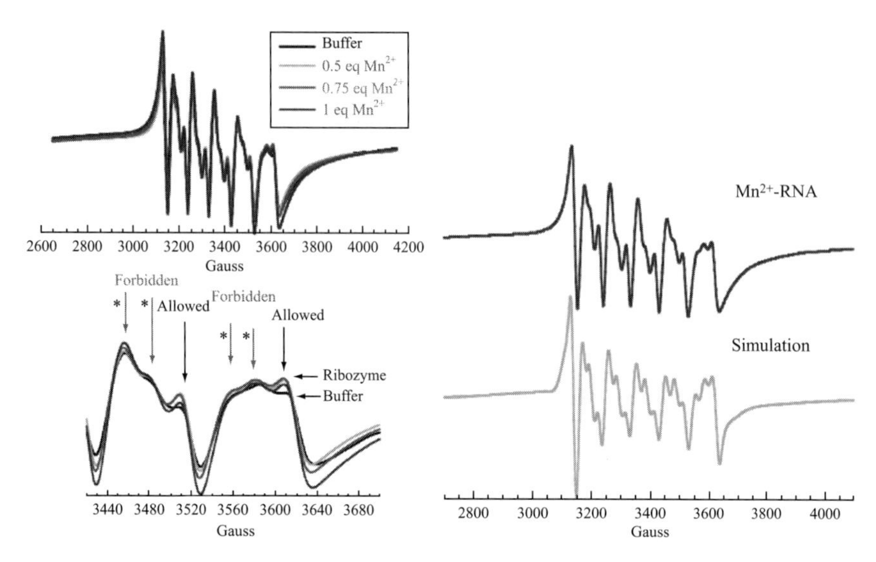

Figure 16.5 Experimental and simulated low-temperature EPR spectra of Mn^{2+} bound to the mHHRz. At left, top: overlay of spectra showing appearance of the distinctive Mn^{2+} EPR feature at substoichiometric Mn^{2+}:RNA ratios. Samples are 250 μM Mn^{2+} and increasing RNA concentrations to achieve 0.5, 0.75, or 1.0 Mn^{2+}: RNA ratio. Buffer: 1 *M* NaCl, 5 m*M* TEA pH 7.8, 20% (v/v) ethylene glycol. EPR spectral parameters as in Fig. 16.4. Left, bottom: expansion of fifth and sixth Mn^{2+} hyperfine lines, showing features due to "forbidden" and "allowed" transitions whose ratios change with different Mn^{2+} environments. Right: experiment (top) and simulation (bottom) for Mn^{2+} EPR spectrum of 0.5 Mn^{2+}:mHHRz using the program EasySpin (see text). Simulation parameters: $g = 2$, $A = 270$ MHz, $D = 520$ MHz, $E = 173$ MHz, D-strain = 1, A-strain = 0. (See Color Insert.)

MHz for the mHHRz, and a reduction of both values to $D = 420$ MHz and $E = 100$ MHz in Mn–GMP (Stich *et al.*, 2007) (Fig. 16.5).

4. ENDOR Spectroscopy to Identify Metal Ligands

Unambiguous identification of the site of a metal ion bound to a nucleic acid requires a signal from the nucleic acid itself. In EPR-based experiments, ligand identification can be based on detecting the nuclear spins from the surrounding environment. Relevant nuclei include ^{31}P from coordinated phosphodiester groups (I=1/2, 100% natural abundance), ^{14}N (I=1, >99% natural abundance) or site-specifically labeled ^{15}N (I=1/2) or ^{13}C (I=1/2) from nucleobases, and ^{1}H (I=1/2, >99% natural abundance) from exchangeable or nonexchangeable proton sites on nucleobases or aqua ligands. The hyperfine interaction between the unpaired electron spin of the metal ion and the nuclear spin is based on distance, orbital overlap, and the properties of the nucleus. In the case of Mn^{2+}, the largest hyperfine interaction from the ^{55}Mn nucleus itself actually splits the EPR signal, but the interactions with nearby environmental nuclei are generally too small to be resolved in regular EPR spectroscopy.

Hyperfine interactions from nuclei within ~6–7 Å of the paramagnetic metal ion can be uncovered using "advanced" EPR techniques that have some similarities to NMR NOESY and COSY methodologies. In ENDOR spectroscopy, the EPR signal is detected while nuclear transitions are directly excited using a separate radiofrequency (RF) input. ENDOR experiments generally monitor a single position in the EPR signal (one magnetic field/microwave frequency), and sweep through RF values. Upon excitation by the RF, transitions between the nuclear sublevels alters the populations of spins giving rise to the EPR signal, eliciting a response in the EPR signal amplitude that is considered the ENDOR response. A hyperfine interaction of an $M_s = \pm 1/2$ feature with a single type of nucleus gives rise to a pair of ENDOR lines at values of $\nu_{ENDOR} = A/2 \pm \nu_n|$, where ν_n is the Larmor frequency for the nucleus n. If the nucleus has $I > 1/2$ and therefore a quadrupole moment, the ENDOR lines will be further split by the nuclear quadrupole interaction. The high-spin $S = 5/2$ Mn^{2+} ion presents a specialty situation where in some cases, hyperfine couplings might be observed from transitions involving the $M_S > 1/2$ sublevels (i.e., between the $M_S = 1/2$ and $M_S = 3/2$, or 3/2 and 5/2 sublevels). In this instance, which happens primarily for data collected on the "wings" of the EPR spectrum, the hyperfine couplings are amplified proportionately to the M_S values (Carmieli *et al.*, 2001; Morrissey *et al.*, 2000; Potapov and Goldfarb, 2006; Tan *et al.*, 1993).

The ENDOR experiment can be performed either in CW mode (CW excitation of both microwave and radio frequencies), or in pulsed-EPR mode (DeRose and Hoffman, 1995; Hoffman, 2003; Thomann and Bernardo, 1993). Both techniques require specialty instrumentation. When available, there are significant advantages to working at a higher magnetic field for ENDOR spectroscopy because of intrinsic increased sensitivity and also because the nuclear Larmor frequencies of the relevant nuclei are much better separated, being proportional to magnetic field strength. At Q-band (~35 GHz, similar to K_a-band), for example, ENDOR signals from ^1H, ^{31}P, and ^{14}N are well-separated, whereas at X-band there can be overlap between ENDOR signals from strongly coupled protons and the other nuclei. Recently, ENDOR spectroscopy of Mn^{2+}-nucleotide and RNA complexes at W-band (90 GHz) has been demonstrated (Bennati *et al.*, 2006; Potapov and Goldfarb, 2006; Schiemann *et al.*, 2007). At W-band, the higher magnetic field allows nearly complete separation of the Mn^{2+} EPR features arising from the $M_S \pm 1/2$ sublevels.

4.1. Protocols for ENDOR spectroscopy of Mn^{2+} in RNA samples

4.1.1. Sample preparation

ENDOR spectroscopy relies on detection of the EPR signal. Samples prepared for low-temperature EPR spectroscopy, described above, are appropriate for ENDOR. The ENDOR spectrum will reflect signals from all of the species that are giving rise to a particular position in the EPR spectrum. If there are different Mn^{2+} sites with overlapping EPR signals, both species will be sampled, and the ENDOR spectrum is additive.

4.1.2. Isotopic substitution

ENDOR signals from ^{31}P, ^{14}N, and ^{1}H might be expected from a metal site in an oligonucleotide. All of these nuclei have ~100% natural-abundance. To identify the site, however, substitution with a different isotope might be desired. Global substitution of a particular nucleobase, such as ^{15}N-labeled guanine, can be achieved via *in vitro* synthesis with ^{15}N-labeled GTP. We performed this procedure with an excess of GMP in the reaction mixture in order to substitute the potentially chelating, ^{15}N-labeled GTP at the 5′-end with a weaker coordinating, and nonisotopically labeled, guanine monophosphate. In order to achieve a site-specific ^{15}N-substituted guanine, the ^{15}N-labeled guanine phosphoramidite was purchased from Cambridge Isotope Laboratories (formerly Spectra Stable Isotopes) and incorporated by a third party. The phosphoramidite is only commercially available in the deoxyribose form, meaning that loss of the 2′-OH at the desired position must be acceptable for RNA function. Literature methods synthesizing the 2′-protected, isotopically labeled ribose phosphoramidite are available.

4.1.3. Deuterium substitution

Substitution of exchangeable protons with ^{2}H is performed as follows. As described above, RNA samples are purified, dialyzed against appropriate buffer, and precipitated in excess EtOH. The resulting pellet is further dried by SpeedVac, and then resuspended in D_2O. The D_2O-suspended sample is lyophilized to dryness. The sample is resuspended again in D_2O and lyophilized, and this procedure is repeated two additional times. The final RNA pellets are resuspended in the appropriate D_2O-containing buffer. Samples are annealed, and then metal ion and cryoprotectant added as described above. Both the metal-ion solutions and the cryoprotectant are prepared in deuterated buffer. In deuterated buffers, because of the altered activity of deuterium, the measured pD is 0.4 higher than the value used in pH (i.e., a pH of 7.5 becomes pD = 7.9).

4.1.4. ENDOR data acquisition

ENDOR spectroscopy is a specialized technique, and details of data acquisition will differ depending on the specific instrument being used. Here, we will provide an overview of the procedure. ENDOR data require liquid He temperatures (10 K and below) in order to slow the electron T_1 relaxation rate sufficiently to allow perturbation of the electron populations by the RF input. In general, the first step is acquisition of an EPR spectrum from the Mn^{2+} ions in the sample. In both CW- and pulsed-EPR modes used in ENDOR spectroscopy, the EPR signal is acquired as an absorption line rather than in the derivative mode that is common to regular EPR spectrometers. The EPR spectrum acquired in this manner shows the six major peaks from the ^{55}Mn hyperfine lines. Broad wings on the sides of these central features are also observed, and are due to features from the higher M_S states. ENDOR data obtained from the central features in theory will sample all possible M_S transitions, but several studies have shown that the $M_S = \pm 1/2$ states will dominate and for the purpose of this chapter, we will focus only on ENDOR features arising from these transitions.

To obtain ENDOR spectra, the magnetic field is then set at a position corresponding to one of the peaks of the six-line feature. This magnetic field **B** will set the nuclear Larmor frequencies, $\nu_N = g_n \beta_n \mathbf{B}$. ENDOR features, to first-order, are observed when the scanning RF reaches values of $\nu \pm = (\nu_N \pm A/2)$, where A is the electron-nuclear hyperfine coupling. Here, the scanning RF, ν_N, and A are generally given in units of MHz (Fig. 16.6).

4.1.5. CW ENDOR spectroscopy

CW ENDOR spectrometers that operate at Q-band exist in laboratories at Northwestern University (Hoffman) and SUNY-Albany (Scholes). A CW ENDOR spectroscopic method that is robust for transition metal ions is performed under conditions of "rapid passage" for the paramagnetic center (Feher, 1956 and reviewed in Doan *et al.*, 2007; DeRose and Hoffman, 1995). In this situation, the temperature is lowered to $\sim$2 K, or supercritical He. At this temperature, the T_1 relaxation is so slow that the standard 100 kHz magnetic field modulation produces an EPR signal with an absorptive lineshape, rather than a derivative. The magnetic field is set to the desired feature in the EPR spectrum, and then the magnetic field is modulated at 100 kHz (the same field modulation that is used in detecting EPR spectra) with a MA desired for the experiment. Very qualitatively, this procedure excites a set of electron spins centered at the field value and, through continuously varying the local magnetic field in a small modulation, provides a relaxation pathway that induces relatively fast relaxation. When those electron spins are induced to change nuclear sublevels via the scanning RF, they are removed from the small packet of excited spins and the magnitude of the EPR signal is decreased. Achieving a high signal:noise ratio for an ENDOR signal acquired in this manner can depend on several intertwined parameters: MA,

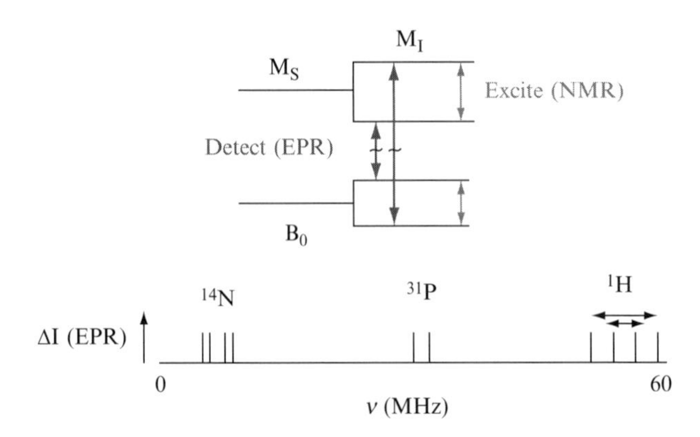

Figure 16.6 Schematic of an ENDOR experiment. Transitions between M_s sublevels and M_I sublevels are individually excited by microwave and RF input, respectively. The resulting ENDOR spectrum is monitored as a perturbation of the EPR signal intensity (y-axis) as a function of RF (x-axis). For each hyperfine-coupled nucleus, a pair of ENDOR features at $\nu_{\pm} = |\nu_N \pm A/2|$ is observed. At Q-band or higher frequencies, $\nu_N > A/2$ (usually) and the ENDOR features are centered at ν_N and separated by values of A. Importantly, this "stick" treatment ignores EPR and hyperfine anisotropy, both of which in addition to normal linebroadening will add a lineshape to the ENDOR features. The ENDOR experiment can be performed by either continuous-wave or pulsed (ESE-detected) methods.

microwave power, RF power, and rate of the RF scan to name a few. This method suffers from some lineshape distortion that depends on the RF sweep conditions. Still, it provides a fairly reliable detection method.

In an ENDOR experiment, RF scans are set for regions corresponding to expected nuclei. For example, to obtain a ^{1}H ENDOR spectrum, the RF is set up to scan a region that is centered at the nuclear Larmor frequency, and sweeps a region ± 10 MHz from that position. In the case of the Mn^{2+}-mHHRz samples, Q-band CW ENDOR was most useful in obtaining ^{1}H ENDOR signals that are centered at the Larmor $\nu_N(^1H)$ value of $\sim$55 MHz (Morrissey *et al.*, 2000). The ^{14}N ENDOR signals appear at a much lower frequency of $\sim$4 MHz and were not of sufficient amplitude against the background to be easily detected. In the case of ^{31}P ENDOR, signals were observed at the appropriate Larmor frequency but a very large peak due to a buildup of "distant" hyperfine interactions, or a sum of many very weak hyperfine interactions, was evident in the CW ENDOR that made lineshape analysis unreliable. As described below, for these latter nuclei it is preferable to use pulsed EPR methods (Fig. 16.7).

4.1.6. Pulsed (ESE) ENDOR spectroscopy

In a pulsed ENDOR (and ESEEM) experiment, the EPR signal is created through application of high-power microwave pulses that result in an electron spin echo (Fig. 16.8). A standard "Hahn" two-pulse echo sequence,

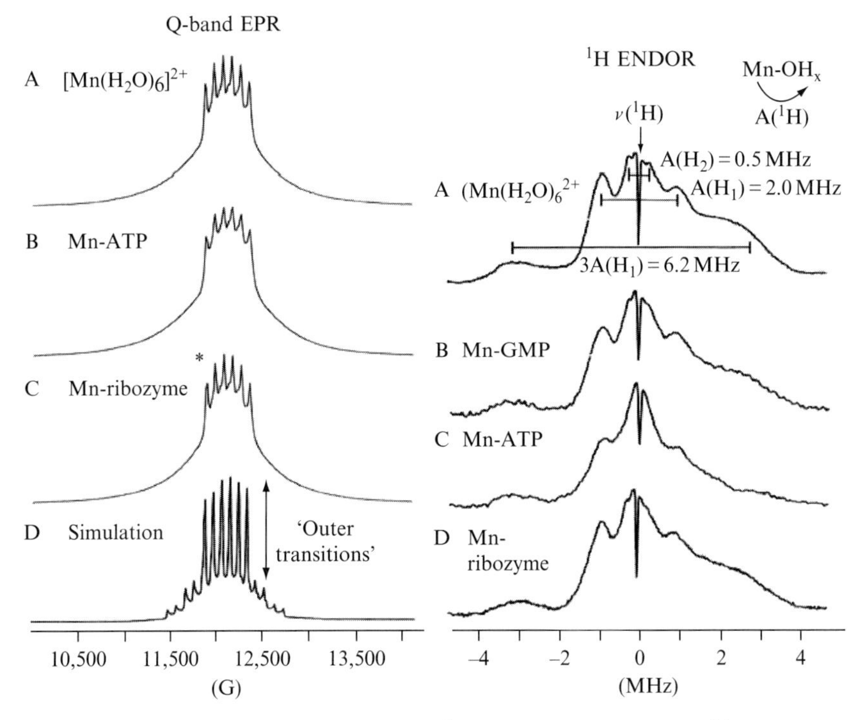

Figure 16.7 Q-band CW EPR (left) and ^{1}H ENDOR (right) of Mn^{2+}-complexes. At left, Q-band EPR spectra obtained in "rapid-passage," giving absorption lineshape. (A–C) All samples are 1 m*M* Mn^{2+} and ligand, and 1.0 *M* NaCl in 5 m*M* TEA, pH 7.9, and 20% (v/v) ethylene glycol. EPR parameters: 2.0 K, 1.7 mW, 0.5 G MA, and 1 scan each. (D) Mn^{2+} EPR spectra were simulated using SimFonia (Bruker). This program is based on perturbation theory and calculates allowed transitions for the situation in which the electronic Zeeman interaction is the dominating interaction. EasySpin (see text) is a more accurate program that includes ZFS and independent broadening functions. Simulation using the following parameters: $g = 2.00$, $A(^{55}\text{Mn}) = 91$ G; $D = 200$ G; line width 10 G, 33.90 GHz microwave frequency. The main six lines are from the "central" $M_S = \pm 1/2$ features, and the broad wings are from the underlying "outer" transitions (see Fig. 16.1). The asterisk indicates the position used for ENDOR spectra. At right, ^{1}H ENDOR spectra obtained for different complexes. Proton hyperfine couplings are assigned to two protons, H1 with A(H1) 2.0 MHz and H2 with $A(\text{H2}) = 0.5$ MHz. The hyperfine coupling of $A = 6$ MHz is assigned to H1 (a value of $\sim 3 \times A(\text{H1})$), but observed from the underlying $M_S = 1/2$–$3/2$ EPR feature which multiplies the observed hyperfine value by a factor of 3. Typical acquisition parameters: 1–10 mW microwave power, 0.5 G 100 kHz field modulation, 20 W applied RF power, 2.0–2.2 K, and 0.25–0.5 MHz/s RF sweep rate, 34 GHz operating frequency, $\sim$50 scans each. Data are taken from Morrissey *et al.* (2000).

a $\pi/2$–τ–π–τ-detect sequence, transfers a packet of electron spins from the z to the xy plane, includes an evolution/mixing time τ, refocuses, and detects the resulting echo. A 3-pulse sequence, $\pi/2$–τ –$\pi/2$–T–$\pi/2$–τ-detect

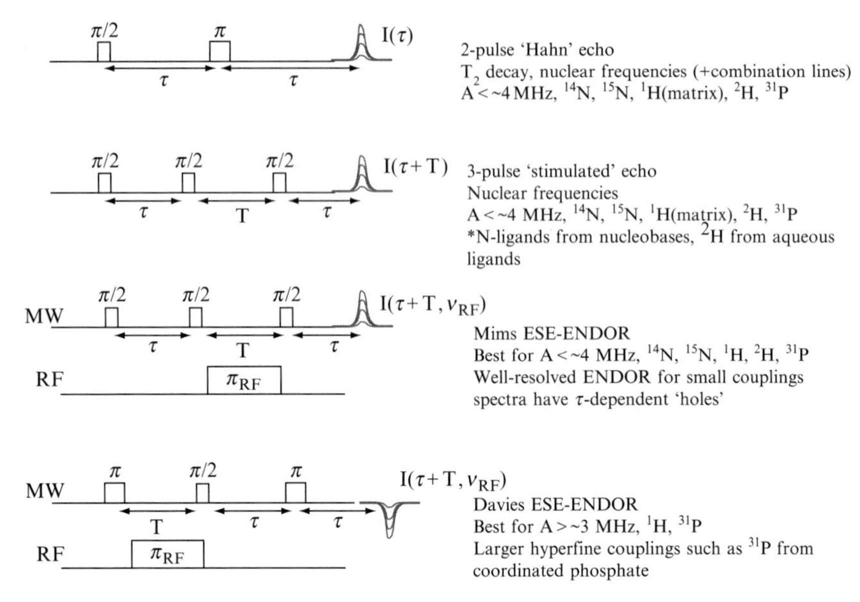

Figure 16.8 Common pulse sequences used in electron spin echo (ESE)-detected EPR, ESEEM, and ENDOR experiments and their (to date) practical uses for Mn^{2+}-RNA experiments.

sequence, stores magnetization in the *z*-axis for a time *T* before bringing it back for detection. In both cases, the amplitude of the echo, monitored across the magnetic field range of resonance, reconstructs an absorption-shaped EPR signal. Sitting at the desired magnetic field position of that signal, microwave pulses in the Hahn or other sequence are used to sculpt out a set of electron spins, and a separate RF pulse is added to create the ENDOR effect. Excitation between nuclear sublevels by the RF pulse moves the electron spins out of the coherence ("packet-shifting"), allowing readout of the ENDOR frequencies on the ESE signal.

The electron spin coherence and resulting electron spin echo required for pulsed ENDOR depends on length and power of the microwave pulse, which will be spectrometer-dependent, and on the interpulse timing. As will be described below for the ESEEM experiment, even in the absence of an RF excitation, hyperfine couplings create modulations on the amplitude of the electron spin echo that depend on both A and ν_N. For this reason, both the ESE-detected EPR signal and ENDOR spectra are sensitive to tau values.

Adequate cryoprotectant is critical for pulsed EPR applications and particularly so for the ENDOR pulse sequences. The ESE experiment is generally performed at 10–20 K. When paramagnetic systems are aggregated, both T_1 and T_2 are shortened. The ESE-ENDOR experiment requires sufficient time to include a long microsecond RF pulse; fast electron spin relaxation will reduce the ESE amplitude at the end of the pulse sequence, lowering the detection limits.

Two different pulsed ENDOR methods, Mims and Davies, are currently in major use with different optimal situations.

Mims ESE-ENDOR. "Mims" ENDOR is a robust method that is particularly useful for detection and accurate lineshape analysis of relatively small hyperfine couplings. In the case of Mn^{2+}, Mims ENDOR is perfect for ^{2}H, ^{15}N, and ^{14}N, all of which exhibit small ($A \lesssim 2$ MHz) hyperfine couplings because of a low magnetogyric ratio and the mainly ionic ligand interactions with Mn^{2+}. The Mims pulse sequence is given in Fig. 16.8. Examples of weak hyperfine couplings from directly coordinated ^{15}N or through-water coordination to phosphate provide example spectra in Fig. 16.9. In the Mims experiment, certain values of tau create holes or blind spots in the ENDOR spectrum when $[(A\tau) = 0, 1, 2, 3......]$. For the practical lower limit on τ of $\sim$120 ns, a "clean" Mims ENDOR can be achieved for $A \lesssim 5$ MHz. For other values of A, the τ-dependence of the spectrum can be valuable but must be considered. Many researchers will turn to the Davies sequence for larger A-values.

Davies ESE-ENDOR. The Davies ENDOR sequence (Fig. 16.8) also has a suppression region but it is for when A is approaching zero. Therefore, suppression effects in Davies ENDOR are somewhat opposite to those in the Mims method, and the two techniques are complementary. When there are overlapping ENDOR features, lineshapes in Davies ENDOR can be biased toward the stronger hyperfine coupling. In Mn^{2+}–RNA interactions, Davies ENDOR has been particularly powerful in measuring ^{31}P hyperfine couplings due to directly coordinated phosphate or phosphodiester groups, with values of $A(^{31}P) \sim 5$–8 MHz that are consistent with expectations based on models (Bennati *et al.*, 2006; Potapov and Goldfarb, 2006; Schiemann *et al.*, 2007; Walsby *et al.*, 2005). An example from the mHHRz is shown in Fig. 16.10. Davies ENDOR would also be appropriate for monitoring proton hyperfine couplings.

5. ESEEM Spectroscopy

ESEEM spectroscopy is a powerful method of detecting hyperfine interactions, and does not require application of an outside RF pulse. There is an upper limit on the detected hyperfine interactions, however, that is based on the fact that the microwave pulse can only excite a limited manifold of the spin system. Effects giving rise to strong ESEEM signals are also most dominant for nonisotropic hyperfine couplings. In general, ESEEM is most sensitive to weak, anisotropic hyperfine couplings. The two- or three-pulse ESE experiment detects electron spin coherences that

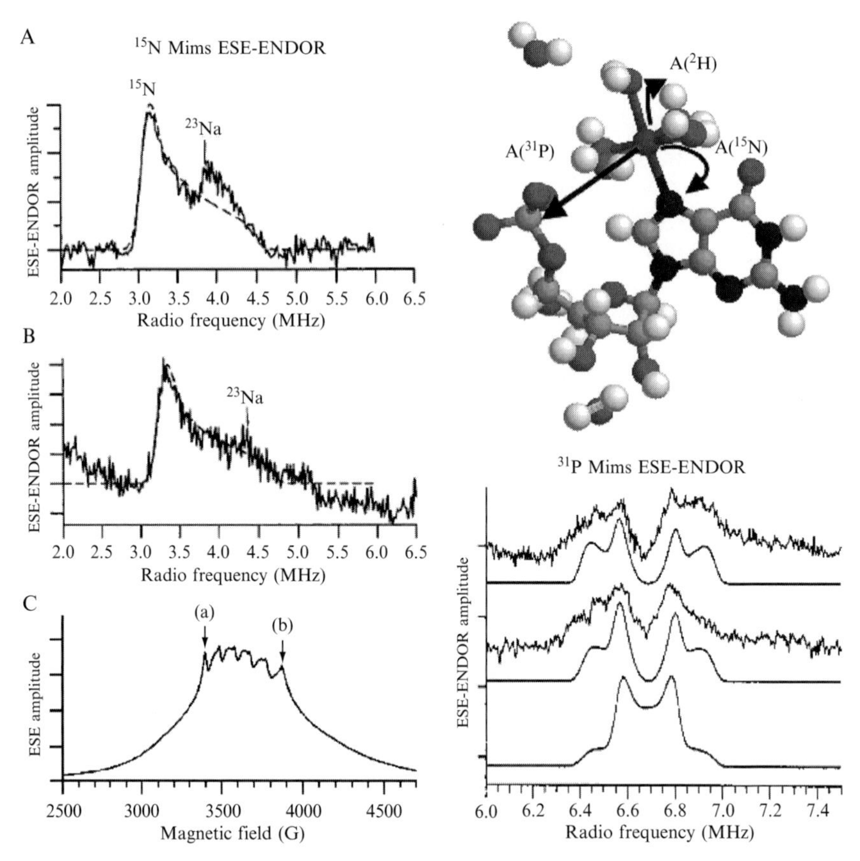

Figure 16.9 Mims ESE-ENDOR in Mn^{2+}-RNA and nucleotide complexes. At left, ^{15}N ENDOR from Mn^{2+}-mHHRz sample globally labeled with ^{15}N-guanine. Only the $\nu_+ = |\nu_N + A/2|$ feature is shown ($A_{iso}(^{15}N) = 4.2$ MHz), and interference from "matrix" ^{23}Na ENDOR is observed. The ESE-detected EPR signal shows the points of data collection. At right, ^{31}P ENDOR features from the hydrogen-bonded monophosphate of Mn-GMP are observed by Mims ESE-ENDOR. Note the <1 MHz resolution of very weak hyperfine coupling resulting from mainly a point-dipole through-space interaction simulated as $A_{iso} = 0.08$ MHz and $A_{dip} = 0.6$ MHz. For comparison, Davies ^{31}P ENDOR from directly coordinated Mn–phosphate interactions are shown in Fig. 16.10. HHRz sample is 1 m*M* mHHRz (^{15}N-guanine):1 m*M* Mn^{2+}, 1.0 *M* NaCl, 5 m*M* TEA pH 7.8, 20% (v/v) ethylene glycol. Mn-GMP sample is 5 m*M* GMP:1 m*M* Mn^{2+} in 10 m*M* sodium cacodylate, pH 7.0 with 0.4 *M* sucrose as cryoprotectant. Data collection: 10.2 GHz (X-band) ESE-ENDOR, 4.2 K, with τ values of 600–1000 ns and a 16 μS RF pulse for ^{31}P ENDOR. Figure adapted from Hoogstraten *et al.* (2002). The Mn–GMP model is based on an in-house X-ray crystal structure and shows the through-water ^{31}P interaction and through-bond ^{15}N interactions observed here by Mims ENDOR. In Fig. 16.11, three-pulse ESEEM analysis of Mn–N interactions is displayed and in Fig. 16.12, ^{2}H couplings from coordinated water ligands are "counted" using three-pulse ESEEM methods.

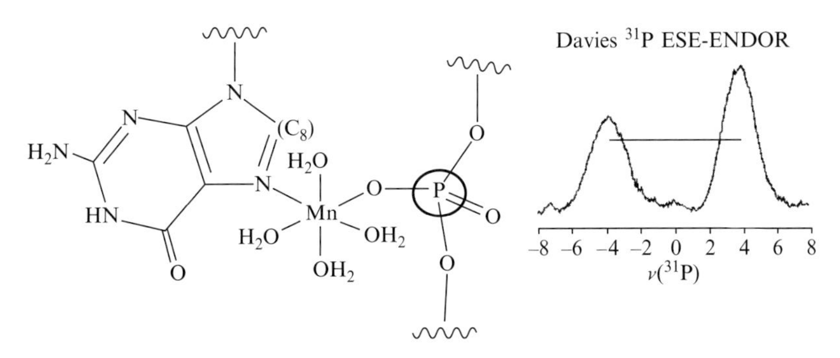

Figure 16.10 "Strong" ^{31}P couplings from Mn^{2+}-coordinated phosphate observed by Q-band Davies ESE-ENDOR. Data obtained on Mn^{2+}-mHHRz samples prepared as in Figs. 16.8 and 16.9 and under conditions described in Walsby *et al.* (2005). The two peaks are split by an observed $A(^{31}P) \sim 8$ MHz. Similar couplings in ^{31}P ENDOR data on the exHHRz have been obtained at W-band by Schiemann *et al.* (2007).

include hyperfine interactions from the surrounding nuclei, and the interpulse timing produces some selectivity for these interactions. Because of this, changes in the interpulse timing will produce different amplitudes of the resulting electron spin echo. A plot of ESE versus interpulse timing, τ or T in a two- or three-pulse experiment respectively, has a modulating amplitude superimposed on the overall decay of the echo height with longer time. Fourier transform (FT) analysis provides the frequencies and amplitudes associated with these modulations.

Like ENDOR, ESEEM frequencies occur at values related to both A and ν_N for every type of hyperfine-coupled nucleus. The amplitude of modulation depends on the number of nuclei, and on the relationship between nuclear-dependent frequencies. In order to produce deep modulation, similar values of $A/2$, ν_N, and (for $I>1/2$ nuclei) the quadrupole parameter e^2qQ are optimal. This situation occurs for ^{14}N coordinated to Mn^{2+}, in X-band spectrometers, making X-band ESEEM very useful for detecting Mn^{2+}-nucleobase coordination. The situation also occurs for the coupling between ^2H and Mn^{2+} in Mn–aqua (deutero) complexes. Interactions with phosphate have also been analyzed by ESEEM spectroscopy. For different types of nuclei coordinated to a Mn^{2+} ion, the resulting ESEEM spectrum is the *product* of the individual spectra. Because of this, ESEEM spectra can be used quantitatively to "count" nuclei, and this effect has been particularly exploited in analysis of metal ion hydration levels in RNA and nucleotide models. Hoogstraten and Britt have provided a methods article on this specific topic (Hoogstraten and Britt, 2002). Practically, for nucleic acids ESEEM has been particularly useful for detecting $^{14,15}N$ from coordinated nucleobases (Hoogstraten *et al.*, 2002; Kisseleva *et al.*, 2005; Morrissey *et al.*, 1999; Schiemann *et al.*, 2003; Vogt *et al.*, 2006) and for quantifying ^2H in ^2H_2O-exchanged samples (Hoogstraten and Britt, 2002; Vogt *et al.*, 2006).

5.1. Protocols for ESEEM spectroscopy of Mn^{2+} in RNA samples

5.1.1. Sample preparation

Sample preparation for ESEEM is the same as that for ENDOR experiments, with similar potential requirements in isotopic labeling. Whereas Q-band or higher frequencies have some advantages for ENDOR spectroscopy, X-band spectroscopy with some tunability (e.g., the 8–18 GHz instrument at UC-Davis) has been very productive for Mn^{2+} ESEEM analysis. This puts the RNA sample requirements at the higher end of $\sim$120 μL, 200–500 μM RNA, with adequate cryoprotection as described above.

5.1.2. ESEEM Spectroscopy

As with all sections, we will focus on some critical parameters for data collection but will not cover details of tuning the instrument, which will be specific to the spectrometer. As with ESE-ENDOR, the first step is to acquire an ESE-detected EPR signal using a two- or three-pulse sequence to generate and detect the electron spin echo amplitude. The magnetic field is then set at different positions of the EPR spectrum; for Mn^{2+}, usually the top of one of the major ^{55}Mn hyperfine lines. At this magnetic field position, two- or three-pulse ESEEM spectra are acquired as echo height versus τ (two-pulse) or $[\tau+T]$ (three-pulse) time. ESEEM data are processed by subtracting a background to account for the decay of the signal, and then by FT. The resulting frequency-dependent data are analyzed in terms of expected modulation frequencies and amplitudes. Data are acquired at different magnetic fields; ν_N is proportionate to magnetic field strength, whereas the hyperfine coupling values are not (assuming an isotropic system, as we are here for Mn^{2+}).

Two-pulse ESEEM. The most basic ESEEM experiment is application of a Hahn echo sequence with increasing values of τ. This resulting plot of echo height versus τ displays an overall decay related to T_2, and is modulated by nuclear frequencies. In a two-pulse experiment, nuclear modulation frequencies arise from fundamental "ENDOR" frequencies and also combinations of them, and result in both positive and negative peaks in the FT that can be complex but informative when carefully analyzed. Typical T_2 values for Mn^{2+} at 10 K are a few microseconds, which gives a short collection time and limits the ability to resolve lower frequency modulations in two-pulse ESEEM experiments. The three-pulse experiment below lengthens this observation window. From the two-pulse ESEEM experiment, an estimate of the T_2 value and the most evident nuclear interactions will be available. Protons will dominate this experiment as a large peak at the $\nu_N(^1H)$ value of $\sim$14 MHz in X-band ESEEM. ^{31}P may also be apparent ($\nu_N(^{31}P) \sim 7$ MHz at X-band magnetic fields), and, if there is coordination to the nitrogen of a nucleobase, strong peaks at frequencies $\lesssim$4 MHz due to ^{14}N.

Three-pulse ESEEM. The three-pulse ESEEM experiment can circumvent short phase (T_2) memories and provide clearer detection of low frequencies, such as those due to ^{14}N or ^{2}H nuclei. The frequencies detected in a three-pulse experiment do not have sum- and combination-features, simplifying the spectrum in comparison with the two-pulse ESEEM experiment. Three-pulse ESEEM can also be set up to suppress strong modulation due to protons by setting $\tau = n(\nu_N)$ ($n = 1,2,3$, etc.). Other modulations might be suppressed in these conditions, though, leading to τ-dependent ESEEM experiments. Analytical solutions to frequencies and amplitudes that are expected in three-pulse ESEEM experiments are available, though simulations of ESEEM data from high-spin Mn^{2+} are complicated by the $S = 5/2$ system.

Hoogstraten and coworkers have published a thorough analysis of Mn^{2+}-ESEEM data in Mn^{2+}-^{14}N GMP, Mn^{2+}-^{15}N GMP, and Mn^{2+} in the natural abundance and ^{15}N-G-labeled mHHRz (Hoogstraten *et al.*, 2002). A combination of Mims ENDOR and three-pulse ESEEM gives self-consistent results concerning a Mn^{2+}-N hyperfine coupling of A_{iso} (^{14}N) $\sim$ 2–3 MHz, $A_{dip}(^{14}N) \sim$ 0.4–0.6 MHz, and a nuclear quadrupole parameter $e^2qQ \sim 2.9$ MHz. With $\nu_N \sim 1.4$ MHz, these values clearly satisfy the condition for strong modulation in ESEEM spectroscopy that the nuclear and Larmor frequencies be very similar in magnitude (Figs. 16.11 and 16.12).

Four-pulse ESEEM and HYSCORE. The interpretation of ESEEM spectra can be complicated when there are overlapping features from the different nuclei, for example, frequencies due to ^{14}N from multiple coordinated ligands. In this instance, the two-dimensional (2D) hyperfine sublevel correlation experiment (HYSCORE) provides assignment of features to the same or different nucleus. HYSCORE uses a four-pulse ESEEM scheme. Data are collected as a function of two different interpulse times, and analyzed as the 2D FT. In a HYSCORE 2D analysis, crosspeaks are present for nuclear hyperfine features arising from the same nucleus. HYSCORE has been applied to the guanine nucleobase coordination to Mn^{2+} in the HHRz, showing the presence of only a single coordinated guanine (Kisseleva *et al.*, 2005; Schiemann *et al.*, 2003). Because it is a full 2D method, data acquisition for HYSCORE is time-intensive and analysis procedures can be complex. In the event of multiple ligands, however, this technique can be highly informative.

6. Summary

In this chapter, we have provided a summary of the ways in which EPR-based methods can be applied toward investigating Mn^{2+} coordination in RNA. Practical details regarding sample preparation, cryoprotectants,

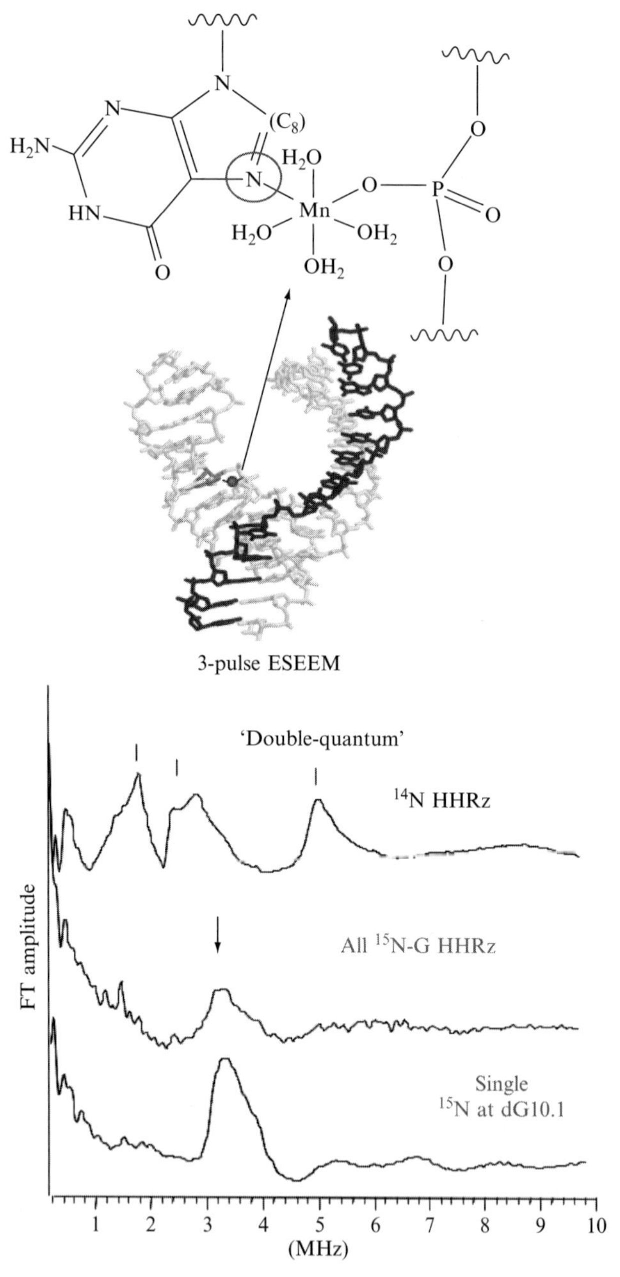

Figure 16.11 Three-pulse X-band ESEEM detection of ^{14}N and ^{15}N from Mn^{2+}-coordinated guanosine in the mHHRz. The three low-frequency peaks in the natural-abundance Mn^{2+}-mHHRz sample are characteristic for weakly coupled ^{14}N with properties as described in the text. When the ribozyme is substituted globally or with

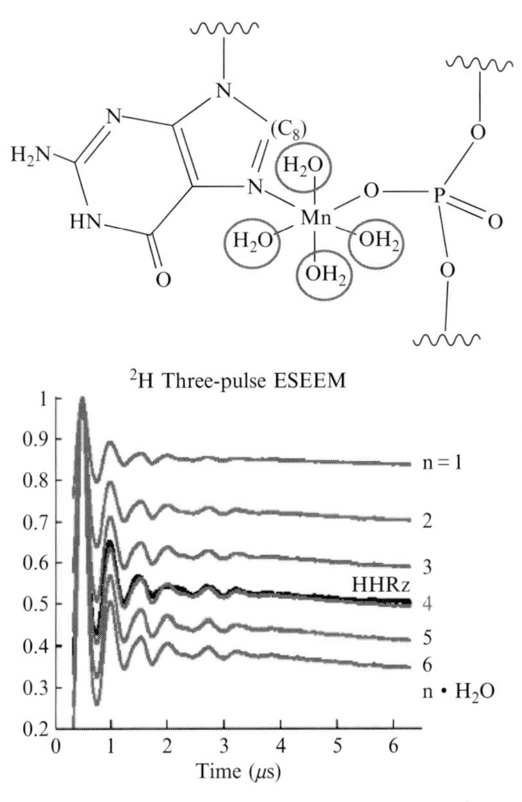

Figure 16.12 Three-pulse X-band ESEEM spectroscopy used for "counting" water ligands in Mn^{2+}-RNA sites by quantifying ^2H ESEEM amplitudes. This experiment, described in detail by Hoogstraten and Britt (2002), relies on the quantitative aspect of ESEEM spectroscopy. Mn^{2+}-RNA samples are exchanged into ^2H_2O, and their three-pulse ^2H ESEEM data obtained. In this case, 0.4 *M* sucrose is chosen as the appropriate cryoprotectant. The data are divided by three-pulse ESEEM of nonexchanged Mn-RNA samples to remove modulation from other nuclei. Resulting traces are compared to traces calculated for different levels of *n* aqua ligands, that is, Mn-$(^2H_2O)_n$. In this example, from the mHHRz, a clear number of four aqueous ligands are obtained. Data taken from Vogt *et al.* (2006). (See Color Insert.)

isotopic substitution are given, as well as basic strategies involved in data accumulation and analysis of EPR, ENDOR, and ESEEM data. Because of the complexities of ENDOR and ESEEM spectroscopic methods, this

---

a site-specific ^{15}N-labeled guanine, the ^{14}N features collapse into a single peak that is similar to that observed by ^{15}N ESE-Mims ENDOR (Fig. 16.9). These data unambiguously identify the site of Mn^{2+} coordination in this RNA. Mn^{2+}-RNA samples ~300–400 μM in same buffer conditions as in previous figures. ESEEM data collected with $\tau = 192$ ns (to suppress ^1H modulation at 3600 Gauss field strength), a 15 ns $\pi/2$ pulse, and 3.2 W microwave power. Data taken from Vogt *et al.* (2006).

chapter is meant to be used as a practical aid in collaborative efforts rather than as stand-alone instruction.

EPR-based methods have been used extensively to characterize metal sites in proteins, providing identities of metal ligands and changes in coordination environment that are correlated to function. At this time, however, there are very few examples of RNAs that have been explored in detail using these methods. Poising an RNA sample such that one or a small subset of Mn^{2+} sites are populated is a requirement for the high-resolution ENDOR and ESEEM methods described here. Judicious choice of supporting cations, such as high concentrations of monovalent ions or mixtures of diamagnetic Mg^{2+} or Ca^{2+} with the paramagnetic Mn^{2+} ions, are potential routes toward selective population of Mn^{2+} in high-affinity RNA sites. Once the appropriate sample is achieved, however, the spectroscopic methods described here are fairly well developed and can be used to unambiguously identify metal sites as well as understand their properties in greater detail. Under-developed, important areas in metal–RNA interactions include the influence of electrostatics on the electronic properties of bound metal ions, and the specificity of hydrogen bonding between metal ligands and the RNA environment. In addition to providing ligand identification and quantitation, EPR-based techniques are uniquely capable of addressing these subtle interactions.

ACKNOWLEDGMENTS

Our pulsed EPR experiments have been carried out as collaborations with members of the laboratory of Prof. R. David Britt at the University of California-Davis, including Charles Hoogstraten (current position: Michigan State University, Department of Biochemistry) and Dr. Simanti Lahiri. Original sample preparation and protocols related to ENDOR and ESEEM spectroscopy of Mn^{2+} in the hammerhead ribozyme were developed by Dr. Susan R. Morrissey, Roxanne Clardy, and Dr. Thomas 'Ed' Horton. The contributions of Dr. Tomek Wasowicz in developing Q-band ENDOR spectroscopy at Texas A&M University are remembered with gratitude. We acknowledge use of protocols and data from the Ph.D. thesis work of Dr. Shannon Burns. Dr. Charles Walsby, and Sarah Tate (Stefanutti) obtained ^{31}P ESE-ENDOR on HHRz samples in the laboratory of Prof. Brian Hoffman at Northwestern University. Unpublished work from the DeRose laboratory is supported by the National Institutes of Health (NIH-GM58096).

REFERENCES

Bennati, M., *et al.* (2006). High-frequency 94 GHz ENDOR characterization of the metal binding site in wild-type Ras center dot GDP and its oncogenic mutant G12V in frozen solution. *Biochemistry* **45,** 42–50.

Britt, R. D. (2003). Electron spin echo methods: A tutorial. *In* "Paramagnetic Resonance of Metallobiomolecules," (J. Telser, ed.), ACS Symposium Series, pp. 16–54.

Carmieli, R., *et al.* (2001). Proton positions in the Mn^{2+} binding site of concanavalin A as determined by single-crystal high-field ENDOR spectroscopy. *J. Am. Chem. Soc.* **123,** 8378–8386.

Cohn, M., and Townsend, J. (1954). A study of manganous complexes by paramagnetic resonance absorption. *Nature* **173,** 1090–1093.

Danchin, A., and Gueron, M. (1970). Cooperative binding of manganese (II) to transfer RNA. *Eur. J. Biochem.* **16,** 532–536.

Das, R., *et al.* (2005). Determining the Mg^{2+} stoichiometry for folding an RNA metal ion core. *J. Am. Chem. Soc.* **127,** 8272–8273.

DeRose, V. J. (2008). Characterization of nucleic acid metal ion binding by spectroscopic techniques. *In* "Nucleic Acid–Metal Ion Interactions," (N. V. Hud, ed.), pp. 154–175. Royal Society of Chemistry, Cambridge.

DeRose, V. J., and Hoffman, B. M. (1995). Protein structure and mechanism studied by electron nuclear double resonance spectroscopy. *Methods Enzymol.* **246,** 554–589.

DeRose, V. J., *et al.* (2003). RNA and DNA as ligands. *In* "Comprehensive Coordination Chemistry II," (J. A. McCleverty and T. J. Meyers, eds)., Elsevier, Oxford.

Doan, P. E., *et al.* (2007). The ups and downs of Feher-style ENDOR. *Appl. Magn. Reson.* **31,** 649–663.

Edwards, T. E., and Sigurdsson, S. T. (2005). EPR spectroscopic analysis of U7 hammerhead ribozyme dynamics during metal ion induced folding. *Biochemistry* **44,** 12870–12878.

Ennifar, E., *et al.* (2003). A crystallographic study of the binding of 13 metal ions to two related RNA duplexes. *Nucleic Acids Res.* **31,** 2671–2682.

Feher, G. (1956). Observation of nuclear magnetic resonances via the electron spin resonance line. *Phys. Rev.* **103,** 834.

Freisinger, E., and Sigel, R. K. O. (2007). From nucleotides to ribozymes—A comparison of their metal ion binding properties. *Coord. Chem. Rev.* **251,** 1834–1851.

Grant, C. V., *et al.* (2000). Solid-state ^{25}Mg NMR of a magnesium(II) adenosine 5′-triphosphate complex. *J. Am. Chem. Soc.* **122,** 11743–11744.

Hoffman, B. M. (2003). Electron-nuclear double resonance spectroscopy (and electron spin-echo envelope modulation spectroscopy) in bioinorganic chemistry. *Proc. Natl. Acad. Sci. USA* **100,** 3575–3578.

Hoogstraten, C. G., and Britt, R. D. (2002). Water counting: Quantitating the hydration level of paramagnetic metal ions bound to nucleotides and nucleic acids. *RNA* **8,** 252–260.

Hoogstraten, C. G., *et al.* (2002). Structural analysis of metal ion ligation to nucleotides and nucleic acids using pulsed EPR spectroscopy. *J. Am. Chem. Soc.* **124,** 834–842.

Horton, T. E., and DeRose, V. J. (2000). Cobalt hexammine inhibition of the hammerhead ribozyme. *Biochemistry* **39,** 11408–11416.

Horton, T. E., *et al.* (1998). Electron paramagnetic resonance spectroscopic measurement of Mn^{2+} binding affinities to the hammerhead ribozyme and correlation with cleavage activity. *Biochemistry* **37,** 18094–18101.

Hunsicker, L. M., and DeRose, V. J. (2000a). Activities and relative affinities of divalent metals in unmodified and phosphorothioate-substituted hammerhead ribozymes. *J. Inorg. Biochem.* **80,** 271–281.

Hunsicker, L. M., and DeRose, V. J. (2000b). Activities and relative affinities of divalent metals in unmodified and phosphorothioate-substituted hammerhead ribozymes. *J. Inorg. Biochem.* **80,** 271–281.

Juneau, K., *et al.* (2001). Structural basis of the enhanced stability of a mutant ribozyme domain and a detailed view of RNA–solvent interactions. *Structure* **9,** 221–231.

Kisseleva, N., *et al.* (2005). Binding of manganese(II) to a tertiary stabilized hammerhead ribozyme as studied by electron paramagnetic resonance spectroscopy. *RNA* **11,** 1–6.

Marathias, V. M., *et al.* (1996). Determination of the number and location of the manganese binding sites of DNA quadruplexes in solution by EPR and NMR in the presence and absence of thrombin. *J. Mol. Biol.* **260,** 378–394.
Markham, G. D., *et al.* (1979). Analysis of EPR powder pattern lineshapes for Mn(II) including 3rd-order perturbation corrections—Applications to Mn(II) complexes with enzymes. *J. Magn. Reson.* **33,** 595–602.
Martell, A. E., and Smith, R. E. (1971). Critical Stability Constants. Plenum, New York.
Milligan, J. F., and Uhlenbeck, O. C. (1989). Synthesis of small RNAs using T7 RNA polymerase. *Methods Enzymol.* **180,** 51–62.
Misra, V. K., *et al.* (2003). A thermodynamic framework for the magnesium-dependent folding of RNA. *Biopolymers* **69,** 118–136.
Morrissey, S. R., *et al.* (1999). Mn^{2+}–nitrogen interactions in RNA probed by electron spin-echo envelope modulation spectroscopy: Application to the hammerhead ribozyme. *J. Am. Chem. Soc.* **121,** 9215–9218.
Morrissey, S. R., *et al.* (2000). Mn^{2+} sites in the hammerhead ribozyme investigated by EPR and continuous-wave Q-band ENDOR spectroscopies. *J. Am. Chem. Soc.* **122,** 3473–3481.
Mustafi, D., *et al.* (2003). Catalytic and structural role of the metal ion in dUTP pyrophosphatase. *Proc. Natl. Acad. Sci. USA* **100,** 5670–5675.
Ottaviani, M. F., *et al.* (1996). Characterization of starburst dendrimers by EPR.4. Mn(II) as a probe of interphase properties. *J. Phys. Chem.* **100,** 11033–11042.
Potapov, A., and Goldfarb, D. (2006). Quantitative characterization of the Mn^{2+} complexes of ADP and ATP gamma S by W-band ENDOR. *Appl. Magn. Reson.* **30,** 461–472.
Prisner, T., *et al.* (2001). Pulsed EPR spectroscopy: Biological applications. *Annu. Rev. Phys. Chem.* **52,** 279–313.
Reed, G. H., and Poyner, R. R. (2000). Mn^{2+} as a probe of divalent metal ion binding and function in enzymes and other proteins. *In* "Metal Ions in Biological Systems," Vol. 37 (H. Sigel and A. Sigel, eds.), pp. 183–207. Marcel Dekker, New York.
Santangelo, M. G., *et al.* (2007). Structural analysis of Cu(II) ligation to the 5′-GMP nucleotide by pulse EPR spectroscopy. *J. Biol. Inorg. Chem.* **12,** 767–775.
Schiemann, O., *et al.* (2003). Structural investigation of a high-affinity MnII binding site in the hammerhead ribozyme by EPR spectroscopy and DFT calculations. Effects of neomycin B on metal-ion binding. *Chembiochem* **4,** 1057–1065.
Schiemann, O., *et al.* (2007). W-band ^{31}P ENDOR on the high-affinity Mn^{2+} binding site in the minimal and tertiary stabilized hammerhead ribozymes. *Appl. Magn. Reson.* **31,** 543–552.
Shi, H. J., and Moore, P. B. (2000). The crystal structure of yeast phenylalanine tRNA at 1.93 angstrom resolution: A classic structure revisited. *RNA* **6,** 1091–1105.
Smith, T. J., *et al.* (2002). Paramagnetic spectroscopy of vanadyl complexes and its application to biological systems. *Coord. Chem. Rev.* **228,** 1–18.
Smoukov, S. K., *et al.* (2002). EPR study of substrate binding to the Mn(II) active site of the bacterial antibiotic resistance enzyme FosA: A better way to examine Mn(II). *J. Am. Chem. Soc.* **124,** 2318–2326.
Stich, T. A., *et al.* (2007). Multifrequency pulsed EPR studies of biologically relevant manganese(II) complexes. *Appl. Magn. Reson.* **31,** 321–341.
Stoll, S., and Schweiger, A. (2006). EasySpin, a comprehensive software package for spectral simulation and analysis in EPR. *J. Magn. Res.* **178,** 42–55.
Tan, X. L., *et al.* (1993). Pulsed and continuous wave electron nuclear double-resonance patterns of aquo protons coordinated in frozen solution to high-spin Mn^{2+}. *J. Chem. Phys.* **98,** 5147–5157.
Thomann, H., and Bernardo, M. (1993). Pulsed electron-nuclear multiple resonance spectroscopic methods for metalloproteins and metalloenzymes. *In* "Methods in Enzymology,"

Vol. 227, Metallobiochemistry, Pt. D (J. F. Riordan and B. L. Vallee, eds.), pp. 118–189. Academic Press, San Diego.

Travers, K. J., *et al.* (2007). Low specificity of metal ion binding in the metal ion core of a folded RNA. *RNA* **13,** 1205–1213.

Vogt, M. (2004). PhD Thesis, Texas A&M University.

Vogt, M., *et al.* (2006). Coordination environment of a site-bound metal ion in the hammerhead ribozyme determined by ^{15}N and ^{2}H ESEEM spectroscopy. *J. Am. Chem. Soc.* **128,** 16764–16770.

Walsby, C. J., *et al.* (2005). Enzyme control of small-molecule coordination in FosA as revealed by P-31 pulsed ENDOR and ESE-EPR. *J. Am. Chem. Soc.* **127,** 8310–8319.

Woody, A. Y. M., *et al.* (1996). Asp537 and Asp812 in bacteriophage T7 RNA polymerase as metal ion-binding sites studied by EPR, flow-dialysis, and transcription. *Biochemistry* **35,** 144–152.

SECTION THREE

RNA THERMODYNAMICS

CHAPTER SEVENTEEN

Optical Melting Measurements of Nucleic Acid Thermodynamics

Susan J. Schroeder* *and* Douglas H. Turner†

Contents

Abstract

Optical melting experiments provide measurements of thermodynamic parameters for nucleic acids. These thermodynamic parameters are widely used in RNA structure prediction programs and DNA primer design software. This review briefly summarizes the theory and underlying assumptions of the method and provides practical details for instrument calibration, experimental design, and data interpretation.

1. Introduction

A theory is the more impressive the greater the simplicity of its premises is, the more different *[sic]* kinds of things it relates, and the more extended is its area of applicability. Therefore, the deep impression which classical

* Department of Chemistry and Biochemistry, University of Oklahoma, Norman, Oklahoma, USA
† Department of Chemistry, University of Rochester, Rochester, New York, USA

Methods in Enzymology, Volume 468
ISSN 0076-6879, DOI: 10.1016/S0076-6879(09)68017-4

thermodynamics made upon me. It is the only physical theory of universal content concerning which I am convinced that, within the framework of the applicability of its basic concepts, it will never be overthrown.

Albert Einstein (Einstein, 1970)

Nucleic acid folding is one area where the basic concepts of thermodynamics have found wide ranging applicability. RNA thermodynamic parameters have applications to diverse areas of study such as rhinovirus evolution and recombination (Palmenberg *et al.*, 2009) antisense therapeutics, for example, Vitravene, which is the first FDA-approved nucleic acid therapeutic and which targets cytomegalovirus in the human eye (Anderson *et al.*, 1996) models of the HIV-1 RNA structural elements (Parisien and Major 2008; Wilkinson *et al.*, 2008) cancer microRNA target specificity (Doench and Sharp, 2004); the mechanisms of RNA interference (Ameres *et al.*, 2007); the mechanism of group I introns (Bevilacqua and Turner, 1991; Narlikar *et al.*, 1997; Pyle *et al.*, 1994); the discovery of noncoding RNAs in genomes (Uzilov *et al.*, 2006; Washietl *et al.*, 2005); and tRNA codon recognition in protein translation (Ogle *et al.*, 2002). In principle, thermodynamics can predict the populations of structures that would be present at equilibrium, although the current knowledge of the sequence dependence of nucleic acid thermodynamics limits the accuracy of such predictions. Much of the known thermodynamics has been measured by optical melting, which has several advantages over the more accurate calorimetric methods. Relatively small quantities of sample are required; the experiments are fast; and the instrumentation is relatively inexpensive. For example, if two 8-mer RNA oligonucleotides with internal loops are predicted to have different stabilities, with only approximately 1 μmol of each RNA and one day of optical melting experiments by a hard-working student, one can determine which internal loop is more thermodynamically stable. (Very few bets in the RNA world can be resolved so quickly!) This chapter provides details on the optical melting methods used most often, and includes both technical aspects and a discussion of the assumptions in interpretation.

2. Instrumentation

UV spectrometers suitable for optical melting experiments are commercially available from Beckman, Cary, and Shimadzu corporations. The primary requirements in a UV spectrometer are good optics; accurate, variable temperature control; and a cell holder for several small cuvettes. This article will discuss details of the Beckman DU800 spectrometer, but the general principles apply to all UV spectrometers. The Beckman DU800 spectrometer specifications for temperature are ± 1 °C from 20 to 60 °C with the DU800 high-performance temperature controller unit, although the instrument range is 13–95 °C. A customized cell holder with chilled water circulation to remove heat from the peltier-controlled cell holder allows accurate ±1 °C temperature control to 0 °C. Dry air or nitrogen gas flowing through the cell chamber prevents condensation on the cells at low temperatures. The microcell holder contains places for six cuvettes and uses the cell transporter unit. Standard Beckman cells have a 1 cm pathlength and a 400 μL volume. Custom quartz cells with pathlengths of 0.1 cm, 0.5 cm, 1.0 cm and volumes of 40 μL, 200 μL, and 400 μL, respectively, in dimensions that fit into the Beckman cell holder can be obtained from Hellma, Inc. and NSG Precision Cells.

3. Calibrations

The Beckman DU800 spectrometer software automatically runs several initialization calibration tests when the instrument is turned on. These tests are run with no samples in the instrument and the lid closed. The initialization tests check the gain, the visible lamp, the light path, the shutter, the filter, the wavelength drive, and the detector performance. Turn the instrument power off when not in use, so that these calibrations are automatically checked every time the instrument is used. In addition, the performance validation tests following the manufacturer's instructions should be run monthly to insure reliable instrument performance. The performance validation checks the wavelength accuracy (±0.2 nm); wavelength repeatability (±0.1 nm); resolution (<1.80 nm); baseline flatness (<0.0010 A); noise at 500 nm (<0.000200 A); and stability at 340.0 nm for 60 min (<0.0030 A drift). ("A" is a unit of absorbance defined by the NIST 930D solid filter at 546 nm.) Additional checks for temperature, absorbance, pathlength, and cell holder alignment can be done manually at installation and as necessary during use.

The temperature can be manually tested with a microprobe, such as the Ertco-eutechnics digital thermometer model 4400. Test the accuracy of the microprobe thermometer in a water bath and compare with an accurate mercury thermometer. Fill six 1-cm cuvettes with double distilled water and seal five cuvettes with Teflon tape and stoppers. Insert the temperature probe into one of the cuvettes and seal with the small stopper around the probe. Check that the probe is directly upright in the cell and does not touch the sides of the cuvette. Keep the lid closed as much as possible while measuring temperature. The temperature of the cell holder can be manually set, and the actual temperature recorded by the software appears in the lower right hand corner of the screen. Check that the temperature is within ±1 °C at all cell positions at several temperatures, for example, 5, 15, 25, 35,

45, and 55 °C. Allow approximately 5 min for equilibration at each temperature. When checking the temperature at higher than 60 °C, take care to note any water evaporation. The temperature measurement will not be accurate if the cell is not full of water. Then check the temperature as if a melting experiment were being run with a heating rate of 0.5 or 1 °C per minute from 0 to 90 °C. Check the temperature in one of the first three and in one of the last three cell positions to insure that the peltier devices embedded in the bottom of the cell holder are accurately matched. The nitrogen flow, the chilled water flow, and the rate of heating can be adjusted so that the actual temperature of the cell matches the recorded temperature. Alternatively, temperature offsets can be included in the data analysis if there is a reproducible difference in temperature between cells, although this is not recommended.

The wavelength accuracy can be measured using a holmium oxide filter in place of the cell holder and scanning wavelength from 200 to 800 nm at a rate of 1200 nm/min. The shape and position of the peaks are the important features of the spectrum rather than the exact intensities. There should be three strong distinct peaks between 440 and 460 nm. Peaks should be clearly distinguished at 241.5, 279.3, 287.6, 333.8, 360.8, 385.8, 418.5, 453.4, 459.9, 536.4, and 637.5 ±0.2 nm (Allen, 2007). Any peaks appearing below 225 nm indicate stray light in the instrument. If this scan does not show the appropriate peaks, then replace the UV bulb following the manufacturer's instructions or troubleshoot other possible problems in the optics.

The absorbance may be checked by measuring the absorbance of a known stock solution, such as 0.00400 g/L of $K_2Cr_2O_7$ in 0.05 *M* KOH, referenced to a 0.05 *M* KOH solution at 25 °C. (Table 17.1) (Gordon and Ford, 1972). Measure the same sample in the same cell in each cell position to check the cell holder alignment. If the absorbance varies more than ±0.0005 A at different cell positions, then rerun the transporter alignment with no samples in the cell holder. The service

Table 17.1 Absorbance for 0.00400 g/L of $K_2Cr_2O_7$ in 0.05 *M* KOH[a]

λ (nm)	A	λ (nm)	A	λ (nm)	A
220	0.446	315	0.046	400	0.396
230	0.171	330	0.149	420	0.124
240	0.295	340	0.316	440	0.054
250	0.496	350	0.559	460	0.018
260	0.633	360	0.830	480	0.004
275	0.757	370	0.987	500	0.000
290	0.428	375	0.991		
300	0.149	390	0.695		

[a] Values are from Gordon and Ford (1972).

diagnostics calibrations can be run by a Beckman service technician to correct the alignment. Use the same solution with a known absorbance in cells of each pathlength and verify the pathlength accuracy using Beer's Law:

$$A = \varepsilon c l \tag{17.1}$$

where A is absorbance; ε is the extinction coefficient; c is concentration; and l is the pathlength.

4. Brief Theory of Optical Melting Experiments

In principle, optical melting curves could be analyzed by a partition function approach in which every base pair is considered separately. This approach, however, would require a global fit of melting data for many sequences and refitting the data when additional sequences are added. Therefore, data are typically analyzed with a two-state model, which assumes that each strand is either completely paired or unpaired. The equilibrium for duplex formation is represented as either a self-complementary or non-self-complementary association (Cantor and Schimmel, 1980; Turner, 2000):

$$2\text{A} \leftrightarrow \text{A}_2 \tag{17.2}$$

$$\text{B} + \text{C} \leftrightarrow \text{B·C} \tag{17.3}$$

The equilibrium for a unimolecular transition, for example, a hairpin, is represented as:

$$\text{D} \leftrightarrow \text{E} \tag{17.4}$$

For self-complementary (Eq. (17.2)) or non-self-complementary equilibria (Eq. (17.3) with equal concentrations of B and C), the equilibrium constant is given by:

$$K = (\alpha/2)/(C_\text{T}/a)(1-\alpha)^2 \tag{17.5}$$

C_T is the total strand concentration:

$$C_\text{T} = [\text{A}] + 2[\text{A}_2] \quad \text{self-complementary} \tag{17.6}$$

$$C_\text{T} = [\text{B}] + [\text{C}] + 2[\text{B·C}] \quad \text{non-self-complementary} \tag{17.7}$$

"a" has a value of 1 for self-complementary duplexes and 4 for non-self-complementary duplexes. α is the fraction of strands in a duplex. For a unimolecular transition,

$$K = \alpha/(1-\alpha) = [\text{E}]/[\text{D}] \tag{17.8}$$

where

$$\alpha = [\mathrm{E}]/([\mathrm{D}] + [\mathrm{E}]) \tag{17.9}$$

Figure 17.1 shows data for the non-self-complementary duplex, 5′-GAGC**GA**CGAC-3′/3′-CUCG**AAG**GCUG-5′. At low temperatures, the strands are in a duplex and the absorbance is low. As the temperature is increased, the duplex dissociates into single strands. The total concentration of strands can be measured using the absorbance at 80 °C and the extinction coefficient for that sequence. The difference in absorbance between duplex and single strands is the hyperchromicity. For self-complementary or non-self-complementary duplexes with equal concentrations of each strand, the melting temperature, T_M, in kelvins or T_m in degrees Celsius, is the point at which the concentrations of strands in duplex and in single strands are equal. The steepness of the transition indicates the cooperativity of the transition. The width and maximum of first derivative of the melting curve can also indicate the cooperativity and melting temperature, although the peak of the derivative curve only occurs at the T_M when the transition is unimolecular (Gralla and Crothers, 1973; Marky and Breslauer, 1987). T_M is most accurately measured by fitting the lower and upper baselines. The melting temperature is measured at several concentrations over a 100-fold range and then plotted versus the concentration in a van't Hoff plot. The van't Hoff equation relates the melting temperature in kelvins (T_M), total strand concentration (C_T), enthalpy (ΔH°), and entropy (ΔS°):

$$1/T_M = (R/\Delta H^\circ)\ln(C_T/a) + \Delta S^\circ/\Delta H^\circ \tag{17.10}$$

where R is the ideal gas constant, 1.987 cal K^{-1} mol^{-1} or 8.3145 J K^{-1} mol^{-1}. The slope of the van't Hoff plot gives the enthalpy change, and the y-intercept gives the ratio of entropy change to enthalpy change. The free energy and equilibrium constant at any temperature can then be calculated using Gibb's relation:

$$\Delta G^\circ = \Delta H^\circ - T\Delta S^\circ, \quad K = e^{-\Delta G^\circ/RT} \tag{17.11}$$

The errors in enthalpy and entropy changes are typically around 10%. Because these errors are correlated, the errors in free energy change are typically 2%.

The melting curve can be fit using Meltwin software (McDowell and Turner, 1996), although other software is available to perform the same mathematical analysis (Draper *et al.*, 2001; Siegfried and Bevilacqua, 2009). The Meltwin software uses seven parameters to fit each curve (Fig. 17.1): the total strand concentration, enthalpy and entropy changes of the transition, the slope and intercept of the lower baseline (double-stranded region), and the slope and intercept of the upper baseline (single-stranded region). The total strand concentration is the only nonfloating parameter and is determined at a high-temperature absorbance point. A Marquardt–Levenberg

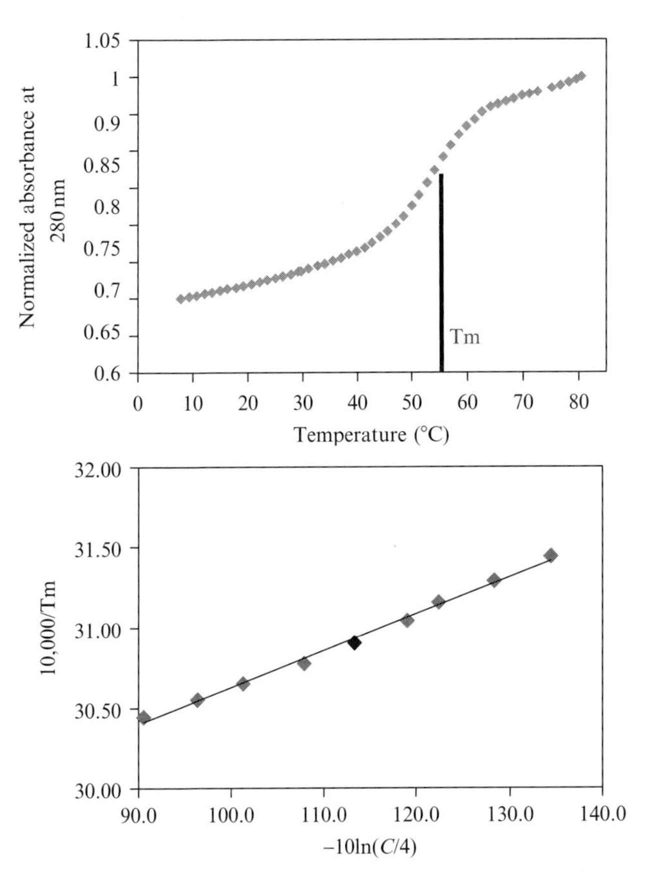

Figure 17.1 Optical melting data. The normalized UV absorbance (280 nm) versus temperature (°C) curve (top). Van't Hoff plot (bottom): $1/T_M$ versus ln (C_T/a).

fitting routine is applied to each curve to find the best parameter values. The parameters and fitting follow an Ising model:

$$A_T = (1 - \alpha)A_{RC} + \alpha A_{ST} \tag{17.12}$$

where A_T is the total absorbance, A_{RC} is the absorbance of the random coil or single strands, A_{ST} is the absorbance of the stacked duplex, and α is the fraction of strands in the duplex conformation. The truncation points for the fitting can be selected by the user. It is important to have enough points in the upper and lower baselines to find a good fit, although with a low T_m the points in the lower baseline may be few in practice. At higher temperatures, it is important to avoid evaporation effects, which cause the absorbance to appear higher than true values, and to exclude such points from the fit. The position of the truncation points of the fit can have a significant impact on the values of the enthalpy parameter.

5. Two-State Assumption

An important assumption in the analysis described above is that only two states exist for the RNA: single strands or duplex. Any intermediates between these two conformations are assumed to be at very low concentration and not to contribute significantly to the absorbance. Also, the duplex must fold into a single conformation with no alternative folds. If the values for the enthalpy change calculated according to the fit of the melting curves and the van't Hoff plot agree within 15%, then the results are consistent with the two-state assumption. Although the exact cutoff value for consistent two-state behavior is a matter of debate and interpretation, if the enthalpy values calculated with the two different fitting methods are not consistent, then the analysis does not provide valid thermodynamic parameters.

6. ΔC_p^o Assumption

The above equations neglect the temperature dependence of ΔH^o and ΔS^o, which for a constant heat capacity, ΔC_p^o, is given by:

$$\Delta H^o(\mathrm{T}) = \Delta H^o_{\mathrm{T_o}} + \Delta C_p^o(T - T_o) \tag{17.13}$$

$$\Delta S^o(\mathrm{T}) = \Delta S^o_{\mathrm{T_o}} + \Delta C_p^o \ln(T/T_o) \tag{17.14}$$

Usually, ΔC_p^o will not be zero because stacking in the single strands is temperature dependent (Holbrook *et al.*, 1999). The best way to determine ΔC_p^o is by isothermal titration calorimetry at different temperatures (Diamond *et al.*, 2001; Mikulecky and Feig, 2006). It is sometimes possible, however, to obtain estimates from optical melting data by plotting the fitted values of ΔH^o from individual melting curves versus T_M and the fitted values of ΔS^o versus $\ln T_M$ (Diamond *et al.*, 2001; Petersheim and Turner, 1983). This is best attempted when the ΔH^o is not large so that there is a large concentration-dependence of T_M. In general, however, the experimental errors in optical melting data do not allow accurate determination of ΔC_p^o (Chaires, 1997). Fortunately, the systematic errors in fitting $\Delta \mathrm{H}^o$ and $\Delta \mathrm{S}^o$ due to neglecting ΔC_p^o compensate each other so that the ΔG^o is a reliable parameter.

7. Experimental Design

When designing sequences to study the thermodynamic stabilities of different noncanonical RNA motifs, consider carefully the stems' stabilities, the expected T_m, the advantages of non-self-complementary or self-complementary systems, and possible alternate folds. When choosing stem duplexes for internal or multibranch loop, the predicted thermodynamic stabilities of each stem should be close so that the duplexes unfold in a two-state manner. The predicted T_m of the folded forms should be as close to 37 °C as possible to minimize the extrapolation for calculating ΔG^o_{37}. Self-complementary systems have the advantages of simplicity and not requiring determination of oligonucleotide concentrations prior to the melt because mixing of equimolar amounts of two different strands is not required. A self-complementary system also enables incorporation of two motifs in one duplex, thus doubling the effect of the motif. This is particularly useful when the free energy increment of the noncanonical motif is small, such as for dangling ends. While non-self-complementary duplexes require accurately mixing equimolar amounts of two strands, they have the potential to "mix and match" different strands and thus enable a wider sequence diversity within a motif using fewer different sequences.

The possibilities of alternate folds should be considered even for short sequences. Sometimes changing a Watson–Crick base pair distant from the intended noncanonical pair can change the propensity for alternate duplex formation. Prediction programs such as mfold (Zuker, 1989) and RNAstructure (Mathews *et al.*, 2004) can generate some suboptimal folds for both unimolecular and bimolecular folds. A simple graphical method for finding alternate folds is to create a grid as shown in Fig. 17.2. Every possible Watson–Crick pairing is assigned an X, potentially stable noncanonical pairs such as GU can be assigned an O. Possible helices appear as a diagonal line of X's and O's. For example, in the sequence on the left in Fig. 17.2, the intended duplex has two GU pairs at the ends of the helix; the thermodynamic stability of GU is often idiosyncratic and depends on the helix position. The only alternative duplexes have positive predicted free energy and are thus unlikely to form. In contrast, the sequence on the right differs only in the order of the two middle GC pairs, but this slight change creates the possibility for several more stable alternative duplexes. The middle GC pairs are unlikely to affect the measurement of the thermodynamic stability of the terminal GU pairs. Keep in mind that these are predicted free energy values; and the idiosyncratic nature of non-Watson–Crick pairs, such as GU pair stabilities, is the reason to continue measuring thermodynamic parameters with optical melting experiments.

One-dimensional proton NMR of 0.2–1 m*M* RNA in 90% H_2O and 10% D_2O, 100 m*M* NaCl, 10 m*M* phosphate, and 0.5 m*M* Na_2EDTA in a 500 MHz spectrometer is a quick way to check that the expected duplex forms. The extensive dialysis and sample preparation necessary for high-quality 2D NMR spectra is often not necessary for a quick 1D proton spectrum. The correct number and chemical shift of imino protons can verify that the proton spectrum is consistent with the intended duplex design. The imino protons in the pairs at the terminus of a duplex may be

	5′G	U	A	U	**C**	G	A	U	G	U
U	O		X			O	X		O	
G		O		O	X			O		O
U	O		X			O	X		O	
A		X		X				X		X
G		O		O	X			O		O
C	X					X			X	
U	O		X			O	X		O	
A		X		X				X		X
U	O		X			O	X		O	
5′G		O		O	X			O		O

5′-GUAUCGAUGU-3′ $\Delta G° = -7.0$ kcal/mol
3′-UGUAGCUAUG-5′

5′-GUAUCGAUGU-3′ $\Delta G° = +0.2$ kcal/mol
3′-UGUACGUAUG-5′

5′-GUAUGCAUGU-3′ $\Delta G° = +1.8$ kcal/mol
3′-UGUACGUAUG-5′

	5′G	U	A	U	**G**	**C**	A	U	G	U
U	O		X		O		X		O	
G		O		O		X		O		O
U	O		X		O		X		O	
A		X		X				X		X
C	X				X				X	
G		O		O		X		O		O
U	O		X		O		X		O	
A		X		X				X		X
U	O		X		O		X		O	
5′G		O		O		X		O		O

5′-GUAUGCAUGU-3′ $\Delta G° = -7.6$ kcal/mol
3′-UGUACGUAUG-5′

5′-GUAUGCAUGU-3′ $\Delta G° = -8.0$ kcal/mol
3′-UGUACGUAUG-5′

5′-GUAUGCAUGU-3′ $\Delta G° = -6.1$ kcal/mol
3′-UGUACGUAUG-5′

Figure 17.2 Plots to facilitate sequence design. The X's represent possible Watson–Crick pairs. The O's represent possible GU pairs. The intended duplex with terminal GU pairs is shown as the central diagonal. Possible alternative duplexes are shown as shorter diagonals. The sequence differences between the left and right examples are highlighted in bold. The predicted free energies at 37 °C of the possible duplexes are calculated according to (Clanton-Arrowood *et al.*, 2008; He *et al.*, 1991; Mathews *et al.*, 1999, 2004; Miller *et al.*, 2008; O'Toole *et al.*, 2005, 2006; Xia *et al.*, 1998). Note that the nearest-neighbor parameter for 5′-GU/3′-UG is known to be context-dependent and shows non-nearest neighbor effects. In this example, the lowest energy possibility for the terminal mismatches was used. The left hand sequence is preferred because the desired pairing is predicted to be considerably more stable than other pairings.

missing or weak due to exchange with water and fraying at the duplex ends. Too many imino protons implies that more than one conformation of the RNA is stable. The imino protons in GC and AU Watson–Crick pairs resonate between 11-13 ppm and 13-15 ppm, respectively. Imino protons in stable mismatches may be protected from exchange with water and then resonate anywhere between 9.5 and 15 ppm depending on the conformation and protonation of the mismatch (Santa Lucia *et al.*, 1991; Schroeder and Turner, 2000). These resonances for non-Watson–Crick pairs can provide strong supporting evidence that different hydrogen-bonded conformations are the basis for the sequence dependence of the thermodynamic stability. For example, wobble GU pairs show imino proton resonances between 10 and 11.5 ppm and a strong NOE between the two imino protons (Schroeder and Turner, 2001).

The buffer used for optical melting experiments is typically 1 *M* NaCl, 10 or 20 m*M* sodium phosphate or sodium cacodylate at pH 7, and 0.5 m*M* Na_2EDTA. One molar NaCl was initially chosen to stabilize short RNA sequences when RNA synthesis was difficult and time-consuming and has become the standard salt concentration. The phosphate or cacodylate buffers maintain a constant pH over a wide temperature range. Because cacodylate anion has an arsenic atom, the buffer can be stored without concerns about bacterial growth. The buffering range of cacodylate is 5.0–7.4 with a pK_a of 6.3. Noncanonical pairs such as $A^+{\cdot}C$ or $C^+{\cdot}C$ are more stable at low pH, and increased stability at low pH provides evidence for the formation of protonated pairs (Santa Lucia *et al.*, 1991). The Na_2EDTA chelates any divalent cations that promote RNA hydrolysis, especially at high temperatures. Other buffers may be used to test the effects of salt on thermodynamic stabilities. A typical buffer that may resemble more physiological salts is 0.15 *M* KCl, 10 m*M* $MgCl_2$, and 20 m*M* sodium cacodylate, pH 7. When using buffers that contain magnesium, however, the samples cannot be diluted and used again for a melt at another concentration because magnesium facilitates hydrolysis at high temperatures.

If a transition is truly two-state, then the thermodynamics should be the same when measured at any wavelength at which the folded and unfolded states have different absorbances (Cantor and Schimmel, 1980). Typically, 260 or 280 nm light is used. For sequences with a high fraction of GC or AU pairs, 280 or 260 nm, respectively, are preferred in order to maximize the hyperchromicity (Fresco *et al.*, 1963). Sometimes local information can be deduced from optical melting curves at other wavelengths. One case is when a nucleotide absorbs outside the region where the rest of the nucleotides absorb. Measurements at 296 nm provide information on global conformation, and the absorption is due to an n to $\pi^\star$ transition (Testa and Gilham, 1993). DNA duplexes with Hoogsteen pairs also have a signature melting profile at 295 nm (Mergny *et al.*, 2005; Miyoshi *et al.*, 2009).

The purity of oligonucleotides for optical melting experiments should be at least 90% as measured by HPLC or by gel electrophoresis on a denaturing gel. The RNA can be diluted to provide 10 concentrations over a 100-fold range. The maximum and minimum accurate absorbance of the Beckman spectrometer is 2.5 and 0.2, approximately a 10-fold range. The use of cuvettes with 0.1 cm and 1.0 cm pathlengths enables another 10-fold range in concentration. A typical RNA duplex concentration range is from 2.5×10^{-4} *M* to 2.5×10^{-6} *M*, and concentration ranges from 60-fold to 150-fold are common. Thus, use Beer's Law and the sequence-dependent extinction coefficient to calculate the necessary dilutions of the RNA samples with optical melting buffer (Cantor and Schimmel, 1980). Unless magnesium is used in the buffer, five sample concentrations can be prepared, run in an optical melting experiment, and then diluted and used again in the next optical melting experiment. Signal-to-noise is usually not a consideration with modern spectrophotometers, but if it is, then the optimum absorbance is 0.434 if the noise is not due to statistical fluctuations in the number of photons hitting the detector (shot noise) (Hammes, 1974; Turner, 1986).

8. Data Interpretation

Analysis of optical melting experiments provides thermodynamic parameters for enthalpy, entropy, and free energy changes for duplex formation. Optical melting experiments, however, do not provide definitive information about the duplex structure, hydrogen bonding, or heat capacity. Although pairing patterns are sometimes inferred from the sequence dependence of RNA motifs, NMR or crystallography is required to provide definitive information about the structure of the RNA duplex. For example, hydrogen-bonded GA and GG pairs are relatively stable thermodynamically in certain contexts (Burkard and Turner, 2000; Schroeder and Turner, 2000; Walter *et al.*, 1994; Wu and Turner, 1996); however, a combination of optical melting experiments and NMR spectroscopy was necessary to determine this. As a consequence of these results, when GA or GG pairs have the potential to form in loops, the possibility of forming a GA or GG pair is often invoked to extend empirical rules to thermodynamic stabilities of unmeasured sequences.

Optical melting studies provide estimates of ΔH° and ΔS° and therefore ΔG° on the basis of the two-state model. The most reliable values for ΔG° are those near the melting temperatures of the experiment. Relative stabilities depend not only on melting temperature but also on the enthalpy change. For example, if duplex A and duplex B have melting temperatures of 70 and 60 °C, respectively, then duplex A is more stable at 70 °C if both RNA concentrations are the same. If the enthalpy of duplex A is larger than the enthalpy of duplex B, however, duplex B may be the more stable duplex at 37 °C. The ΔG° at the temperature of interest is the true measure of relative stability.

9. Error Analysis

There are many possible sources of error in optical melting studies. SantaLucia and Turner (SantaLucia and Turner, 1997) list two sources of random error: (1) signal-to-noise ratio of the absorbance measurements and (2) variations in sample preparation, and four sources of systematic error: (a) incorrect calibration, (b) a non-two-state transition, (c) incorrect choice of baselines, and (d) neglect of $\Delta C_{\mathrm{p}}^{\circ}$.

Sampling errors in $1/T_{\mathrm{M}}$ versus ln C_{T} plots and in fitted data provide measures of the random errors. Equations for calculating sampling error are given by SantaLucia and Turner (1997) and Xia *et al.* (1998). The random errors from $1/T_{\mathrm{M}}$ versus ln C_{T} plots are typically on the order of 3%, 3%, and 1% for ΔH°, ΔS°, and ΔG° near the T_{M}, respectively (Xia *et al.*, 1998). Random errors from averaging values from fitting curves are typically two- to threefold larger. In a study of 51 Watson–Crick complementary duplexes, the errors in ΔS° were about 13% larger than in ΔH° because the uncertainty in ΔS° depends on more terms (Xia *et al.*, 1998). The error in ΔG° is smaller than that in ΔH° and ΔS° because the errors in ΔH° and ΔS° are highly correlated in a compensating manner. Random errors in T_{m} are typically 1–2 °C.

The magnitudes of systematic errors are difficult to estimate. Optical melting results from different laboratories provide one measure. For three DNA sequences reported by separate laboratories (Breslauer *et al.*, 1986; SantaLucia *et al.*, 1996; Sugimoto *et al.*, 1996), differences were 6%, 6%, 3%, and 1 °C for ΔH°, ΔS°, ΔG_{37}°, and T_{M}, respectively (SantaLucia and Turner, 1997). Values of ΔG_{50}° for four sequences measured by substrate inhibition in a group I ribozyme reaction (Narlikar *et al.*, 1997) differed an average of 8% from those measured by optical melting (Pyle *et al.*, 1994).

When comparing the free energies of two short RNA duplexes, 0.5 kcal/mol is a reasonable rule of thumb for estimating a significant difference in thermodynamic stabilities. The calculated error in duplex free energy is typically ± 0.2 kcal/mol when using Meltwin software. Non-nearest neighbor effects due to the length of the duplex stems or the position of the mismatch in a helix can be approximately 0.5 kcal/mol (Kierzek *et al.*, 1999; Schroeder and Turner, 2000). The rules for predicting thermodynamic stabilities of duplexes include terms with values less than 0.5 kcal/mol, however. These terms are calculated from linear regression analysis of typically 50–200 duplex free energies, and the terms are justified by statistical significance.

10. Summary

Optical melting experiments can provide a large number of thermodynamic measurements from which generalized rules for predicting nucleic acid stabilities can be derived. These thermodynamic measurements and rules form the core of most RNA and DNA structure prediction algorithms. Contrafold (Do *et al.*, 2006), a new algorithm based on computational conditional training methods and databases of known RNA structures (Griffiths-Jones *et al.*, 2003, 2005), calculated base stacking energies with the same rank order as the nearest neighbor parameters measured by optical melting experiments. This result supports the use of thermodynamic parameters and free energy minimization to predict the structure of functional RNA conformations. Thus, thermodynamic analysis and optical melting experiments are useful tools for exploring the unknown landscapes of the RNA world amidst a flood of sequencing information, low free energy valleys, and peaks of activation energies in ribozymes.

Acknowledgements

The authors thank Koree Clanton-Arrowood, Nic Hammond, Biao Liu, and Mai-Thao Nguyen for critical reading of the manuscript. We thank all our students who remind us that teaching and learning are a dynamic equilibrium with arrows that go both ways. D. H. Turner is supported by NIH grant #GM22939. S.J. Schroeder is supported by NSF grant #0844913 and grants from the Oklahoma Center for the Advancement of Science and Technology Plant Science Research Program, the Pharmaceutical Research and Manufacturers of America Foundation, and the American Cancer Society Institutional Research Grant to the Oklahoma Health Science Center.

References

Allen, D. (2007). Holmium oxide glass wavelength standards. *J. Res. Natl. Inst. Stand. Technol.* **112,** 303–306.

Ameres, S. L., Martinez, J., and Schroeder, R. (2007). Molecular basis for target RNA recognition and cleavage by human RISC. *Cell* **130,** 101–112.

Anderson, K., Fox, M., Brown-Driver, V., Martin, M., and Azad, R. (1996). Inhibition of human cytomegalovirus immediate-early gene expression by an antisense oligonucleotide complementary to immediate-early RNA. *Antimicrob. Agents Chemother.* **40,** 2004–2011.

Bevilacqua, P. C., and Turner, D. H. (1991). Comparison of binding of mixed ribose-deoxyribose analogues of CUCU to a ribozyme and to GGAGAA by equilibrium dialysis: Evidence for ribozyme specific interactions with 2′OH groups. *Biochemistry* **30,** 10632–10640.

Breslauer, K. J., Frank, R., Blocker, H., and Marky, L. A. (1986). Predicting DNA duplex stability from the base sequence. *Proc. Natl. Acad. Sci. USA* **83,** 3746–3750.

Burkard, M. E., and Turner, D. H. (2000). NMR structures of r(GCAGGCGUGC)2 and determinants of stability for single guanosine-guanosine base pairs. *Biochemistry* **39,** 11748–11762.

Cantor, C. R., and Schimmel, P. R. (1980). *Biophysical Chemistry* W.H. Freeman and Company, New York.

Chaires, J. B. (1997). Possible origin of differences between van't Hoff and calorimetric enthalpy estimates. *Biophys. Chem.* **64,** 15–23.

Clanton-Arrowood, K., McGurk, J., and Schroeder, S. J. (2008). 3′-Terminal nucleotides determine thermodynamic stabilities of mismatches at the ends of RNA helices. *Biochemistry* **47,** 13418–13427.

Diamond, J. M., Turner, D. H., and Mathews, D. H. (2001). Thermodynamics of three-way multibranch loops in RNA. *Biochemistry* **40,** 6971–6981.

Do, C., Woods, D., and Batzglou, S. (2006). CONTRAfold: RNA secondary structure prediction without physics-based models. *Bioinformatics* **22,** e90–e98.

Doench, J. G., and Sharp, P. A. (2004). Specificity of microRNA target selection in translational repression. *Genes Dev.* **18,** 504–511.

Draper, D. E., Bukham, Y. V., and Gluick, T. C. (2001). Thermal methods for the analysis of RNA folding pathways. *Curr. Protoc. Nucleic Acid Chem.* **11.3.**

Einstein, A. (1970). Autobiographical Notes on Albert Einstein: Philosopher-Scientist. Cambridge University Press, London.

Fresco, J. R., Klotz, L. C., and Richards, E. G. (1963). A new spectroscopic approach to the determination of helical secondary structure in ribonucleic acids. *Cold Spring Harbor Symp. Quant. Biol.* **28,** 83–90.

Gordon, A., and Ford, R. (1972). *The Chemist's Companion: A Handbook of Practical Data, Techniques, and References* Wiley, New York.

Gralla, J., and Crothers, D. M. (1973). Free energies of imperfect nucleic acid helices III. Small internal loops resulting from mismatches. *J. Mol. Biol.* **78,** 301–309.

Griffiths-Jones, S., Bateman, A., Marshall, M., Kharna, A., and Eddy, S. R. (2003). Rfam: An RNA family database. *Nucleic Acids Res.* **31,** 439–441.

Griffiths-Jones, S., Moxon, S., Marshall, M., Kharna, A., Eddy, S. R., and Bateman, A. (2005). Rfam: Annotating non-coding RNAs in complete genomes. *Nucleic Acids Res.* **33,** D121–D124.

Hammes, G. G. (1974). Temperature-jump methods. *In* "Techniques of Chemistry, Vol. 6," (G. G. Hammes, ed.), Wiley-Interscience New York.

He, L., Kierzek, R., SantaLucia, J. Jr., Walter, A. E., and Turner, D. H. (1991). Nearest-neighbor parameters for GU mismatches. *Biochemistry* **30,** 11124–11132.

Holbrook, J., Capp, M., Saeker, R., and Record, M. T. Jr. (1999). Enthalpy and heat capacity changes for formation of an oligomeric DNA duplex: Interpretation in terms of coupled processes of formation and association of single-stranded helices. *Biochemistry* **38,** 8409–8422.

Kierzek, R., Burkard, M. E., and Turner, D. H. (1999). Thermodynamics of single mismatches in RNA duplexes. *Biochemistry* **38,** 14214–14223.

Marky, L. A., and Breslauer, K. J. (1987). Calculating thermodynamic data for transitions of any molecularity from equilibrium melting curves. *Biopolymers* **26,** 1601–1620.

Mathews, D. H., Sabina, J., Zuker, M., and Turner, D. H. (1999). Expanded sequence dependence of thermodynamic parameters improves prediction of RNA secondary structure. *J. Mol. Biol.* **288,** 911–940.

Mathews, D. H., Disney, M. D., Childs, J. L., Schroeder, S. J., Zuker, M., and Turner, D. H. (2004). Incorporating chemical modification constraints into a dynamic programming algorithm for prediction of RNA secondary structure. *Proc. Natl. Acad. Sci. USA* **101,** 7287–7292.

McDowell, J. A., and Turner, D. H. (1996). Investigation of the structural basis for thermodynamic stabilities of tandem GU mismatches: Solution structure of (rGAG-GUCUC)2 by two-dimensional NMR and simulated annealing. *Biochemistry* **35,** 14077–14089.

Mergny, J.-L., Li, J., Lacroix, L., Amrane, S., and Chaires, J. B. (2005). Thermal difference spectra: A specific signature for nucleic acid structures. *Nucleic Acids Res.* **33,** e138.

Mikulecky, P., and Feig, A. (2006). Heat capacity changes associated with nucleic acid folding. *Biopolymers* **82,** 38–58.

Miller, S., Jones, L. E., Giovannitti, K., Piper, D., and Serra, M. J. (2008). Thermodynamic analysis of 5' and 3' single- and 3' double nucelotide overhangs neighboring wobble terminal base pairs. *Nucleic Acids Res.* **36,** 5652–5659.

Miyoshi, D., Nakamura, K., Tateishi-Karimata, H., Ohmichi, T., and Sugimoto, N. (2009). Hydration of Watson–Crick base pairs and dehydration of Hoogsteen base pairs inducing structural polymorphism under molecular crowding conditions. *J. Am. Chem. Soc.* **131,** 3522–3531.

Narlikar, G., Kosla, M., Usman, N., and Herschlag, D. (1997). Quantitating tertiary binding energies of 2'OH groups on the P1 duplex of the *Tetrahymena* ribozyme: Intrinsic binding energy in an RNA enzyme. *Biochemistry* **36,** 2465–2477.

Ogle, J., Murphy, F., Tarry, M., and Ramakrishnan, V. (2002). Selection of tRNA by the ribosome requies a transition from an open to a closed form. *Cell* **111,** 721–732.

O'Toole, A. S., Miller, S., and Serra, M. J. (2005). Stability of 3' double nucleotide overhangs that model the 3' ends of siRNA. *RNA* **11,** 512–516.

O'Toole, A. S., Miller, S., Haines, N., Zink, M. C., and Serra, M. J. (2006). Comprehensive thermodynamic analysis of 3' double nucleotide overhangs neighboring Watson–Crick terminal base pairs. *Nucleic Acids Res.* **34,** 3338–3344.

Palmenberg, A., Spiro, D., Kuzmickas, R., Wang, S., Djikeng, A., Rather, J., Fraser-Liggett, C., and Liggett, S. (2009). Sequencing and analyses of all known human rhinovirus genomes reveal structure and evolution. *Science* **324,** 55–59.

Parisien, M., and Major, F. (2008). The MC-Fold and MC-Sym pipeline infers RNA structure from sequence data. *Nature* **452,** 51–55.

Petersheim, M., and Turner, D. H. (1983). Base-stacking and base-pairing contributions to helix stability: Thermodynamics of double-helix formation with CCGG, CCGGp, CCGGAp, ACCGGp, CCGGUp, and ACCGGUp. *Biochemistry* **22,** 256–263.

Pyle, A. M., Moran, S., Strobel, S. A., Chapman, T., Turner, D. H., and Cech, T. R. (1994). Replacement of the conserved GU with a GC pair at the cleavage site of the *Tetrahymena* ribozyme decreases binding reactivity and fidelity. *Biochemistry* **33,** 13856–13863.

SantaLucia, J. Jr., Kierzek, R., and Turner, D. H. (1991). Stabilities of consecutive A·C, C·C, G·G, U·C, and U·U mismatches in RNA internal loops: Evidence for stable hydrogen-bonded U·U and C·C$^+$ pairs. *Biochemistry* **30,** 8242–8251.

SantaLucia, J. Jr., and Turner, D. H. (1997). Measuring the thermodynamics of RNA secondary structure formation. *Biopolymers* **44,** 309–319.

SantaLucia, J. Jr., Allawi, H. T., and Seneviratne, P. A. (1996). Improved nearest-neighbor parameters for predicting DNA duplex stability. *Biochemistry* **35,** 3555–3562.

Schroeder, S. J., and Turner, D. H. (2000). Factors affecting the thermodynamic stability of small asymmetric internal loops in RNA. *Biochemistry* **39,** 9257–9274.

Schroeder, S. J., and Turner, D. H. (2001). Thermodynamic stabilities of internal loops with GU closing pairs in RNA. *Biochemistry* **40,** 11509–11517.

Siegfried, N. A., and Bevilacqua, P. C. (2009). Thinking inside the box: designing, implementing, and interpreting thermodynamic cycles to dissect cooperativity in RNA and DNA folding. *Methods Enzymol.* **455,** 365–393.

Sugimoto, N., Nakano, S., Yoneyama, M., and Honda, K. (1996). Improved thermodynamic parameters and helix initiation factor to predict stability of DNA duplexes. *Nucleic Acids Res.* **24,** 4501–4505.

Testa, S. M., and Gilham, P. T. (1993). Analysis of oligonucleotide structure using hyperchromism measurements at long wavelengths. *Nucleic Acids Res.* **21,** 3907–3908.

Turner, D. H. (1986). Temperature jump methods. *In* "Techniques of Chemistry, Vol. 6," (C. Bernasconi, ed.), Wiley-Interscience, New York.

Turner, D. H. (2000). Conformational changes. *In* "Nucleic Acids: Structures, Properties, and Functions," (V. A. Bloomfield, D. M. Crothers, and I. Tinoco Jr., eds.) pp. 259–334. University Science Books, Sausalito, CA.

Uzilov, A. V., Keegan, J. M., and Mathews, D. H. (2006). Detection of non-coding RNAs on the basis of predicted secondary structure formation free energy change. *BMC Bioinformatics* **7,** 173–203.

Walter, A. E., Wu, M., and Turner, D. H. (1994). The stability and structure of tandem GA mismatches in RNA depend on closing base pairs. *Biochemistry* **33,** 11349–11354.

Washietl, S., Hofacker, I. L., and Stadler, P. F. (2005). Fast and reliable prediction of noncoding RNAs. *Proc. Natl. Acad. Sci. USA* **102,** 2454–2459.

Wilkinson, K. A., Gorelick, R. J., Vasa, S. M., Guex, N., Rein, A., Mathews, D. H., Giddings, M. C., and Weeks, K. M. (2008). High-throughput SHAPE analysis reveals structures in HIV-1 genomic RNA strongly conserved across distinct biological states. *PLoS Biol.* **6,** e96.

Wu, M., and Turner, D. H. (1996). Solution structure of r(GCGGACGC)2 by two-dimensional NMR and the iterative relaxation matrix approach. *Biochemistry* **35,** 9677–9689.

Xia, T., SantaLucia, J. Jr., Burkard, M. E., Kierzek, R., Schroeder, S. J., Jiao, X., Cox, C., and Turner, D. H. (1998). Thermodynamic parameters for an expanded nearest-neighbor model for formation of RNA duplexes with Watson–Crick base pairs. *Biochemistry* **37,** 14719–14735.

Zuker, M. (1989). On finding all suboptimal foldings of an RNA molecule. *Science* **244,** 48–52.

CHAPTER EIGHTEEN

Analyzing RNA and DNA Folding Using Temperature Gradient Gel Electrophoresis (TGGE) with Application to *In Vitro* Selections

Durga M. Chadalavada *and* Philip C. Bevilacqua

Contents

Abstract

Gel electrophoresis is a ubiquitous separation technique in nucleic acid biochemistry. Denaturing gel electrophoresis separates nucleic acids on the basis of length, while native gel electrophoresis separates nucleic acids on the basis of both shape and length. Temperature gradient gel electrophoresis (TGGE), in which a temperature gradient is present across the gel, combines the advantages of denaturing and native gel electrophoresis by having native gel-like properties at low temperatures and denaturing gel-like properties at high temperatures. We describe here the techniques of perpendicular and parallel TGGE and some of their applications. Isolation of stable and unstable RNA and DNA sequences from combinatorial libraries is accomplished with TGGE-SELEX, while thermodynamic

Department of Chemistry, The Pennsylvania State University, University Park, Pennsylvania, USA

Methods in Enzymology, Volume 468
ISSN 0076-6879, DOI: 10.1016/S0076-6879(09)68018-6

characterization of an RNA tertiary motif is performed by perpendicular TGGE-melts. Specific examples are chosen from the literature to illustrate the methods. TGGE provides a powerful biophysical approach for analyzing RNA and DNA that complements more traditional methodologies.

1. Introduction

The native fold of a biopolymer is essential to its biological function. The fraction of molecules that occupy the native fold is dictated by the difference in free energy between the native and unfolded states. This free energy difference is often measured experimentally by denaturing the biopolymer by chemical or thermal means, and both denaturing approaches have been coupled to gel electrophoresis as ways of analyzing nucleic acid stability. In this section of the chapter, we provide definitions and surveys of some of the applications of denaturing gradient gel electrophoresis (DGGE) and temperature gradient gel electrophoresis (TGGE) to biophysical characterization of RNA and DNA in order to motivate the subsequent two sections on method development.

Separation of long dsDNA molecules that differ by as little as one base pair was initially demonstrated by chemical denaturation using DGGE (Fischer and Lerman, 1983), wherein a linear gradient of urea or formamide is prepared in the gel perpendicular to the direction of the electric field (Fischer and Lerman, 1979; Myers *et al.*, 1985). While this method met with some initial successes, it suffered from difficulty in pouring gradients in a stable and reproducible manner.

Subsequently, thermal denaturation through TGGE was developed as a more robust method for the investigation of the melting behavior of long DNAs and RNAs (Rosenbaum and Riesner, 1987; Thatcher and Hodson, 1981). In general, a thermal gradient is more reproducible and easier to fine tune than a denaturing gradient. The TGGE apparatus used in our laboratory was adapted from Wartell *et al.* (1990), as described in Section 2.2.

TGGE can be used to separate a mixture of identical length nucleic acids on the basis of their relative thermodynamic parameters. As such, TGGE can provide information about mutational variation, mismatches in base-pairing, and helix-coil transitions. In the case of simple helices, single base changes can be detected (Bevilacqua and Bevilacqua, 1998; Zhu and Wartell, 1997, 1999). For example, Wartell *et al.* (1990) detected AT to TA transversions in long dsDNAs using vertical TGGE, and determined the thermal stabilities of short DNA fragments that differ by only a single unpaired base (Ke and Wartell, 1995). In addition, RNA hairpin oligonucleotides differing by only 0.3 kcal/mol in ΔG^0_{37} have been separated in our laboratory by parallel TGGE (Bevilacqua and Bevilacqua, 1998), as described in Section 2.4. Similarly, Kang *et al.* (1994) used a horizontal

TGGE setup to analyze two highly cooperative, temperature-dependent conformational transitions in a dsRNA, and thermal stabilities of base pair stacking interactions and single mismatches in long RNA have been measured by TGGE (Zhu and Wartell, 1997).

TGGE can also be applied to analyze conformational transitions and sequence variations of biological RNAs and protein–RNA interactions. It was shown by TGGE that circular viroid RNA undergoes one highly cooperative transition to a slower migrating band (Riesner *et al.*, 1989). In addition, these investigators reported distinct denaturation curves for viral DNA genomes that differ only slightly in their base composition or sequence from TGGE. Crameri *et al.* (1995) successfully analyzed genomic reassortants of two serotypes of bluetongue virus (BTV) using TGGE, where traditional SDS-PAGE methods were found to be ineffective. Sequence variants of other dsRNA viruses have also been successfully analyzed for minor nucleotide sequence changes (Steger *et al.*, 1987).

More recently, the bacterial composition of water bodies has been examined by TGGE of 16S rRNA genes and revealed complex band patterns suggestive of high levels of diversity of bacterial species in this habitat (Beier *et al.*, 2008). In another study, PCR and TGGE were combined for tracing the strain types, contamination routes, and sources of *Listeria monocytogenes* (Tominaga, 2007). Other applications for TGGE include examining melting profiles of individual DNA topoisomers in the presence of DNA binding drugs (Danko *et al.*, 2005).

In our research, we have combined the ability of TGGE to separate mixtures of RNA or DNA with *in vitro* selection (or SELEX) (Ellington and Szostak, 1990; Robertson and Joyce, 1990; Tuerk and Gold, 1990) and identified rare thermodynamically stable and unstable sequences (Bevilacqua and Bevilacqua, 1998; Nakano *et al.*, 2002; Proctor *et al.*, 2002; Shu and Bevilacqua, 1999). In addition, the Cech lab has applied TGGE to the thermodynamic characterization (Guo *et al.*, 2006; Szewczak *et al.*, 1998) and selection (Guo and Cech, 2002) of RNA tertiary motifs, and in some cases RNA tertiary motifs with enhanced stability have been found to have favorable crystallographic properties (Guo *et al.*, 2004). An advantage of TGGE, as discussed below, is that it can be used in the presence of modest amounts of monovalent and divalent ions, as required for the folding of RNA secondary and tertiary structures.

2. Temperature Gradient Gel Electrophoresis

2.1. Basic principles

We begin this section with an overview of the basic principles of TGGE and then provide more detailed descriptions of the instrumentation and setup. TGGE is achieved by sandwiching a polyacrylamide gel encased in

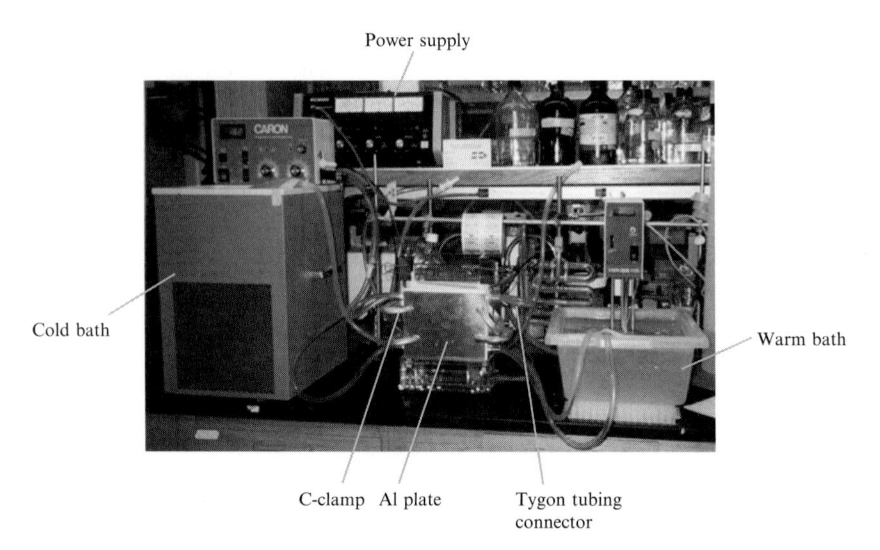

Figure 18.1 Setup for TGGE experiment. In the center of the photo is the electrophoresis apparatus. Not visible are the two aluminum blocks that have water flowing through them (see Fig. 18.2), or the polyacrylamide gel in glass plates sandwiched between these blocks. Behind the back aluminum block is a foam piece used to seal the apparatus and prevent leakage of the buffer from the upper reservoir. In addition, a piece of closed cell foam is placed outside of each block in order to minimize temperature fluctuations, and a thin aluminum plate is placed in the front of the assembly for support during clamping. These pieces are held together by four C-clamps (shown). This photograph is of a perpendicular TGGE experiment, as can be deduced from the orientation of the Tygon tubing connectors (compare Fig. 18.2), as well as from the tracking dyes entering the gel, which are slanted from left to right. To the left of the apparatus is a refrigerated water bath, used to provide cold water, and to the right side is a warm bath.

glass plates between two aluminum blocks that have a temperature gradient established perpendicular or parallel to the electric field (Figs. 18.1 and 18.2). Perpendicular TGGE is used to obtain a full melting profile of the RNA, while parallel TGGE is used to separate mixtures of RNAs, which can have very similar properties (Fig. 18.3). In general, perpendicular TGGE is conducted by loading the sample in a single wide lane that spans the width of the gel, while parallel TGGE is conducted in individual narrow lanes. As described below, we have used perpendicular TGGE to estimate the temperatures over which an RNA library melts, and then performed parallel TGGE over this temperature range to separate the members of the library according to stability.

In both types of TGGE, RNA in the unfolded form is in an extended conformation and therefore has slower electrophoretic mobility (Fig. 18.3A, right), while RNA in the folded form is compact and has faster

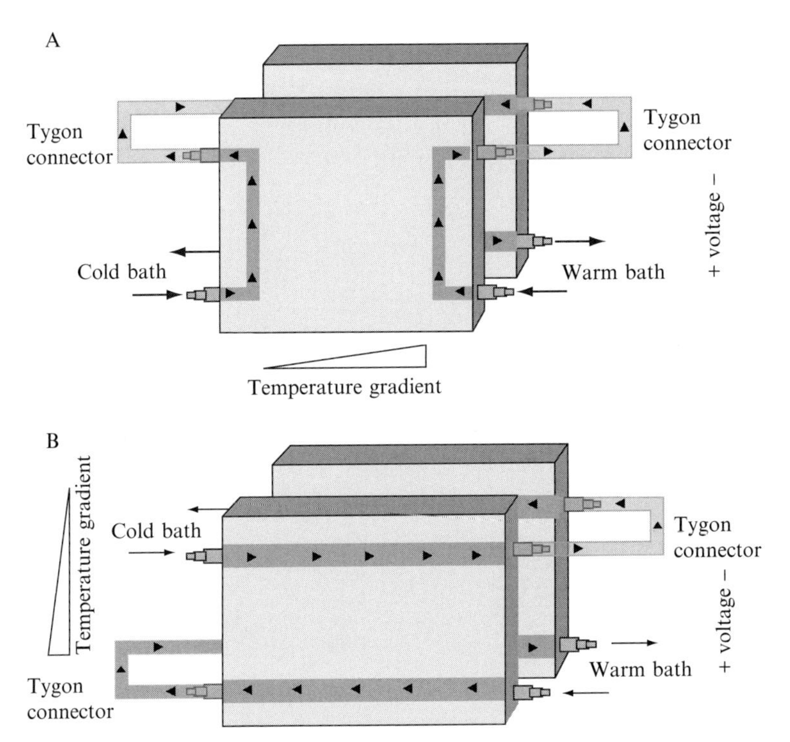

Figure 18.2 Schematics of aluminum blocks used for water circulation. Shown are the channels drilled into the aluminum blocks and the directions of water flow. In both cases, two external water baths are required, as are two Tygon connectors to direct the flow from the front block to the back block. (A) Perpendicular TGGE schematic. Here, water from the cold bath regulates the temperature of the left side of the blocks, while water from the warm bath regulates the temperature of the right side of the blocks. Water enters from the bottom of the blocks to ensure good flow. The voltage gradient is shown and is perpendicular to the direction of the temperature gradient. In general, the sample is loaded evenly into a single lane that spans the width of the gel. (B) Parallel TGGE schematic. Here, water from the cold bath regulates the temperature of the top of the blocks, while water from the warm bath regulates the bottom of the blocks. The voltage gradient is shown and is parallel to the direction of the temperature gradient. In general, the samples are loaded into individual lanes.

electrophoretic mobility (Fig. 18.3A, left). The decrease in electrophoretic mobility as the RNA unfolds is the basis for TGGE separations.

As described below, electrophoresis is in practice limited to water baths set near 4 and 65 °C. In order to have the RNA melt in the center of this transition, which is ideal for thermodynamic characterization and separation, one can use partially denaturing conditions in which a modest amount of urea (2–4 *M*) is present uniformly throughout the gel (Bevilacqua and Bevilacqua, 1998; Wartell *et al.*, 1990). The amount of urea needed is

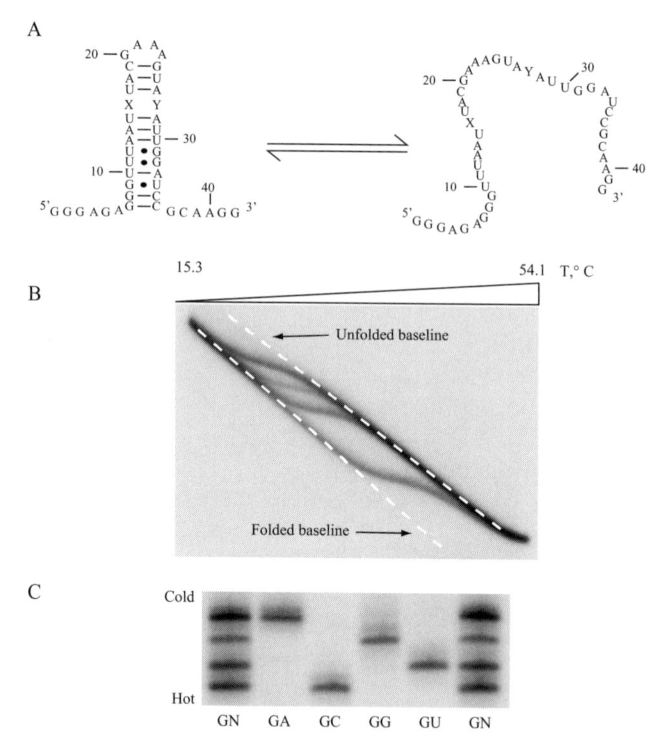

Figure 18.3 Separation of model RNA hairpins. (A) Equilibrium between folded and unfolded forms of model hairpins. Watson–Crick base pairs are denoted with a dash, and wobble base pairs with a dot. The letters "X" and "Y" denote positions of substitution with various base pairs and mismatches. (B, C) Separation of a simple RNA library. RNA oligonucleotides were 5′-^{32}P-labeled, with XY from panel A substituted as follows: X = G, and Y = N, a mix of A, C, G, and U. Buffer was 1×TBEK$_{50}$/4 *M* urea. (B) Perpendicular TGGE data. The temperature in the gel ranged from 15.3 to 54.1 °C, as determined with a thermometer inserted into the gel (see text). Shown in dashed lines are the folded and unfolded baselines, which are linear except at the very outer part of the gel where the gradient is lost. Distinct melting curves are seen for the four members of the library. (C) Parallel TGGE data. Separation of the library is revealed in the first and last lanes. Positions of GA, GC, GG, and GU substitutions are indicated. Low- and high-temperature baths were set to 18 and 32 °C. All three panels are adapted from Bevilacqua and Bevilacqua (1998) with permission from the American Chemical Society.

determined empirically for a given sequence. It is possible to extrapolate thermodynamic parameters back to 0 *M* urea (Bevilacqua and Bevilacqua, 1998), as is typically done for protein folding (Pace, 1975); however, in many applications, such as *in vitro* selections, only relative ranking of thermodynamic stabilities of various RNAs is needed. In such instances, actual thermodynamic parameters are obtained by standard methods such as UV melting or calorimetry. We have shown that urea does not alter the relative stability of various RNA constructs either in the gel or in solution (Bevilacqua and Bevilacqua,

1998). In some instances, the melting temperature of the RNA has also been tuned by limiting the amount of Mg^{2+} (a renaturant) in the gel (Szewczak *et al.*, 1998), which is the counterpart of increasing the amount of urea (a denaturant) in the gel.

2.2. Instrumentation

Rosenbaum and Riesner (1987) introduced a horizontal temperature gradient system to study the structural transitions of a biopolymer. Based on these studies, Wartell *et al.* (1990) designed a vertical TGGE apparatus to detect single base pair substitutions in DNA. We designed and performed biophysical experiments on model RNA and DNA oligonucleotides using a vertical TGGE apparatus similar to that of Wartell and coworkers (Bevilacqua and Bevilacqua, 1998).

A note about availability of TGGE apparatuses is in order. The TGGE apparatus that we designed and use is not commercially available. However, the interested reader can design his/her own apparatus based on the detailed description in the Wartell *et al.* (1990) study, as well as the experimental sections below. In addition, at the time of writing this article, TGGE apparatuses are commercially available from Biometra (horizontal unit) and BioRad (vertical unit), although we have not used these.

The basic TGGE apparatus used in our studies closely resembles a typical vertical gel electrophoresis setup and is shown fully assembled in Fig. 18.1. During the TGGE run the glass plates sandwiching the gel are subjected to high temperatures and sharp temperature gradients, which stresses the plates; thus, thicker plates are used (see Section 2.5 for details on plates). In general, thinner gels were used for analytical applications, typically with perpendicular TGGE, while thicker gels were used for preparative applications, typically with parallel TGGE. The glass plates containing the gel are sandwiched between two aluminum blocks, each of which has two channels for circulating water. Aluminum is a good material for the blocks because it is an excellent conductor of heat.

Channels in the block are machined such that the temperature gradient can either run from left to right, generating a temperature gradient perpendicular to the electric field, or from top to bottom, generating a temperature gradient parallel to the electric field (Fig. 18.2). Convenient lower and upper temperatures for the externally circulating water baths were found to be 4 and 65 °C, respectively. Temperatures lower than 4 °C led to excessive condensation on the apparatus and difficulty in holding a steady temperature, while temperatures higher than 65 °C tended to crack the glass plates.

For reproducible and interpretable results, it is imperative that the temperature gradient in the gel itself be known. We measured the temperature in the perpendicular TGGE apparatus by directly inserting a temperature probe into a 3 mm thick gel to a depth of 2 cm below the top of the aluminum blocks. An Omega HH21 microprocessor thermometer fitted with a 0.02 in. diameter, 6 in. long thermocouple probe (Omega) was used. It is important that the power supply be turned off during this measurement otherwise the probe will deteriorate.

Prior to using the probe, we calibrated it by taking readings in an ice bath and a boiling water bath whose temperature was determined using the atmospheric pressure in the room that day. The probe was allowed to equilibrate before taking a reading, and readings were made at regular distances across the gel. It was found that the gradient was linear across the gel, although the temperatures in the gel were different than those on the baths. As an example, setting the cold and hot baths to 4 and 60 °C, respectively, led to readings of 15.3 and 54.1 °C in the gel at the edges of the plate (Bevilacqua and Bevilacqua, 1998). In addition, linear diagonal migration of the tracking dyes and the baselines of the RNA was observed, which was also consistent with a linear gradient, although exceptions occurred at the very outer parts of the gel where gradient was lost (e.g., Fig. 18.3B, high temperature). Loss of linearity of the gradient at the outer part of the gel is formally analogous to the "smiling" that is often observed in conventional electrophoresis. For a 2 h run, there was no fluctuation in the temperatures measured at various points inside of the gel, consistent with a sustained gradient despite potential Joule heating of the gel. In the next two sections, we discuss advantages and disadvantages of parallel and perpendicular TGGE.

2.3. Perpendicular TGGE

In a perpendicular TGGE setup, the temperature gradient is perpendicular to the electric field with the channels in the aluminum blocks keeping the (arbitrarily) left side of the apparatus cool and right side warm (Fig. 18.2A). Typically, for a large temperature gradient, the water bath on the left is set to 4 °C and the bath on the right is set to 65 °C, although adjusting the temperature of the two water baths closer to each other can improve the resolution of the gel. The samples used in these experiments were applied to a single, wide lane of 15 cm, typically on a 0.5 mm thick analytical gel to get optimal separation. As mentioned above, urea is included in the gel stocks to ensure that the RNA melts near the center of this convenient temperature zone.

The migration of the sample through the gel is followed by the inclusion of tracking dyes such as bromophenol blue and xylene cyanol. A 1-in. area is present above and below the aluminum plates, which allows the user to visually monitor the progress of the gel run. When the lower tracking dye appears below the aluminum plate, the gel run is terminated.

An example of such a perpendicular TGGE run is shown in Fig. 18.3B, where a library containing four different sequences was well separated according to independently determined relative stabilities. Sloping upper and lower base lines are due to a general increase in electrophoretic mobility with temperature, and are also observed for RNAs that do not undergo a transition (Szewczak *et al.*, 1998), as well as for the tracking dyes alone. Although perpendicular TGGE can be used for both preparative and analytic purposes, it is most often used to observe melting transitions of an RNA mixture. These melting profiles provide useful information on the overall stability of an RNA population. For example, a smeary melting transition would be indicative of an RNA mixture with differing T_{M}s, while a sharp transition would suggest that RNAs in a mixture have similar thermodynamic properties. Information obtained from perpendicular TGGE gels is then applied to set up conditions for parallel gels for enriching individual RNAs within a larger pool. All analytical gels are dried and exposed to a PhosphorImager plate to examine the melting transitions.

2.4. Parallel TGGE

In a parallel TGGE setup, the temperature gradient is established in the same direction as the electric field, with the channels in the aluminum blocks keeping the top of the apparatus cold and the bottom warm (Fig. 18.2B). The samples are normally loaded into individual 10 mm wide lanes on a 1.5 mm thick preparative gel to facilitate recovery of sample, as required for *in vitro* selections. Thus, as compared to perpendicular TGGE, the sample volumes are smaller. Parallel gels are typically run over a shallower temperature gradient than perpendicular gels (e.g., a 15 °C gradient, as in Fig. 18.3C), with the water bath settings determined from the lower and higher temperature limits of the melting transition obtained by perpendicular TGGE, determined by interpolation. The same concentration of urea is used in the parallel and perpendicular gels.

After completion of the gel run, preparative parallel TGGE gels were visualized by autoradiography and the RNA bands of interest were excised, eluted, and purified by a standard crush and soak procedure (Shu and Bevilacqua, 1999). The RNA was concentrated by ethanol precipitation and either used directly in the next selection cycle or characterized further. Parallel TGGE offers the advantage that multiple lanes can be run on a single gel, allowing simultaneous examination of mixtures and individual components. As with perpendicular TGGE, migration of the sample through the gel is followed by the observation of tracking dyes.

The thermodynamic behavior of model RNA hairpins has also been analyzed using parallel TGGE (Fig. 18.3C). As mentioned previously, RNA in the unfolded form is open and has slower electrophoretic mobility. Thus, as a mixture of RNAs with different stabilities migrates through the parallel TGGE gel, the least stable ones open earlier, are retarded, and finish higher in the gel.

2.5. Preparing and running TGGE experiments

In this subsection, we provide details on how to perform a TGGE experiment. Section 3 will provide details on preparing RNA and DNA samples for TGGE and *in vitro* selections. We have attempted to present this section in a general fashion so that it will be of utility to users of a wide range of TGGE apparatuses, whether commercial or home-made (see Section 2.2).

In general, we used an 8% polyacrylamide gel in the vertical orientation, with an acrylamide to bis-acrylamide ratio of 29:1. The electrophoresis buffer present in the gel and the running buffer is 1×$TBEK_{50}$ (100 m*M* Tris, 83 m*M* Boric Acid, 1 m*M* Na_2EDTA, 50 m*M* KCl), pH 8.77 at 20 °C, and 4 *M* urea. This buffer is a composite of TBE, which is an excellent buffer for electrophoresis of nucleic acids, and a modest amount of KCl, which stabilizes the folded form of the RNA or DNA. Higher amounts of KCl were avoided because of associated Joule heating, and KCl was chosen over NaCl because the former is more prevalent in cells.

Most buffers have a temperature-dependence to their pK_as, which was of potential concern for a TGGE experiment. We measured the pH dependence of 1×$TBEK_{50}$ and found it to be −0.014 ΔpH/°C, providing a pH of 8.85 at the gel temperature of 15.3 °C, and a pH of 8.30 at the gel temperature of 54.1 °C. These values are outside the standard pK_a values of the bases, especially when they are base paired, and are low enough to avoid alkaline denaturation of the helix (Bevilacqua *et al.*, 2004; Moody *et al.*, 2005). An additional concern with pH is that the electrochemistry of KCl leads to changes in pH during the run, which is addressed by recirculating the running buffer as described below.

To facilitate analytical and preparative applications, either 0.5 or 1.5 mm thick spacers are used between the glass plates, which are 3/16 in. thick in order to tolerate the heat gradients. The shorter glass plate is 7 3/8 in. × 8 1/2 in., while the longer plate is 7 3/8 in. × 10 in. After the gel is poured and polymerized, the TGGE apparatus is assembled. To assure good thermal contact with the aluminum blocks, the outsides of the glass plates are cleaned of any tape and acrylamide residue and thoroughly dried. The back aluminum block is set on a plexiglass piece in the lower buffer reservoir with a foam piece behind it to prevent leakage of the buffer from the upper reservoir. Next, the gel itself is positioned, followed by the front aluminum block and another piece of close cell foam, which helps insulate the apparatus. The user should make sure that the proper set of aluminum blocks is in place (i.e., two of the same blocks for parallel TGGE or two of the same blocks for perpendicular TGGE). Lastly, a thin aluminum plate is placed over the foam for support and the entire apparatus is secured with four C-clamps (see Fig. 18.1 for final setup).

The two water baths are connected to the aluminum blocks with Tygon tubing, which is secured with copper wiring. Note that the connections are different for perpendicular and parallel baths and are provided in the schematics in Fig. 18.2. We suggest putting a flow monitor in line with each bath to assure adequate flow. In addition, temperature readings of the water as it returns to each bath should be taken and confirmed to equal the set temperature of the bath to assure that the flow rate is sufficient.

Prior to loading of samples, gels are preelectrophoresed for 30 min at 350 volts, with water baths running. This is done to equilibrate both ions in the gel and the temperature gradient. To permit the RNAs to fold into the proper hairpin conformation, at the start of each experiment the appropriate mixture of RNAs is renatured in TE by heating for 3 min at 90 °C followed by cooling on the bench for at least 10 min (see Section 3.1 for further details on preparing and folding the nucleic acids). Samples are then mixed with TGGE loading buffer (final concentrations: 1×$TBEK_{50}$, 10% glycerol, 0.05% xylene cyanol, and 0.05% bromophenol blue).

For perpendicular TGGE, 150 μL of sample is loaded into a single well across the top of the gel. For parallel TGGE, 2 μL of sample is loaded into each 10 mm wide lane for the analytical (0.5 mm thick) gels, and 20 μL for the preparative (1.5 mm thick) gels. Each gel is run at constant voltage (350 V) until the bromophenol blue migrates just below the aluminum plate (approximately 2 h). Every 60 min during electrophoresis, the upper and lower reservoirs are emptied, buffers mixed, and reservoirs refilled. This is done because the electrochemistry of the 50 m*M* KCl leads to opposing changes in pH in the two buffer reservoirs. In addition, Cl_2 gas is generated in these runs, therefore adequate ventilation is needed. For analytical gels, the gel is dried on Whatman 3 *M* paper and analyzed using a PhosphorImager (Molecular Dynamics).

3. Experimental Design and Application of TGGE to RNA and DNA

RNA folds into a wide variety of stable structures whose folds and populations are dictated by the intramolecular interactions and thermodynamic stability of a particular fold. Thermodynamic parameters for many RNA structural features have been determined by melting of individual RNA oligonucleotides, which has helped to improve RNA structure prediction (Mathews and Turner, 2006; Xia *et al.*, 1998). However, there are many other RNA structural motifs for which thermodynamic parameters are not yet known; moreover, many of these are too complex to permit study of every possibility. Thus, a high throughput approach wherein the identity of especially stable and unstable RNA folds within a given motif is easily determined would prove valuable.

To address this issue, we prepared short RNA hairpins in a combinatorial library using TGGE. As described above, TGGE of these short RNA hairpins is sensitive to small differences in free energy and is capable of resolving a simple RNA library into stable and unstable members (Bevilacqua and Bevilacqua, 1998). In this section, we discuss how these RNA and DNA libraries were designed and provide an overview of the *in vitro* selection experiments.

3.1. Separation of RNA and DNA secondary structures

The general secondary structure of the RNAs used in TGGE is provided in Fig. 18.3A. These RNAs are prepared by T7 transcription from hemiduplex DNA templates, purified by denaturing PAGE, and stored in TE buffer (10 mM Tris, 1 mM Na_2EDTA, pH 7.5) at -20 °C (Bevilacqua and Bevilacqua, 1998; Bevilacqua and Cech, 1996; Milligan and Uhlenbeck, 1989). RNAs are then dephosphorylated using calf-intestinal phosphatase and 5′-^{32}P labeled with polynucleotide kinase and [γ-^{32}P]ATP to facilitate visualization during electrophoresis. These RNAs are repurified by denaturing PAGE and eluted with a crush and soak protocol. Protocols for transcribing and labeling the RNAs are widely available (see references above and elsewhere) and so are not reproduced here. However, the reader may benefit from a few comments on sequence design.

The RNA hairpins are designed to facilitate RT-PCR reactions necessary for *in vitro* selection experiments, to minimize alternative conformations upon folding, and to provide a good change in electrophoretic mobility upon melting (Bevilacqua and Bevilacqua, 1998). The single-stranded tails on either end of the helical region are present to minimize the influence of the additional nucleotides added by T7 polymerase on RNA folding and to provide binding sites for the RT PCR primers that allow the stem to be subsequently invaded (Proctor *et al.*, 2002; Shu and Bevilacqua, 1999). The GGGAGA stretch at the 5′-end of the RNA is an optimal T7 polymerase start sequence, and the opposing GCAAGG stretch on the 3′-end tail is designed to not interact with the 5′-end.

The effect of potential alternative conformations upon folding was assessed by putting the sequence through an RNA structure prediction algorithm (mfold) with a window size of zero, which shows all suboptimal structures (Jaeger *et al.*, 1990). The larger the difference in free energy between the optimal and next optimal sequence, the more likely the optimal fold is to be unique. The next most stable structure is 5.1 kcal/mol in free energy away from the optimal fold, which is a population enhancement of at least 4000-fold (Bevilacqua and Bevilacqua, 1998). The loop sequence is a phylogenetically conserved GAAA, which is present to encourage native folding of the RNA. A *Bam*H1 cloning site is present near the 3′-end of the RNA for cloning purposes, although it could have been introduced in the final round of selection with the bottom strand PCR primer (Bevilacqua and Bevilacqua, 1998).

The RNAs shown in Fig. 18.3A gave reasonable hypomobilities of 11–21% depending on salt and urea conditions, which facilitated separations. A simple library was made by placing a G at position X and N at position Y in these RNAs. This led to four transitions by perpendicular and parallel TGGE, and each band migrated with one of the four discretely prepared RNAs: GA, GC, GG, or GU (see Fig. 18.3B and C). Extraction of thermodynamic parameters from the perpendicular TGGE experiment was not attempted because of the presence of partially denaturing conditions of 4 *M* urea. Rather, the perpendicular TGGE experiments were used to demonstrate separation and ranking of RNAs according to relative stability by parallel TGGE and to provide optimal temperatures for parallel TGGE.

It is really parallel TGGE that is powerful for selections because it spatially separates the RNA on the basis of their relative thermodynamic stabilities (see Sections 2.3 and 2.4). We note that perpendicular TGGE can also be used to provide thermodynamic parameters, as described in Section 3.4 below.

3.2. TGGE-SELEX of RNA

We were interested in determining the identities of the unusually stable and unstable RNAs separated by parallel TGGE, and so turned to *in vitro* selection, or SELEX (Ellington and Szostak, 1990; Robertson and Joyce, 1990; Tuerk and Gold, 1990). We refer to this combined methodology as TGGE-SELEX, and have applied it to both RNA triloops and tetraloop libraries (Proctor *et al.*, 2002; Shu and Bevilacqua, 1999). The RNA library, which is provided here for the triloops library (Fig. 18.4A), was similar to that of the RNA model oligonucleotides. The library for tetraloop selections was quite similar to the triloops library, with small differences indicated below. As noted in the Introduction, TGGE-SELEX has also been applied to the selection of stable RNA tertiary structures (Guo and Cech, 2002), which have provided high-resolution crystal structures (Guo *et al.*, 2004). The basic approach is similar to that described in this section, although perpendicular TGGE is typically conducted, and the interested reader is referred to those papers.

One technical issue in studying triloops was encouraging triloops in the selection over pentaloops (Shu and Bevilacqua, 1999). This was accomplished by placing complementary bases at positions 19:23 of the library through the preparation of CG, GC, UA, AU, GU, and UG 19:23 sublibraries; the AU and GU were made by a single RU sublibrary, as were the UA and UG, where "R" is a mixture of purines. The various sublibraries were mixed in the following ratio: CG, GC, UR, RU 1:1:2:2 to encourage equal representation of each sequence in the initial library. For the tetraloop selections, a six-nucleotide library without sublibraries was used, which was

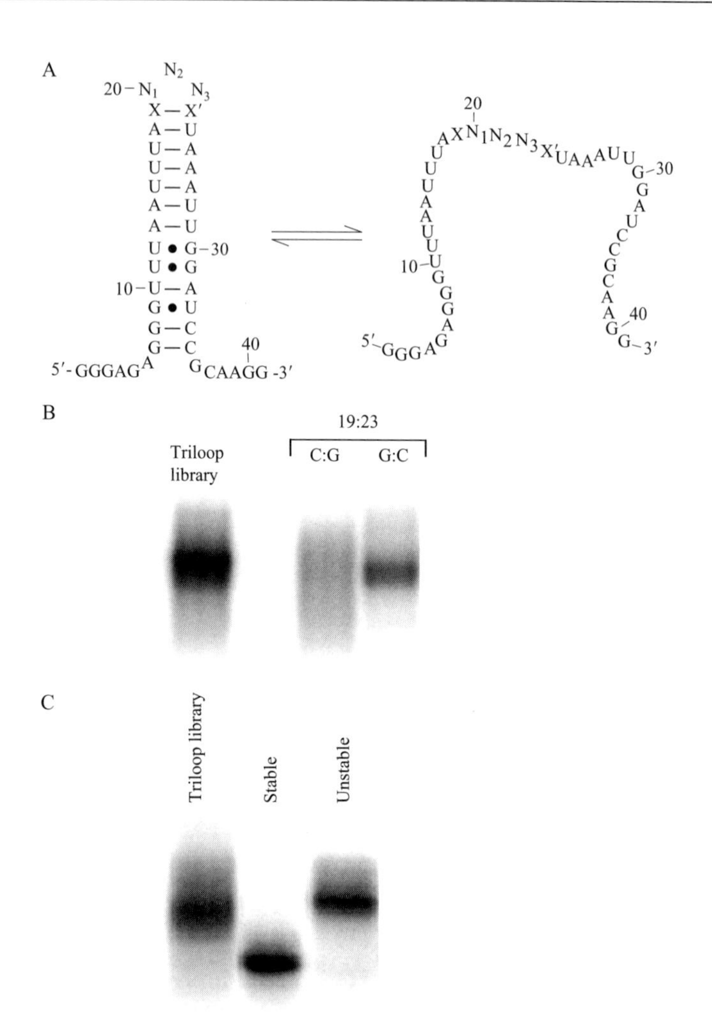

Figure 18.4 Selection of a triloop library. The triloop library was a mixture of 384 different sequences (6 closing base pairs and 64 triloops) and designed similar to the constructs in Fig. 18.3A. (A) Equilibrium between folded and unfolded forms of RNA hairpins. Watson–Crick base pairs are denoted with a dash, and wobble pairs with a dot. The letter N denotes a mix of A, C, G, and U. The letters X and X, denote complementary bases. (B) Parallel TGGE of the triloop library and CG and GC closing base pair sublibraries. The temperature baths were set at 27 and 37 °C. Triloop sublibraries have complete randomization at positions 20–22 and complementarity at positions 19 and 23 as indicated in the figure. (C) Results of selections for stable and unstable populations. Stable and unstable samples were enriched by 3–4 rounds of selection. "Stable" and "Unstable" signify high- and low-mobility bands, respectively. Parallel TGGE gel under standard conditions, with temperature baths set at 30 and 37 °C. Shown here are results of one of two related selection methods described in the original study (Shu and Bevilacqua, 1999). Adapted from Shu and Bevilacqua (1999) with permission from the American Chemical Society.

based on the assumption that tetraloops with a selected closing base pair would be more stable than hexaloops with an AU closing base pair—an assumption that was borne out by the actual study (Proctor *et al.*, 2002).

The parallel TGGE for the selections was conducted in a very similar fashion to the aforementioned separation of RNA secondary structures in Section 3.1. One interesting experiment was comparison of the full triloop library to the CG and GC sublibraries. As shown in Fig. 18.4B, the CG sublibrary had a much broader range of melting behavior giving a smear that migrated further down the gel, where stable RNAs migrate, which foreshadowed their prevalence in the final selected RNAs (Shu and Bevilacqua, 1999).

Selection was conducted according to standard selection approaches described in detail elsewhere (Bevilacqua *et al.*, 1998; Ellington and Szostak, 1990; Proctor *et al.*, 2002; Robertson and Joyce, 1990; Shu and Bevilacqua, 1999; Tuerk and Gold, 1990). We focus here on those features that are peculiar to TGGE-SELEX. Of particular note is that the RNA is removed by excising the appropriate region of the gel with a razor blade: below the main band of the library for stable RNAs and above the main band of the library for unstable RNAs. Three rounds of selection led to a clear enrichment of stable bands that migrated faster and of unstable bands that migrated slower than the main band (Fig. 18.4C). An important control for TGGE-SELEX is to fractionate the final RNAs and the initial library under completely denaturing conditions (either of high temperature and/or high urea), which should lead to a single band (no smear) of identical mobility across the various selected and unselected species. This control assures that separation is due to genuine differences in thermal stability and not to artifacts such as deletion of residues during PCR.

The isolated RNA bands of unusual stability were cloned and sequenced, as described elsewhere (Bevilacqua *et al.*, 1998; Shu and Bevilacqua, 1999). Of more specialized importance to TGGE-SELEX is postselection characterization of the stabilities of the selected RNAs. Because the partially denaturing TGGE methods only rank the relative stabilities of the RNAs, it is important that another method be available for determining thermodynamic parameters, especially free energy parameters if the results are to be used in structure prediction algorithms. We characterized the most frequent stable and unstable hairpins found in our selections and tabulated their free energies; as expected, RNAs isolated as stable were exceptionally stable as judged by UV melts, while those isolated as unstable were unstable. These results serve as additional controls that the selections worked as expected (Shu and Bevilacqua, 1999).

Other useful application of the TGGE-SELEX procedure has been identification of new folding motifs of stable RNAs, such as YNMG tetraloops (Proctor *et al.*, 2002), and subsequent structural work (Du *et al.*, 2003; Theimer *et al.*, 2003). One item of note is that the selections on tetraloops, for which there were representatives already in the literature, failed to turn up the common GNRA tetraloops, although they did give the common (and slightly *more* stable) UNCG and related tetraloops. The reason for this may be that GNRA sequences are also stable in DNA (Tanikawa *et al.*, 1991), whereas UNCGs are not. In fact, a d(cGAAAg) sequence (lower case is the closing base pair) could be recovered only upon providing longer tails (see Section 3.3). It may be of interest to repeat the tetraloop selections with longer tails in an effort toward exhaustiveness in the RNA tetraloop selections.

3.3. TGGE-SELEX of DNA

It is also possible to perform TGGE-SELEX on DNA libraries (Nakano *et al.*, 2002). The actual experiment is quite similar to the one performed with RNA libraries with the key exceptions that (1) there is no transcription, and (2) the top strand of the DNA must be isolated from the bottom strand. We achieved the latter by biotinylating the bottom strand primer, capturing the PCR product on streptavidin beads, and isolating the nonbiotinylated top strand by eluting in 0.15 *M* NaOH, which alkaline denatures the dsDNA without releasing the biotinylated strand (Bock *et al.*, 1992). Other methods of isolating the top strand have also been developed such as putting a 2′-hydroxyl into the bottom strand primer, cleaving it by alkali, and PAGE-purifying the longer strand; however, the above method proved robust in our hands and avoided the need for a subsequent gel purification step.

In these experiments, the top strand primer was radiolabeled to facilitate tracking of the DNA throughout the selection procedure, especially in the TGGE gels (Nakano *et al.*, 2002). One other technical issue of note is that because the selected DNA is thermodynamically stable, the cloning and sequencing can prove troublesome. We found that more reliable sequences were obtained by using a bottom strand primer for sequencing; since this strand is the reverse complement of the stable DNA, it is less likely to be especially stable. In addition, all cloning steps were performed at 30 °C rather than 37 °C, which tends to provide fewer mutations. On a related note, we also found that it was necessary to have longer primer binding sites, achieved by lengthening the tails in the library from 6 to 12 nt, in order to recover the known stable d(cGAAAg) stable tetraloop motifs from the selection (Nakano *et al.*, 2002). To date, TGGE-SELEX of DNA has only been applied to tetraloop libraries, for which it gave several motifs of stable DNAs whose stability we determined by UV melts. Some of these sequences were known stable DNA families, which gave confidence in the method, while others were new families, some of which had interesting biological function (Habig and Loeb, 2002; Nakano *et al.*, 2002).

3.4. Perpendicular TGGE melts

In the experiments described above, TGGE was used primarily to rank the relative stabilities of RNA libraries, not to determine thermodynamic parameters. However, it is also possible to use perpendicular TGGE to extract thermodynamic parameters in a method similar to UV melts. One case that serves to illustrate this point is the perpendicular TGGE melts of the P4–P6 tertiary domain from the *Tetrahymena* ribozyme (Szewczak *et al.*, 1998). These experiments were done in the absence of urea, precluding need for urea extrapolations. Instead, the position of the melting transition was controlled by adjusting the concentration of $MgCl_2$.

The data were treated by assuming a two-state model for the tertiary unfolding of the P4–P6 domain involving compact and extended forms of the RNA, and applying a formalism identical to that used in UV melting curves (Szewczak *et al.*, 1998). The fraction folded as a function of temperature, $\alpha(T)$, was calculated as:

$$\alpha(T) = \frac{D_x(T) - D_U(T)}{D_F(T) - D_U(T)} \qquad (18.1)$$

where $D_x(T)$ is the vertical distance x from the well that the RNA migrated at a given temperature, and $D_U(T)$ and $D_F(T)$ are the vertical distances from the well for the unfolded and folded baselines, respectively. The extended nature of the RNA was confirmed by comparison to a base paired mutant that prevents compaction; in general, such controls are valuable in confirming assumptions about the states being probed.

Once $\alpha(T)$ is determined, $K(T)$ is calculated from the standard relationship:

$$K(T) = \alpha(T)/[1 - \alpha(T)] \qquad (18.2)$$

and enthalpy is determined from the standard van't Hoff relationship,

$$\Delta H = -R\frac{\partial \ln K(T)}{\partial 1/T} \qquad (18.3)$$

Entropy, free energy, and T_M follow from standard thermodynamic relationships.

What advantages do perpendicular TGGE melts offer over standard UV melts? First, the observable is change in electrophoretic mobility rather than hypochromicity. The former depends on changes in shape, while the latter depends on changes in base stacking. It is possible to have a folding transition that involves large shape changes but small stacking changes, especially for tertiary folding changes; in these cases, TGGE should be a more powerful approach. In addition, it is possible to melt two (or more) RNA samples simultaneously in a given perpendicular TGGE experiment, assuming the RNAs have different electrophoretic mobilities. By melting the two samples simultaneously, one is assured that the mixture of samples is exposed to identical experimental conditions. Such an approach was used in the case of P4–P6 melting (Szewczak *et al.*, 1998). Lastly, one can envision selecting for large enthalpy changes using perpendicular TGGE-separated libraries, since the larger the enthalpy change the steeper the melting transitions (Puglisi and Tinoco, 1989).

ACKNOWLEDGMENT

This work was supported by NSF Grant 0527102.

REFERENCES

Beier, S., Witzel, K. P., and Marxsen, J. (2008). Bacterial community composition in Central European running waters examined by temperature gradient gel electrophoresis and sequence analysis of 16S rRNA genes. *Appl. Environ. Microbiol.* **74,** 188–199.

Bevilacqua, J. M., and Bevilacqua, P. C. (1998). Thermodynamic analysis of an RNA combinatorial library contained in a short hairpin. *Biochemistry* **37,** 15877–15884.

Bevilacqua, P. C., and Cech, T. R. (1996). Minor-groove recognition of double-stranded RNA by the double-stranded RNA-binding domain from the RNA-activated protein kinase PKR. *Biochemistry* **35,** 9983–9994.

Bevilacqua, P. C., George, C. X., Samuel, C. E., and Cech, T. R. (1998). Binding of the protein kinase PKR to RNAs with secondary structure defects: Role of the tandem A–G mismatch and noncontiguous helixes. *Biochemistry* **37,** 6303–6316.

Bevilacqua, P. C., Brown, T. S., Nakano, S., and Yajima, R. (2004). Catalytic roles for proton transfer and protonation in ribozymes. *Biopolymers* **73,** 90–109.

Bock, L. C., Griffin, L. C., Latham, J. A., Vermaas, E. H., and Toole, J. J. (1992). Selection of single stranded DNA molecules that bind and inhibit human thrombin. *Nature* **355,** 564–566.

Crameri, G. S., Wang, L. F., and Eaton, B. T. (1995). Differentiation of cognate dsRNA genome segments of bluetongue virus reassortants by temperature gradient gel electrophoresis. *J. Virol. Methods* **51,** 211–219.

Danko, P., Kozak, A., Podhradsky, D., and Viglasky, V. (2005). Analysis of DNA intercalating drugs by TGGE. *J. Biochem. Biophys. Methods* **65,** 89–95.

Du, Z., Yu, J., Andino, R., and James, T. L. (2003). Extending the family of UNCG-like tetraloop motifs: NMR structure of a CACG tetraloop from coxsackievirus B3. *Biochemistry* **42,** 4373–4383.

Ellington, A. D., and Szostak, J. W. (1990). *In vitro* selection of RNA molecules that bind specific ligands. *Nature* **346,** 818–822.

Fischer, S. G., and Lerman, L. S. (1979). Length-independent separation of DNA restriction fragments in two-dimensional gel electrophoresis. *Cell* **16,** 191–200.

Fischer, S. G., and Lerman, L. S. (1983). DNA fragments differing by single base-pair substitutions are separated in denaturing gradient gels: Correspondence with melting theory. *Proc. Natl. Acad. Sci. USA* **80,** 1579–1583.

Guo, F., and Cech, T. R. (2002). Evolution of *Tetrahymena* ribozyme mutants with increased structural stability. *Nat. Struct. Biol.* **9,** 855–861.

Guo, F., Gooding, A. R., and Cech, T. R. (2004). Structure of the *Tetrahymena* ribozyme: Base triple sandwich and metal ion at the active site. *Mol. Cell* **16,** 351–362.

Guo, F., Gooding, A. R., and Cech, T. R. (2006). Comparison of crystal structure interactions and thermodynamics for stabilizing mutations in the *Tetrahymena* ribozyme. *RNA* **12,** 387–395.

Habig, J. W., and Loeb, D. D. (2002). Small DNA hairpin negatively regulates *in situ* priming during duck hepatitis B virus reverse transcription. *J. Virol.* **76,** 980–989.

Jaeger, J. A., Turner, D. H., and Zuker, M. (1990). Predicting optimal and suboptimal secondary structure for RNA. *Methods Enzymol.* **183,** 281–306.

Kang, J., Harders, J., Riesner, D., and Henco, K. (1994). TGGE in quantitative PCR of DNA and RNA. *Methods Mol. Biol.* **31,** 229–235.

Ke, S. H., and Wartell, R. M. (1995). Influence of neighboring base pairs on the stability of single base bulges and base pairs in a DNA fragment. *Biochemistry* **34,** 4593–4600.

Mathews, D. H., and Turner, D. H. (2006). Prediction of RNA secondary structure by free energy minimization. *Curr. Opin. Struct. Biol.* **16,** 270–278.

Milligan, J. F., and Uhlenbeck, O. C. (1989). Synthesis of Small RNAs Using T7 RNA Polymerase. *Methods Enzymol.* **180,** 51–62.

Moody, E. M., Lecomte, J. T., and Bevilacqua, P. C. (2005). Linkage between proton binding and folding in RNA: A thermodynamic framework and its experimental application for investigating p*K*a shifting. *RNA* **11,** 157–172.

Myers, R. M., Fischer, S. G., Maniatis, T., and Lerman, L. S. (1985). Modification of the melting properties of duplex DNA by attachment of a GC-rich DNA sequence as determined by denaturing gradient gel electrophoresis. *Nucleic Acids Res.* **13,** 3111–3129.

Nakano, M., Moody, E. M., Liang, J., and Bevilacqua, P. C. (2002). Selection for thermodynamically stable DNA tetraloops using temperature gradient gel electrophoresis reveals four motifs: d(cGNNAg), d(cGNABg), d(cCNNGg), and d(gCNNGc). *Biochemistry* **41,** 14281–14292.

Pace, C. N. (1975). The stability of globular proteins. *CRC Crit. Rev. Biochem.* **3,** 1–43.

Proctor, D. J., Schaak, J. E., Bevilacqua, J. M., Falzone, C. J., and Bevilacqua, P. C. (2002). Isolation and characterization of a family of stable RNA tetraloops with the motif YNMG that participate in tertiary interactions. *Biochemistry* **41,** 12062–12075.

Puglisi, J. D., and Tinoco, I. Jr. (1989). Absorbance melting curves of RNA. *Methods Enzymol.* **180,** 304–325.

Riesner, D., Steger, G., Zimmat, R., Owens, R. A., Wagenhöfer, M., Hillen, W., Vollbach, S., and Henco, K. (1989). Temperature-gradient gel electrophoresis of nucleic acids: Analysis of conformational transitions, sequence variations, and protein-nucleic acid interactions. *Electrophoresis* **10,** 377–389.

Robertson, D. L., and Joyce, G. F. (1990). Selection *in vitro* of an RNA enzyme that specifically cleaves single-stranded DNA. *Nature* **344,** 467–468.

Rosenbaum, V., and Riesner, D. (1987). Temperature-gradient gel electrophoresis: Thermodynamic analysis of nucleic acids and proteins in purified form and in cellular extracts. *Biophys. Chem.* **26,** 235–246.

Shu, Z., and Bevilacqua, P. C. (1999). Isolation and characterization of thermodynamically stable and unstable RNA hairpins from a triloop combinatorial library. *Biochemistry* **38,** 15369–15379.

Steger, G., Po, T., Kaper, J., and Riesner, D. (1987). Double-stranded cucomovirus associated RNA 5: which sequence variations may be detected by optical melting and temperature-gradient gel electrophoresis? *Nucleic Acids Res.* **15,** 5085–5103.

Szewczak, A. A., Podell, E. R., Bevilacqua, P. C., and Cech, T. R. (1998). Thermodynamic stability of the P4–P6 domain RNA tertiary structure measured by temperature gradient gel electrophoresis. *Biochemistry* **37,** 11162–11170.

Tanikawa, J., Nishimura, Y., Hirao, I., and Miura, K. (1991). NMR spectroscopic study of single-stranded DNA fragments of d(CGGCGAAAGCCG) and d(CGGCAAAAGCCG). *Nucleic Acids Symp. Ser.* **25,** 47–48.

Thatcher, D. R., and Hodson, B. (1981). Denaturation of proteins and nucleic acids by thermal-gradient electrophoresis. *Biochem. J.* **197,** 105–109.
Theimer, C. A., Finger, L. D., and Feigon, J. (2003). YNMG tetraloop formation by a dyskeratosis congenita mutation in human telomerase RNA. *RNA* **9,** 1446–1455.
Tominaga, T. (2007). Rapid determination of multi-locus sequence types of *Listeria monocytogenes* by microtemperature-gradient gel electrophoresis. *J. Microbiol. Methods* **70,** 471–478.
Tuerk, C., and Gold, L. (1990). Systematic evolution of ligands by exponential enrichment: RNA ligands to bacteriophage T4 DNA polymerase. *Science* **249,** 505–510.
Wartell, R. M., Hosseini, S. H., and Moran, C. P. Jr. (1990). Detecting base pair substitutions in DNA fragments by temperature-gradient gel electrophoresis. *Nucleic Acids Res.* **18,** 2699–2705.
Xia, T., SantaLucia, J. Jr., Burkard, M. E., Kierzek, R., Schroeder, S. J., Jiao, X., Cox, C., and Turner, D. H. (1998). Thermodynamic parameters for an expanded nearest-neighbor model for formation of RNA duplexes with Watson–Crick base pairs. *Biochemistry* **37,** 14719–14735.
Zhu, J., and Wartell, R. M. (1997). The relative stabilities of base pair stacking interactions and single mismatches in long RNA measured by temperature gradient gel electrophoresis. *Biochemistry* **36,** 15326–15335.
Zhu, J., and Wartell, R. M. (1999). The effect of base sequence on the stability of RNA and DNA single base bulges. *Biochemistry* **38,** 15986–15993.

CHAPTER NINETEEN

Studying RNA–RNA and RNA–Protein Interactions by Isothermal Titration Calorimetry

Andrew L. Feig

Contents

Abstract

Isothermal Titration Calorimetry (ITC) provides a sensitive and accurate means by which to study the thermodynamics of RNA folding, RNA binding to small molecules, and RNA–protein interactions. The advent of extremely sensitive instrumentation and the increasing availability of ITC in shared facilities have made it increasingly valuable as a tool for RNA biochemistry. As an isothermal measurement, it allows analysis at a defined temperature, distinguishing it from thermal melting approaches (UV melting and differential scanning calorimetry, for instance) that provide thermodynamic information specific to

Department of Chemistry, Wayne State University, Detroit, Michigan, USA

Methods in Enzymology, Volume 468
ISSN 0076-6879, DOI: 10.1016/S0076-6879(09)68019-8

the melting temperature. Residual structures at low temperature in the unfolded state and heat capacity changes lead to potential differences between thermodynamic values measured by ITC and those derived from melting studies. This article describes how ITC can be put to use in the study of RNA biochemistry.

1. Introduction

Isothermal Titration Calorimetry (ITC) is a methodology that directly measures the heat (q) taken up or given off by a reaction. Since heat is a universal readout for any reaction, no extrinsic labeling of the sample is required. Reactions are followed over a series of small injections such that in the earliest injections, reactions go to completion due to the large molar excess of the titrate (material in the sample cell) relative to the titrant (material in the syringe being added to the cell), but by the end of the titration, no binding occurs because all of the titrate is already in a complex with its binding partner. A single ITC titration allows one to measure an affinity constant for an interaction (K_a), the reaction enthalpy (ΔH) and the stoichiometry (n). This differs from thermal melting studies where for bimolecular interactions, one often collects data over a range of concentrations to obtain comparable thermodynamic data. Since K_a is related to the Gibbs Free Energy change (ΔG), one can solve indirectly the entropy change (ΔS). If reactions are carried out over multiple temperatures, heat capacity changes (ΔC_P) of binding can also be determined.

The most common instruments used for the ITC analysis of biological materials are power-compensation calorimeters that measure the power consumption required to keep a sample cell and a control cell at constant temperature during the titration (Wiseman *et al.*, 1989). As energy is released by an exothermic reaction, less power is required to hold the cells at the same temperature and a negative deflection from the baseline is observed. If a reaction is endothermic, a positive deflection occurs. Since practically all biological interactions have at least a small enthalpic component, most binding reactions can be studied in this manner. This chapter will lead the reader through the sample preparation, binding reaction, and data analysis required for the characterization of an RNA binding to another species. Examples of such experiment include two RNAs binding each other to form duplexes or tertiary structures (Mikulecky and Feig, 2006a; Mikulecky *et al.*, 2004; Reymond *et al.*, 2009; Takach *et al.*, 2004; Vander Meulen *et al.*, 2008), RNAs binding to proteins (McKenna *et al.*, 2006; Niedzwiecka *et al.*, 2004; Recht and Williamson, 2004; Recht *et al.*, 2008), or small molecule ligands (Bernacchi *et al.*, 2007; Gilbert and Batey, 2009; Li *et al.*, 2004). Examples of all three classes of experiments have been measured successfully by ITC.

2. Required Materials

- VP-ITC or ITC_{200}
- Loading syringe
- Thermovac station for degassing samples
- Titrant solution
- Titrate solution
- 5% solution of Top Job or Mr. Clean (for cell cleaning)
- ddH_2O

3. Instrumentation

A schematic diagram of an ITC instrument is shown in Fig. 19.1. This schematic is based on the VP-ITC produced by MicroCal (now a subsidiary of GE Healthcare) but is in principle equivalent to the ITC_{200} (MicroCal), and nano-ITC2G (TA instruments). The auto-ITC_{200} (MicroCal) has the added attachment of an autoloader to facilitate higher throughput for compound screening. The titrant is loaded into the sample chamber and the sample is sealed with a special injection syringe. The paddle-shaped needle of this syringe doubles as the stirring mechanism within the cell. Injections occur through the action of a screw-driven plunger allowing accurate delivery of as little as 1 μL for the VP-ITC and 0.1 μL for the ITC_{200}. Specific instrument parameters for data collection are described below and include settings for temperature, injection volume, injection speed, and stirring speed.

Most casual and first-time ITC users will take advantage of a multiuser facility that has an ITC. This situation poses a significant problem for most RNA experiments. The issue derives from the training and calibration routine popularized by MicroCal involving the binding of CMP to Ribonuclease A (Wiseman *et al.*, 1989). While this is a terrific standard for labs working on proteins and small molecules, it involves loading the sample cell with concentrated nuclease solutions that will readily destroy any RNA sample. For users considering the use of such a facility, it is essential that you speak with the facility manager and inquire about the training and calibration protocols well in advance of running your experiment. Alternative protocols are in use today including titrations involving Ni(II) binding to histidine (Salim and Feig, 2009), Ba(II) binding to 18-Crown-6 (Gilbert and Batey, 2009), or tris base with nitric acid (Baranauskiene *et al.*, 2009). We strongly urge facilities to consider switching to one of these more benign reactions as a training standard. Whether or not RNase A has been in the sample cell recently, it is prudent to test the cell for nuclease

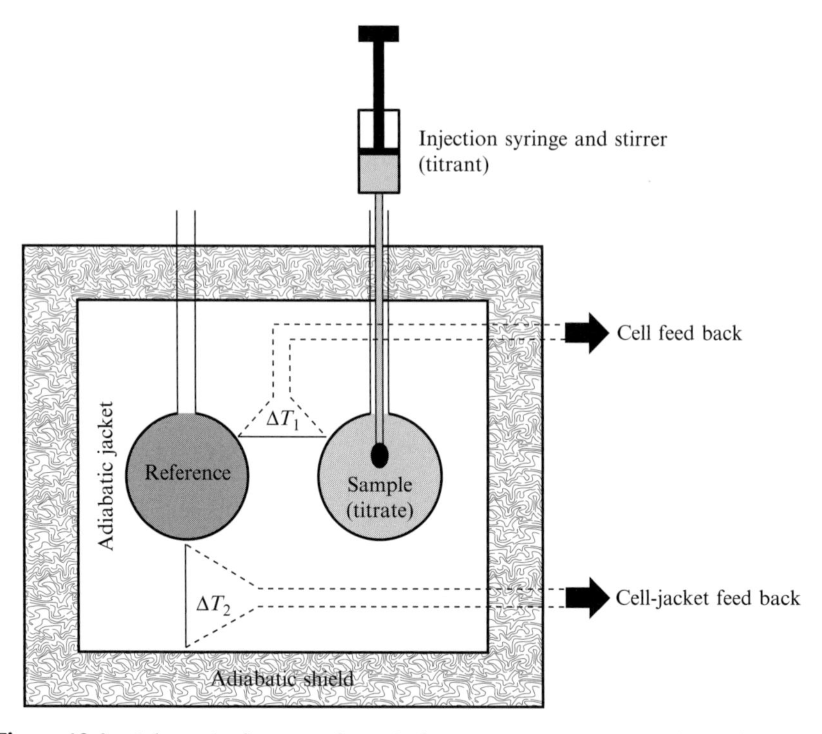

Figure 19.1 Schematic diagram of a typical power compensation isothermal titration calorimeter such as the VP-ITC or the ITC_{200}. (Reprinted with permission from Salim and Feig (2009). Copyright, Elsevier, Inc.)

contamination in advance of any experiment. Typically, this involves loading an RNA sample in the cell, incubating it for several hours or overnight, followed by PAGE analysis. If sample integrity is compromised, one should perform the stringent decontamination protocols described below.

4. Sample Considerations and Preparation

One of the major drawbacks of ITC is the sample sizes required for analysis. While this has been coming down in recent years, it is still quite significant and limits certain applications, especially with naturally isolated RNAs or those containing synthetic modifications. On the VP-ITC instrument with a 1.44 mL sample cell, a typical titration requires approximately 1.8–2.0 mL of titrate at a concentration of 1–2 μM. The titrant will typically be at a concentration 10–20 times that of the titrate. Even though the injection syringe only holds 240 μL, one realistically needs about 400 μL to fill it in a bubble-free manner. The newer ITC_{200} has a 200 μL sample cell and 40 μL injection syringe. Sample requirement is thus about sevenfold smaller in this instrument.

While this concentration range is often a good place to start, one may adjust this up or down depending on the specific interaction being studied. The limitations are dependent on the experimental enthalpy of the interaction and the binding constant. When binding enthalpy is low, the experiment becomes heat limited. This problem can be resolved by increasing the sample concentration or the injection volume as either will increase the heat evolved per injection. Ideally, one uses a parameter called the c-value (defined in Eq. (19.1)) to set the concentration of titrate where n is the stoichiometry of the interaction, K_B is the affinity constant and M_T is the concentration of the titrate in the cell. The c-value must lie between 1 and 1000 for ITC data to be meaningful and optimally between 10 and 100 (Tellinghuisen, 2005). Theoretical studies have shown that the optimal concentration ratio (R_m) of titrant (B_T) to titrate (M_T) for ITC is defined in Eq. (19.2) and is between 15 and 30 for simple binding systems with 1:1 stoichiometry (Tellinghuisen, 2005).

$$c = n \times K_B \times M_T \tag{19.1}$$

$$R_m = \left(\frac{B_T}{M_T}\right) = \frac{6.4}{c^{0.2}} + \frac{13}{c} \tag{19.2}$$

The process of making RNA goes beyond the scope of this discussion. However, one does need to be sure that large amounts of high-quality material are available, be it T7 transcribed or synthetic. It should be clean of impurities (typically HPLC, FPLC, or gel purified depending on size) and properly folded. Specific artifacts to watch out for and avoid include the formation of dimeric species due to the high concentrations required in the titration syringe and alternative folds. Such species are particularly prevalent in the analysis of hairpins when annealed at high concentrations. It is strongly recommending that native gel analysis be performed on the material prior to ITC to ensure that material is free from these types of defects as the energy associated with unfolding an alternative conformer will adversely affect the thermodynamic values measured and the presence of inactive conformers can impact the measured stoichiometry of the reaction.

The size of the RNAs to be used also dictates to a great extent what must happen just prior to the experiment. A critical issue for collecting high-quality ITC data is the buffer-match between titrate and titrant solutions. The best way to ensure that both samples are in identical buffer conditions is to dialyze them against the same reservoir overnight prior to the experiment. This process ensures rigorously identical buffer conditions and works well if the RNAs are long, but becomes problematic for short RNAs, like the 6–10 nt species often used to study of duplex formation.

In the latter case, the final steps in the purification are often either HPLC purification and ethanol or $LiClO_4$/Acetone precipitation. In these cases, special care should be taken to remove as much salt as possible during workup as mismatch between the ionic strength of the samples leads to significant heats of dilution. This problem is mitigated somewhat by using moderate to high salt (100 m*M* or greater) binding buffers but can still be a significant source of problems.

Knowledge of the exact concentration of the RNAs is also very important, as any errors will result in a nonintegral stoichiometry. Readers should be aware that approximations based on base composition are useful for very short RNAs or for the determination of ballpark concentrations, but the errors get significant for longer and more structured nucleic acids. It is best to measure exact extinction coefficients for folded RNAs and use those for determining the sample concentrations.

5. Cleaning the Sample Cell and Titration Syringe

Maintaining a clean sample cell is essential for the proper and effective operation of the ITC. While RNAs are typically quite soluble and not particularly prone to fowling of the cell, one is often measuring the binding of the RNA to a protein or small molecule that may have less solubility and greater potential for deposition onto cell walls. On a multiuser instrument, we always recommend performing stringent cleaning prior to beginning a multiday experiment as it is sometimes unclear how careful the previous user has been. This precaution can prevent accidental degradation of precious RNA samples. Three separate cleaning protocols are described below.

5.1. Gentle cleaning before and after runs

Standard cleaning involves rinsing the cell with degassed ddH_2O, washing it with 100 mL 5% Mr. Clean or Top Job and then flushing the cell exhaustively with up to 1 L of ddH_2O. This procedure can be run overnight as well, soaking the instrument in 5% Mr. Clean followed by flushing the instrument with water the next day. It is essential to remember to also flush and rinse the titration syringe and loading syringes.

5.2. Stringent cleaning and nuclease contamination issues

The preliminary test for all RNA experiments on a new instrument should always be a stability control experiment. If degradation occurs within the instrument, stringent cleaning and decontamination are essential. Sample degradation will dramatically erode data quality and the degradation process itself is exothermic leading to heat evolution over time that affects baseline stability and changes the sample concentration during the course of the experiment. Stringent cleaning of the ITC cell involves filling the sample cell with a 5% (v/v) solution of Contrad-70 and letting it incubate for several hours, often at elevated temperature (up to 65 °C). This step is reasonably effective at removing minor nuclease contamination (such as left over from a protein titration of a previous user) although repeated treatment is sometimes necessary for more significant contamination (experiments involving nuclease, for instance). Stringent cleaning is also required periodically to remove buildup from the cell walls, especially after the use of protein samples.

Protein buildup in the cell can be removed by trypsin or proteinase K treatment of the cell. Our protocol involves filling the cell with a degassed solution of 1 mg/mL solution of proteinase K in 50 m*M* Tris buffer, pH 7.5, and incubating it in the cell for at least 1 h (but can be left in the cell overnight). The enzyme solution is then removed from the cell and the cell is rinsed with 500 mL ddH_2O followed by the gentle cleaning cycle described above.

Serious nuclease contamination occurs if the RNase A calibration routine has been used in the instrument. In these cases, as stated above, it can be extremely difficult to fully decontaminate the instrument. After proteinase K treatment, the cell and titration syringe should be filled with neat RNaseZap (Ambion) and allowed to incubate overnight. After rinsing the cell with deionized water, follow the gentle cleaning protocol above and retest the cell for contamination. This procedure may have to be repeated two or three times before sample integrity is maintained.

6. Collecting Titration Data

Data collection in an ITC titration is a two-step process involving collection of information on the background heat and the heat of a reaction itself. Sources of background heat include buffer mismatch, heats of dilution, heats of mixing, etc. and accounting for this energy is an important part of the data analysis. One of the two protocols for the background measurement can be used, the selection of which dictates the way in which the primary data will be collected. Our preferred method is a procedure involving just a single titration where we collect additional data at the end of a titration after all binding is complete. The second method (which is probably more common in the field of biocalorimetry) involves performing a second independent titration where the titrant is added to a buffer solution in the absence of titrate. We prefer the single titration method because it saves time and material relative to performing an independent series of background injections, but it cannot be used in all cases. In particular, if there is a significant nonspecific binding above and beyond a specific binding event, one must collect the background titration separately (two titration method).Standard data collection parameters for a typical RNA experiment are listed in Table 19.1.

6.1. Choosing who should be titrated into whom

Concern about who should be titrated into whom is pretty common. The simple answer is that for well-behaved systems, it should not matter and one should get equivalent data from either direction. In practice, however, it sometimes does make a big difference. One sample must be 15–20 times more concentrated than the other. For protein or small molecule binding, this fact sometimes leads to solubility problems. For those dealing with RNA–RNA titrations, problems can be manifested in terms of alternative folding problems like dimerization equilibria. In general, if one binding partner has lower solubility, consider using this material as the titrate initially. The other consideration is one of nonspecific binding. This manifests itself in terms of nonlinear baselines at the end of the titration and noninteger stoichiometries.

6.2. Data collection procedure using the single titration method

Degas 2 mL of 1–2 μM solution of RNA 1 (titrate), 300 μL of a 15–30 μM solution of RNA 2 (titrate) and some excess buffer using the thermovac aparatus. Rinse the cell with titration buffer and empty completely. Fill the sample cell with titrate being careful not to bend the cell-loading syringe or to introduce bubbles into the cell. Then load the syringe with the titrate solution. Save the excess solutions from loading the cell and use this material to determine the exact concentration of the reagents in the cell and the syringe. Carefully affix the titration syringe into the top of the sample cell and prepare the instrument by setting the data collection parameters. The small initial injection of 1–2 μL will not be used in the data analysis. This step prevents an artifact associated with the mechanical backlash of the screw mechanism in the drive syringe from which under delivers titrant in this first injection (Mizoue and Tellinghuisen, 2004). Plan the experiment to ensure that 5–10 data points are collected after complete saturation is achieved, as these will be used to define the background heat for the reaction. For a 1:1 stoichiometry experiment, this will often require that you titrate out to a molar ratio of $\sim$2.5. The integrated injection heats from the last few injections will be fit to a linear model and extrapolated through the entire dataset as the background correction.

Table 19.1 Typical experimental parameters for an ITC experiment[a]

Experimental parameter	Setting	Comments
Cell temperature	2–80 °C	See specific concerns regarding high-temperature experiments in text
No of injections	25–35[b]	Based on single titration method; fewer injections can be used for the two-titration method
Injection volume	First Inj.: 1–2 μl; Rest of Inj.: 7–12 μl	First small injection is used to account for syringe backlash; data from this injection should be omitted from data analysis
Injection spacing	250–500 s	Typical spacing is about 300 s; should be lengthened if titration fails to adequately return to baseline prior to commencing the next injection. Spacing can be shortened if long flat baselines are present between each injection
Titrate/titrant concentration	Cell: 1.8 μM; Syringe: 42 μM	Can be optimized depending on reaction heat; this example comes from an RNA–RNA experiment with ΔH of ~40 kcal/mol and a 1:1 stoichiometry
ITC equilibration options	Automatic	
Reference power	25–30 μCal/s	Setting a much higher value than is required may affect the sensitivity of the instrument
Initial delay	60–100 s	Ensure stable baseline prior to first injection
Stirring speed	270–310 rpm	Faster stirring speeds may lead to high-frequency noise in the data
Feedback mode	Fast	
Filter period	2 s	For slow reactions, the filter period can be increased to improve data quality

[a] Based on use with a VP-ITC from MicroCal. Specific parameter settings might need to be adjusted for use with other instruments.

[b] A terminal ratio of about 3 is obtained for a 1:1 interaction (i.e., the final excess ratio of the titrant over the titrate).

6.3. Data collection procedure using the two-titration method

This procedure is similar to the single titration method except that less titrant is used in the initial titration with a terminal molar ratio of approximate 1.5–2 instead of 2.5–3. After the initial data collection is complete, empty and rinse the cell and refill the instrument with buffer. Then, using the exact same instrument protocol as for the first titration, titrate the titrant into buffer. This will serve as the background titration and integrated heats of injection from this run are subtracted from the experimental run. Note that in this case, the titrant is clean but dilute at the end of the background titration and can be recovered so long as no degradation occurred.

If the experiment has worked well, data should look like that shown in Fig. 19.2. Specific things to look for in the raw data are: (i) low noise in the heat versus time curve; (ii) clean, well-shaped peaks; (iii) level baseline after each injection (injection i should not tail into injection $i+1$); (iv) clear evidence of saturation behavior; (v) two or more points with maximal heat evolution at the beginning of the titration; (vi) several points within the transition region; (vii) small integrated heat from injections after saturation was achieved. Problems with (i–iv) typically derive from data collection parameter issues such as stirring speeds, injection volumes, and injection rates. Data quality problems with (v–vii) usually derived from issues related to titrate or titrant concentrations. In general, when working with small and precious samples with low heat output (small molecule binding to RNAs), the best data often results from using fewer injections. The limit of this is actual a titration involving just a single injection which gives very accurate ΔH parameters but provides no information on binding affinity. If the reaction being studied has a large ΔH (such as most RNA duplex formation experiments), then a larger number of small volume injections often works well in our hands.

7. Data Processing and Analysis

Data analysis is typically performed using the Origin software supplied by the manufacturer. The automated routines allow for numerical integration of the peaks and subtraction of background heat from either the one-titration or two-titration methods. Note that the heat from injection i (q_i) is given by the formula in Eq. (19.3) where n is the reaction stoichiometry, F is the fractional saturation of the reaction, M_T is the titrate concentration, ΔH is the reaction enthalpy and V is the cell volume. For a simple binding equilibrium when the concentration of titrant (B_T) is known, one can solve for the fractional saturation at any point along the titration based on Eqs. (19.4) and (19.5). Equation (19.5) is then fit by nonlinear least squares methods to determine n, K_B and ΔH (Wiseman *et al.*, 1989).

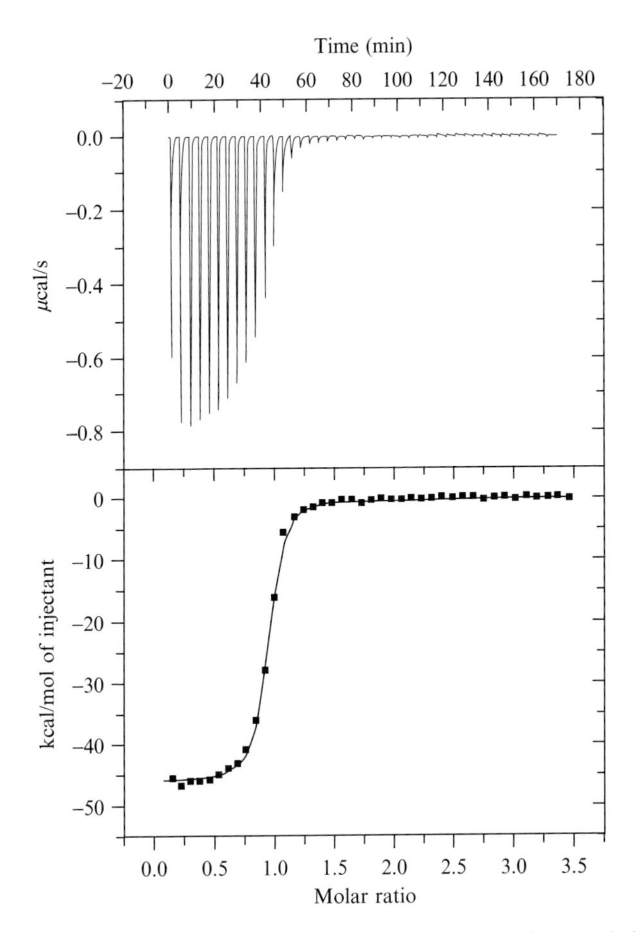

Figure 19.2 Example ITC data from an experiment measuring the association of two single-stranded 7-mer RNAs. Data collected at 15 °C in 50 m*M* HEPES, pH 7.5, and 1 *M* NaCl. Top panel. Power versus time curve. Bottom panel. Integrated injection enthalpy plotted versus the mole ratio of the reactants. $\Delta H = -46.1 \pm 0.2$ kcal/mol, $K_a = 4.3 \times 10^{-7}$ *M* and $n = 0.91 \pm 0.1$ (Reprinted with permission from Takach *et al.* (2004). Copyright, American Chemical Society.)

$$q_i = nFM_T\Delta HV \qquad (19.3)$$

$$F^2 - F\left\{1 + \frac{B_T}{nM_T} + \frac{B_T}{nK_BM_T}\right\} + \left\{\frac{B_T}{nM_T}\right\} = 0 \qquad (19.4)$$

$$q_i = nM_T\Delta H\left(\frac{V}{2}\right)\left\{X - \sqrt{\left(X^2 - \frac{4B_T}{nM_T}\right)}\right\} \qquad (19.5)$$

where

$$X = 1 + \frac{B_T}{nM_T} + \frac{1}{nK_BM_T}$$

More complicated binding models involving sequential parallel binding equilibria are available if necessary, but knowledge of the system involved should be used to determine if their use is warranted.

8. Special Considerations

8.1. The importance of pH and proton association/dissociation reactions

The experimental pH of a titration is of special interest in some ITC studies. Due to the polyanionic nature of RNA and the typical positive charge of RNA binding proteins, there is the potential for changes in protonation state to occur during binding. While this occasionally occurs in simple RNA folding interactions where intrinsic pK_as are altered, it is common in RNA–protein interactions. One must pay attention to it in the case of ITC because it leads to an enthalpic artifact that one must correct for if it occurs. If a proton is lost or taken up during the binding event, the reaction is accompanied by (de)protonation of the buffer. This results in a very specific artifact in which the measured enthalpies become buffer dependent. This problem can be solved by either shifting the pH of the experiment to make the proton inventory zero or remeasuring the thermodynamic values in a second buffer at the same pH. In this case, if one knows the ionization enthalpies of both buffers, one can calculate the corrected enthalpy change of the reaction (ΔH_{corr}) by accounting for the ΔH_{ion} and the number of protons released or taken up (Δn) using Eqs. (19.6) and (19.7). Tables of the protonation enthalpies (ΔH_{ion}) for common biological buffers can be found in the literature (Feig, 2007; Fukada and Takahashi, 1998).

$$\Delta H_{obs1} = \Delta H_{corr} + \Delta H_{ion1}\Delta n \tag{19.6}$$

$$\Delta H_{obs2} = \Delta H_{corr} + \Delta H_{ion2}\Delta n \tag{19.7}$$

8.2. Temperature-dependent phenomena

Measurement temperature is a major concern with the evaluation of thermodynamic parameters of RNA interactions by ITC. While the ability to set the experimental temperature is one of the major strengths of ITC relative to thermal melting analysis, it is also one of the greatest weaknesses if sufficient care is not taken to recognize and account for potential artifacts.

There are two main issues at play. The first has to do with residual structure. When RNA folding thermodynamics is studied by thermal melting, the unfolded or unbound form is a high-temperature state whereas the structured form is the low-temperature state. ITC, on the other hand, is an isothermal measurement as the name implies. Both the bound and free states are at modest temperature and that means that the uncomplexed form of the RNA may exhibit significant residual structure (Feig, 2007; Holbrook *et al.*, 1999; Mikulecky and Feig, 2006a,b). Thus, the heats that are measured in this experiment are really the sum of the typically unfavorable unfolding of whatever residual structure exists in the initial state followed by the typically favorable free energy associated with folding into the final bound state. Depending on the RNAs being studied and the extent of the folding of the starting materials, this can lead to very complex temperature dependencies yielding highly nonlinear heat capacity changes (Feig, 2007; Mikulecky and Feig, 2006b).

A second effect occurs at high temperature. As we know from thermal melting analysis, as one approaches the T_M for a given RNA folding transition, the amount of product that forms at equilibrium is temperature dependent. In ITC, one relies on the fact that all of the titrant can and will bind the titrate. However, if a fraction of the material is thermally unfolded at a given temperature, this material may not bind, affecting the fractional saturation F at equilibrium, Eq. (19.4). Under these conditions, if one studies an RNA by ITC at the high temperature (a good working definition of high temperature for this discussion is within 20 °C of the melting transition), the data might need to be corrected for the percentage of unfolded material. Note, there are many systems that might be of interest for study with T_Ms in the range of 50–60 °C. For these systems, this effect already leads to measureable deviations at 37 °C, so care must be taken to know the T_M behavior of any systems being studying.

9. Conclusions

ITC can be put to excellent use in the analysis of RNA folding and RNA-binding interactions. Care must be taken, however, to design the experiments to respect the idiosyncrasies of this biopolymer and understand how these thermodynamic values differ from those obtained by thermal melting. With due care and increasingly available instruments, ITC is likely to be of exceptional value to RNA scientists interested in fundamental folding and binding phenomena.

REFERENCES

Baranauskiene, L., *et al.* (2009). Titration calorimetry standards and the precision of isothermal titration calorimetry data. *Int. J. Mol. Sci.* **10,** 2752–2762.

Bernacchi, S., *et al.* (2007). Aminoglycoside binding to the HIV-1 RNA dimerization initiation site: Thermodynamics and effect on the kissing-loop to duplex conversion. *Nucleic Acids Res.* **35,** 7128–7139.

Feig, A. L. (2007). Applications of isothermal titration calorimetry in RNA biochemistry and biophysics. *Biopolymers* **87,** 293–301.

Fukada, H., and Takahashi, K. (1998). Enthalpy and heat capacity changes for the proton dissociation of various buffer components in 0.1 *M* potassium chloride. *Proteins* **33,** 159–166.

Gilbert, S. D., and Batey, R. T. (2009). Monitoring RNA–ligand interactions using isothermal titration calorimetry. *Methods Mol. Biol.* **540,** 97–114.

Holbrook, J. A., *et al.* (1999). Enthalpy and heat capacity changes for formation of an oligomeric DNA duplex: Interpretation in terms of coupled processes of formation and association of single-stranded helices. *Biochemistry* **38,** 8409–8422.

Li, T. K., *et al.* (2004). Drug targeting of HIV-1 RNA.DNA hybrid structures: Thermodynamics of recognition and impact on reverse transcriptase-mediated ribonuclease H activity and viral replication. *Biochemistry* **43,** 9732–9742.

McKenna, S. A., *et al.* (2006). Uncoupling of RNA binding and PKR kinase activation by viral inhibitor RNAs. *J. Mol. Biol.* **358,** 1270–1285.

Mikulecky, P. J., and Feig, A. L. (2006a). Heat capacity changes associated with DNA duplex formation: Salt- and sequence-dependent effects. *Biochemistry* **45,** 604–616.

Mikulecky, P. J., and Feig, A. L. (2006b). Heat capacity changes associated with nucleic acid folding. *Biopolymers* **82,** 38–58.

Mikulecky, P. J., *et al.* (2004). Entropy-driven folding of an RNA helical junction: An isothermal titration calorimetric analysis of the hammerhead ribozyme. *Biochemistry* **43,** 5870–5881.

Mizoue, L. S., and Tellinghuisen, J. (2004). The role of backlash in the "First injection anomaly" in isothermal titration calorimetry. *Anal. Biochem.* **326,** 125–127.

Niedzwiecka, A., *et al.* (2004). Thermodynamics of mRNA 5′ cap binding by eukaryotic translation initiation factor eIF4e. *Biochemistry* **43,** 13305–13317.

Recht, M. I., and Williamson, J. R. (2004). RNA tertiary structure and cooperative assembly of a large ribonucleoprotein complex. *J. Mol. Biol.* **344,** 395–407.

Recht, M. I., *et al.* (2008). Monitoring assembly of ribonucleoprotein complexes by isothermal titration calorimetry. *Methods Mol. Biol.* **488,** 117–127.

Reymond, C., *et al.* (2009). Monitoring of an RNA multistep folding pathway by isothermal titration calorimetry. *Biophys. J.* **96,** 132–140.

Salim, N. N., and Feig, A. L. (2009). Isothermal titration calorimetry of RNA. *Methods* **47,** 198–205.

Takach, J. C., *et al.* (2004). Salt-dependent heat capacity changes for RNA duplex formation. *J. Am. Chem. Soc.* **126,** 6530–6531.

Tellinghuisen, J. (2005). Optimizing experimental parameters in isothermal titration calorimetry. *J. Phys. Chem. B Condens Matter Mater. Surf. Interfaces Biophys.* **109,** 20027–20035.

Vander Meulen, K. A., *et al.* (2008). Thermodynamics and folding pathway of tetraloop receptor-mediated RNA helical packing. *J. Mol. Biol.* **384,** 702–717.

Wiseman, T., *et al.* (1989). Rapid measurement of binding constants and heats of binding using a new titration calorimeter. *Anal. Biochem.* **179,** 131–135.

Author Index

L

M

N

O

P

Q

R

S

T

Subject Index

D

E

K

L

M

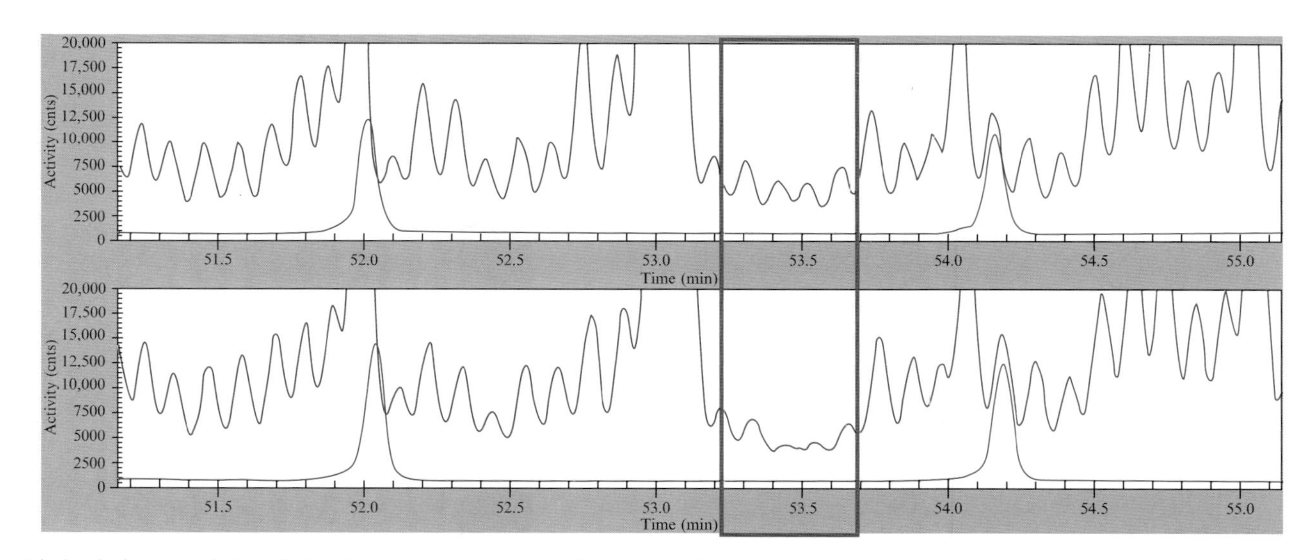

Inna Shcherbakova and Somdeb Mitra, Figure 2.2 Capillary separation of the products of the •OH-induced cleavage of the RNA backbone. Peak profiles of Cy5-labeled cDNAs corresponding to cleaved unfolded (upper panel) and folded (lower panel) *T. thermophila* ribozyme are shown by blue lines. Red peaks corresponding to the Beckman® size standard ladder are used to correlate the peaks with the RNA nucleotides. The peaks that demonstrate significant decrease in reactivity upon folding are enclosed within gray contour.

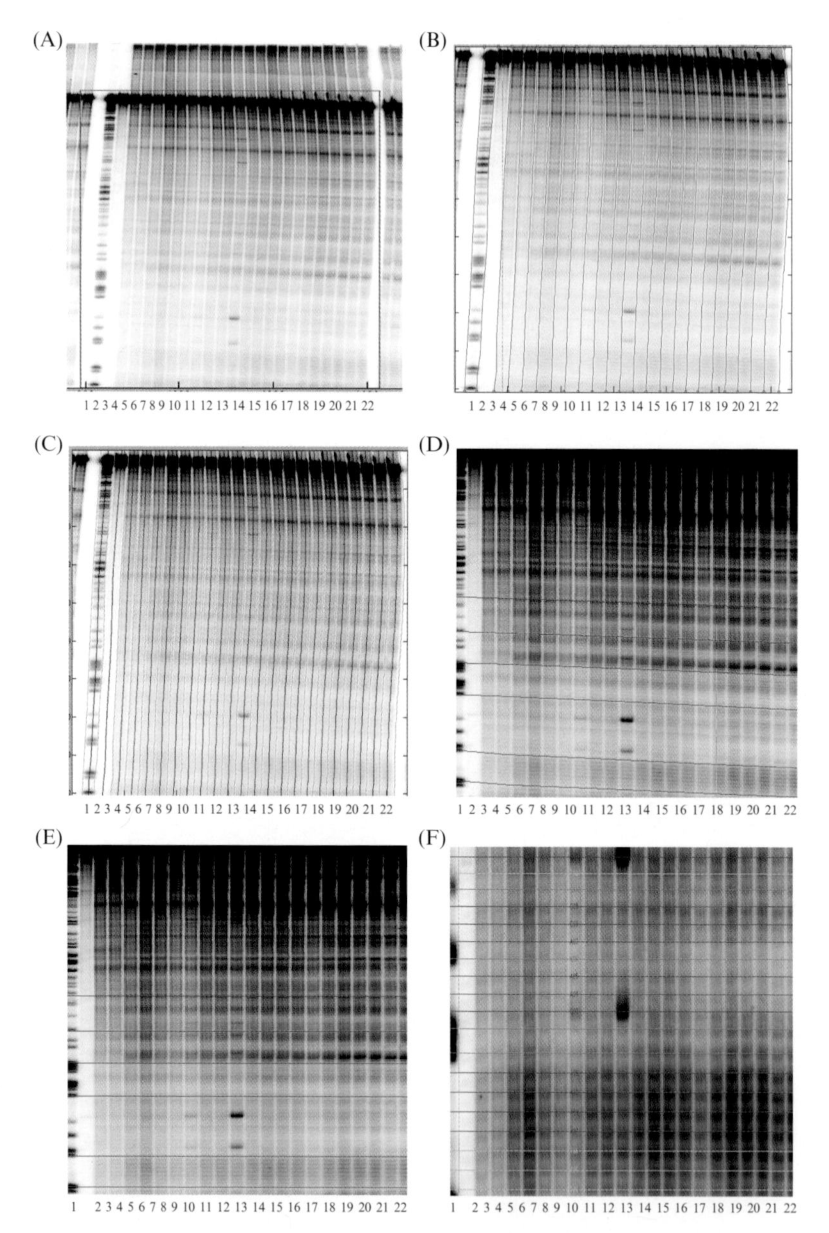

Katrina Simmons *et al.*, Figure 3.3 Illustration of the gel rectification procedure implemented in SAFA. (A) Screenshot of the main SAFA window with a gel image loaded; red box delineates the user-defined cropping boundary for subsequent analysis. (B) Screenshot of the lane definition procedure. Red line is manually drawn that marks the boundaries of each lane. Once a few lanes have been manually defined, SAFA is able to "guess" the outline of the following lanes, which is done by typing the letter "G." (C) Screenshot of the completed and recorded defined lanes that occur after the user types "Z" or "Q." The color of the lines changes from red to green and a dotted line

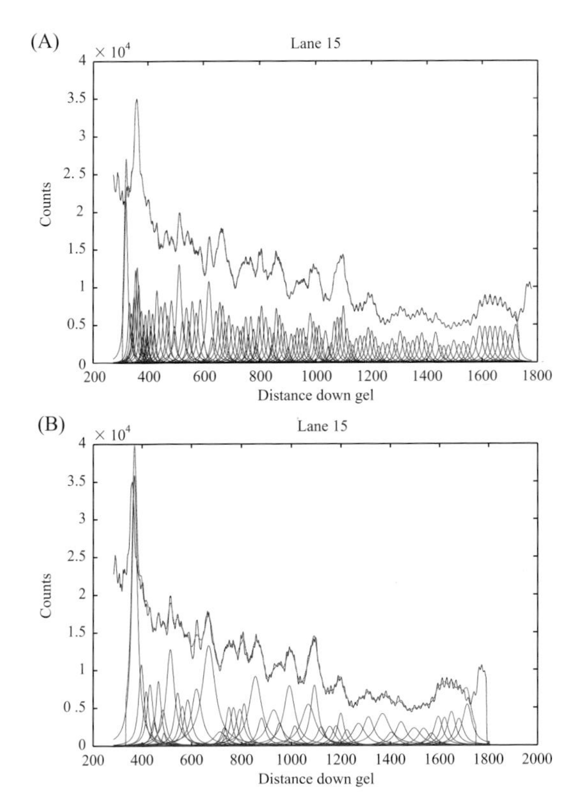

Katrina Simmons *et al.*, Figure 3.4 SAFA performs a peak-fitting procedure to determine the relative band intensity in the gel image. When the fitting procedure has converged, SAFA will plot the results as shown here. Red lines represent individual Lorentzians that when summed yield the lane profile shown in blue. (A) A successful and accurate single-peak fitted graph for Lane 15 of the gel shown in Fig. 3.1. (B) Visual inspection of the fitted profile can be used to determine if the user was not accurate in assigning bands (Fig. 3.3F). In this example, there are large differences in the peak widths and irregularly spaced peaks. These are signs of inaccurate results.

---

marks the middle of the lane when the procedure in completed. (D) Screenshot of the horizontal alignment procedure. The gel is aligned horizontally to ensure the bands are parallel. Similar to the lane definition process, the user identifies bands across the gel image with a horizontal line. Only one line should be drawn per band. The left button of the mouse begins the line and marks sequential points throughout the row and the right button ends the line. The number of rows to mark is at the users' discretion; however, the more rows used will result in a more accurate an adjustment of the image. (E) Screenshot of the finished horizontally aligned gel. (F) Screenshot of the band assignment procedure.

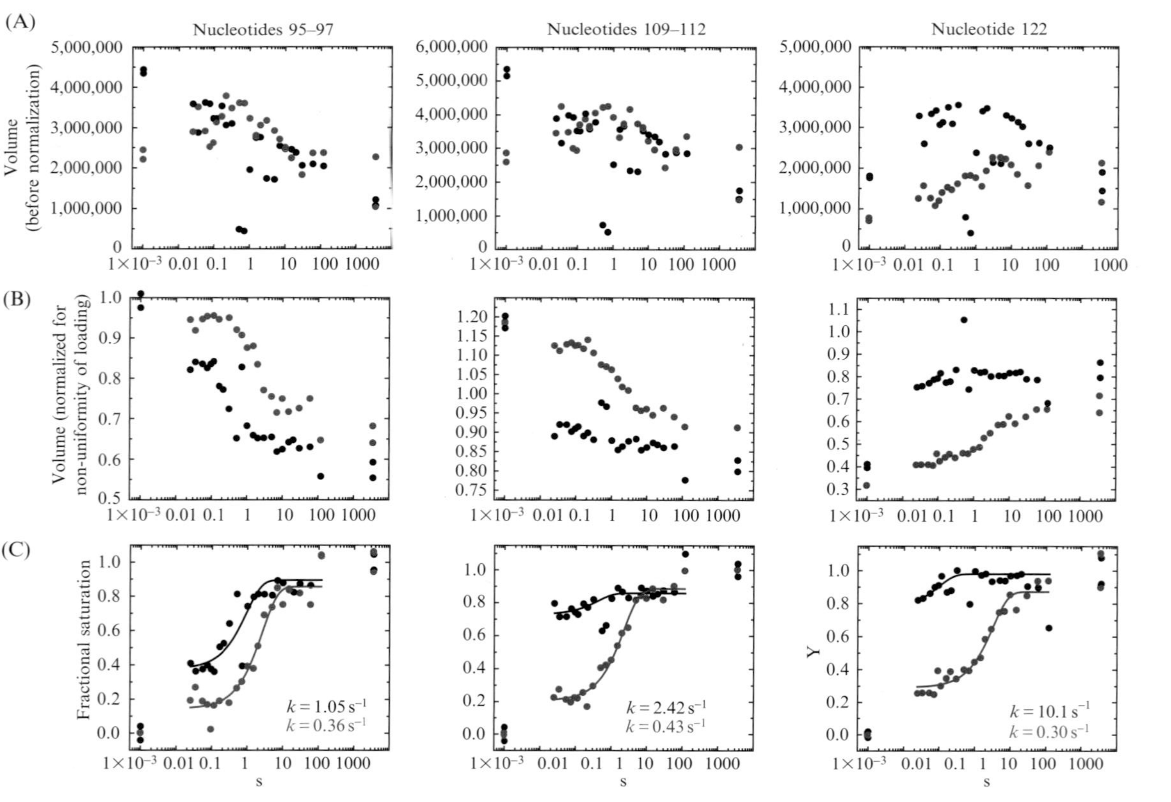

Katrina Simmons *et al.*, Figure 3.5 Time–progress curves following SAFA quantification. The wild-type data are shown in black and the L5b mutant in red. (A) When the data that are not normalized are plotted as a function of time, it is difficult to identify any trends in the changes in reactivity. (B) When invariant residue normalization is applied (in this case based on invariant residues 113–115), much clearer changes in reactivity are revealed. (C) When further normalization procedures are applied based on setting the initial and final fractional saturation to zero for $t = 0$ and one for the final state, we are able to compare the data between experiments and determine rate constants for the transitions.

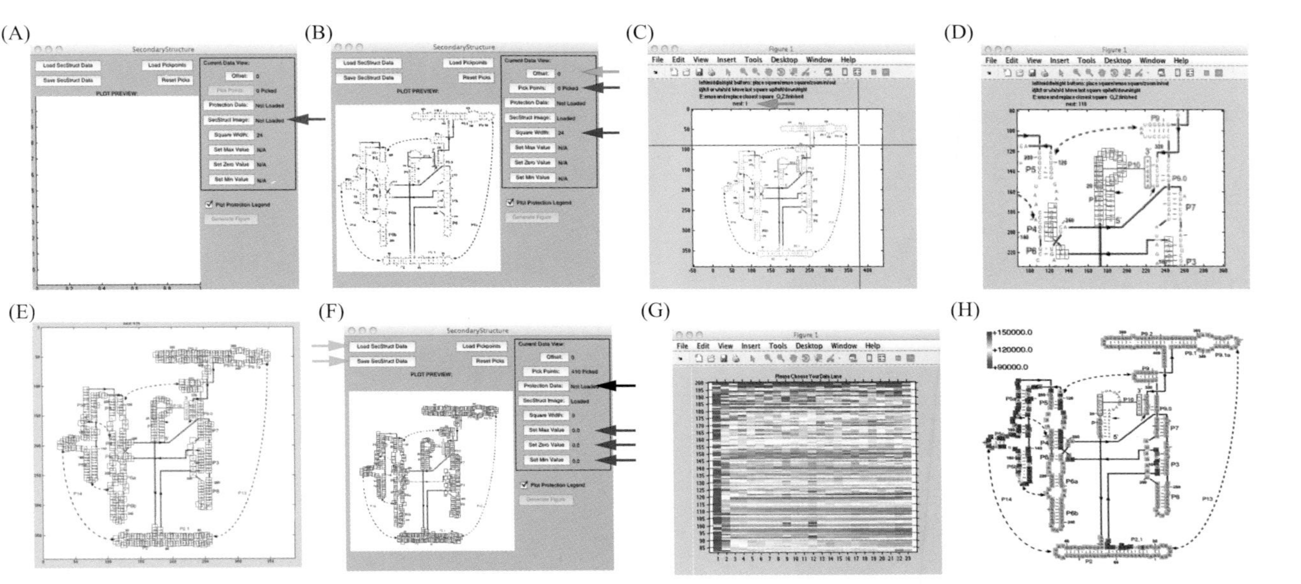

Katrina Simmons *et al.*, Figure 3.6 The data visualization component allows users to map their data onto secondary structure diagrams. (A) Screenshot of the starting window that opens when *Visualize Data → Secondary Structure Plot* is selected. Selecting the *SecStruct Image* button (highlighted by red arrow) allows the user to load a JPEG, GIF, or TIFF image of their secondary structure. (B) Once the image is loaded, it will appear in the window and the next step is to pick points on the image. The *Pick Points* button indicated by blue arrow begins the procedure where the user defines the location of each individual nucleotide on the diagram. Magenta arrow indicates the *Square Width* button, which determines the size of the box that is placed over the selected residues, and green arrow indicates the *Offset* button that allows the user to account for any offset in the start nucleotide of the nucleic acid. (C) A new window appears when the *Pick Points* button is pushed. Indications for keyboard shortcuts are printed above the upper *X*-axis along with the next box number indicated by orange arrow. (D) An example of pick points procedure on the *T. thermophila* group I intron secondary structure diagram. (E) An example of a completed procedure on the same figure. Each nucleotide in the diagram is covered by a square. (F) Once the procedure of picking points is completed, the image and corresponding boxes are saved and can also be reloaded. The boxes change color from red to blue signifying the points have been saved.

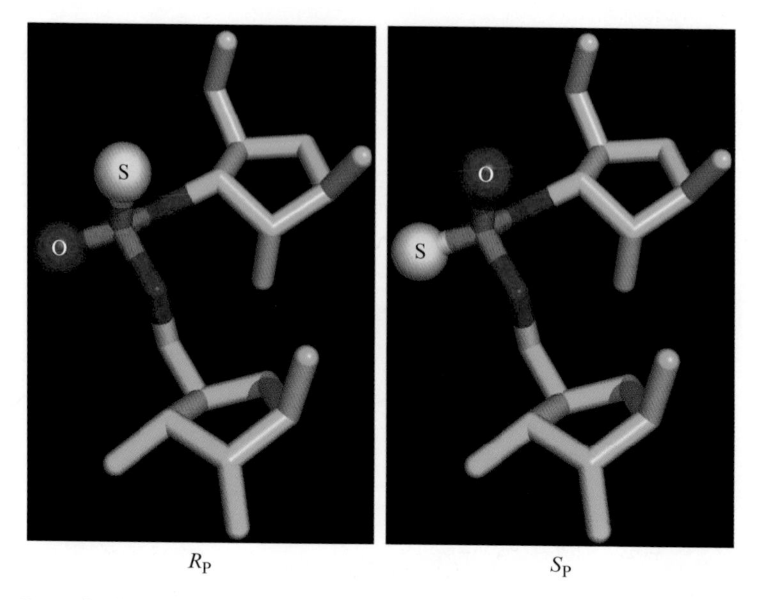

John K. Frederiksen and Joseph A. Piccirilli, Figure 14.1 RNA phosphorothioate diastereomers. Sulfur sizes and bond lengths not drawn to scale.

The light blue arrows mark the *Load SecStruct Data* and *Save SecStruct Data* buttons that allow the user to save their progress and load their data. Black arrow highlights the *Protection Data* button that initiates the next step in the process where the user chooses the data to be displayed on the figure. The purple arrows indicate the *Set Max/Zero/Min Value* buttons that allow the user to adjust these values. (G) Representation of the SAFA quantified data for the gel image shown in Fig. 3.1 in a color plot allowing the user to choose which lane to plot on the secondary structure. (H) Mapping of Lane 15 onto the secondary structure of the *T. thermophila* group I intron indicating nucleotides that are protected in blue and cleaved by •OH radicals in red.

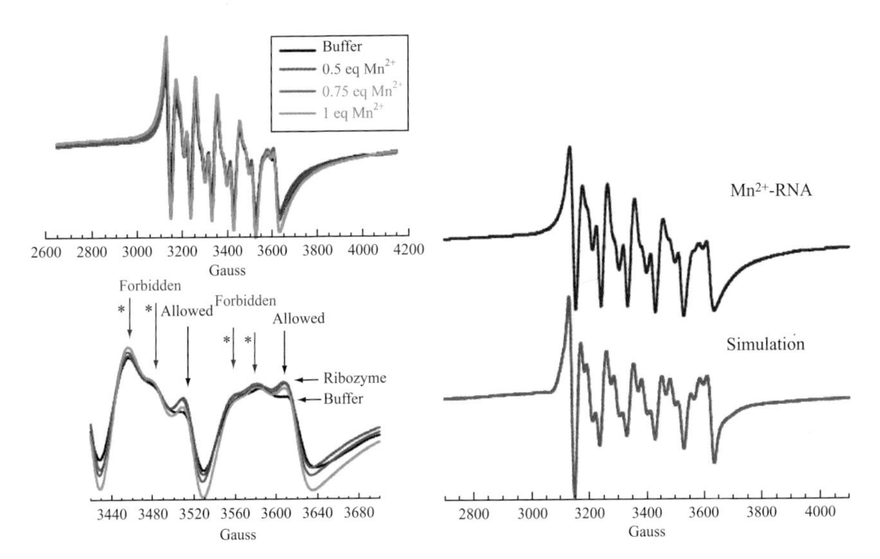

Laura Hunsicker-Wang *et al.*, Figure 16.5 Experimental and simulated low-temperature EPR spectra of Mn^{2+} bound to the mHHRz. At left, top: overlay of spectra showing appearance of the distinctive Mn^{2+} EPR feature at substoichiometric Mn^{2+}:RNA ratios. Samples are 250 μM Mn^{2+} and increasing RNA concentrations to achieve 0.5, 0.75, or 1.0 Mn^{2+}:RNA ratio. Buffer: 1 M NaCl, 5 mM TEA pH 7.8, 20% (v/v) ethylene glycol. EPR spectral parameters as in Fig. 16.4. Left, bottom: expansion of fifth and sixth Mn^{2+} hyperfine lines, showing features due to "forbidden" and "allowed" transitions whose ratios change with different Mn^{2+} environments. Right: experiment (top) and simulation (bottom) for Mn^{2+} EPR spectrum of 0.5 Mn^{2+}:mHHRz using the program EasySpin (see text). Simulation parameters: $g = 2$, $A = 270$ MHz, $D = 520$ MHz, $E = 173$ MHz, D-strain = 1, A-strain = 0.

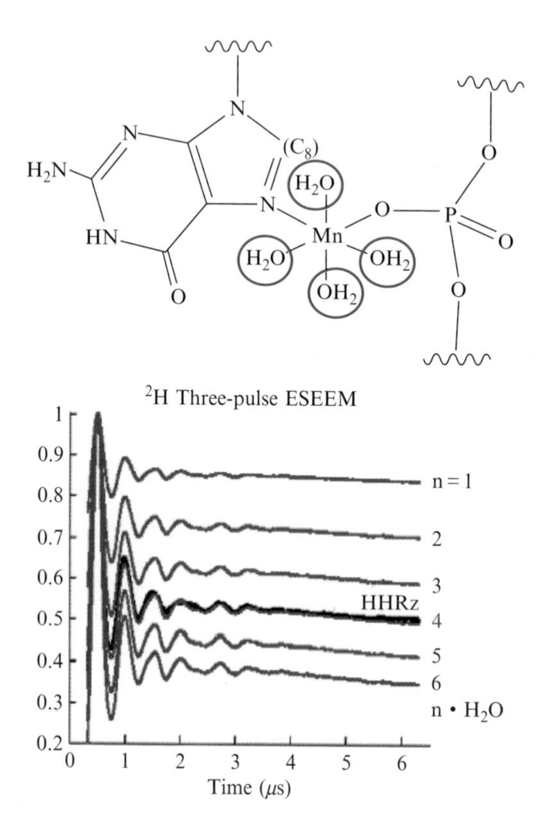

Laura Hunsicker-Wang *et al.*, Figure 16.12 Three-pulse X-band ESEEM spectroscopy used for "counting" water ligands in Mn^{2+}-RNA sites by quantifying ^{2}H ESEEM amplitudes. This experiment, described in detail by Hoogstraten and Britt (2002), relies on the quantitative aspect of ESEEM spectroscopy. Mn^{2+}-RNA samples are exchanged into $^{2}H_2O$, and their three-pulse ^{2}H ESEEM data obtained. In this case, 0.4 *M* sucrose is chosen as the appropriate cryoprotectant. The data are divided by three-pulse ESEEM of nonexchanged Mn-RNA samples to remove modulation from other nuclei. Resulting traces are compared to traces calculated for different levels of *n* aqua ligands, that is, Mn-$(^{2}H_2O)_n$. In this example, from the mHHRz, a clear number of four aqueous ligands are obtained. Data taken from Vogt *et al.* (2006).